T0178209

Understand Mathematics, Understand Computing

Arnold L. Rosenberg • Denis Trystram

Understand Mathematics, Understand Computing

Discrete Mathematics That All Computing Students Should Know

 Springer

Arnold L. Rosenberg
College of Information
and Computer Science
University of Massachusetts
Amherst, MA, USA

Denis Trystram
Grenoble INP
Université Grenoble Alpes
Saint Martin d'Hères, France

ISBN 978-3-030-58378-1   ISBN 978-3-030-58376-7   (eBook)
https://doi.org/10.1007/978-3-030-58376-7

This Springer imprint is published by the registered company Springer Nature Switzerland AG
The registered company address is: Gewerbestrasse 11, 6330 Cham, Switzerland

Addition   Roger Trystram, 1973

To my beloved wife, Susan, and my children Paul and Rachel for decades of patience during my extended episodes of "cerebral absence."

To Professor Oscar Zariski, who changed my life by introducing me to "real" mathematics.

With love to my wife Cécile and my children Alice and Noé who help me to stay connected to the "real" world.

To all past students who inspired this book. To all future students who will read this book.

Contents

Manifesto

The technologies that enable both the hardware and software systems of modern computers have grown in complexity at least as fast as they have in performance. For decades, it has been impossible to design computing systems without using tools whose underpinnings are rooted in *discrete mathematics*. For too long, though, these tools have been used to create widgets which are then assembled into coherent systems using only cleverness and untutored ingenuity. The net result is that many of the computing systems which impact our daily lives are neither well-understood nor well-controlled. We read daily about breaches of security and/or privacy that can neither be controlled nor repaired because the victimized system is too complex to be understood. The lesson is clear: **We must stop being satisfied with mere** *knowledge* **of how to make a system take in inputs and emit outputs. We must strive for** *understanding* **of every system's structure and behavior.** There are myriad examples of successful such computing systems: compilers and databases and electronic circuits begin a long list. Notably, many success stories trace back to the development of relevant new discrete mathematics.

This book has a simple but fundamental goal—to endow readers with *operational* conceptual and methodological understanding of the discrete mathematics that can be used to study, and understand, and perform computing. We want each reader to *understand* the elements of computing, rather than just *know* them. Thereby, the interested reader will be able to *develop concepts* and *invent techniques and technologies* for expanding the capabilities of the hardware that performs computations and the software that controls the hardware. We stress the word *operational:* **We want the readers' level of understanding to enable them to** *"do"* **mathematics.**

The bold goal of reaching the point of "doing" mathematics is within the reader's grasp. We brace the spine and embolden the spirit by invoking no less an authority than the great mathematician Leopold Kronecker. We read in the historical tome [13] of Kronecker's assurance that "God made the integers; all else is the work of man." Therefore, in order to achieve our goal of "doing" mathematics, we need only stand on the shoulders of (mathematical) giants—there is no need to insolently attempt to wrest fire from Olympus!

Preface

We live in a world of increasing professional and social diversity. Computing curricula have responded to these changes by dramatically changing focus and scope.

Early computation-oriented academic programs focused narrowly on the principles and practices related to designing and using digital computers. Discrete mathematics was "automatically" prepended to such programs, in recognition of the fact that every aspect of the design and use of computers built upon mathematical concepts and tools. Today, these traditional programs have been joined by a broad range of curricula which focus on the computational concerns of an equally broad range of *consumers* of the craft of computing: from those interested in fields that touch upon technology, such as scientists or engineers, to those interested in nontechnological pursuits, such as scholars in the humanities or social studies, to practitioners in professions such as business, law, or medicine, and so on, all the way to members of the general population whose computing-related interests are avocational (e.g., hobbyists). Within this new world, many aspiring users of computers need only very specialized knowledge of mathematics in order to compute successfully. But, the amount and direction of the specialization can vary greatly based on a student's, or an academic program's, focus; therefore, unless all or most students have a broad background in mathematics, some students may lack the technical tools necessary to conceive of and to pursue truly bold novel computational ideas.

The number of nontraditional students seeking technical education is growing, and the ways in which their nontraditional goals and backgrounds manifest themselves is increasing. We have striven to keep this diversity in mind while writing this text.

- We have included a broad range of material, in subject and level and depth.

  - Some topics will be appropriate only for students with specific goals/interests. For instance:

    We expect only aspiring hardware specialists study the design of carry-free digital adders (Appendix E).

We expect only students with interest in foundational topics to even glance at topics such as "bijective" number systems (Section 10.2.2).

- Some topics will be studied casually by most students but much more deeply by aspiring specialists. A topic such as graph coloring (Section 13.1) should be on the menu for most students, because of the extensive range of "real" applications which benefit from the topic. We would, however, expect only students with specialized interests—such as task-scheduling—to delve deeply into the subject, say, by studying algorithms for assigning colors to the vertices of general graphs.

- An extreme example of the preceding point relates to rather specialized subtopics that the general student will likely never encounter. While all students will benefit from a cultural/historical introduction to the Fibonacci numbers (Section 9.2.2), only a self-selected few will want to delve into the mathematical depths of the topic, as in Sections 9.2.2.3 and C.3.

• We have tried throughout to expound on the fundamentals of mathematical ideas, unadorned by superficial (from a mathematical perspective) details that arise in specific applications.

• We have addressed certain basic notions (e.g., "number") in layers of increasing detail, as we guide the reader along a journey from student to apprentice.

Our goal is to endow each reader with an *operational* understanding of an activity that we—hopefully evocatively—call "*doing*" mathematics. We want the reader to be able to recognize situations—especially, but not exclusively, relating to the world of computing—wherein mathematical reasoning and analysis can make a positive difference. Of course, a large component of enabling this activity requires imparting mathematical knowledge relating to the fundamental concepts and tools and intellectual artifacts that we have inherited from our mathematical forebears. Another part, though, requires transmitting the understanding of how to use the knowledge effectively and creatively.

We have tried to design the text so that each chapter adds new computing-related mathematical concepts and techniques to the reader's toolkit. We have designed each chapter to develop mathematical tools *in multiple ways*, with the hope that each reader, no matter what her goals, background, and interests, will learn how to think mathematically and to achieve an *operational* understanding of both the activity of computing and the systems that enable that activity.

Our approach toward achieving our goal is embodied in the previous paragraph's promise to develop mathematical tools "*in multiple ways*".

• *We often present several fundamentally different proofs for the same result.*

With tongues only mildly impinging on our cheeks, we encapsulate the philosophy underlying this practice in the following "self-evident truth", or, "axiom".

The Conceptual Axiom. *One's ability to think deeply about a complex concept is enhanced by having more than one way to think about the concept.*

An extreme illustration of this philosophy resides in our multiple derivations of the sum of the first n positive integers. The reader will encounter (in Chapter 6) derivations that adopt each of the following worldviews:

1. The problem is a "textual" one: One writes out the summation symbolically and manipulates the resulting string (as in elementary school).

2. One views each positive integer k as a height-k unit-width rectangle: One sums by calculating the combined area of the rectangles.

3. One views each positive integer k as a collection of k tokens: One sums by aggregating and manipulating tokens into a perspicuous configuration.

4. One views the process of summation "combinatorially"—as the process of determining the number of ways of selecting two items from a set of n items.

5. One views the problem geometrically, by drawing a figure and calculating its area.

6. One views integers via their familiar numerals (say, in decimal notation): One sums by replicating, rearranging, and counting the number of occurrences of the various integers.

While the *number* of derivations of this result exceeds our norm, the *fact* of exploiting multiple viewpoints is a hallmark of our approach.

- *We organize discrete mathematical topics in conceptual layers.*

Our treatment of the fundamental topic "number" illustrates this approach well.

1. We describe the number system that we have inherited from our mathematical forebears in Chapter 4. The treatment in that chapter is mostly discursive but it is peppered with a number of important results and computational insights.

2. In Chapter 8, we take a careful look inside our number system, addressing topics such as prime numbers—and their use within encoding schemes and security. In order to do justice to this deeper layer of material, we insert Chapters 5 and 6, which respectively cover arithmetic and summations, between Chapters 4 and 8.

3. Finally, we turn in Chapter 10 to the issue of how to represent numbers in order to compute with them. Here again, we insert auxiliary material—in this case, Chapter 9, which covers recurrences—between number-oriented chapters, to lay necessary groundwork.

- *We continually point out mathematical insights into* computing-related *concepts, tools, and systems.*

While this book is unquestionably a text on *mathematics*, we never lose sight of the computational motivation for our study. Traces of this motivation are visible, for instance, in:

- Chapters 4, 8, and 10, where we develop material concerning numbers and their representations that underlies much of the world of computing:

  › how to efficiently encode complicated structures as integers

    There are applications to security and cryptography within this subject.

  · how numbers and numerals can model *self-referentiality* in languages

    The consequences of self-referentiality abound throughout philosophy, linguistics, and the foundations of computing. The property leaves its mark on computing-related topics ranging from programming languages and compilers to the inherent complexity of broad families of computations.

  · how to design an adder that does not expend time "rippling carries"

    This is basically an engineering-oriented concern, but its resolution involves important mathematical ideas.

- Chapter 9, where we discuss in detail material about recurrences that is of crucial importance in the design and analysis of algorithms

- Chapters 12 and 13, where we discuss aspects of graphs and networks that are particularly relevant to topics such as:
  · social networks
  · the interconnection networks of parallel computer architectures
  · the design of integrated electronic circuits

• *We describe explicit connections between the mathematics we develop here and the computational topics that have informed our choices of material.*

Section 1.4 discusses how to excerpt from this book within courses that cover a broad range of computation-related topics, from *Algorithms* to *Digital Design* to *Programming Systems* to *Social Media*, and beyond.

• *We sprinkle stories into the text, in the hope of inspiring the reader, while stimulating mathematical thinking and understanding.*

As we remarked at the end of our *Manifesto*, mathematics is done—created, analyzed, applied—by humans. It is both inspiring and illuminating to realize the variety of backgrounds, goals, and motivations that our mathematical forebears had. The digressions also provide momentary diversions to help the reader approach the ensuing material reinvigorated.

• *We present a number of short, informal introductory essays on advanced topics.*

We expect that much of this material will go beyond the level of mathematical achievement that most readers will start out with. We hope that the citations we provide to accompany these essays will inspire further reading, and often even further mathematical exploration.

In furtherance of our hope to excite the reader about further involvement with mathematics, we annotate the citations in our *References* section with tags that identify the roles that individual references play within the mathematical story that we are telling.

- Citations tagged "[H/C]" are identified as being of **H**istorical and/or **C**ultural interest. They expose how our mathematical forebears thought about the subjects being discussed and, thereby, how mathematical thinking has evolved since their writing. Some of these references are, in fact, modern histories; others are (translations of) original texts. It can be fascinating to read about Leonardo Pisano's (*that's Fibonacci's real name*) interest in the demographics of rabbits and about John Arbuthnot's curiosity about the relative birth rates of the sexes in the London of his day.

- Citations tagged "[T/R]" point to advanced and/or specialized **T**extbooks and **R**eference books. They will lead the interested reader beyond the introductory level of this text. For instance, our description in Appendix E of "carry-free" addition can be a stepping stone to a study of *computer arithmetic*.

- Untagged citations identify "modern" expositions of the material we describe in the text. They also usually will lead the interested reader beyond the introductory level of this text. Indeed, they are often quite technical companions to quite dramatic human-interest stories.

- *The text offers many (clearly labeled) opportunities for enrichment within the body of the text and within the exercises that accompany each chapter.*

  - Throughout the text, the symbols "⊕" and "⊕⊕" identify sections that are either more advanced than the body of the text or targeted at a more specialized audience.
  - In similar ways, the symbols "⊕" and "⊕⊕" identify exercises that are more challenging than their untagged kin. We provide hints for the single-⊕ problems, and we provide solutions for the double-⊕ problems, in Appendix H: *Solutions to Selected Exercises*.

- *Many of the single-⊕ and double-⊕ exercises provide fodder for Socratic teaching, group work, special projects, and independent study.*

The Authors

Research. The authors have a long cumulative history of research on

MATHEMATICS:	Discrete math and Mathematical logic
COMPUTER SCIENCE:	Algorithms, Computer architecture, Computation theory, Data analytics, Information retrieval, Resource management, Task scheduling
COMPUTER ENGINEERING:	Digital logic design, Energy conservation, Fault tolerance, Testing of digital logic

They have cumulatively published more than 450 research articles in elite journals and conferences, as well as several research books and a few textbooks on subjects relating to mathematics and/or computing.

Teaching. Over a cumulative span approaching three-quarters of a century, the authors have taught courses in mathematics and mathematical logic, as well as a broad range of topics in computer science and engineering at the following institutions (alphabetically):

Duke University, University of Grenoble, University of Massachusetts, University of Paris, New York University (NYU), Polytechnic Institute of New York, The Technion (Israel Institute of Technology), Faculté des Sciences de Tunis, University of Toronto, Yale University

The courses taught by the authors have included:

Algorithms, Computation and Complexity, Approximation Theory, Computer Architecture, Mathematical Logic, Research Methodology, VLSI Design
in addition to numerous specialized seminars

We owe a tremendous debt to the colleagues and friends and to the students, both undergraduate and graduate, who have enriched our lives over the decades. We do not attempt to list their names for fear of inadvertently omitting some.
We do, however, proudly express our gratitude to the colleagues who have shared their wisdom by commenting on portions of the book. Alphabetically:

Raphael Bleuse	Daniel Cordeiro	Fanny Dufossé
Christoph Durr	Giorgio Lucarelli	Fabrizio Luccio
Clément Mommessin	Geppino Pucci	Peter Rodgers
Ramesh Sitaraman	Nguyen Kim Thang	Jean-Marc Vincent

The exposition has benefited greatly from their input. Of course, whatever imperfections remain are solely our responsibilities.

Paraphrasing the greats over the centuries who have acknowledged a debt to their predecessors,[1] we have truly "seen farther by standing on the shoulders of giants"!

Arnold L. Rosenberg   Denis Trystram
Falmouth, MA, USA   *Grenoble, FRANCE*

---

[1] See R.K. Merton, *On the Shoulders of Giants: A Shandean Postscript*. The Free Press, NY.

*Mathematics is the foundation on which
the edifice of Computing stands*

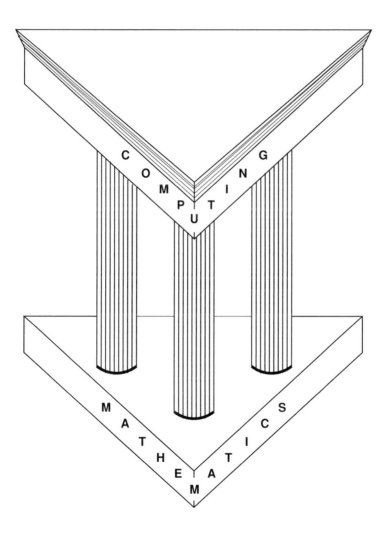

Chapter 1
Introduction

*Je n'ai fait cette lettre-ci plus longue que parce que je n'ai pas eu
le loisir de la faire plus courte.*
(*I have only made this letter longer because I have not had the time
to make it shorter.*)

Blaise Pascal (*The Provincial Letters* [Letter 16, 1657])

1.1 Why Is This Book *Needed?*

How much mathematics does an aspiring computing professional need—and at what
level of expertise? We believe that the answer to this pedagogically fundamental
question is time dependent.

The early generation. In the early days of computing, all aspects of the field were
considered the domain of the "techies"—the engineers and scientists and mathematicians who designed the early computers and figured out how to use them to
solve a range of (mostly compute-intensive) problems. Back then, one expected every computing professional to have a mastery of many mathematical topics.

Children of the early generation. Times—and the field of computing—changed.
The "techies" were able to craft a variety of sophisticated tools that opened up the
world of computing to the general population. Even people at the lower levels of the
educational edifice were able to use imagination and ingenuity, rather than theorems
and formulas, to produce impressive software artifacts of considerable utility.

The modern generation. Pendulums were made to swing. As we note in our *Manifesto*, we now encounter almost daily problems that arise from unanticipated consequences of the often-unstructured ingenuity that produced various software arti-

A. L. Rosenberg, D. Trystram, *Understand Mathematics, Understand Computing*,
https://doi.org/10.1007/978-3-030-58376-7_1

facts. Many would say that we need a renewed commitment to technical discipline that will endow artifacts with:

- *understandable structure*, so that we can determine *what* went wrong when something *does* go wrong.
- *sustainability*, so that changes, which are inevitable in complex artifacts, will not create new problems (especially ones we do not know how to solve)
- *controllability*, so that "smart" artifacts do not become modern instances of Dr. Frankenstein's monster.[1]

While we certainly need not return to the era of the "techies", it is indisputable that we do need a larger contingent of computing professionals who have *bona fide* expertise that enables the needed technical discipline. This text is devoted to the mathematics that underlies the needed computing and science and engineering.

1.2 Why Is *This* Book Needed?

There are many introductory texts on discrete mathematics. What separates this text from its siblings is the plan we have implemented to accommodate our intended audience of aspiring computing professionals.

We begin by recognizing that we live in a world of ever-increasing professional and social diversity.

Historically, computing curricula began as predominantly technically-oriented studies within a school of science or engineering. The evolution of the computing field has led us to move beyond that historical curricular worldview. Modern computing-oriented curricula offer a variety of educational trajectories, including:

- the traditional path, which emphasizes science, engineering, and mathematics,
- paths which emphasize subfields of the humanities or social studies,
- paths which emphasize the *practice* of computing, either in a general setting or within a focused applied field such as business or finance or law or . . . .

The preceding reality has led to a phenomenal broadening of the audience for computing-related curricula, hence for some level of mathematics education. It has also engendered two far-reaching changes in the way we think about the mathematical component of a computing-related education.

1. Many of today's aspiring computer professionals—particularly those targeting the newer subareas of computing—arguably need only specialized knowledge of mathematics. Educators in computing-related fields must, consequently, serve a large population of students in a manner that accommodates the students' quite diverse needs and aspirations.

---

[1] See Mary Shelley (English author), *Frankenstein; or, The Modern Prometheus.*

2. The fields of *information* and *computing* have developed an unprecedented level of overlap. As recently as the mid-twentieth century, these fields were viewed as largely separate concerns: computing was concerned with manipulating discrete objects; information was concerned with transmitting sequences of bits. As parallel computing machines were developed, beginning in the 1960s, computing practitioners had to start paying more attention to the specific ways that information flowed among the processors of a parallel machine, and between these processors and the devices in which data was stored. The development of the Internet, around the 1990s, focused yet more attention on information flow. The marriage between the fields of computing and information dissemination was completed by the end of the twentieth century. Orchestrating the way that information flowed and spread became a *bona fide* specialty within the areas of mathematics that hitherto had focused only on computing. The introduction of information into computing had revolutionary aspects. Information could be replicated at a speed and to a volume that was unmatched by the objects studied in any physical science.[2] Flowing information enabled the construction and operation of the kind of virtual universe hinted at by the computation-theoretic notion of *nondeterminism* (see, e.g. [91]), which had hitherto been viewed as a "pure" mathematical idea. The mathematics that students learn had to adapt so that students learned how to think about information, most especially within the context of computing.

We have sought to keep the increasing diversity of students and of approaches to computing in mind as we have written this text. We have included a broad range of material, in both subject and level. As noted in the *Preface*, we have tried to accommodate the different backgrounds of our expected readership, while leading them all to a level of mathematical sophistication that will enable them to *understand*—and to *do*—mathematics.

In the remainder of this chapter we describe our approach to the text:

- We describe both our strategy for selecting and developing mathematical topics and our tactical organization of topics into chapters, sections, etc.

- We describe the various types of problems that we have included as *exercises* at the end of each chapter, ranging from "practice"-oriented problems to rather challenging ones, whose anticipated level of difficulty is signaled by the symbols "⊕" (challenging) and "⊕⊕" (really challenging).

  We provide hints for the single-⊕ problems along with the problems' specifications; and we provide complete solutions for the double-⊕ problems, in Appendix H: *Solutions to Selected Exercises*.

- We describe the more advanced *enrichment*-oriented material which we have included for the ambitious reader, in several guises:

---

[2] Of course, there are areas within the physical sciences that generate "astronomical" amounts of data—*astronomy* is such an area. But the data accumulated in, say, astronomical studies could safely be warehoused for more relaxed later processing. The data flying around social media have to be dealt with in real time.

   – as sections and subsections in the core text: Enrichment sections are signaled by the prependage of the symbols "$\oplus$" (less specialized or advanced) or "$\oplus\oplus$" (more specialized or advanced) to the title. These sections are typically a bit off the beaten curricular track, because of either specialized subject matter or more advanced argumentation.

   – as appendices: The chapters will also be signaled by "$\oplus$" or "$\oplus\oplus$" in their titles. You should keep in mind that these chapters are being "doubly signaled" by their non-inclusion in the core text. They are likely too specialized or too advanced for the typical introductory mathematics course.

1.3 The Structure of This Book

In our hope of endowing each reader with an operational understanding of how to "*do*" mathematics, we want the reader

- to recognize situations—especially relating to computing—wherein mathematical reasoning and analysis can make a positive difference;
- to identify the mathematical tools that are appropriate for these situations.

1.3.1 Our Main Intellectual Targets

This book is devoted to covering the discrete-mathematics underpinnings of the endeavor of computing: from the design and implementation of devices that perform the actions necessary to compute to the design of the processes that control the devices—including whatever communications are needed among processes and among (sub)devices. Our exposition aims at a number of intellectual targets.

1. *Fundamental concepts*

    Examples:

    - sets—and their embellishments: tuples, arrays, tables, etc.—as embodiments of *object*

    - numbers—and their operational descriptions, numerals—as embodiments of *quantity*

    - graphs—in their many, varied, forms—as embodiments of *connectivity* and *relationship*

    - algebras and functions—adding operations to sets, numbers, and graphs—as embodiments of *structured dynamism* and *computing* and *process*

    We thereby expand the scope of what can be thought about "mathematically".

2. *Fundamental representations*

   Examples:

   - representing and thinking about numbers via many metaphors: slices of pie, tokens arranged in stylized ways, characteristics of geometrical figures (such as rectangles or circles), textual objects

   - understanding the strengths and weaknesses of various positional number representations (e.g., what can you tell about a number by studying its representation?)

   - using grouping and/or replication to represent relationships among objects

   - viewing interrelated objects via many structures: tables, tuples, graphs, pictures, geometric drawings

   We thereby expand the universe of conceptual paradigms that one can use while thinking "mathematically".

3. *Fundamental tools/techniques*

   Examples:

   - using induction to extrapolate from simple examples to complex ones

   - "hopping" between the discrete and continuous mathematical worlds (as one does, e.g., when using integration to approximate summation)

   - using the conceptual tools of asymptotics to argue *qualitatively* about *quantitative* phenomena

   - "hopping" between *"real"-world* mathematical reasoning and the *idealized, and often impractical* reasoning embodied in formal system of logic

   We thereby elucidate and expand upon the conceptual tools that one has access to when one *does* mathematics.

4. *Beyond the fundamentals*

   Our goal in writing this text is not just to expound on the basic notions that enable one to study the phenomenon of computation and its accompanying artifacts. We strive additionally to inspire each reader to dig deeper into the lore of one or more of these notions. To this end, we have presented many basic notions via multiple explanations and numerous exemplars.

   Examples:

   - The ability to "encode" tuples of objects by using the underlying objects themselves lies at the base of some of the pillars of computing. We expound on two approaches to such encodings, one based on *prime numbers* and one on *pairing functions*. We provide pointers to literature that shows how to apply encoding techniques to obtain nonobvious solutions for "real" computing problems. Two such problems are: *efficiently storing arrays/tables whose dimensions can change dynamically* and *tracking the computing output of agents in a collaborating team (such as a volunteer computing project).*

- The use of recursion as a control structure in programming/computing has a purely mathematical analogue, namely, *recurrences*. Recurrences are extremely useful in analyzing the correctness and efficiency of recursively specified computations. Every student gains at least some familiarity with the most basic versions of recurrences. We introduce variations on the theme of these familiar recurrences, which offer approaches to more complicated computational situations. We illustrate the use of recurrences in crafting a quite approachable analysis of an apparently dauntingly complex *token game*.

- Many phenomena that appear to arise from arcane, inherently computational sources are actually rather easily understood applications of purely mathematical phenomena. We spend considerable time expounding on such phenomena and their applications. *Satisfiability problems* are one phenomenon of this sort; certain *specialized number systems* are another.

1.3.2 Allocating Our Targets to Chapters

1.3.2.1 Chapter 2: "Doing" Mathematics

We strive to acclimate the reader to mathematics as a living discipline and an integral part of the world of computing, by beginning the book with a chapter entitled *Doing Mathematics*. The chapter provides a toolkit for mathematical reasoning, mainly by focusing on the practicalities of mathematical reasoning, but most importantly by expounding on the concept of a mathematical proof. We survey the most commonly encountered techniques for crafting such proofs, and we illustrate each with several examples.

Part of understanding proofs is being aware of what makes the endeavor of proving things difficult. We provide both mathematical and historical background relating to important intellectually challenging topics, including, notably, how to reason about objects that are infinite and about objects that are finite but very large (and, possibly, growing).

1.3.2.2 Chapter 3: Sets and Their Algebras

Sets are the stem cells of mathematics. They begin as the most primitive mathematical objects, but as soon as one endows them with operations—for creating more inclusive ("bigger") sets, for selecting more exclusive ("smaller") subsets, for enabling the structure of tupling—they quickly afford one a powerful substrate for doing almost all of mathematics.

The power of tupling. Tupling enables one to isolate entities such as *relations*, which then enable one to impose and identify *order* within, and among, sets. Tupling enables one to isolate *functions*, which then enable one to *encode* sets as other, seemingly unrelated, ones—*example*: encoding computer programs as positive integers.

Algebras: Operations and the laws that govern them. The operations within any collection of operations inevitably obey certain "laws" as they interact; for instance, operation ∘ may (or may not) be *commutative*, i.e., obey the relation

$$x \circ y = y \circ x$$

it may (or may not) be *associative*, i.e., obey the relation

$$x \circ (y \circ z) = (x \circ y) \circ z$$

The combination of a set S (of "objects") and a set of operations on S, together with the laws that govern the operations, is an *algebra*—and there are myriad algebras that play important roles in our (daily and mathematical) lives.

- *Boolean algebras* focus on sets and operations on sets.

- A special class of Boolean algebras is the class of *Propositional logics*. These algebras underlie computing-related topics that range from *digital logic*, the basis of all computer design, to the *satisfiability problems* that play a major role in aspects of *complexity theory* and *artificial intelligence* (*AI*).

- *Numerical algebras* govern our daily lives, by enabling us to perform crucial basic functions such as counting and calculating (using arithmetic).

1.3.2.3 Chapters 4, 8, 10: Numbers and Numerals

The first mathematical concepts that children learn about usually involve numbers. We spend much of our early lives expanding our number-based knowledge. We proceed from counting to manipulating numbers by adding and subtracting and multiplying and dividing them. We progress from using fingers and toes as "names" of numbers to using a variety of numeral-forming schemes.

For many of us, "mathematics" *means* "numbers and arithmetic": We never get to explore the powerful, and beautiful, mathematical concepts and tools which have enabled many of the great advances of science and engineering. Even fewer of us get to explore the conceptual extremities of mathematics which led to the field's being dubbed *"the queen of the sciences"* by the great nineteenth-century German mathematician Carl Friedrich Gauss.

While mathematics is assuredly much more than "just" numbers and arithmetic, one could spend one's life fruitfully exploring nothing beyond these topics. The three chapters we describe in this subsection introduce the reader to numbers and arithmetic "in layers".

- *The basic objects and properties of our number system.*

  Chapter 4 introduces the subject by means of a "biography" of our number system as it has evolved over the millennia. The system's evolution has been stimulated by the need for greater explanatory and manipulatory power over the phenomenally expanding knowledge base exposed by science and technology.

- *Building the integers and building with the integers.*

  Chapter 8 is devoted to looking both inward and outward at the most easily intuited component of our number system, the *integers* (or, "counting numbers"). Looking inward, we discover the *prime numbers* (familiarly, "the primes"). A theorem from antiquity—which has been honored with the name *The Fundamental Theorem of Arithmetic*—exposes the primes as the "building blocks" of the entire set of integers. Looking outward, we introduce the use of the primes as the basis of a variety of *coding schemes* for a broad range of structures. Both the strengths and the weaknesses of prime-based encoding schemes have inspired the development of many other important such schemes. These encoding schemes lead to numerous nonobvious applications, ranging from storage schemes for dynamic data structures to provably secure encryptions of various structures.[3]

- *Operational number representations and their consequences.*

  In Chapter 10 we shift our focus from *numbers*, the objects that we count with, to *numerals*, the *names* that we use to represent and manipulate—i.e., compute with—numbers. The importance of using an appropriate numeral system for a particular application can be illustrated by the design of *digital adders*— circuits that implement the operation of addition. Digital adders that execute the "carry-ripple" process which we all learned in elementary school take roughly n steps to add a pair of n-digit numbers. We could, in contrast, design digital adders that embody a nonstandard, nonobvious, *signed-digit* representation scheme and that require only a fixed constant number of steps to add a pair of n-digit numbers—independent of the length n. (Adders that use signed-digit schemes incur heavy costs that standard adders do not, which is why we generally use standard adders.)

We spread our coverage of numbers and numerals over a *noncontiguous* sequence of three chapters because we need additional material to progress from one number-oriented chapter to the next. Specifically, we employ concepts and tools from Chapters 5 and 6 (which cover *arithmetic* and *summation*, respectively) in essential ways within Chapter 4. Similarly, we need material from Chapter 9 (which covers *recurrences*) as we develop Chapter 10.

1.3.2.4 Chapters 5 and 6: Arithmetic and Summation

Numbers are important in our daily lives only when we *use* and *manipulate* them.

---

[3] We provide in Chapter 8 a brief discussion of the relationship between encoding and encryption.

Chapter 5 discusses *arithmetic*, the basic operations—addition, multiplication, etc.—that we use to manipulate numbers and the laws that these operations obey. The chapter then moves on to complex operations on numbers—polynomials, exponentials, and logarithms. It closes with pointers to topics for advanced study—including topics that are central to modern computer applications such as big data. The chapter's treatment of polynomials is far-reaching:

- The chapter begins with basic facts about these special functions.

- The chapter then introduces the centuries-old study of solving—or being unable to solve—single-variable polynomials using *radicals*—square roots, cube roots, etc. (The latter topic is discussed only informally because of its advanced nature.) The well-known *quadratic formula* is the simplest instance of solving polynomials by radicals: The two solutions to the generic polynomial equation

$$ax^2 + bx + c = 0$$

are

$$x = \frac{-b + \sqrt{b^2 - 4ac}}{2a} \quad \text{and} \quad x = \frac{-b - \sqrt{b^2 - 4ac}}{2a}$$

- The chapter's treatment of polynomials then digresses to present an extremely important result about two-variable polynomials, Newton's famous *Binomial Theorem*. This result enables one to go easily between polynomials in a certain family and their roots. The Theorem's "smallest" instances provide the following equations.

$$(x+y)^2 = x^2 + 2xy + y^2$$
$$(x+y)^3 = x^3 + 3x^2y + 3xy^2 + y^3$$

- The chapter's treatment of polynomials closes with a short, informal discussion of *Hilbert's Tenth Problem*, an advanced topic of immense mathematical import. In one line: The work on this Problem demonstrates that the single topic of discovering integer roots of arbitrary polynomials with integer coefficients—or proving that no such roots exist—captures (read: *encodes*) the full complexity of performing arbitrary computations!

The chapter finally introduces the operations of taking exponentials and logarithms. These are mutually inverse operations, in the sense embodied in the equations

$$x = \log_b(b^x) = b^{\log_b(x)}$$

A fundamental insight here is that the arithmetical system based on these functions is *almost* identical to our conventional arithmetic—but with multiplication replacing addition and division replacing subtraction. (*This is the basic insight underlying the slide rules that were techies' pocket calculators for many decades*.). The qualifier "almost" hints at the adjustments needed to accommodate the impossibility of dividing by 0. This pair of operations plays a fundamental role in the field of information theory, whose importance to computing cannot be overstated.

Chapter 6 is basically a *tools* chapter. It studies how to break a complex operation into simple constituents. It is devoted to analyzing a broad variety of families of summations, providing exact solution-sums for many and approximate sums for others. All of the finite summations we study have the general form

$$S = s_1 + s_2 + \cdots + s_n \tag{1.1}$$

Each s_i is termed a *summand* of S.

Explanatory note

In the interest of precision, it is convenient to reserve the word *summation* for the formal mathematical expression (1.1) and to reserve the word *sum* for the result of evaluating the summation. Thus, for instance, if each summand in expression (1.1) is an integer, then the sum of summation S is an integer.

In situations where we allow summations to have infinitely many terms—for instance, with the famous summation of inverse powers of 2

$$1 + \frac{1}{2} + \frac{1}{4} + \frac{1}{8} + \cdots + \frac{1}{2^k} + \cdots$$

(whose sum is 2), we allow this pattern to continue without end. We include here *arithmetic summations*, in which all adjacent summands, s_{i+1} and s_i, have a common *difference*, and *geometric summations*, in which all adjacent summands, s_{i+1} and s_i, have a common *ratio*. We discuss also powerful techniques for estimating the solution-sums of summations of consecutive terms of a "smooth" function.

The mathematics we exploit to derive exact and/or approximate sums for many of the classes of summations we study provides us with the opportunity of looking at numbers and their summations in many quite distinct ways—from textual to pictorial to geometric, and beyond. For this reason, Chapter 6 is extremely important as the reader gains traction in the endeavor of "doing mathematics".

1.3.2.5 Chapter 7: The Vertigo of Infinity

This chapter deals with mathematical objects whose very size—ranging from *finite but very large* to truly *infinite*—is difficult to reason about, for one of two reasons.

1. Humans have developed the important ability to reason abstractly by blurring descriptions. We notice, e.g., the ways in which all phenomena that experience *linear* growth differ from phenomena that experience *quadratic* growth, and that both classes differ from phenomena that experience *cubic* growth—and so on.

Additionally, we often encounter finite objects whose sizes can change dynamically or can be known only approximately. Consider, e.g., social networks: they expand and contract in unpredictable manners, so their population statistics cannot be known exactly.

We require a language for talking—and rigorously reasoning—about blurry distinctions and approximately known objects. We also need a formal analogue of arithmetic for "calculating" the statistics of such objects.

The topic of *asymptotics* (Section 7.1) fills both needs, for a large variety of dynamic mathematical phenomena. Asymptotics thereby enables *reasoning qualitatively about inherently quantitative phenomena*.

2. We often have to reason about objects that are actually infinite. We encounter such situations, e.g., in areas that combine, in some way,

- mathematics—e.g., classes of numbers or of functions

- logic—e.g., expressions with special properties, such as theorems or proofs

- linguistics—e.g., sentences that share (syntactic or semantic) characteristics

Since antiquity, we have been confronted by infinite objects that behave very differently from the finite objects of daily discourse. As just two examples:

- Why does a shot arrow reach its target? The following seemingly cogent argument shows that it does not.

  The argument states—correctly—that after the arrow traverses one-half the distance to the target, it still has one-half the distance to go; after traversing half of the remaining distance, it still has one-quarter of the original distance to go. Continuing, the arrow will have a never-ending shrinking distance that it has yet to traverse: one-eighth, then one-sixteenth, then one-thirty-second, and so on. How, then, can the arrow ever reach the target?

- We have had to adapt to the fact that "provable" and "true" are distinct concepts within most realistic systems of logic, despite their intuitive coincidence.

In a slightly more sophisticated vein, we all know that there are "equally many" odd integers as all integers—even though the former set is obtained by discarding half of the latter set's elements.

Section 7.2.1 provides the background necessary to cope with these puzzles and navigate the hazard-laden unfamiliar world of the infinite.

1.3.2.6 Chapter 9: Recurrences

The functions discussed in Chapter 5 are described by static "closed-form" expressions. In contrast, Chapter 9 is devoted to a family of *computational procedures—recurrences*—which calculate the value of a function f on an argument n in terms of the values of f on smaller arguments. The patterns that underlie recurrent computations have been known for millennia to occur throughout nature—in the growth patterns of many plants and in the demographics of many animals, among other places. The mathematics that describes such computations is as elegant and aesthetic as it is useful. Among the enormous variety of recurrent patterns that one can identify, we select three computationally important ones to focus on.

1. *Linear recurrences.* Recurrences of the form

$$f(n) = af(bn+c) + dn + e \qquad (1.2)$$

(where a,b,c,d,e are constants) are a welcome friend when one analyzes the costs of executing a broad range of algorithms, using cost measures such as time, memory usage, power requirements, and amount of communication.

We introduce two simple, yet nontrivial uses of recurrence (1.2) in the analysis of algorithms. We refer the reader to an algorithms text such as [36] for details.

- The *binary search algorithm*, which determines whether an item x occurs within a given ordered list of n items, begins by partitioning the list in two (so $b = 1/2$ and $c = 0$ in (1.2)). It then compares x with the list's middle item(s)—there is one middle item when n is odd, two when n is even—(so $d = 0$, and e is the fixed constant cost of the comparison(s)). Based on the outcome of the comparison, the algorithm recurses with a binary search on the half of the list that could contain x (so $a = 1$).

- The *merge-sort algorithm* builds on a natural n-step algorithm for merging two n-item lists of "order-comparable" items (i.e., for every pair, x,y, of distinct list items, either $x < y$ or $y < x$). The algorithm first merges adjacent odd-even pairs of items, to end up with sorted two-item lists. It then merges adjacent two-item lists, to end up with sorted four-item lists. It continues thus with eight-item lists, then sixteen-item lists, ..., until it finally merges the "top-level" two $n/2$-item lists to achieve the goal of a sorted n-item list. The performance of this algorithm, as measured by operation-count, is given by recurrence (1.2) with $a = 2$, $b = 1/2$, $c = 0$, $d = 1$, and $e =$ (the fixed cost of a two-item comparison).

The centerpiece of our discussion of linear recurrences is the so-called *Master Theorem*, which uses geometric summations to generate explicit—rather than recurrent—expressions for the values of a function f on an arbitrary argument n.

2. We highlight two *bilinear* recurrences[4] of especial computational importance.

- The *binomial coefficient* articulated as "n choose k", and commonly denoted by one of the following three notations

$$\binom{n}{k} \quad \text{or} \quad \Delta_{n,k} \quad \text{or} \quad C(n,k)$$

---

[4] The expression at the heart of a *bilinear* recurrence involves a linear combination of two variables.

plays a central role in a broad range of mathematical domains. Two rather distinct examples:

- $C(n,k)$ is the number of ways to select k items from a set of n items. For instance, $C(4,2) = 6$, because the two-element subsets of $\{a,b,c,d\}$ are:

$$\{a,b\}, \ \{a,c\}, \ \{a,d\}, \ \{b,c\}, \ \{b,d\}, \ \{c,d\}$$

Analyzing $C(n,5)$ will reveal, for example, why the "three-of-a-kind" deal beats the "two-pair" deal in the game of poker.

- These numbers play a prominent role in evaluating arithmetic summations; e.g., we derive the following equation (in several ways) in later chapters:

$$1 + 2 + 3 + \cdots + n \ = \ C(n+1,2)$$

The *family* of binomial coefficients is defined by the bilinear recurrence

$$C(n+1,k+1) = C(n,k) + C(n,k+1) \ \text{ for any } \ k \geq 0$$
$$C(n,0) = 1$$
$$C(n,1) = n$$

- The sequence of *Fibonacci numbers*, named for the Pisano mathematician known by the nickname "Fibonacci", are defined by the recurrence

$$F(n+1) = F(n) + F(n-1)$$
$$F(0) = F(1) = 1$$

Fibonacci invented the famous sequence named for him as he observed the populations of consecutive generations of progeny produced by a pair of rabbits; the sequence also arises in the structure of many plants; it further is said to have a semi-religious aspect in the architecture of ancient Greek temples. Aside from its important descriptive role, the sequence plays a significant role in the analysis of algorithms and as the basis for a nonstandard system of numerals.

Variations on the preceding three families of recurrences provide supplemental material in this chapter.

1.3.2.7 Chapter 11: The Art of Counting: Combinatorics, with Applications to Probability and Statistics

Certain subfields of mathematics have been known for centuries as "the art of counting".[5] This chapter introduces three related areas of discrete mathematics that are

---

[5] German mathematician Gottfried Leibniz used the phrase as the title of his 1666 doctoral thesis.

based on that art. Indeed, one can argue that the following chain of applications, while grossly oversimplified, is not misleading.

Statistics   can be viewed as applied *Probability*
Probability   can be viewed as applied *Combinatorics*
Combinatorics can be viewed as applied *Counting*

These are important connections. Elements of probability and statistics infuse every area of endeavor—from science to finances to informatics, and beyond—where computing is involved. In order to function successfully in today's world, one needs statistical and probabilistic literacy—a command of the foundations and the operational rules of "the art of counting".

- This chapter begins by developing the basic *rules of counting:* How many strings of length n can one form using c characters? How many subsets does a set of k elements have? We discover that many seemingly distinct problems of this type are actually *encodings* of one another!

- The chapter moves on to concepts related to *grouping and arrangement* using mechanisms such as permutations, combinations, and derangements. Variations of these themes allow us to develop *selection*-based concepts, such as: In how many deals of five playing cards do all cards have the same suit?

- We develop the elements of *discrete* (or, *combinatorial*) *probability*. The probability (or, *likelihood*) of an event is defined as the ratio of the number of ways that the event occurs, divided by the total possible number of outcomes. For instance, the probability of achieving the result "7" when rolling two dice is the ratio of the number of rolls that produce "7" (which is 6), divided by the total number of rolls (which is 36).

  We illustrate the elements of probability by simple, fun, examples such as assessing the likelihoods of various deals in the card game *poker* and of various rolls of a pair of dice in the game *craps*.

  The most provocative portion of the chapter discusses the value of probability-related concepts in two situations where one might not expect to encounter them. Both situations are described via puzzles which, while lacking in intrinsic importance, provide good platforms for developing significant aspects of probabilistic thinking. The first puzzle, in Section 11.2.2, arose from a television game show that was popular a few decades ago. We demonstrate in this instance how a knowledge of probabilities can enhance one's likelihood of making good guesses in the presence of partial information. The second puzzle, in Section 11.2.3, asks how many people one must assemble in order to expect (at least) two to have the same birthday. Not surprisingly, this second puzzle is widely known as *the birthday puzzle*.

- Finally, we illustrate the *statistical* way of thinking by carefully analyzing two ways for deriving the likelihood of achieving a specific sum—such as 6—when rolling *three* dice. This example will enable us to generalize from the probabilities of specific events to (statistical) *distributions* of these probabilities.

Even if one intends to "do" statistics mainly with the aid of preprogrammed packages (or apps), it is valuable to understand what the numbers produced by an app mean—and what the numbers *do not* mean! Anyone who aspires to designing and/or executing and/or analyzing experiments *must* understand crucial notions such as *randomness* and should be conversant with the most common statistical distributions. *Lives can depend on such knowledge!*

1.3.2.8 Chapters 12 and 13: Graphs and Related Topics

Graphs are arguably the most important representational tool in all of mathematics— most certainly in the subfields related to computation. In their most basic form, graphs represent any binary relation; a brief sampler:

- the structure of a family, as exposed by the parent-child relation;
- the structure of an electronic or a communication circuit, where certain pairs of entities have the right to intercommunicate—perhaps only unidirectionally.

The structure represented by a graph can expose not only the fact that certain pairs of entities can intercommunicate, but also the number of inter-entity "links" that must be traversed to achieve communication.

Even with this rudimentary discussion, one can intuit the range of real-life problems that can be modeled using graphs. Many such problems use a graph to represent entities that can "talk" to one another, in some sense. A variety of associated questions could be of the form, "Who knew what when?"

Among the representational roles that graphs play in computational settings are as models that expose dependencies among the objects that the graph interconnects. A common relation studied in computer applications exposes that task A in a program depends on input from task B—so that B must be performed *before* A. The notion of *graph coloring* is exceedingly important in scheduling and related applications: In its conceptually simplest form, one colors the tasks of a dependency graph in such a way that like-colored tasks can be executed concurrently—because they are computationally independent. The challenge in this scenario is to color a given task-graph with *as few colors as possible*. This is a computationally difficult task in general, but we expose some of the sophisticated mathematics that has been developed in order to study the graph-coloring problem. *Path problems* in graphs provide another entry to a large variety of scheduling applications. Of particular interest are problems that require some object (e.g., a datum) to be passed around within a graph in a manner that achieves a "coverage" goal, e.g., so that the object encounters all of the graph's entities or traverses all of the graph's inter-entity links.

If *binary* relations are not adequate for a person's modeling needs, then the expanded notion of *hypergraphs* can be used to model relations beyond binary, even those in which the "arity" of the relation varies from one related group of items to the next; a brief sampler:

- the structure of a family, as exposed by two relations: the parent-child relation and the sibling relation;

- the structure of a bus-connected communication setup: entities on a single bus can all "hear" one another;

- social networks in which aggregations of "friends" have special intercommunicating privileges.

1.4 Using the Text in Courses

We have tried to design this book to meet the mathematical needs of a broad range of readers, in terms of both preparation and goals. We hope that the reader will use the book as the first step in a lifelong journey. Toward this end, we have designed the book to fill multiple needs for both students and instructors:

- as a *textbook* for a beginning university discrete math course

- as a *source of mathematical preliminaries* for a broad range of beginning and intermediary university courses that benefit from background in mathematical topics

- as a *tutorial* on how mathematicians model and think about the worlds of mathematics, of science, of nature, and of people

- as a *launchpad* into further study of the historical, social, and technical aspects of mathematics

1.4.1 Resources

1.4.1.1 References

We have provided several types of references, which are clearly marked, either by labels or by their positioning within the text.

SOURCE MATERIAL FOR ALL COVERED TOPICS   We provide pointers to references for all topics that we cover in this text—both concepts and results. By following these pointers, the reader will have access to the original presentations of covered topics. (Of course, we have worked toward finding a single level of discourse and a single "voice" for the cited material. But it is often beneficial to experience the multiple voices in the cited references.)

We have tried to make this book more valuable by supplementing the "standard" citations to modern technical source materials with two other genres of citation.

Historical and/or cultural sources.   We provide many references to the *original* work on many important topics. The entries in our bibliography to such references are tagged with the label **[H/C]**, for "**H**istorical or **C**ultural".

- By perusing historical sources such as (naming[6] just a diverse few):

  - J. Arbuthnot: An argument for Divine Providence taken from the constant regularity observed in the births of both sexes

  - J. Backus et al.: The FORTRAN automatic coding system

  - T. Bayes, R. Price: An essay towards solving a problem in the doctrine of chance

  - G. Boole: *An Investigation of the Laws of Thought*

  - A. de Moivre: *The Doctrine of Chances*

  - Euclid: *Elements*

  - J. Nash: Equilibrium points in *n*-person games

  - I. Newton: *Philosophia Naturalis Principia Mathematica* (Referred to popularly as *Principia Mathematica*)

  - A.M. Turing: On computable numbers, with an application to the Entscheidungsproblem

  - J. von Neumann, O. Morgenstern: *Theory of Games and Economic Behavior*

  the reader can gain a sense of "where it all began". Some of these sources are challenging to read—for example, Euclid's verbal exposition in *Elements* certainly differs from that in a modern text (although his drawings will just as certainly have some familiarity)—but one cannot escape the feeling that one is observing greatness!

- Most of the preceding citations reference the original treatises on various subjects, but we must never forget our indebtedness to the—usually anonymous—workers in the vineyards of mathematics who made "minor" advances *and* to the greats of ages past who ensured that the many contributions of these workers would be available to following generations. A stalwart among these transmitters of knowledge was the great calculator al-Khwārizmī (*eponym of the word "algorithm"*) whose monumental *Liber Abaci* transmitted much mathematical wisdom from Asia to the then-backward lands of Europe.

- Some of the historical/cultural references read like an adventure story—albeit of a very intellectual sort:

---

[6] Complete citations appear in *References*.

– There is no better way to absorb the drama and excitement of the mathematical journey than to read of Hilbert's celebrated 23-problem agenda for twentieth-century mathematics and then to read of the blockbuster studies that exploded some of the goals of that agenda:

– The epochal results of K. Gödel in the 1930s which showed that mathematics could *provably* never capture *all* truths.

– The blockbuster solution of Hilbert's famous *Tenth Problem* in the 1970s. In fact, you can read the *firsthand account* by M. Davis, Y. Matiyasevic, and J. Robinson describing the hills and valleys of their lengthy journey toward this solution.

– Highlighting a different twentieth-century adventure: You can read the *firsthand account* by K. Appel and W. Haken of the journey that led to their proof of the celebrated *Four-Color Theorem*. This result answers a question that originated in the nineteenth century: How many distinct inks suffice to color the countries on *any* map of the world in such a way that bordering countries received distinct colors? Easily (as we shall see in Section 13.1), the answer can be no smaller than four; Appel and Haken show that four colors suffice! Adding to the drama of Appel and Haken's accomplishment: *This was the first proof of a major mathematical theorem in which a digital computer played a* mathematical *role rather than just a* calculational *role.*

ADVANCED/SPECIALIZED TEXTS AND REFERENCE BOOKS. The entries in our bibliography to such references are tagged with the label **[T/R]**, for "**T**extbook or **R**eference". We provide pointers to a long list of excellent texts that "pick up" where this introductory text "leaves off". These texts cover a broad range of mathematical topics, including: Algorithms, Coding Theory, Combinatorics, Computation Theory, Computer Arithmetic, Graph Theory, Mathematical Logic, Number Theory, Probability Theory, and Statistics.

1.4.1.2 Cultural asides

Mathematics is invented by humans. Motivated by superficial curiosity, humans exploit their innate powers to detect patterns. Motivated by a deeper form of curiosity, humans develop conceptual tools in order to explain and spread understanding of the observed patterns.

It is our continuing goal to lend the reader a historical perspective on the way that progress has been made in major areas in mathematics. By exposing the human side of what have turned out to be fruitful pursuits of curiosity, we strive to motivate readers to launch their own journeys of discovery. Some illustrative examples of fruitful curiosity:

Archimedes's quest to count the number of grains of sand on beaches; Leonardo Pisano's (Fibonacci) curiosity about the demographics of rabbits; John Arbuthnot's curiosity about the relative frequencies of male vs. female births in humans.

We strive to help the reader learn how to *do mathematics* by describing the entire lifespans of several mathematical ideas and results. We exhort the readers to:

1. cultivate the ability to observe possible patterns in the world

2. develop facility with mathematical concepts so as to describe observed patterns with precision and rigor

3. hone their analytical skills so that they can pursue the implications of the patterns that they have either verified or refuted.

The stories within our text are intended to foster the first step in this program. Our inclusion of multiple proofs for many results, developed from quite distinct viewpoints, is intended to foster both the second and third steps in the program.

1.4.1.3 Exercises

Continuing with the pedagogical philosophy that motivates all aspects of this book, we have included a spectrum of exercises to supplement the text. Each is annotated by the lesson it is supposed to teach/reinforce.

- At the lowest rungs in the ladder—a physical metaphor seems consistent with the noun "exercise"—are the practice problems that reinforce the reader's command over the concepts and tools we develop.

  These exercises will expose possible places that need further study and, we hope, suggest digressions that the reader will find motivating and inspiring.

- In the intermediate range of the ladder are the exercises that require the reader to "do" some original mathematics—for which the text to that point has adequately prepared the reader.

  Each of these exercises is tagged with the symbol "$\oplus$" and is accompanied by a hint.

- As the reader reaches the upper rungs of the ladder, she will encounter exercises that really require mathematical insights.

  Each of these exercises is tagged with the symbol "$\oplus\oplus$" and is annotated by

  - an explanation of why the exercise was included

  - a hint at how to get started

  - a sketch of a solution, in Appendix H: *Solution Sketches for $\oplus\oplus$ Exercises.*

- *Many of the single-⊕ and double-⊕ exercises provide opportunities for Socratic teaching, group work, and independent study.*

Explanatory/historical note.

The fifth-century BCE Athenian philosopher Socrates is the eponym of a teaching paradigm which is equally rewarding for the instructor and the instructed.

WIKIPEDIA™ defines Socratic teaching as "a form of cooperative argumentative dialogue between individuals, based on asking and answering questions to stimulate critical thinking and to draw out ideas and underlying presuppositions."

Within the context of teaching mathematics, Socratic teaching calls for discussions in which participants cooperate toward joint discoveries. Instructor-guided cooperative solving of problems is a major focus in these discussions.

1.4.1.4 Appendices

Mathematics is done by humans! The authors are human mathematicians. We each have our own favorite tidbits that we would like to share with you. Some of these tidbits come from our own research; others come from studies that are kindred to those we have engaged in ourselves; yet others just strike our personal fancies. We have prepared a buffet of such tidbits for the interested readers' amusement. Shorter tidbits appear (clearly identified as "*enrichment*") within the core text; longer ones appear within the appendix that follows the core text. None of the tidbits requires mathematics that reaches far beyond the topics covered in the text. Our hope is that even skimming an appendix that deals with a topic of interest might kindle a flame that will motivate some readers to make mathematics a part of their lives.

1.4.2 Paths Through the Text for Selected Courses

The material in this Introductory chapter and in the early parts of Chapters 2–13 is designed to help the reader decide which topics are appropriate for her desired course of study. The development within each chapter will help the reader fine-tune a path through a topic, deciding whether the chapter's content should become the reader's

- *casual acquaintance*

  A reader who is unlikely to need detailed information about graphs might, for instance, want only passing familiarity with the basic concepts of graph theory (the first half of Chapter 12).

- *good friend*

  A reader who is interested in the architecture of communication networks and/or parallel architectures—as a user—might supplement the first half of Chapter 12 with the second half of that chapter, which gives some detail about graphs that have proven useful in such specialties.

- *intimate friend*

  A reader who expects to use graph-theoretic modeling in sophisticated endeavors such as task-scheduling or designing compilers or analyzing complex data might add Chapter 13 to her study list—or perhaps even use this material as a springboard to more advanced material such as appears within the enrichment content.

To assist the reader in selecting appropriate paths through the text, we close this chapter with the following course-chapter table, which focuses on eleven courses that appear in many computation-oriented university curricula. To help the reader focus on the table, we provide the following capsule summary of the course content in the chapters or sections that head the columns of the table.

Chapter	Topic
2	Introduction to mathematical reasoning
	Overview of proof techniques
3.2-3.3	Sets and structured sets
3.4	Boolean algebra and logic
4	Numbers 1: Introduction to our number system:
	the integers, the rationals, the reals, the complex numbers
5	Introduction to arithmetic (operations and their laws)
6	Summation: Exact and approximate evaluation of finite and infinite series
7.1	Calculating with infinite objects: asymptotics
7.2	Reasoning about infinite objects and systems: paradoxes
8	Numbers 2: Prime numbers; Pairing functions; Finite number systems;
	Countability; Number-based encoding
9	Recurrences: Solving recurrences; Applications of recurrences
10	Numbers 3: Numerals: Representations of numbers:
	Uncountability; Recognizing integers, rationals via numerals
11	Combinatorics; Discrete probability; introduction to Statistics
12.1-12.2	Graphs 1: Basic properties and definitions; Graph-based models; Trees
12.3	Graphs 1: "Named" classes of graphs and their areas of application
13	Graphs 2: Vertex- and edge-coloring, with applications;
	Path- and cycle-detection and selection

Course/Chapter	2	3.2	3.4	4	5	6	7.1	7.2	8	9	10	11	12.1	12.3	13
		3.3											12.2		13
Everyone	•	•	•	•	•		•	•					•		
Algorithm Design and Analysis	•	•	•	•	•	•	•	•	•	•	•	•	•	•	•
Artificial Intelligence	•	•		•			•	•			•		•		
Computational Science	•	•		•	•		•	•		•	•		•		
Computer Architecture	•	•	•	•		•	•	•		•	•		•	•	•
Cryptography and Cryptology	•	•	•	•	•	•	•	•	•	•	•	•	•		
Data & Information Retrieval	•	•	•			•	•	•					•		•
Digital Logic	•	•	•	•			•	•		•	•	•	•	•	•
Networks	•	•		•		•	•	•			•		•	•	•
Social Media	•	•	•				•	•					•	•	
Systems & Program Models	•	•	•		•		•	•					•		•
Theory: Complexity, Computation	•	•	•	•	•	•	•	•	•	•	•	•	•		•

Of course, the *extent* and *depth* with which a reader studies a relevant chapter will depend on the reader's detailed goals.

Chapter 2
"Doing" Mathematics: A Toolkit for Mathematical Reasoning

> ... *where a Mathematical reasoning can be had, it's as great*
> *folly to make use of any other, as to grope for a thing in*
> *the dark, when you have a candle standing by you.*
>
> John Arburhnot (*Of the Laws of Chance*, 1692)

Mathematics can be viewed as a discipline that crafts formal models of real phenomena or structures and then uses the models to reason rigorously about reality. This is a daunting charter for the neophyte who aspires to "do" mathematics. How does one observe and isolate the features of a real phenomenon within a formal model? What, in fact is a *formal* model? Once one has such a model, how does one use it to reason *rigorously* about the piece of reality that the model is intended to capture?

This chapter constitutes our attempt to answer the preceding daunting questions. Of course, details and examples will largely await the technically deeper chapters of this text. What we present here merely lays a foundation by providing intuition and motivation and examples. This chapter "amuses the palate"[1] in preparation for the introductory mathematical repast that we offer the reader throughout this text. Before we present the hors d'oeuvres that we hope will entice the reader, we "set the table" for our feast.

It is surprising to most non-mathematicians that much of "doing" mathematics—a subject that the brilliant nineteenth-century German mathematician Carl Friedrich Gauss touted as "queen of the sciences"[2], can be viewed as *pattern-matching*—albeit of a monumentally sophisticated variety. Mathematicians are trained to understand pieces of reality to a depth that allows them to understand how apparently unrelated concepts *A* and *B* can be conceptualized via the same *abstract representation*, and to analyze (computationally, in our bailiwick) the advantages of exploiting

---

[1] ... in the sense of the French *amuse-bouche*.

[2] See, e.g., Wolfgang Sartorius von Waltershausen, *Gauss zum Gedächtniss* (1856).

© Springer Nature Switzerland AG 2020
A. L. Rosenberg, D. Trystram, *Understand Mathematics, Understand Computing*,
https://doi.org/10.1007/978-3-030-58376-7_2

such representations. It will be useful to the reader to keep the "mathematics-as-pattern-matching" metaphor in mind while reading (from) this book, all the better to enjoy the many instances of pattern-matching that populate its pages.

This text is devoted to describing and explaining the practice of mathematics within the world of computing. By means of plentiful examples, we hope to convince the reader of the importance of mathematics in one's quest to master the technology and the methodology of computing. By means of extensive explanations—we often verify the same fact many times, from multiple, orthogonal vantage points—we provide the reader tools to recognize mathematical aspects of computational settings and phenomena and technology, and we provide guidelines for using those tools effectively.

We begin with two motivational aphorisms.

REGARDING MATHEMATICAL MODELS.
  Entia non sunt multiplicanda praeter necessitatem.
  (Explanations and arguments should employ as few concepts as possible.)
                                                    William of Occam (fourteenth century)

Occam's Razor, known also as *The Principle of Parsimony*, urges the reader to keep every (philosophical) argument as simple as possible. This quest for simplicity is essential as one seeks perspicuous mathematical models of complex computational phenomena and as one reasons about the resulting models.

REGARDING MATHEMATICAL ARGUMENTATION.
  I mean the word proof not in the sense of the lawyers, who set two half proofs equal to a whole one, but in the sense of a mathematician, where $\frac{1}{2}$ proof $= 0$, and it is demanded for proof that every doubt becomes impossible.
                                                    Carl Friedrich Gauss,
                                                    letter to Heinrich Wilhelm Matthias Olbers (14 May 1826)

Without being prescriptive regarding the *form* of the arguments that mathematicians accept as "proofs", Gauss articulates here the essential required characteristic of such arguments.

The task of crafting mathematical models requires knowledge of the domain being modeled. We delegate this topic to the subject areas that motivate the reader's interest in mathematics. Our interest resides solely in what one does with the domain knowledge that one has access to. In particular, our goal is always to enable the crafting of mathematical arguments that are rigorous yet accessible.

Without further ado, let us begin to explore the craft of mathematical reasoning.

2.1 Rigorous Proofs: Theory vs. Practice

Contrary to the all-too-common view of mathematical argumentation as arcane strings of symbols that must be manipulated in rigid ways, mathematics is a vibrant, evolving system of thinking whose evolution is influenced by the ever-changing objects that are being thought about and by the ever-changing population that are doing the thinking. That said, mathematics does have its metaphorical Cesium atom against whose internal vibrations we can—*in principle*—measure any mathematical argument. Indeed, the quest for such a standard occupied much of the early nineteenth century.

Historical note.

We do not go back earlier than the nineteenth century in our quest for rigor in mathematical proofs because formal notions of "rigor" are, historically, a relatively recent phenomenon. Indeed, many "proofs" predating the nineteenth century fall below the standard that we now expect even of students—usually by failing to resolve—or even mention—every crucial issue raised in a proposed argument.

Perhaps the most famous example of omitted details relates to the famous "last theorem" of the great French polymath Pierre de Fermat. During the 1630s, Fermat made (in Latin, of course) the following claim in the margin of a copy of the classic *Arithmetica*, by third-century Greek mathematician Diophantus:

> *It is impossible to separate a cube into two cubes, or a fourth power into two fourth powers, or in general, any power higher than the second, into two like powers. I have discovered a truly marvelous proof of this, which this margin is too narrow to contain.*

This claim translates in modern parlance to the following well-known result.

Theorem 2.1 (Fermat's Last Theorem). *There do not exist positive integers a, b, and c such that $a^k + b^k = c^k$ for any positive integer $k > 2$.*

Fermat attributed the absence of details to the absence of room to supply them in the margin.

It is, indeed, possible that the small margins in Fermat's copy of *Arithmetica* were the only cause for his omitting a detailed argument. Subsequent history, though, suggests that the real cause for the omission was the absence in the seventeenth century of the mathematics needed to prove the theorem. A complete proof of the theorem was published only in 1995, by the English mathematician (Sir) Andrew Wiles, using highly novel mathematics whose development occupies an entire issue of the journal *Annals of Mathematics* [114].

The ultimate development of a "Cesium atom" for mathematical proofs resulted from seminal philosophical developments by mathematical logicians in the nineteenth century. Not surprisingly, the rigidity of style and the length of these ultimate-standard proofs made them quite unfriendly for humans to either craft or understand. It's time to remember that mathematics is an endeavor by and for humans. Just as we allow a carpenter to use a tape measure rather than a scored platinum bar while

building a house, we allow humans to employ human-palatable modes of argumentation as they build theorems.

We turn now to a description of the mathematical analogue of the Cesium atom
that we aspire to and a discussion of the mathematical analogues of tape measures
that we actually work with.

2.1.1 Formalistic Proofs and Modern Proofs

This section provides a rather informal discussion of rather formal topics. The presentation here is introductory and motivational; it is expanded and made rigorous in
the remainder of the book. The reader will find in Section 3.4.1 a rigorous presentation of much of the coming material.

2.1.1.1 Formalistic proofs, with an illustration

The purely formalistic view of mathematical discourse is devoid of intuition or im-
agery. Each discourse is just a sequence of syntactically valid *statements*. Certain
designated statements are *axioms*—meaning "self-evident truths". There are also
rules of inference—formal rewriting rules—which enable one to create new statements from sequences of pre-existing ones and to append these new statements to
the discourse. Within this world:

- A *proof* is a sequence of statements.

  – The first statement must be an axiom.

  – Each subsequent statement must be either an axiom or the result of applying
    a rule of inference to the statements that already appear in the sequence.

- A *theorem* is the last statement of a proof.

We are raised mathematically to view a "theorem" as some kind of holy grail, a term
that is reserved for the most important of proved assertions. Within the formalistic
setting, though, *a theorem is any assertion that is proved.*

Explanatory note.

To expand on the metaphorical pedestal that we reserve for *theorems*:
We typically employ some sort of hierarchy to classify assertions that we prove.
There is no generally accepted hierarchy, but the following categories of proved
assertions will resonate with at least some mathematical practitioners.

> *theorem*:   an assertion having more than local import
> *proposition*: an assertion of local import
> *lemma*:   a stepping stone toward a proposition or theorem
>   — often of little intrinsic import
> *corollary*:   a consequence of the proof of a more important assertion

---

One finds the formalistic point of view we have just introduced expounded in detail in, e.g., the classic *Principia Mathematica* by British logicians Alfred North
Whitehead and Bertrand Russell [113].

We now illustrate the formalistic world that we have sketched—in a form that
we hope renders it more palatable to humans—by peeking ahead into Section 2.2,
specifically Section 2.2.1. We describe and apply the well-known rule of inference
called the *Principle of Finite Induction*. Stated in human-consumable terms, the
Principle asserts the following.

The Principle of Finite Induction
Let $\mathbf{P}(n)$ *be an assertion involving the positive integer n.*
   if *one can prove the assertion*
      $\mathbf{P}(1)$
   and *one can prove the assertion*
      $\mathbf{P}(m)$ **implies** $\mathbf{P}(m+1)$
   then *one can* infer *the assertion*
      for all n $\mathbf{P}(n)$ *— which is often written symbolically:* $(\forall n)\mathbf{P}(n)$

This is an opportune moment to digress and introduce the important mathematical notion of *an unknown*, which is often used but less often acknowledged.

As soon as all of us mastered the elements of *arithmetic*—adding numbers, multiplying numbers, etc.—we were introduced to the rudiments of *algebra*. Suddenly,
we encountered expressions that had the same form as in arithmetic, but some numbers were represented by letters or words rather than by strings of digits. These
letters and words exemplify the conceptual shorthand of an *unknown* (although we
likely never heard that term at the time).

An unknown is essentially a number that is wearing a mask or, alternatively, is a
generic number: We are told that n is a number, but we are not told *which* number. In
fact, we are often told, cryptically, that n's identity is not relevant to the discussion.
It is not a huge leap to use unknowns to refer to generic instances of mathematical
objects other than just numbers. Here are some sample sentences that use unknowns.

Let 2n be an even integer.
(The unknown here is *n*.)

Let p be an arbitrary prime number.
(The unknown here is p.)

Let f be a numerical function of two variables.
(The unknown here is f.)

Let v be an arbitrary vertex of graph $\mathcal{G}$.
(Unknowns v and $\mathcal{G}$ here range over different domains!)

For any integers m and n, $m+n = n+m$.
(The two unknowns here range over the same domain.)

The last example illustrates a common use of unknowns: to discuss a universally true property of the objects within a mathematical domain.

The notion of *unknown* is distinct from the notion of *variable*. The former notion refers to a single object that is generic within a certain domain. The latter notion refers to a placeholder which can be instantiated with any object from a certain domain. The somewhat subtle difference expressed here will hopefully be clarified by comparing the earlier sentence

"For any integers m and n, $m+n = n+m$"

to the following quite different way of expressing the commutativity law for addition (which we shall revisit in Section 5.2).

Consider the two-variable numerical function $f(m,n) = m+n$.
For all variables m and n, $f(m,n) = f(n,m)$.

In fact, variables are a special kind of unknown which is used specifically when discussing functions. We shall talk more about them beginning in Section 3.3.4.

Back to proofs by induction:

A sample induction: *Verifying the summation formula for the first n integers.* We use the Principle of Finite Induction to verify the well-known formula for the sum of the first n positive integers.

Proposition 2.1 *For all $n \in \mathbb{N}$,*

$$S_n \stackrel{def}{=} 1+2+\cdots+(n-1)+n = \frac{1}{2}n(n+1) \tag{2.1}$$

Notation: the deep triviality.

Well-chosen notation can free one's mind from nonessentials, thereby enabling deep intellectual advances. The history of mathematics is filled with examples of major advances which are enabled—sometimes even triggered—by shifts in notation. The introduction of Hindu-Arabic numerals, the invention of positional numerals, and the discovery of algebraic notation are but three notational blockbusters among many over the millennia.

Not all notational advances are as consequential as the three just mentioned. Some less-monumental advances created compact notations which enabled a reader to discover—and a writer to disclose—the essential message in an expression more quickly and more clearly. We shall encounter many such advances throughout our journey, but three are about to come in quick succession, so we sequester them here with this message.

1. "$\overset{\text{def}}{=}$" is a shorthand for "equals, by definition".

   For illustration: "$X \overset{\text{def}}{=} Y$" is read "$X$ *equals* Y, by definition."

2. The symbol "$\square$" signals the end of a proof.

   The symbol is usually articulated "Q.E.D.", for "*Quod Erat Demonstrandum*" [Latin for "which was to be proved"].

3. "$(\forall x)$" is shorthand for the phrase "for all x".

   "$(\exists x)$" is shorthand for the phrase "for some x" or "there exists all x".

Proof. For every positive integer m, let $\mathbf{P}(m)$ be the proposition

$$1 + 2 + \cdots + m \;=\; \frac{1}{2}m(m+1)$$

We proceed according to the Principle's prescribed format of an inductive argument.

Because $\frac{1}{2}(1+1) = 1$, proposition $\mathbf{P}(1)$ is true.

Let us assume, for the sake of induction, that proposition $\mathbf{P}(m)$ is true, and consider the summation

$$1 + 2 + \cdots + m + (m+1)$$

Because $\mathbf{P}(m)$ is true, we replace the first part of this expression, and we obtain by direct calculation:

$$\begin{aligned}
1 + 2 + \cdots + m + (m+1) &= \frac{1}{2}m(m+1) \,+\, (m+1) \\
&= (m+1)(1 \,+\, \frac{1}{2}m) \\
&= (m+1)\frac{m+2}{2}
\end{aligned}$$

We have thus shown that

- **P**(1) is true

- *and* **P**($m+1$) is true whenever **P**(m) is true

The Principle of Finite Induction now assures us that **P**(n) is true for all positive integers n. □

Intuition. *Why does finite induction work?*

This question is actually a rather deep one, as one might expect from its focus on a *principle of reasoning*—i.e., on a *meta*-mathematical concept. But there is a simple *mathematical* theorem about numbers that at least lends one intuition about why induction works, even if it does not really address the meta-question. We state this result now, even though we have not yet developed its formal setting: You are already familiar with an informal version of the setting.

Proposition 2.2 *The set S defined as follows contains all of the positive integers.*

- *The integer 1 belongs to S.*

- *If the integer i belongs to S, then so also does the integer i + 1.*

We defer the proof of Proposition 2.2 to Section 4.3.1.2, where we develop the mathematical tools needed for the proof.

The intuitive relevance of Proposition 2.2 to the Principle of Finite Induction is that the construction of the set *S* in the proposition mimics the "construction" by the Principle of the set of indexed propositions **P**(i) for which the inductive proposition holds. The relevance is valid only at an intuitive level because the Principle talks *about* mathematics, not *within* mathematics!

2.1.1.2 Modern proofs, with an illustration

We are not overly uncomfortable with a (slightly cosmetized) formal proof of a result such as Proposition 2.1, because the result is so simple and because its statement is purely mathematical. Our discomfort will probably grow significantly, even with purely mathematical assertions, as the complexity of the asserted proposition grows, because the austere framework of mathematical logic does not capture the more discursive way that humans—even mathematicians!—think. A very helpful guidebook from the rigid world of logical notation to the more free-flowing way that mathematicians think about purely mathematical assertions can be found in the aptly titled book *Logic for Mathematicians* [94].

However, sooner or later—in fact, by the very next chapter(!)—we are going to encounter assertions that require some modeling to turn them into mathematical statements. Our goal in this book is to reason about computing-related phenomena, and such phenomena seldom come prepackaged in purely mathematical formats.

Even an "almost purely" mathematical assertion such as the following requires a modicum of free-flowing formulation before we obtain a directly provable format.

> *Any comparison-based algorithm that determines whether a given item occurs in an ordered list of n items must, in the worst case, employ* $\log_2 n$ *comparisons.*

Where should we begin our analysis? Putting aside technical jargon such as "$\log_2 n$" and "comparison" and "comparison-based algorithm", how does one organize one's reasoning in order to *perspicuously* and *convincingly* prove such an assertion? (If the proof is not both *perspicuous* and *convincing*, then what good is it?)

We clearly need some human-oriented guidance to augment the austere formalistic approach! In order to guide the upcoming discussion, let us keep in mind the following proposition—which is even less pre-packaged mathematically than our preceding example. We provide a "modern" rigorous proof of the upcoming assertion after our discussion.

Friends and strangers at a party. While the following result is worded here in an anthropomorphic, "homely" fashion, it is a quite-serious instance of a widely applicable genre of *inevitable subgraph* problem. Each such problem has the form: If you have n entities that relate to one another in such-and-such a way, then some $m < n$ of the entities must relate to one another in so-and-so a way. We shall encounter more such problems in Chapters 12 and 13.

Proposition 2.3 *In any gathering of six people, at least one of the following assertions is true.*
A. *There is a group of three people who know each other.*
B. *There is a group of three people none of whom knows either of the others.*

We want to garner intuition about how to prove Proposition 2.3. Let us begin with some discussion about what our goal should be. If we cannot reduce the provable world to sequences of assertions, then what is our goal? Using evocative terms, the twentieth-century French mathematician René Thom tells us

> *Est rigoureuse toute démonstration, qui, chez tout lecteur suffisamment instruit et préparé, suscite un état d'évidence qui entraîne l'adhésion.*

> [A demonstration is *rigorous* if it would convince every adequately educated and prepared reader.]

Thom transports the notion "proof" from the domain of the formal and absolute into the domain of humans—much as Kronecker did for the notion "number" (see our Preface). This importation is expanded upon—in regard to the problem of proving the correctness of programs—in the thought-provoking essay [46]. These sources propose that the prime endeavor of a mathematician—the proving of interesting, valuable theorems—is a *social exercise*. The rigid formalism that characterized the proofs of the late nineteenth and early twentieth century is henceforth replaced within the world of the practicing mathematician by a free-form, vibrant system of thought that, when convenient,

- allows one to represent the number n, as convenient, by, e.g.,

  - a numeral in some positional number system

  - a set (usually imagined) of n balls or widgets or ...

  - a unit-width rectangle that is n units high;

- freely "mixes and matches" different modes of argumentation, even within a single proof;

- freely invokes a highly tested computer program to check mind-numbing proliferations of clerically verifiable details.

We provide (for now) just a single instance of modern argumentation, which we employ to prove Proposition 2.3. The primary technical mechanism—a modern form of "rule of inference"—which we exploit in the proof is the widely applicable *Pigeonhole Principle*. There are numerous other "friends and strangers at a party" problems and numerous ways of solving such problems.

The Pigeonhole Principle
If one places $n + 1$ items (the pigeons) into n boxes (the pigeonholes), then at least one box must end up with more than one item.

Proof (Proposition 2.3). Let us observe a gathering of six indistinguishable people, named P_1, P_2, P_3, P_4, P_5, P_6. Focus on an arbitrary person, say P_5. (This choice "sounds" more arbitrary than P_1—but, of course, it is not.)

Now, there are five people, namely, P_1, P_2, P_3, P_4, P_6, each of whom P_5 either *knows* or *doesn't know*. Some three of these five people must "lie on the same side of the *knows/doesn't-know* fence." This follows from the pigeonhole principle: we have *two* boxes—namely, "*knows*" and "*doesn't know*"—and *five* pigeons—namely, the people P_1, P_2, P_3, P_4, P_6. Any way of putting the pigeons into the boxes will place at least three people into some one of the boxes.

Say, with no loss of generality, that P_5 *knows* P_1, P_2, and P_3.

Explanatory note.

Why can we claim that the selected situation—"P_5 knows P_1, P_2, and P_3"—can be assumed "with no loss of generality"?

One should *always* ask this question about such a claim! In the current case, the claim follows from the following facts.

The names that we use to refer to the six assembled people are just for our expository benefit. The names carry no inherent meaning related to the *Friends and Strangers* problem. You can repeat our argument while choosing arbitrary replacements for P_1, P_2, P_3, P_5, with no change to the logical outcome.

You can also interchange the "*knows*" and "*doesn't-know*" labels. The underlying logic will not change, although the conclusions regarding options A and B in the statement of the proposition will clearly "flip".

Having decided that P_5 *knows* P_1, P_2, and P_3, we now consider the implications of the possible relations between each of the three pairs of people chosen from $\{P_1, P_2, P_3\}$, namely, the pairs $\{P_1, P_2\}$, $\{P_1, P_3\}$, and $\{P_2, P_3\}$. There are precisely two logical possibilities.

- Some two of P_1, P_2, P_3 know each other—say, with no loss of generality, P_1 and P_2. In this case, P_1, P_2, and P_5 form a trio of people who know one another (option A in the statement of the proposition).

- No two of P_1, P_2, P_3 know each other. In this case, P_1, P_2, and P_3 form a trio of people none of whom knows either of the others (option B in the statement of the proposition).

This disjunction completes the proof.   □

We remark finally that nothing in the proof precludes the possibility that *both* option A *and* option B are true!

Are you convinced? If you are not, then you can contact one (or both) of the authors, and we shall gladly supply more details. This is the essence of the social modality of proof. At the "end of the day", we allow the totality of the readers to vote on whether we, the authors, have discharged our assertion-proving duties. The *Friends and Strangers* problem is simple enough that we do not expect a deluge of mail requesting more details, but for other, more complex, assertions, there could, indeed, be a need for further details. You can be certain, for example, that Wiles's proof of Fermat's Last Theorem (Theorem 2.1 in Section 2.1) engendered a vigorous discussion which stretched far beyond the world of mathematicians!

The current volume is dedicated to trying to overcome people's resistance to mathematical analysis and argumentation, by expounding on a modern, human-friendly—but no less rigorous—methodology for "thinking mathematically", especially with regard to computation-related matters. We attempt to describe proof systems and methods that the reader can comfortably develop facility with.

Our avenue for promoting mathematical *understanding* rather than just rote knowledge does not promote any specific formalism. Instead, throughout this text, we develop multiple proofs for the topics of discourse. We employ, when appropriate, multiple representations—textual, pictorial, symbolic, discursive—of the objects being discussed and the relationships being exposed; we employ, when appropriate, multiple modes of analysis and argumentation. We hope that our offering multiple ways to think about individual situations will enhance the likelihood that every reader will relate to an approach that is congenial to their way of thinking. By practicing crafting proofs and analyses involving more and more topics, the reader will begin to find it increasingly easy to state informally what ultimately needs to be analyzed rigorously and to intuit how to embark on the path toward such rigor.

2.1.1.3 Some elements of rigorous reasoning

We close our discussion of proof methodology by highlighting a number of ideas for the reader's consideration. These ideas range from pointing out common pitfalls to suggesting ways to develop intuition.

A. Distinguishing name from object

A common, fundamental stumbling block in the road toward cogent reasoning arises from the failure to distinguish *names* from the *objects* they denote. Two prime examples within the world of computing reside in the following distinctions that are often missed.

- A *function* is a special genre of infinite set of argument-value pairs; see Section 3.3.4. *A function is not a program*—even when the program computes the values of the function. Indeed, a program that computes a function can/should be viewed as a *name* for the function.

  Note that the often-used view of a function as a *rule* for assigning values to arguments should be avoided, because it suggests—**erroneously**—that an implementable such rule always exists!

- A number is an abstract notion that denotes a quantity. *You cannot touch a number; you cannot compute with a number!* As we discuss at length in our three chapters about numbers—Chapters 4, 8, and 10—it is *numerals*, i.e., *number representations*, that we manipulate in order to compute.

Instances of the *name-vs.-object* distinction—such as the two just noted—must always be in the forefront of the mind of a person while "doing" mathematics.

B. Quantitative reasoning

Readers who aspire to "do" mathematics must understand the foundational distinction between a quantity's *growing without bound* and the quantity's *being infinite*. Within this theme, they should appreciate situations such as the following. Every integer, and every polynomial with integer coefficients, is finite, but there are infinitely many integers and infinitely many polynomials. Readers should be able to verify (cogently but not necessarily via any particular formalism) assertions such as the following.

- Let us be given polynomials with positive coefficients: $p(x)$ of degree a and $q(x)$ of degree $b > a$, where a, b need not be integers. There must exist a constant $X_{p,q}$ (i.e., a constant that depends on the forms of polynomials p and q) such that

$$(\forall x > X_{p,q}) \quad p(x) < q(x)$$

Thus, polynomials having bigger degrees eventually *majorize*—i.e., have larger values than—polynomials having smaller degrees.

- Continuing with polynomial q of degree b: For any real number $c > 1$, there exists a constant $Y_{c;q}$ (i.e., a constant that depends on the polynomial q and the constant c) such that:

$$(\forall x > Y_{c;q}) \quad c^x > q(x)$$

Thus, exponential functions eventually *majorize* polynomials.

(Of course, the preceding examples are just simple illustrations of the kinds of principles that the reader should keep an eye out for.)

C. The elements of *empirical* reasoning

> *Scientific knowledge consists in the search for truth, but it is not the search for certainty–All human knowledge is fallible and therefore uncertain.*
>
> Karl Popper, *Search for a Better World and Essays from Thirty Years*

Experimentation (or, empiricism) is an invaluable guide both in the sciences and in mathematics, but these two domains of inquiry acquire quite different kinds of guidance from the outcomes of experiments.

The goal of *science*—especially physical science and biological science—is always to better understand some aspect of the universe—usually in order to explain and predict the evolution of observable phenomena. The scientific method seeks to objectively explain natural phenomena by the use of reproducible demonstrations. The process is always the same: observe and experiment with the goal of generating hypotheses and, thereby, predictions—and then study the consequences of these hypotheses in order to test and evaluate them. A theory is accepted as valid only in proportion to how well it fits with observed facts and predicts not-yet-observed facts related to the target phenomena. Dissonant results usually lead to a new theory which enriches understanding. This new theory can be either *revolutionary*, as, e.g., when the phlogiston theory of combustion was replaced by an understanding of oxidation, or *evolutionary*, as when Newton's theory of gravitation evolved into the more comprehensive theory of relativity. We are still awaiting a theory that provides a uniform description at both the sub-atomic and cosmic scales; some believe that quantum mechanics will play a role in this synthesis.

Mathematics is a science which seeks to learn "absolute" truths—not just evolving approximations to such truths—and it strives to accomplish this via pure reasoning. As such, mathematics does not really fit into the previous description of empirical sciences. Yet, empiricism plays an indispensable role in the process of "doing" mathematics!

While empirical reasoning does not convey the certitude that formal reasoning does, one cannot overemphasize how useful it is in preparing the ground for formal reasoning. Every practicing mathematician develops intuition to prepare for an attempted formal proof via activities such as

- trying to solve small instances of quantitative assertions of interest

- drawing pictures that caricature phenomena of interest

- playing with continuous versions of discrete phenomena of interest

Such experimentation can lead to the insight(s) needed to develop formal proofs. These insight(s), though, sometimes take years, or even decades, to "ripen".

Many instances of quick-ripening intuitions involve summations. In this arena, experimentation helps one to *guess* the right expressions—and to detect small mistakes, because any form of mathematics does not tolerate any mistake! Proposition 2.4, which we encounter later in this chapter, provides an example of quick-ripening intuition: It is not obvious *a priori* that each perfect square n^2 is the sum of the first n odd numbers, but it is! Indeed, Chapter 6 is a treasure trove of the fruits of quick-ripening intuition.

Throughout this text, we discuss numerous results that required years, or decades, or even centuries to prove.

- One of the best-known such "slow-cookers" is *Fermat's Last Theorem* (Theorem 2.1), which we discussed earlier. Its smallest instance asserts that the sum of perfect cubes, $a^3 + b^3$, cannot itself be a perfect cube, c^3.

- Another well-known "slow-cooker" is the *Four-Color Theorem* (Theorem 13.2). This result asserts that any map of the world can be colored using four colors in such a way that entities sharing a border get different colors. Adding to the interest in this result is that it was the first (significant) theorem that enlisted the aid of a computer program as a "collaborator" in the developing its proof.

- The final example we cite here is *Hilbert's-Tenth Theorem* (Theorem 5.4), so named for its position in the 23-item to-do list for twentieth-century mathematics promulgated by the German mathematician David Hilbert. This momentous result, which is less well known but no less consequential than our other two sample "slow-cookers", can be viewed as asserting that the problem of solving polynomial equations using integers (whole numbers) completely captures the essence of complex computations.

The proofs of long-maturing results such as the preceding three are beyond the scope of this text, but the stories behind them usually have lessons that any aspiring practitioner of the mathematical arts can benefit from.

To sum up: empirical exploration can be extremely useful in the endeavor of "doing" mathematics. Indeed, as we explore phenomena of immense complexity or even phenomena that inherently involve immensely large numbers, empirical studies are the only known avenue toward deep understanding.

2.2 Overview of Some Major Proof Techniques

2.2.1 Proof by (Finite) Induction

The version of the Principle of Finite Induction that we enunciated in Section 2.1.1.1 is often termed *weak* induction, to distinguish it from the *strong* form of the Principle that we enunciate now. The qualifiers "weak" and "strong" are somewhat unfortunate, because a proof constructed using either version of the Principle is as valid as a proof constructed using the other version. In fact, the two versions differ mainly in their ease of use in various situations.

> **The *Strong* Principle of Finite Induction**
> *Let* $\mathbf{P}(n)$ *be an assertion involving the positive integer n.*
> > **if** *one can prove the assertion*
> > > $\mathbf{P}(1)$
> > **and** *one can prove the assertion*
> > > $\big[(\forall\, k \leq m)\mathbf{P}(k)\big]$ **implies** $\mathbf{P}(m+1)$
> > **then** *one can* infer *the assertion*
> > > $(\forall\, n)\mathbf{P}(n)$

2.2.1.1 Two more sample proofs by induction

In Section 2.1.1.1, we specified the Principle of Finite Induction and exemplified its use by verifying the formula (2.1) for the sum of the first *n* integers (Proposition 2.1). We now provide two more sample applications of the Principle.

The inductive proof of the following result complements the constructive proofs of the same result in Proposition 6.3.

Proposition 2.4 *For all $n \in \mathbb{N}^+$, the nth perfect square is the sum of the first n odd integers. Symbolically:*

$$n^2 \;=\; 1 \,+\, 3 \,+\, 5 \,+\cdots+\, (2n-3) \,+\, (2n-1)$$

Proof. For every positive integer *m*, let $\mathbf{P}(m)$ denote the assertion

$$m^2 \;=\; 1+3+5+\cdots+(2m-1).$$

We proceed according to the standard format of an inductive argument.

Base case. Assertion $\mathbf{P}(1)$ is true because $1 \cdot 1 = 1$.

Inductive hypothesis. Assume, for the sake of induction, that assertion $\mathbf{P}(m)$ is true for all positive integers strictly smaller than *n*.

Inductive extension. Consider now the summation

$$1+3+5+\cdots+(2n-3)+(2n-1)$$

Because $\mathbf{P}(n-1)$ is true, we know that

$$
\begin{aligned}
1+3+\cdots+(2n-3)+(2n-1) &= \big(1+3+\cdots+(2n-3)\big)+(2n-1) \\
&= \big(1+3+\cdots+(2(n-1)-1)\big)+(2n-1) \\
&= (n-1)^2+(2n-1)
\end{aligned}
$$

By direct calculation, we observe that

$$(n-1)^2+(2n-1) \;=\; (n^2-2n+1)+(2n-1) \;=\; n^2.$$

Because n is an arbitrary positive integer, we conclude that $\mathbf{P}(n)$ is true whenever

$$[\mathbf{P}(1) \text{ is true}] \quad and \quad [\mathbf{P}(m) \text{ is true for all } m < n]$$

The Principle of (Finite) Induction tells us that $\mathbf{P}(n)$ is true for all $n \in \mathbb{N}^+$.   □

Our final sample inductive proof complements the constructive proofs of the same result in Proposition 6.4(b).

Proposition 2.5

$$S(n) \;=\; 1 + \frac{1}{4} + \frac{1}{16} + \frac{1}{64} +\cdots+ \frac{1}{4^k} +\cdots+ \frac{1}{4^n} \;=\; \frac{4}{3}\left(1 - \frac{1}{4^{n+1}}\right).$$

Proof. Since this is our third proof by induction, we can be a bit sketchier than earlier. Let $\mathbf{P}(m)$ denote the assertion

$$1 + \frac{1}{4} + \frac{1}{4^2} +\cdots+ \frac{1}{4^m} \;=\; \frac{4}{3}\left(1 - \frac{1}{4^{m+1}}\right).$$

Base case. Assertion $\mathbf{P}(0)$ is true because $S(0) = \frac{4}{3}\cdot\frac{3}{4} = 1$.

Inductive hypothesis. Assume that $\mathbf{P}(m)$ is true for all $m < n$.

Inductive extension.

$$
\begin{aligned}
S(n) &= S(n-1) + \frac{1}{4^n} \\
&= \frac{4}{3}\left(1 - \frac{1}{4^n}\right) + \frac{1}{4^n} \\
&= \frac{4}{3} + \frac{1}{4^n}\cdot\left(1 - \frac{4}{3}\right) \\
&= \frac{4}{3} - \frac{4}{3}\cdot\frac{1}{4^{n+1}}
\end{aligned}
$$

This extends the induction, hence completes the proof.   □

2.2.1.2 A *false "proof"* by induction: The critical base case

There is an old adage about the value of learning from one's mistakes. Let's observe that in action. We intentionally *mis*-use the Principle of Finite Induction to craft a fallacious "proof" of the following absurd "fact" about horses that are monochromatic—i.e., each horse has a single color.

Non-Proposition. *All monochromatic horses have the same color.*

Non-Proof. We craft an argument that follows the form of an induction. Focus on a set S of monochromatic horses.

Base case. If there is only a single monochromatic horse in set S, then all monochromatic horses in the set are the same color.

Inductive hypothesis. Say, for induction, that if set S contains no more than n monochromatic horses, then all monochromatic horses in S are the same color.

Inductive extension. Let us be given a set S' that consists of $n+1$ monochromatic horses. If we remove a single horse from S', then the remaining set, call it S'', consists of n monochromatic horses. By our inductive hypothesis, all of the horses in S'' have the same color.

Now remove a single horse from S'' and replace it with the horse that we just removed from the $(n+1)$-horse set S'. We now have a new n-horse set, call it S'''. Because S''' contains only n horses, we can once again invoke the inductive hypothesis to conclude that all horses in S''' have the same color.

Finally, let us reunite all of the horses—which reconstitutes the $(n+1)$-horse set S'. Because the relation "has the same color" is *transitive*—which, loosely speaking, says that things that are both equal to a third thing are equal to each other—we may conclude, by induction, that all of the horses in set S' have the same color.   □-NOT!

Well—we all know that all monochromatic horses do *not* share the same color. Where has our argumentation gone wrong? In a word, the *base case* that we used was not adequate for the "proof" we were crafting. In detail, we argued that when we remove a horse from set S' and then remove another (different!) horse from set S', we still have a (third!) horse left to compare those two horses to. *This means that we must have started with at least* three *horses in set S'!* Said differently, this means that $n+1$ *must be no smaller than* 3, or, equivalently, *n must be no smaller than* 2. The base case of our induction must, therefore, be valid for sets that contain *two* horses! Since the same-color "assertion" is clearly absurd for such sets, the base case of our induction is *false*—which invalidates the entire argument!

This analysis brings us to the critical issue of how to select the "small" cases that comprise the base of a valid inductive argument.

By definition, an inductive proof is required to cover all cases ("for all positive integers, …"). It accomplishes this by explicitly treating some "small" cases individually and then showing that the inductive rule at the heart of the argument—which resides implicitly within the inductive assumption—enables one to generate all non-"small" cases from the "small" cases.

In some domains, it is rather simple to determine what cases provide an adequate set of "small" cases. This is usually true, for example, when one solves numerical recurrences, as we shall do in Chapter 9, because the required "small" cases often coincide with a target recurrence's boundary conditions. (Situations whose structures are founded upon sequences such as the Fibonacci numbers and the binomial coefficients often require multiple "small" instances for the base case of an induction.)

In contrast, the challenge of verifying assertions about, say, graphs, as we do in Chapters 12 and 13, is quite often not at all easy. For newly discovered propositions, we may not know *a priori* how the extension step of an inductive argument reduces the graph size n until we understand relevant characteristics of general n-vertex graphs. Because of this, when one argues about complex objects such as graphs, it is a good practice to "play" with several small graph sizes initially. Hopefully, in addition to giving your inductive argument a robust base case, such "playing" will enable you to develop valuable intuition for the general case of the induction.

2.2.1.3 The Method of Undetermined Coefficients

Proofs by induction are important tools for *verifying* the correctness of alleged results—but induction by itself is not a tool for *discovering* new results. *(You cannot verify assertion **P**(n) if you do not know exactly what the assertion is asserting.)*

There are many techniques of analysis that yield *approximations* to the values of important quantities—but not exact values. Within the purely mathematical realm that we study in this text, the summation technique of Section 6.3.1 provides an important example. This technique tells us, for instance, that the sum of the first n positive integers grows *quadratically* with n, but it does not provide the exact formula (2.1). Similarly, the technique tells us that the sum of the first n perfect squares grows *cubically* with n, but it does not provide the exact formula for the sum.[3] Moving beyond the purely mathematical realm, there are many "exhaustive-search" computational heuristics[4] whose approximate values for a variety of optimization problems have been found to be rather good in practice. The drawback with such techniques is that one often must let these heuristics run for a *very* long time before they provide values that are even close to optimal.

The *Method of Undetermined Coefficients*, which we describe now, can sometimes refine approximate results that one has somehow discovered (say, via a heuristic) to provide exact results which one can then verify via induction. We illustrate

---

[3] For the curious: The exact formula for the sum of the first n perfect squares is :

$$1 + 4 + \cdots + (n-1)^2 + n^2 \;=\; \frac{1}{3}n^3 + \frac{1}{2}n^2 + \frac{1}{6}n \qquad (2.2)$$

You will be asked to verify this as an exercise.

[4] *Particle swarm optimization* [65] and *simulated annealing* [66] are two popular such heuristics.

the method by deriving a formula for the sum of the first n integers—under the following assumptions:

- We know that this sum has the form of a quadratic (degree-2) polynomial in n.
- We do not know the coefficients of the polynomial.

Say that we know from some external source—perhaps Section 6.3.1—that *for all positive integers n*:

$$S(n) \overset{\text{def}}{=} 1 + 2 + 3 + \cdots + n = c_2 n^2 + c_1 n + c_0 \qquad (2.3)$$

for some numbers c_0, c_1, c_2.

By evaluating formula (2.3) at the first few positive integers n, we find that:

$$S(1) = 1 \text{ so that } c_2 + c_1 + c_0 = 1 \qquad (2.4)$$
$$S(2) = 3 \text{ so that } 4c_2 + 2c_1 + c_0 = 3 \qquad (2.5)$$
$$S(3) = 6 \text{ so that } 9c_2 + 3c_1 + c_0 = 6 \qquad (2.6)$$

We can now do arithmetic on these equations (adding and/or subtracting pairs of them) to obtain new equations which will help us home in on the values of c_2, c_1, and c_0.

- We combine Eq. (2.4) and Eq. (2.5) to get

$$3c_2 + c_1 = 2 \qquad (2.7)$$

- We combine Eq. (2.5) and Eq. (2.6) to get

$$5c_2 + c_1 = 3 \qquad (2.8)$$

- We combine Eq. (2.7) and Eq. (2.8) to get

$$2c_2 = 1 \qquad (2.9)$$

This last equation, Eq. (2.9), tells us that $\boxed{c_2 = \tfrac{1}{2}}$. Given this, our first two original equations become

$$S(1) = 1 \text{ so that } c_1 + c_0 = 1/2 \qquad (2.10)$$
$$S(2) = 3 \text{ so that } 2c_1 + c_0 = 1 \qquad (2.11)$$

By combining these new equations we find that $\boxed{c_2 = \tfrac{1}{2}}$. Finally, by employing this value in Eq. (2.10), we find that $\boxed{c_0 = 0}$.

Using the preceding analysis, beginning with the initially undetermined coefficients, we have now determined that

$$S(n) = \frac{1}{2}n^2 + \frac{1}{2}n$$

(Of course, we already verified this formula in Proposition 2.1.)

With more (calculational) work, but no new (mathematical) ideas, one can use the Method to derive an explicit expression for the sum of the first n kth powers, i.e., the quantity $S_n^{(k)}$, for any positive integer k. One's starting point could be the knowledge—obtained, say, from Section 6.3.1—that

$$S_n^{(k)} = c_{k+1}n^{k+1} + c_kn^k + \cdots + c_2n^2 + c_1n + c_0$$

2.2.2 Proof by Contradiction

The importance of this proof technique has been recognized since antiquity, under *contradictio in contrarium* or, perhaps less accurately, *reductio ad absurdum*.

2.2.2.1 The proof technique

Suppose that we want to prove some proposition P. Think, as a concrete example, of P as the (false) assertion:

There are two distinct additive identities *for the numbers, call them* 0 *and* $0'$.

This proposition means in detail that:

(a) $0 \neq 0'$. **(b)** *For all numbers x, both of the following assertions are true.*

$$x + 0 = 0 + x = x$$
$$x + 0' = 0' + x = x$$

Note that we are asserting that 0 *and* $0'$ *are both left and right additive identities.*

The basic strategy of a proof by contradiction is the following:

Beginning with Assertion P—i.e., assuming that P is true—craft a logically valid chain of correct logical assertions

$$P \text{ implies } P_1 \text{ implies } P_2 \text{ implies } \cdots \text{ implies } P_m$$

where Assertion P_m is known to be false (i.e., is *a contradiction*).

We then infer that Assertion P is false. (Of course, there is a parallel situation in which we begin by assuming that P is false, and infer after the chain of assertions that P is true.)

The chain of reasoning for our simple Assertion P is encapsulated symbolically in the following chain of equations

$$0 = 0 + 0' \quad \text{because } 0' \text{ is a } \textit{right} \text{ additive identity}$$
$$= \quad 0' \quad \text{because } 0 \text{ is a } \textit{left} \text{ additive identity}$$

We can rewrite the preceding chain to yield an explicit contradiction, as follows.

Assume, for contradiction, that Assertion P is true—i.e., that 0 and $0'$ are *distinct* additive identities; i.e., $0 \neq 0'$. Then:

1. Because $0'$ is a *right* additive identity, it follows that $0 + 0' = 0$.

2. Because 0 is a *left* additive identity, it follows that $0 + 0' = 0'$.

3. Combining the previous statements, we infer that $0 = 0'$, because the relation "=" is *transitive*, i.e., because numbers that are equal to the same number are equal to each other.

Statement 3 asserts that $0 = 0'$, which contradicts P's contention that $0 \neq 0'$. The principle of Proof by Contradiction now states (finally) that P is false. □

2.2.2.2 Another sample proof: There are infinitely many prime numbers

We develop a nontrivial proof by contradiction to establish a result which is traditionally attributed to the Greek mathematician Euclid, one of the patriarchs of mathematics. We say that an *integer n* (known also as a *counting number* or a *whole number*) is *prime* if n is (exactly) divisible only by 1 and by n itself.

Proposition 2.6 *There are infinitely many prime numbers.*

Proof. Let us assume, contrarily, that there are only finitely many primes. Say, in particular, that the following r-element sequence enumerates all (and only) primes, in increasing order of magnitude:

Prime-Numbers $= \langle P_1, P_2, \ldots, P_r \rangle$

where

- $P_1 = 2$

- $P_2 = 3$

- $P_i < P_{i+1}$ for all $i \in \{1, 2, \ldots, r-1\}$.

We verify the *falseness* of the alleged exhaustiveness of the sequence **Prime-Numbers** by analyzing the positive integer

$$n = 1 + \prod_{i=1}^{r} P_i = 1 + (P_1 \times P_2 \times \cdots \times P_r).$$

We make three crucial observations.

1. *The number n is not divisible by any number in the sequence* **Prime-Numbers**.
   To see this, note that for each P_k in the sequence,

   $$\frac{n}{P_k} = \frac{1}{P_k} + \prod_{i \neq k} P_i.$$

   Because $P_k \geq 2$, we see that n/P_k obeys the inequalities

   $$\prod_{i \neq k} P_i \; < \; \frac{n}{P_k} \; < \; 1 + \prod_{i \neq k} P_i.$$

   The discreteness of the set $\mathbb{Z}$—see Section 4.3.1.2.C—implies that n/P_k is not an integer, because it lies strictly between two adjacent integers.

2. Because of assertion 1, if the sequence **Prime-Numbers** actually contained *all* of the prime numbers, then we would have to conclude that *the number n is not divisible by any prime number*.

3. The Fundamental Theorem of Arithmetic (Theorem 8.1) implies that *every positive integer except 1 is divisible by (at least one) prime number*.

We thus have a chain of assertions that lead to a mutual inconsistency: On the one hand, integer n exceeds 1 and has no prime divisor. On the other hand, no positive integer greater than 1 can fail to have a prime divisor!

Let us analyze how we arrived at this uncomfortable place.

- At the front end of this uncomfortable string of assertions we have the assumption that there are only finitely many prime numbers. We have (as yet) no substantiation for this assertion.

- At the back end of this uncomfortable string of assertions we have the (*rock solid*) Fundamental Theorem of Arithmetic.

- In between these two assertions we have a sequence of assertions, each of which follows from its predecessors via irrefutable rules of inference.

It follows that the *only* brick in this edifice that could be faulty—i.e., the only assertion that could be false—is the assumption that there are only finitely many prime numbers. Since this assumption leads to an inconsistent set of assertions, we must conclude that the assumption is false! This classical proof by contradiction thus informs us that there are infinitely many prime numbers.   □

2.2.3 Proofs via the Pigeonhole Principle

The proof technique we discuss now builds on an observation that is almost embarrassingly obvious—yet its simplicity is exceeded by its importance as a source of strikingly surprising results.

2.2.3.1 The proof technique

The technique, known variously as *the Pigeonhole Principle* or *Dirichlet's Box Principle* (after the German mathematician Peter Gustav Lejeune Dirichlet), exploits the fact that if one has n objects (they're the pigeons) and $m < n$ boxes (they're the pigeonholes), then any way of putting pigeons into boxes must place at least two pigeons into the same box.

2.2.3.2 Sample applications/proofs

Choosing a pair of matching socks. You have n pairs of socks, the socks in each pair having a distinct color (one pair of red socks, one pair of blue socks, ...). Since you wake up "very slowly", you want to grab some number of unpaired socks that is certain to yield at least one pair of same-color socks. Clearly, if you grab any $n + 1$ socks (the pigeons), the pigeonhole principle guarantees that you will have at least one monochromatic pair, because there are only n distinct sock-colors (the boxes).

Finding birthday-mates. You are attending a conference and wander into a lecture that has 367 attendees (including you). It is certain that at least two attendees share the same birthday: There are 366 possible birthdays (the boxes for a leap year) and 367 birthday-possessors (the pigeons). You can now add the punchline! (We will uncover much more interesting information about birthday-mates in Chapter 11.)

2.3 Nontraditional Proof Strategies

2.3.1 Pictorial Reasoning

2.3.1.1 The virtuous side of the method

We present in this section a special type of nontraditional proof which complements well the more traditional proofs that build on algebra, analysis, number theory, ... The underlying idea here is to employ representations of (natural) numbers via geometrical or pictorial patterns—construing the term "pattern" very broadly; e.g., we might represent the number n as a set of n tokens or as a geometric object one of whose dimensions is n. We remark that this manner of reasoning has its origins in antiquity:

> Let no one who is ignorant of geometry enter here.
>
> Plato, Greek philosopher (fourth century BC)
written in front of his Academy, in Athens.

We develop two sample proofs—and hope that the reader will be inspired to add to this collection.

1. *Expanding the expression* $(n+1)^2$. In this example, we employ basic geometrical objects—in this case the surfaces of squares and rectangles. Let us focus on the following well-known equation:

$$(n+1)^2 = n^2 + 2n + 1. \tag{2.12}$$

This equation can be validated as a special case of the *restricted Binomial Theorem* (Theorem 6.1), which we prove algebraically in Section 5.3.2. The equation can be validated using a highly perspicuous pictorial argument based on Fig. 2.1. The

Fig. 2.1 A geometric proof of the identity $(n+1)^2 = n^2 + 2n + 1$.

figure tells its tale by exhibiting four rectangles that make up an $(n+1) \times (n+1)$ square; the area of this square is, of course, $(n+1)^2$. This large square is made up of four rectangles.

- Reading across the top of the figure, we encounter a darkly shaded $n \times n$ square (whose area is n^2) and an unshaded $n \times 1$ rectangle (whose area is n).

- Reading across the bottom of the figure, we encounter an unshaded $1 \times n$ rectangle (whose area is n) and a lightly shaded 1×1 square (whose area is 1).

The lesson from the figure is that $(n+1)^2$, which is the area of the composite square,

is the sum of

$\qquad\qquad n^2$ (the area of the darkly shaded square)

plus

$\qquad\qquad 2n$ (the combined areas of the unshaded rectangles)

plus

$\qquad\qquad 1$ (the area of the lightly shaded square)

The preceding geometric validation does not "look like" the proofs whose highly structured forms we struggled with in high school. But: (1) Our geometrical proof is as valid as any algebraic one. (2) It can be a lot more fun to come up with. (3) It may trigger reasoning that could lead to new discoveries.

2. *Solving a quadratic equation.* We now develop a more complex example of pictorial reasoning. The spirit of the following proof comes from the 12th-century mathematician al-Khwārizmī, whose seminal work [2] was indispensable in introducing mathematical and computational ideas to Europe. The following proof builds on his work on the second-degree equation

$$x^2 + 10x = 39 \qquad (2.13)$$

Explanatory/Cultural note

You may remark that we used the letter n for the unknown of Example 1, and we are using the letter x for the unknown of the current example. From a *strict* perspective no message is being sent via these notational choices. More subtly, though, many discrete mathematicians *tend* to employ middle letters—i, j, m, n—for integer-valued quantities and late letters—x, y, z—for general numerical quantities. Indeed, this practice was incorporated into the *FORTRAN* programming language [10].

The methodology that al-Khwārizmī employed to solve quadratic polynomial equations such as (2.13) builds solutions in terms of the areas of squares and rectangles and employs both algebraic and geometric representations of the left-hand sides of the target equations. For equation (2.13) specifically, al-Khwārizmī found the perspicuous geometric representation of the left-hand side depicted in Fig. 2.2. In the figure, the shaded perfect square on the left and the unshaded perfect square

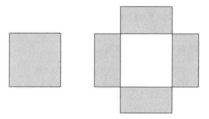

Fig. 2.2 The left-hand side of Eq. (2.13) depicted geometrically via the shaded perfect square x^2 and (i.e., added to) four identical shaded rectangles of area $\frac{5}{2}x$ each.

on the right both have side x, hence, area x^2. Of the four shaded rectangles around the unshaded perfect square, two have dimensions $\frac{5}{2} \times x$, and two have dimensions $x \times \frac{5}{2}$. The figure's representation of the left-hand side of equation (2.13) is matched by the following rewriting of the left-hand side of (2.13):

$$x^2 + 10x = x^2 + \left(4 \times \frac{5}{2}\right)x$$

We finally turn to solving the equation.

1. The area of the entire shaded surface in Fig. 2.2 is given by the right-hand side of equation (2.13), that is 39.

2. Hence, 39 is also the area of the "cross" on the right-hand side of Fig. 2.2 which would be created by moving the shaded $x \times x$ square into the unshaded $x \times x$ square.

3. Finally, 39 is also the area of the unshaded "cross" in Fig. 2.3. (The color scheme

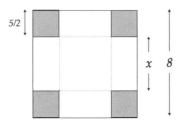

Fig. 2.3 The last step of the geometric solution of the equation $x^2 + 10x = 39$. The four shaded squares augment the "cross" of Fig. 2.2, whose area is 39, to the big square, whose area is 64.

in this figure has been changed to highlight the new corner squares.)

4. As in Fig. 2.3, let us append to the area-39 unshaded "cross" four small shaded squares whose addition augments the "cross" to the big, bi-color square of Fig. 2.3.

5. We see from Fig. 2.2 that each of the small shaded squares has dimensions $\frac{5}{2} \times \frac{5}{2}$, hence, area $25/4$.
   This means that the area of the large square is

$$39 + \left(4 \times \frac{25}{4} \right) = 39 + 25 = 64.$$

We now have the information needed to compute a value of x that satisfies equation (2.13). Consider the large square of Fig. 2.3.

- On the one hand, the side-length of this square is

$$x + (2 \times 5/2) = x + 5$$

- On the other hand, the side-length is

$$x = 8$$

Since all sides of the square are identical, we infer that

$$x = 8 - 5 = 3$$

Indeed, we could—but do not need to!—verify this value algebraically. (It might be a useful exercise for the reader at this point.)

An important caveat: The pictorial/geometric method which we have just employed is not "complete"—it yields only the *positive* solution for x. As we discuss at length in Sections 5.3.1.1 and 5.3.1.2—see, in particular, Theorem 5.1—there exists another solution for x, because its defining polynomial is *quadratic* (i.e., of degree 2). In the light of the defining equation's intended goal—namely, to determine the area of a plot of land—the single (positive) solution for x that al-Khwārizmī's tools yield was sufficient. That said, there are situations wherein one has need of both roots of a quadratic polynomial equation, and the algebraic solution methods will produce both solutions—although not always in an intuitive way.

Returning to equation (2.13): the second, negative, solution for x is $x' = -13$, as is witnessed by the following polynomial factorization:

$$x^2 + 10x - 39 = (x-3)(x+13)$$

There is a happy coincidence in this specific example: Because 64 is the square of -8 as well as of $+8$, *for this example*, the negative solution-value for x can also be obtained via our geometric argument:

$$x - 5 = -8 - 5 = -13$$

We close our discussion by noting that pictorial/geometric arguments can be fun to develop and can trigger unexpected discoveries! We highly recommend them as fodder for "mathematical doodling".

2.3.1.2 Handle pictorial arguments with care

We have just mentioned one limitation of pictorial mathematical arguments, namely their limitation to certain types of problem solutions. We now discuss an even more important reason to be careful when employing such arguments. This reason is described most eloquently in terms of a paradox which is associated with the nineteenth-century English mathematician and fiction writer Lewis Carroll.

Focus on a standard 8×8 chessboard. Let us "mutilate" the chessboard in the following way.

1. Cut the board into four pieces in the manner specified schematically in Fig. 2.4.

2. Reassemble the four pieces into a rectangle in the manner depicted in Fig. 2.5.

On the one hand, Fig. 2.4 supports assessing the *area* of the chessboard based on its conventional form, as an 8×8 array of unit-side squares. Thus viewed, the board has area $8 \cdot 8 = 64$.

On the other hand, Fig. 2.5 supports viewing the chessboard in its cut-up-and-reassembled form. Viewed in this way, the board has the area of a rectangular 5×13 array of unit-side squares. Hence, within this view, the board has area $5 \cdot 13 = 65$.

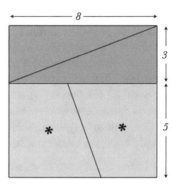

Fig. 2.4 A standard chessboard cut into four pieces

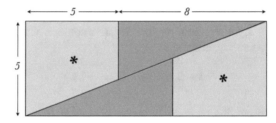

Fig. 2.5 The four pieces of the chessboard reassembled into a 5×13 rectangle

Obviously, something is wrong with our pictorial reasoning—but what?

We can expose the problem by examining the southwest-northeast diagonal of the rectangle in Fig. 2.5 very carefully. In the standard discussion of this "paradox", the diagonal is drawn in a way that suggests that it is a line that *bisects* the 5×13 rectangle into two congruent (hence, equal-area) right triangles, and that this line is the shared hypotenuse of these triangles. However, if one draws the figure in great detail, then one finds that the diagonal does not quite bisect the rearranged unit-side squares of the 8×8 square! In fact—see Fig. 2.6—the pair of curves that actually bisect the rearranged 8×8 square *do not form a single line!* These two curves are really close to the diagonal, but they are separated by a *very small space*—indeed a space whose aggregate area equals the area of a unit-size square of the original 8×8 chessboard.

The moral of this tale. Pictures can deceive as well as illuminate! *Use pictures for inspiration—but always to verify that you are really seeing what you think you are!*

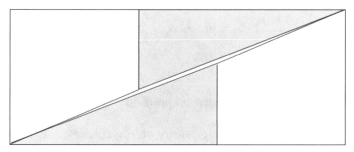

Fig. 2.6 Looking carefully at the border between the northwestern and southeastern halves of the 5×13 rectangle

2.3.2 Combinatorial Intuition and Argumentation

Combinatorics is a branch of discrete mathematics that specializes in the following operations (among others): selecting objects from a given set; arranging and rearranging the objects; counting the number of ways that one can arrive at specified target configurations. Chapter 11 introduces the foundations of this subfield, following with its applications to the kindred fields of probability and statistics.

2.3.2.1 Summation as a selection problem

Sometimes combinatorial argumentation can be used in unexpected ways. We illustrate that fact here by combinatorially deriving an explicit expression for the summation $S_{n-1} = 1 + 2 + \cdots + (n-1)$.

Our summation begins, unexpectedly, by counting the number of ways—call it $C(n, 2)$—of selecting two items from a set of n items. We describe this selection in a somewhat unconventional way.

- The first integer of the two we are selecting can be chosen in $n - 1$ ways, corresponding to the $n - 1$ elements of the set

$$\{1, 2, \ldots, (n-2), (n-1)\}$$

- If the bigger integer chosen was k, then we can select the second, smaller integer in $k - 1$ ways, from among the integers smaller than k.

We thereby observe the following summation, which yields the sum S_{n-1}.

$$C(n, 2) = (n-1) + \sum_{k=2}^{n-1} C(k-1, 1)$$

$$= (n-1) + \sum_{k=2}^{n-1} (k-1)$$

$$= (n-1) + \sum_{k=1}^{n-2} k$$

$$= S_{n-1}$$

2.3.2.2 Summation as a rearrangement problem

Our next proof technique combines geometry with the combinatorial rearrangement of a configuration of objects. The second of these techniques owes an intellectual debt to Guido Fubini, whose eponymous *Principle* mandates *looking at configurations from multiple points of view, to gather multiple intuitions.*

We apply this two-stage reasoning paradigm to the problem of evaluating the summation $S_n = 1 + 2 + \cdots + n$.

The first idea in our reasoning is to represent each positive integer k as a horizontal sequence of k tokens, depicted as darkened circles in Fig. 2.7. Using such a

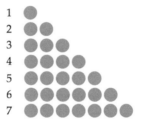

Fig. 2.7 Representing the first n positive integers using tokens. In this illustration, $n = 7$.

representation, we can naturally view the summation S_n as a *triangular pattern* of tokens. How can we use this representation to evaluate S_n without just counting and summing? We call upon an intuition attributed to the great nineteenth-century mathematician Carl Friedrich Gauss—who famously had this intuition as a preteenager. We remark that if we *replicate* the triangular pattern in the figure and *flip* it, then we can configure the resulting triangles—see Fig. 2.8—in the rectangular pattern depicted in Fig 2.9. Next, we observe two important properties of the rectangular pattern that we have achieved in Fig 2.9.

- The number of tokens in the rectangular pattern is precisely $2S_n$.

  This is a pictorial version of the seminal intuition that underlies Gauss's (text-based) evaluation of the summation S_n; see Proposition 6.1.

- The rectangular pattern has $n+1$ columns, each having n tokens.

  This means that the rectangular pattern contains $n \cdot (n+1)$ tokens.

Combining these observations, we conclude that

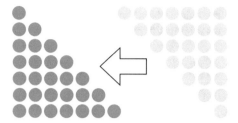

Fig. 2.8 Putting two copies of a triangle together to form a rectangle

Fig. 2.9 The two copies of a triangle have become an $n \times (n+1)$ rectangular array of tokens

$$2S_n = n \cdot (n+1) \quad \text{so that} \quad S_n = \frac{1}{2} n \cdot (n+1)$$

2.3.3 The Computer as Mathematician's Assistant

Computers have played an essential role in the process of "doing" mathematics from the earliest days when computers had enough power to explore mathematical terrain, searching and verifying.

- Some of the calculations performed by these "silicon assistants" have a game-like quality to them:

  - Calculate more digits of π than anyone else has done

  - Calculate a bigger Mersenne prime than anyone else has

- Yet other calculations represent attempts to hone intuition for what will someday morph into serious attempts at proofs:

  - Search for an odd perfect number

  - Determine the relative frequencies of the decimal digits in the (infinite) decimal expansion of π

- Finally, some of these calculations represent attempts to contribute essential detail to already serious attempts at proofs: These are the calculations that interest us here.

Many significant theorems languished as conjectures for decades, some even for centuries, awaiting their eventual elevation to theoremhood. While many of these eventual theorems languished in mathematical limbo awaiting the development of new mathematics—Fermat's Last Theorem (Theorem 2.1) is a prime example— yet others endured their long waits because of a veritable "mountain" of cases that needed (often minor) calculational verification.

The latter category of conjecture can increasingly be tackled with the aid of the ever more powerful computers that are being developed with ever-increasing frequency. When used in this way, computers are clearly "junior" partners to their mathematician-users. When the domain of discourse within a conjecture is discrete—i.e., does *not* involve real numbers "to infinite precision" nor any continuity-related notions—then it is always possible—at least in principle—to eliminate the computer from the "team" by transforming such a verification into a classical formal proof. The role of the computer in the "team" is solely to assist the mathematician—by performing calculations that are too long and/or too numerous and/or too complicated for humans to perform reliably and efficiently.

The renowned Four-Color Theorem (Theorem 13.2) was a landmark in the use of computers as partners to humans in proving theorems. The Theorem asserts— ignoring a few technical details—that any map on Earth (or on a topologically similar planet) can be colored using four colors in such a way that distinct countries that share a border receive distinct colors. As we discuss in detail in Section 13.1.2, this result languished in mathematical limbo for many decades before a pair of American mathematicians, Kenneth Appel and Wolfgang Haken, used a computer to help resolve the thousands of cases that had to be verified-via-calculation in the proof.

The mathematical world erupted in controversy over this first-ever use of a computer in an essential way to construct a mathematical proof. It took years—plus the result-replicating program of a team of Japanese mathematicians—before the global mathematical community accepted the claim by Appel and Haken that Theorem 13.2 had, indeed, been proved!

There may be many conjectures currently in mathematical limbo that could be elevated to the status of theorem if some team of mathematicians would feel free to employ computers to offload needed but onerous computational drudgery.

2.4 Exercises: Chapter 2

Throughout the text, we mark each exercise with 0 or 1 or 2 occurrences of the symbol $\oplus$, as a rough gauge of its level of challenge. The 0-$\oplus$ exercises should be accessible by just reviewing the text. We provide *hints* for the 1-$\oplus$ exercises;

Appendix H provides *solutions* for the 2-⊕ exercises. Additionally, we begin each exercise with a brief explanation of its anticipated value to the reader.

1. **Verifying some summations via Finite Induction**
   LESSON: Practice using Induction plus some algebraic calculation

   Use Finite Induction to verify that the following summation formulas hold for all positive integers n.

   a. Summing the first n fixed powers of integers.

      i. $S_2(n) \overset{\text{def}}{=} 1 + 2^2 + \cdots + n^2 = \frac{1}{6}n(n+1)(2n+1)$

      ii. $S_3(n) \overset{\text{def}}{=} 1 + 2^3 + \cdots + n^3 = \frac{1}{4}n^2(n+1)^2$

      Note that the sum of the first n integers is quadratic (degree 2) in n (Proposition 2.1); the sum of the first n squares of integers is cubic (degree 3) in n; the sum of the first n cubes of integers is quartic (degree 4) in n. Do you detect the pattern? *New mathematics is born from such detected patterns!* We shall verify this pattern in Chapter 6.

   b. Summing the first n powers of a fixed base number (we use base 2).

      i. $\sum_{i=0}^{n} 2^i = 1 + 2 + 4 + \cdots + 2^n = 2^{n+1} - 1$

      ii. $\sum_{i=0}^{n} i2^i = 2 + 8 + 24 + \cdots + n2^n = n2^{n+1} + 2$

      Note that exponentials grow so fast (as functions of n) that the sum of terms is just a bit larger than the largest term! We shall learn, in detail, how to discover this fact in Chapter 6.

2. ⊕ **Meeting people at a party**
   LESSON: Experience using the Pigeonhole Principle

   You are attending a cocktail party that is populated by n couples. Attempting to create a warm atmosphere, the host requests that each attendee shake the hand of every attendee that he or she does not know.

   Prove that some two attendees shake the same number of hands.

3. **An elementary result about encoding**
   LESSON: Practice using Proof by Contradiction

   Here is a simple way to encode an ordered pair of positive integers as a single positive integer.

   Prove the following assertion using a proof by contradiction.

   Proposition 2.7 *Let a, b, c, d be (not necessarily distinct) positive integers. If $2^a 3^b = 2^c 3^d$, then $a = c$ and $b = d$.*

4. ⊕ **Bi-colored necklaces in tubes**
   LESSON: The nature of gradual transitions—a discrete analogue of *Rolle's Theorem* (Michel Rolle, 1691)

Purely mathematical, continuous, version: You start at point A, carrying w_1 units of load. You move toward point B, shedding load at a steady pace. *If you reach point B with w_2 units of load, then at some point in your journey, you had $\frac{1}{2}(w_1 + w_2)$ units of load.*

Discrete version in a "real" setting. You have a *necklace* composed of $2n$ jewels: $2a$ black jewels and $2b$ white jewels. For illustration, the necklace in Fig. 2.10 has $n = 6$, $a = 5$, and $b = 1$. In part (a) of the figure, the necklace is unadorned; in part

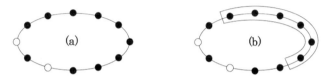

Fig. 2.10 (a) A necklace having 12 jewels: 10 black and 2 white. (b) The necklace in a tube.

(b), the necklace appears within a length-n *tube* which isolates one string—i.e., half-necklace—of n jewels from the complementary string.

Prove the following.

Proposition 2.8 *For any bi-colored necklace of the form described—i.e., with even numbers of jewels, black jewels, and white jewels—there is a way to position the tube so that inside the tube and outside the tube, there are equally many jewels, equally many black jewels, and equally many white jewels.*

Hint: Slide the tube around the necklace, and count both black and white jewels at each step. How can these numbers change in a single step?

5. **There is only one 1 (as a multiplicative identity)**
   LESSON: There is precisely one multiplicative identity for the integers.

   Prove the following assertion by contradiction.

   Proposition 2.9 *The integer 1 is the unique multiplicative identity for the integers. In other words, if the positive integer a satisfies* either *the equation*

   $$a \times x = x$$

   or *the equation*

   $$x \times a = x$$

   for all positive integers x, then $a = 1$.

6. ⊕ **Using *geometric* intuition to sum inverse powers of** 4
   LESSON: Exploiting geometric intuition toward a sophisticated end.

We turn to a variant of Proposition 2.5. This time, we focus on the entire infinite summation

$$S = \frac{1}{4} + \frac{1}{4^2} + \cdots + \frac{1}{4^k} + \cdots$$

and we do so from a geometric point of view.

We prove in Proposition 6.4(b) that this infinite summation converges to the value $\frac{1}{3}$. A simple way to see this is to multiply the summation S term by term by the fraction $1/4$. We observe—just by inspection—that the resulting product, which clearly has the value $S/4$, equals $S - 1/4$, so that $S = 1/3$.

There is a charming geometric argument which mimics the preceding conversion of the original summation into a quarter-valued version of itself, followed by a summarizing calculation. This argument exposes in a compelling way the *self-similarity* which underlies the argument. We get you started on this argument in Fig. 2.11. Observe that we can iterate the partitioning process in the figure on

Fig. 2.11 Partitioning a unit-area isosceles triangle into 4 sub-triangles, each of area $1/4$.

any one of the four sub-triangles of the original triangle—because all four sub-triangles are identical to each other (except for placement and orientation) and are similar to the original triangle.

We have now described the process at a very high level. Your assignment is to flesh out the details. *Prove that the four sub-triangles*

- *are similar to one another*

- *are similar to the original triangle*

- *have an area which is $1/4$ that of the original triangle*

Assemble these facts into an evaluation of the sum S.

Chapter 3
Sets and Their Algebras: The Stem Cells of Mathematics

> *Les grands mathématiciens ont, de tout temps, été ceux qui ont su substituer les idées au calcul.*
> *(The great mathematicians have always been those who knew how to substitute thinking for calculating.)*
>
> Attributed to Peter Gustav Lejeune Dirichlet

3.1 Introduction

This chapter studies three of the most basic concepts that underlie mathematics:

1. *Sets*: "pure" objects which have no structure or apparent capability of operating on anything, nor of being operated on

2. *Structured sets*: sets whose objects have structure that connotes their relationships with other objects

3. *Algebras*: sets, with or without structure, which are enriched by operations that manipulate either the sets or their objects.

We describe these concepts informally in this introductory section.

Basic sets. (Section 3.2) Sets are probably the most basic object of mathematical discourse. Sets exist to have *elements*, or *members*; these are the entities that *belong to* the set. Despite the conceptual simplicity of the notion "*set*", that notion is surprisingly difficult to specify formally: Philosophers have been debating the nature of the notion for millennia. Yet, the intuitive grasp of the concept that *everyone* develops just in the course of living is, surprisingly, adequate for almost all intellectual endeavors regarding the concept. So, we take the basic definition of "set" as given, and we begin our journey into the wondrous world of mathematics from that point.

© Springer Nature Switzerland AG 2020
A. L. Rosenberg, D. Trystram, *Understand Mathematics, Understand Computing*,
https://doi.org/10.1007/978-3-030-58376-7_3

We begin to develop the rudiments of science and mathematics by imposing structure upon the sets of interest and assembling a repertoire of operations to manipulate them.

As soon as our mathematical progenitors developed a repertoire of operations on sets, they observed patterns regarding how various operations interact with one another. They recognized that the patterns which persisted—i.e., which held in all situations—are actually *"laws"* that govern interactions among the operations. Two "laws" that the reader has certainly encountered assert that one can add a set of numbers in a variety of distinct ways without affecting the sum:

- *The commutative law.*[1]

$$(\forall x_1, x_2) \left[x_1 + x_2 \ = \ x_2 + x_1 \right]$$

- *The associative law.*

$$(\forall x_1, x_2, x_3) \left[\left(x_1 + (x_2 + x_3) \right) \ = \ \left((x_1 + x_2) + x_3 \right) \right]$$

Explanatory note.

Laws: in society, in science, in mathematics.
The word "law" plays at least three mutually inconsistent roles in our lives.

1. We read one day in a newspaper about a new "law" that has been enacted by a governmental entity. The next day, we read about a "law"—perhaps the one that was just enacted—that has been amended or even abrogated. These "laws" have finite lifetimes which are at the whim of governmental entities.

2. We learn in school about certain "laws" of physics—the "law" of gravity, the "law" of relativity, the "law" of conservation of mass, to name just a few. As time goes by and new science is discovered, some of these "laws" get amended—we learn, for instance, that mass and energy are not conserved: they are, in fact, interchangeable. Indeed, some "laws" are discovered to have been *false*—the tale of *phlogiston* comes to mind. These "laws" are *approximations to/predictors of reality* which survive until new "laws" are discovered that are better approximations/predictors.

3. We either learn about or discover certain mathematical facts that become known as *"laws"*. We discover, for instance, that one can add a list of numbers in any order without changing the sum (the commutative law which we just discussed). *These "laws" are immutable!* They do not rely on human experience up to any point in time but rather on logical reasoning about specific constants of life: numbers, geometrical shapes, etc. These are the "laws" that we will begin to uncover in this and subsequent chapters.

---

[1] The symbol "∀" is a shorthand for the phrase "for all".

Having clarified our intent, we shall no longer place the word "law" in quotes.

Structured sets. (Section 3.3) The first operations on sets which we study endow sets with enough structure to talk about aspects of "real life". One can view these operations as the mathematical analogue of "systems programs" within the realm of computers.

- We formulate a rigorous analogue of the intuitive concept of *relation*. We can thenceforth talk about relations such as "parent-child", "set-subset", "numbers and their squares", and we can study how some of these relations behave like— or unlike—some others.

- We use various kinds of structure in sets to create complex objects that have sub-objects and sub-· · ·-sub-objects, to arbitrary depths.

- We identify the notion *function*, which is among the central concepts of mathematics, and we identify valuable genres of function (one-to-one, onto, . . .).

Once we have these notions, we can begin to formulate *mathematical models* for real-life entities and situations. Two features that will stand out: how naturally the formal notions capture the intuitive notions of the vernacular; how technically simple the formal notions are—which facilitates using them in sophisticated arguments and analyses.

Algebras. (Section 3.4) Once one has a set, plus operations on the set which obey certain laws, one has an *algebra*. There are several natural algebras whose objects are sets. We discuss two of them at some length.

- *Boolean Algebras.* We focus first on the algebra that is built upon sets and the most basic operations on sets: *union* and *set difference*. The first of these operations, denoted $S \cup T$, combines the membership of its argument sets, S and T; the second, denoted $S \setminus T$, excludes from set S all members of set T. These two operations provide a *basis* for the algebra of sets, in the sense that one can combine these two operations in many different ways to craft an immense repertoire of operations on sets. As an exercise, the reader should define the following two operations from union and set difference: *intersection* ($S \cap T$), which isolates all elements that sets S and T share, and *symmetric difference* ($S + T$), which isolates all elements that belong to precisely one of S and T.

  Importantly, we can now begin to study the laws that these operations obey. One of the central topics in this study is the class of algebras on sets which are named in honor of the nineteenth-century English mathematician George Boole. Boole is generally credited with inventing these *Boolean* algebras.

- *Propositional Logic.* It is difficult to explain the meaning of the Boolean set operations without using "logical" terms such as *and*, *or*, and *not*. These terms fall naturally within the domain of the simplest variety of *mathematical logics*— the *logic of propositions*, or, more familiarly, *Propositional Logic*. This branch of logic studies how the *truth-values*, TRUE and FALSE, of elementary logical expressions combine via operators such as AND, OR, and NOT to produce a truth-value for any complex logical expression.

Propositional Logic is the simplest genre of mathematical logic because it does not deal with *quantifiers*—such as "THERE EXISTS" ($\exists$), or "FOR ALL" ($\forall$)—or with *modalities*—such as "EVENTUALLY" or "FROM SOME MOMENT ON".

Propositional Logic is a fascinating special form of Boolean Algebra for at least two reasons which are rather technical and highly consequential.

- *From a mathematical perspective:* Propositional Logic enjoys a special algebraic property known as *free*-ness, which enables one to prove theorems in Propositional Logic using *truth tables*. This ability means that one can prove theorems within this special system of logic by a form of "symbolic evaluation" rather than by struggling with the axioms-plus-rules-of-inference that many of us found so mysterious and onerous in high school geometry.

- *From a computational perspective:* There is a genre of Propositional logical expression which can model a genre of *computation* so faithfully that it can serve as a foundation for a theory that explains the *complexity of computation*—why is it harder to compute some functions than others?

This will be an exciting chapter to start our study of mathematics with.

3.2 Sets

3.2.1 Fundamental Set-Related Concepts

The reader certainly knows informally what a set is and recognizes that some sets are finite while others are infinite. Continuing to speak informally—a formal treatment will follow in later chapters—here are a few illustrative finite sets:

- the set of words in this book

  We do not know how big this set is, but you as a reader likely have a better intuitive feel than we as authors.

- the set of characters in any *JAVA* program

  Note that while this set is surely finite, we are not so confident about the number of seconds that a given program will run!

- the set consisting of *you*

  Paraphrasing the iconic television figure Mister Rogers, "You are unique." This set has just one element.

- the set of unicorns in New York City

  We will not argue with you about this, but we suspect that this is the *empty set* $\emptyset$, which has zero members.

Some familiar infinite sets are the sets of:

- *nonnegative integers*

- *positive integers*

- *all integers*

- nonnegative *rational numbers*—which are quotients of integers

- nonnegative *real numbers*—which can be viewed as the numbers that admit infinite decimal (or binary, or octal, or hexadecimal, or . . .) expansions,

- *complex numbers*—which can be viewed as ordered pairs of real numbers,

- *all* finite-length binary strings (or ternary, or quaternary, or . . .).

  A *binary string* is a sequence of 0s and 1s. When discussing computer-related matters, one often calls each 0 and 1 in a binary string a *bit* (for *binary digit*). The term "bit" leads to the term *bit string* as a synonym of *binary string*. A *ternary string* is a sequence of 0s, 1s, and 2s; a *quaternary string* is a sequence of 0s, 1s, 2s, and 3s; and so on.

Our assumption about your prior experience with sets notwithstanding, we begin the chapter by reviewing some basic concepts concerning sets and operations on sets.

As noted earlier, sets were created to contain members/elements. We denote the fact that element *t* *belongs to* or *is an element of* set *T* by the notation $t \in T$. Contrarily, we denote the fact that element *t* *does not belong to* or *is not an element of* set *T* by the notation $t \notin T$.

A *subset* of a set *T* is a set *S* each of whose members belongs to *T*. The subset relation occurs in two forms.

1. The *strong* form of the subset relation, denoted $S \subset T$, asserts that every element of *S* is an element of *T*, but *not* conversely; i.e., *T* contains (one or more) elements that *S* does not. When $S \subset T$, we call *S* a *proper* subset of *T*.

2. The *weak* form of the subset relation, denoted $S \subseteq T$, is defined as follows:

$$[S \subseteq T] \quad \text{means:} \quad \left[either \;\; [S = T] \;\; or \;\; [S \subset T] \right].$$

For any *finite* set *S*, we denote by $|S|$ the *cardinality* of *S*, which is the number of elements in *S*. Finite sets having three special cardinalities are singled out with special names. The limiting case of finite sets is the *empty set*; this set, which we denote by $\emptyset$, is *unique*. We say that $\emptyset$ is *characterized* by the equation $|\emptyset| = 0$, meaning that the equation can be used as a definition of the set. (The empty set is often a limiting case of set-defined entities.) If $|S| = 1$, then we call *S* a *singleton*; and if $|S| = 2$, then we call *S* a *doubleton*. (One could, of course, continue with "tripletons" and "quadrupletons", etc., but people tend not to do this.)

It is often useful to have a convenient term and notation for *the set of all subsets of a set S*. This bigger set—we shall see before long that it contains $2^{|S|}$ elements when S is finite—is denoted $\mathscr{P}(S)$ and is called the *power set* of S.[2] Note carefully the two set-relations that we are talking about here:

> *If set T is a subset of set S, then T is an element of the set $\mathscr{P}(S)$.*

You should satisfy yourself that the biggest (i.e., *most populous*) and smallest (i.e., *least populous*) elements of $\mathscr{P}(S)$ are, respectively, the set S itself and the empty set $\emptyset$.

3.2.2 Operations on Sets

Focus on two sets, S and T, as depicted schematically in Fig. 3.1.

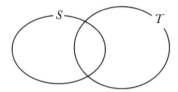

Fig. 3.1 Two (overlapping) sets, S and T

Historical/Cultural note.

Pictorial representations of sets such as those in Fig. 3.1 and its kindred illustrations are called *Venn diagrams*, in honor of John Venn, who employed them in his 1880 paper, "On the diagrammatic and mechanical representation of propositions and reasonings", which appeared in the *Philosophical Magazine and Journal of Science*.

We denote by:

- $S \cap T$ the *intersection* of S and T; see Fig. 3.2.

    The elements of $S \cap T$ belong to *both* S and T:

$$[s \in S \cap T] \quad \text{means} \quad \Big[[s \in S] \textbf{ and } [s \in T]\Big]$$

- $S \cup T$ the *union* of S and T; see Fig. 3.3 .

    The elements of $S \cup T$ belong either to S, or to T, *or to both*:

$$[s \in S \cup T] \quad \text{means} \quad \Big[[s \in S] \textbf{ or } [s \in T] \textbf{ or } [s \in S \cap T]\Big]$$

---

[2] The name "power set" arises from the relative cardinalities of S and $\mathscr{P}(S)$ for finite S.

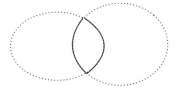

Fig. 3.2 Two operations on S and T. The region within the bold border is their *intersection*. The region within the dotted border—excluding the bold lines—is their *symmetric difference*

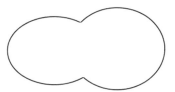

Fig. 3.3 Union of sets S and T

To emphasize the qualifier "or to both", this operation is sometimes called *inclusive union*.

- $S \setminus T$ the *(set) difference* of S and T; see Fig. 3.4.

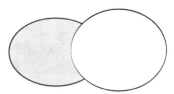

Fig. 3.4 Difference of sets S and T (in grey)

The elements of $S \setminus T$ belong to S but *not* to T:

$$[s \in S \setminus T] \quad \text{means} \quad \Big[[s \in S] \text{ and } [s \notin T] \Big]$$

(Particularly in the United States, one often encounters the notation "$S - T$" instead of "$S \setminus T$".)

We illustrate the preceding operations with the sets $S = \{a, b, c\}$ and $T = \{c, d\}$:

$$S \cap T = \{c\}$$
$$S \cup T = \{a, b, c, d\}$$
$$S \setminus T = \{a, b\}$$

In many situations where sets are being studied, the sets of interest will be subsets of some fixed "universal" set U.

Explanatory note.

We use the term "universal" contextually, in the sense of a "universe of discourse". We are *not* using the term in the absolute, self-referencing sense of a set U that contains all sets as members—"self-referencing" because an absolute universal set U would perforce contain itself as a member. As we shall discuss in Section 7.2.2.2, the absolute, self-referencing construct has been shown by the British philosopher/logician Bertrand Russell to lead to the mind-bending paradox known eponymously as *Russell's Paradox* [95, 113].

Given a universal set U and a *subset $S \subseteq U$*, we observe the set-inequalities

$$\emptyset \subseteq S \subseteq U$$

When we study a context within which there exists a universal set U, we include *(set) complementation* within our repertoire of set-related operations.

- $\overline{S} \overset{\text{def}}{=} U \setminus S$ is the *complement* of set S (relative to the universal set U); see Fig. 3.5.

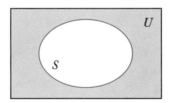

Fig. 3.5 Complement of set S (in grey) relative to the universal set U

$\overline{S} = U \setminus S$ is the set of all elements of U that do not belong to S. For instance, the set of *odd positive integers* is the complement of the set of *even positive integers*, relative to the set of *all positive integers*.

We note a number of basic identities involving sets and operations on them.

- $S \setminus T = S \cap \overline{T}$,
- If $S \subseteq T$, then
  1. $S \setminus T = \emptyset$,
  2. $S \cap T = S$,
  3. $S \cup T = T$.

Note, in particular, that[3]

$$[S = T] \text{ iff } \Big[[S \subseteq T] \text{ and } [T \subseteq S]\Big] \text{ iff } \Big[(S \setminus T) \cup (T \setminus S) = \emptyset\Big]$$

The operations union, intersection, and complementation—and operations formed from them, such as set difference—are called the *Boolean (set) operations*, so named for the nineteenth-century English mathematician George Boole. There are several important identities involving the Boolean set operations. Among the most useful (hence, the most frequently invoked) are the two *laws* named for the nineteenth-century English mathematician Augustus De Morgan:

$$\text{For all sets } S \text{ and } T: \quad \begin{cases} \overline{S \cup T} = \overline{S} \cap \overline{T} \\[2mm] \overline{S \cap T} = \overline{S} \cup \overline{T} \end{cases} \tag{3.1}$$

Elementary logic can be used to verify these laws. To spell out just one case: An element s that belongs to neither S nor T (so that $s \in \left(\overline{S \cap T}\right)$) cannot belong to S or to T, hence cannot belong to the union of S and T.

(Algebraic) Closure. We end this section with a set-theoretic definition that one encounters in *many* contexts.

- Let $\mathscr{C}$ be a (finite or infinite) collection of sets.

- Let S and T be elements of collection $\mathscr{C}$.[4]

- Let $\circ$ be an operation on sets—so that $S \circ T$ is a set.

We say that collection $\mathscr{C}$ is *closed* under the operation $\circ$ if whenever sets S and T (which could be the same set) both belong to $\mathscr{C}$, the set $S \circ T$ also belongs to $\mathscr{C}$.

As a concrete example of the use of the notion of closure, we note the following. De Morgan's laws tell us that a collection $\mathscr{C}$ of finite sets is closed under the operation of intersection whenever it is closed under the operations of union and complementation.

3.3 Structured Sets

The power of set-theoretic concepts to model complex aspects of reality increases immeasurably when we consider sets whose elements enjoy even modest structure. We begin our discussion of structured sets by adding a new (binary) set operation to our earlier repertoire.

---

[3] "iff" is the common abbreviation for the mathematical phrase, "if and only if".

[4] Note that $\mathscr{C}$ is a set whose elements are sets.

Given (finite or infinite) sets S and T we denote by $S \times T$ the *direct product* of S and T, which is the set of all *ordered pairs* whose first coordinate is an element of set S and whose second coordinate is an element of set T. The direct product operation is often called the *Cartesian* product, because of the notion's origin in the formulation of Analytical Geometry by the French mathematician-philosopher René Descartes.

We illustrate the direct product operation in two ways, in order to capture all of its subtleties. Textually, if $S = \{a, b, c\}$ and $T = \{c, d\}$, then

$$S \times T = \{\langle a,c \rangle, \langle b,c \rangle, \langle c,c \rangle, \langle a,d \rangle, \langle b,d \rangle, \langle c,d \rangle\}$$

Pictorially, Fig. 3.6 provides a schematic illustration of $S \times T$.

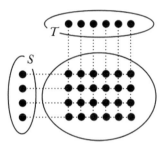

Fig. 3.6 The Direct/Cartesian product of sets S and T

3.3.1 Binary Relations: Sets of Ordered Pairs

The direct-product operation on sets affords us a simple, yet powerful, formalization of the notion of *binary relation*. Given (finite or infinite) sets S and T, a *relation* ρ *on S and T* (in that order) is any subset

$$\rho \subseteq S \times T.$$

When $S = T$, we often call ρ a *binary relation* **on** *(the set)* S. The qualifier "*binary*" indicates that relation ρ establishes links between *pairs* of elements of set S. (By extension, a *ternary* relation would establish links among *triples* of elements of S, and so on for larger arities.)

Relations are so common that we use them in every aspect of our lives without formally acknowledging the rules that govern them. The relations "equal to", "less than", and "greater than or equal to" are simple examples of binary relations on the integers. These same relations apply also to other familiar number systems such as the rational and real numbers; only the relation "equals," though, holds (in the natural way) for the complex numbers. (You will see much more about these sets of numbers in Chapters 4, 8, and 10.) Some subset of the three relations "is a parent of", "is a child of", and "is a sibling of", are binary relations that may apply to (the set of people constituting) your family. The preceding relations all apply to a single set S; a familiar relation for which the sets S and T are distinct is the relation "A is taking course X", which is a relation on

$$(\text{the set of all students}) \times (\text{the set of all courses})$$

By convention, when we deal with a binary relation $\rho \subseteq S \times T$, we often use *infix* notation: We write "$s\rho t$" in place of the more stilted "$\langle s,t \rangle \in \rho$". For instance, we (almost always) write "$5 < 7$" in place of the strange-looking (but formally correct) "$\langle 5,7 \rangle \in <$".

The following operation on relations occurs in many guises, in almost all areas of mathematics. Let ρ and ρ' be binary relations on a set S. The *composition* of relations ρ and ρ' (in that order) is the relation

$$\rho'' \stackrel{\text{def}}{=} \Big\{ \langle s,t \rangle \in S \times S \mid (\exists u \in S)\big[[s\rho u] \text{ and } [u\rho't]\big] \Big\}$$

(Note that we use both of our notations for relations in this equation.)

Enrichment note.

Even at this early stage in our study, we are using concepts that appear in "real-life" applications. The operation of *composing relations* is an essential feature of *relational databases*, as described in the original source on the subject, [32].

When we discuss a relation $\rho \subset S \times T$, it is important to be able to express the assertion that elements $s \in S$ and $t \in T$ are *not* ρ-related. We already have the elaborate notation, "$\langle s,t \rangle \notin S \times T$", but it would be good to have a streamlined notation also. Several notations have been developed for this purpose, as the following table suggests.

Relation	Notation	Negation
set membership	$\in$	$\notin$
equality	$=$	$\neq$
less than (strong)	$<$	$\not< \text{ or } \geq$
less than (weak)	$\leq$	$\not\leq \text{ or } >$
greater than (strong)	$>$	$\not> \text{ or } \leq$
greater than (weak)	$\geq$	$\not\geq \text{ or } <$
generic	ρ	$\neg\rho \text{ or } \bar{\rho}$

(3.2)

Several special classes of binary relations are so important that we single them out immediately, in the upcoming subsections.

3.3.2 Order Relations

A binary relation ρ on a set S is a *partial order relation*, or, more briefly, is a *partial order* if ρ is transitive. This means that, for all elements $s, t, u \in S$,

$$\text{if} \quad s\rho t \quad \text{and} \quad t\rho u \quad \text{then} \quad s\rho u \tag{3.3}$$

The qualifier "partial" warns us that some pairs of elements of S do not occur in relation ρ. Number-related orders supply an easy illustration. Given any two *distinct* integers, m and n, one of them must be less than the other: i.e., either $m < n$, or $n < m$. In contrast, if we consider *ordered pairs* of integers, then there are pairs of pairs that are not related by the "less than" relation in any natural way. For instance, even though we may agree that—by a natural extension of the number-ordering relation "less than"—the ordered pair $\langle 4, 17 \rangle$ is "less than" the ordered pair $\langle 22, 19 \rangle$, we may well not agree on which of the ordered pairs $\langle 4, 22 \rangle$ and $\langle 19, 17 \rangle$ is "less than" the other.

In many domains, order relations occur in two "flavors", *strong orders* and *weak orders*. For many such relations ρ—consider, e.g., "less than" on the integers—the weak version is denoted by underscoring the strong version's symbol. This will be our convention. Just as $\leq$ denotes the weak version of $<$, and $\geq$ denotes the weak version of $>$, we shall denote the weak version of a generic order ρ by the compound symbol $\underline{\rho}$. Strong and weak versions of an order relation ρ (denoted, respectively, ρ and $\underline{\rho}$) are distinguished by their behavior under simultaneous membership. For illustration, instantiate the following template with ρ being "$<$" (less than) and with $\overline{\rho}$ being "$>$" (more than):

For a strong order ρ: **if** $[s \, \rho \, t]$, **then** $[t \, \overline{\rho} \, s]$
For the weak version $\underline{\rho}$ of ρ: **if** $[s \, \underline{\rho} \, t]$ **and** $[t \, \underline{\rho} \, s]$, **then** $[s = t]$.

It is important to note that negating a strong relation (e.g., $<$) yields a weak relation ($\geq$), while negating a weak relation (e.g., $\leq$) yields a strong relation ($>$). Keep this in mind when we study Propositional Logic in Section 3.4.1. We shall note there that the following transformations

$$\text{weak order} + \text{negation} \longrightarrow \text{strong order}$$

$$\text{strong order} + \text{negation} \longrightarrow \text{weak order}$$

result from Proposition 3.5, the logical analogue of De Morgan's Laws (3.1).

3.3.3 Equivalence Relations

A binary relation ρ on a set S is an *equivalence relation* if it enjoys the following three properties:

1. ρ is *reflexive*: $(\forall s \in S)$ $[s\rho s]$
2. ρ is *symmetric*: $(\forall s, s' \in S)$ $[[s\rho s']$ iff $[s'\rho s]]$
3. ρ is *transitive*: $(\forall s, s', s'' \in S)$ if both $[s\rho s']$ and $[s'\rho s'']$, then also $[s\rho s'']$

Sample familiar equivalence relations are:

- The equality relation $=$ on a set S.

  This relation relates each $s \in S$ with itself (i.e., $s = s$) but with no other element of S.

- The relations $\equiv_{12}$ and $\equiv_{24}$ on integers, where[5]

  1. $n_1 \equiv_{12} n_2$ if and only if $|n_1 - n_2|$ is divisible by 12.

  2. $n_1 \equiv_{24} n_2$ if and only if $|n_1 - n_2|$ is divisible by 24.

  We use relation $\equiv_{12}$ (without formally acknowledging it) when we specify the time using a 12-hour clock; we use relation $\equiv_{24}$ when we specify the time using a 24-hour clock.

Closely related to the notion of an equivalence relation on a set S is the notion of a *partition* of S—i.e., a nonempty collection of subsets $S_1, S_2, \ldots$ of S that are

  - *mutually exclusive*: for distinct indices i and j, $S_i \cap S_j = \emptyset$;
  - *collectively exhaustive*: $S_1 \cup S_2 \cup \cdots = S$.

We call each set S_i a *block* of the partition.
One verifies the following Proposition easily.

Proposition 3.1 *A partition of a set S and an equivalence relation on S are just two ways of looking at the same concept.*

Proof (Sketch). To get an equivalence relation from a partition. Given any partition $S_1, S_2, \ldots$ of a set S, define the following relation ρ on S:

  $s\rho s'$ if, and only if, s and s' belong to the same block of the partition.

We claim that relation ρ is an equivalence relation on S. Specifically, the *collective exhaustiveness* of the partition ensures that each $s \in S$ belongs to some block of the partition, while the *mutual exclusivity* of the partition ensures that s belongs to only one block.

To get a partition from an equivalence relation. Focus on any equivalence relation ρ on a set S. For each $s \in S$, denote by $[s]_\rho$ the set

$$[s]_\rho \stackrel{\text{def}}{=} \{s' \in S \mid s\rho s'\}$$

we call $[s]_\rho$ the *equivalence class of s under relation ρ*.

The equivalence classes under ρ form a partition of S. Specifically, ρ's *reflexivity* ensures that the equivalence classes collectively exhaust S; ρ's symmetry and transitivity ensure that equivalence classes are mutually disjoint. $\square$

---

[5] As usual, $|x|$ is the *absolute value* or *magnitude* of the number x: If $x \geq 0$, then $|x| = x$; if $x < 0$, then $|x| = -x$.

The *index* of the equivalence relation ρ is its number of classes—which can be finite or infinite. Henceforth, we conform to common usage and use the symbol $\equiv$, possibly embellished by a subscript or superscript, to denote an equivalence relation.

Let $\equiv_1$ and $\equiv_2$ be two equivalence relations on a set S. We say that the relation $\equiv_1$ *is a refinement of* or *refines* the relation $\equiv_2$ precisely if each block of $\equiv_1$ is a subset of some block of $\equiv_2$. We leave the verification of the following basic result as an exercise.

Theorem 3.1. *The equality relation $=$ on a set S is the* finest *equivalence relation on S, in the sense that $=$ refines every equivalence relation on S.*

3.3.4 Functions

3.3.4.1 The basics: definitions and generic properties

One learns early in school that a function from a set S to a set T is a rule that assigns a unique value from T to every value from S. Simple examples illustrate that this notion of function is more restrictive than necessary. Think, e.g., of the operation *division* on integers. We learn that division, like multiplication, is a function that assigns a number to a given pair of numbers—yet we are warned immediately not to "divide by 0": The quotient upon division by 0 is "undefined", So, division is *not quite* a function of the same sort as addition or multiplication, which both *do* conform to the notion envisioned by our initial definition of "function". In more homely terms, note that, in contrast to an expression such as "$4 \div 2$", which should lead to the result 2 in every programming environment,[6] expressions such as "$4 \div 0$" can lead to wildly different results in different programming environments. Since "wildly different" is anathema in any mathematical setting, mathematicians have dealt with situations such as one encounters with the operation of division by broadening the definition of "function" in a way that behaves like our initial simple definition under "well-behaved" circumstances and that extends the notion in an intellectually consistent way under "ill-behaved" circumstances. Let us begin to get formal.

A *(partial) function from set S to set T* is a relation $F \subseteq S \times T$ that is *single-valued;* i.e., for each $s \in S$, there is *at most* one $t \in T$ such that sFt. We traditionally write "$F : S \to T$" as shorthand for the assertion, "F is a function from the set S to the set T"; we also traditionally write "$F(s) = t$" for the more conservative (but correct) "sFt". (The single-valuedness of F makes the nonconservative notation safe.) We often call the set S the *source* or *domain* of function F, and we call set T the *target* or the *range* of function F. In situations when there is always a (perforce, unique) $t \in T$ for each $s \in S$, then we call F a *total* function.

You may be surprised to encounter functions that are not total, because most of the functions you deal with daily are *total*. Our mathematical ancestors had to do

---

[6] We are, of course, ignoring demons such as round-off error.

some fancy footwork in order to make your world so neat. Their choreography took two complementary forms.

1. *They expanded the target set T on numerous occasions.*

   As just two instances:

   - They appended both 0 and the negative integers to the preexisting positive integers in order to make subtraction a total function.

   - They appended the *rational numbers* to the preexisting integers in order to make division (by nonzero numbers!) a total function.

   The *irrational algebraic* numbers, the *nonalgebraic real* or *transcendental* numbers, and the *nonreal complex* or *imaginary* numbers were similarly appended, in turn, to our number system in order to make certain (more complicated) functions total. (Chapter 4 expands on this brief history of our number system.)

2. *They adapted the function.*

   For programming systems, in particular, literal undefinedness is anathema—a computer must always do *something* in response to a command—so programming languages typically have chosen (sometimes artificial) ways of making functions total, via devices such as "integer division" (so that odd integers can be "divided by 2") in addition to various ploys for accommodating "division by 0".

The (twentieth-century) inventors of *Computation Theory* insisted on a theory of functions on nonnegative integers (or some transparent encoding thereof). The price for such "pureness" is that they had to allow functions to be literally undefined on some arguments. Thus the Theory renders such functions as "division by 2" and "taking square roots" as *nontotal*: Both are defined only on subsets of the positive integers (the even integers and the perfect squares, respectively).

Three special classes of functions merit explicit mention. For each, we give both a down-to-earth name and a more scholarly Latinate one. (Both types of names are actually used.)

1. A function $F : S \to T$ is *one-to-one* or *injective* if for each $t \in T$, there is at most one $s \in S$ such that $F(s) = t$;

   Example:

   - "multiplication by 2" is injective: If you are given an even integer $2n$, you can always respond with the integer n.

   - "integer division by 2" is not injective—because performing the operation on arguments $2n$ and $2n+1$ yields the same answer (namely, n).

   An injective function F is called an *injection*.

   Importantly, each injection F has a *functional inverse*, which is commonly denoted F^{-1}, and which is defined as follows.

For each $t \in T$: $F^{-1}(t) = \begin{cases} s & \text{if there is an } s \in S \text{ such that } F(s) = t \\ \textit{undefined} & \text{if there is no } s \in S \text{ such that } F(s) = t \end{cases}$

Because F is *injective*, there is at most one element $s \in S$ such that $F(s) = t$. In other words, an element $t \in T$ can occur in the range of F only because of a single element $s \in S$ in the domain of F. This means that

- the preceding definition of F^{-1} is a valid definition—i.e., the notion "functional inverse of an injection" is *"well defined"*.

- F^{-1} is a (partial) function $F^{-1} : T \to S$ whose domain is the range of F.

2. A function $F : S \to T$ is *onto* (or, *surjective*) if for each $t \in T$, there is at least one $s \in S$ such that $F(s) = t$;

   Example:

   - Two surjective functions on the nonnegative integers:
     - "subtraction of 1" is surjective, because "addition of 1" is a total function.

     - "taking the square root" is surjective because the operation of squaring is a total function.

   - Two functions on the nonnegative integers that are *not* surjective:
     - "addition of 1" is not surjective, because, e.g., 0 is not "1 greater" than any nonnegative integer.

     - "squaring" is not surjective, because, e.g., 2 is not the square of any integer.

   A surjective function F is called a *surjection*.

3. A function $F : S \to T$ is *"one-to-one, onto"* or *bijective* if for each $t \in T$, there is precisely one $s \in S$ such that $F(s) = t$.

   A bijective function F is called a *bijection*. When F is a bijection from S onto T, we often write $F : S \leftrightarrow T$.

Finally, we present two terms that appear frequently in discussions of functions and their source and target sets.

Focus on a function f that maps a set A *into* a set B; symbolically, we write $f : A \to B$. The *image* of the set A under f is the subset of B

$$\text{IMAGE}(A) \stackrel{\text{def}}{=} \{f(a) \mid a \in A\}$$

The *preimage* of the set B under f is the subset of A

$$\text{PREIMAGE}(B) \stackrel{\text{def}}{=} \{a \in A \mid f(a) \in B\}$$

For any set S, a bijection $f : S \leftrightarrow S$ is called a *permutation* of set S. Note that the set S plays a lot of roles here. S is:

- the source of function f
- the target of function f
- the image of set S under function f
- the preimage of set S under function f

While the operation of *(functional) composition*, as introduced in Section 3.3.1, is important for general binary relations, it is a daily staple with relations that are functions! Let us be given two functions on a set S

$$F : S \to S \quad \text{and} \quad G : S \to S$$

The *composition* of F and G, *in that order*, is the function

$$F \circ G : S \to S$$

defined as follows.

$$\text{For each } s \in S \quad F \circ G(s) = G(F(s)). \tag{3.4}$$

The unexpected change in the orders of writing F and G on the two sides of Eq. (3.4) results from the existence of two historical schools that both contributed to the formulation of this material. One school cleaved to the tradition of *abstract algebra*. They wanted expressions to be written with *all* operators, including the composition operator $\circ$, in *infix* notation—i.e., in the form "$s\rho t$". Another school, which could be called the *applicative algebraic school*, wanted to view functions as "applying" to their arguments. Both notations have significant advantages in certain contexts, so both have survived. It is a good idea for neophyte readers to be prepared to encounter both notations—but they have to keep their eyes open regarding the relative orders of F and G.

Before we proceed with new material, let us take a moment to verify that the important operation of composition behaves the way one would want and expect it to—by preserving the type of function being composed.

Proposition 3.2 *Let us be given functions $F : S \to S$ and $G : S \to S$ on the set S, together with their composition $F \circ G$, as defined in Eq. (3.4).*
(a) *$F \circ G$ is a function on S.*
(b) *If F and G are injections, then so also is $F \circ G$.*
(c) *If F and G are surjections, then so also is $F \circ G$.*
(d) *If F and G are bijections, then so also is $F \circ G$.*

Proof. We prove each of the four assertions by invoking the underlying definitions.
(a) Because F and G are functions on S, for each $s \in S$, there is at most one $t_1 \in S$ such that $F(s) = t_1$ and at most one $t_2 \in S$ such that $G(s) = t_2$. Hence, we identify three possibilities.

1. *if F* is defined at $s \in S$ *then $F(s) \in S$ is unique*
   and if G is defined at $F(s) \in S$ *then $G(F(s)) = F \circ G(s) \in S$ is unique*
2. *if F* is not defined at $s \in S$ *then $F \circ G$ is not defined at $s \in S$*
3. *if F* is defined at $s \in S$ *then $F(s) \in S$ is unique*
   and if G is not defined at $F(s) \in S$ *then $F \circ G$ not defined at $s \in S$*

Hence, for each $s \in S$, there is at most one $t \in S$ such that $F \circ G(s) = t$; in other words, $F \circ G$ is a function on S.

(**b**) Focus on any $s \in S$. Because F is an injection on S, there exists at most one $t \in S$ such that $F(t) = s$. Because G is an injection on S, there exists at most one $u \in S$ such that $G(u) = t$. Thus, there exists at most one $u \in S$ such that $F \circ G(u) = s$. Hence, $F \circ G(u)$ is an injection on S.

(**c**) Focus on any $s \in S$. Because F is a surjection on S, there exists $t \in S$ such that $F(t) = s$. Because G is a surjection on S, there exists $u \in S$ such that $G(u) = t$. This means, however, that $F \circ G(u) = s$. Because $s \in S$ was arbitrary, it follows that $F \circ G(u)$ is a surjection on S.

(**d**) If each of F and G is a bijection on S, then each is an injection on S, and each is a surjection on S. Then Part (b) tells us that $F \circ G$ is an injection on S, and Part (c) tells us that $F \circ G$ is a surjection on S. Hence, $F \circ G$ is a bijection on S. □

3.3.4.2 ⊕ The Schröder-Bernstein Theorem

This section is devoted to a remarkable theorem which states that *bijections ($f : S \leftrightarrow T$)* and *paired injections ($g : S \to T$ and $h : T \to S$)* always travel together. In detail,

For all sets S and T,

  if there exist both *an injection from S to T* and *an injection from T to S*

  then there exists a bijection between S and T.

Historical note

The celebrated theorem of this section is usually attributed to Ernst Schröder and Felix Bernstein, but it is often attributed alternatively to (Georg) Cantor and Bernstein. As this complex attribution suggests, the theorem has a rather complicated history. Picking just a few high points of this history: The theorem was first stated without proof by Cantor in [27]. Roughly a decade later, Schröder produced a flawed proof in [99]. As reported in [44], Schröder soon thereafter provided a correct proof—as, independently, did Bernstein.

We shall refer to the theorem by its customary name, the *Schröder-Bernstein Theorem*, for that is the most likely name the reader will encounter.

The Schröder-Bernstein Theorem can be immensely useful when reasoning about a large variety of mathematical scenarios that involve correspondences between the elements of two sets. We encounter such a scenario, for example, when we prove

Lemma 10.3 in Section 10.4. Acknowledging the importance of the result, we now provide a formal statement and a somewhat informal sketch of its proof.

Theorem 3.2 (The Schröder-Bernstein Theorem). *For any sets S and T, if there exist both an* injection

$$F^{(S \to T)} : S \to T$$

and an injection

$$F^{(T \to S)} : T \to S$$

then there exists a bijection

$$F^{(S \leftrightarrow T)} : S \leftrightarrow T$$

Proof (Sketch). We give the highlights of a perspicuous proof from [107].

Let $S_0 = S$ and $T_0 = T$. We set up a sequence of indices to orchestrate the upcoming argument: We have an index i for each integer $i \in \mathbb{N}$, and we have a special index ω which takes care of the single case that the integer indices do not.

For each index $i \in \mathbb{N}$, define

$$S_{i+1} = F^{(T \to S)}(T_i) \quad \text{and} \quad T_{i+1} = F^{(S \to T)}(S_i)$$

We are repeatedly using here the notational convention that for any set A and function f defined on A, the *image* of set A under function f is the set

$$f(A) = \{f(a) \mid a \in A\}$$

Because each of our injections, $F^{(S \to T)}$ and $F^{(T \to S)}$, maps its domain-set *into* its range-set (and not necessarily *onto*), we observe that increasing indices lead to nested progressions of sets, in the sense that

$$\text{Each} \quad S_{i+1} \subseteq S_i \quad \text{and each} \quad T_{i+1} \subseteq T_i$$

We now define two new sequences of sets:

- The sets U_ω and $U_0, U_1, U_2, \ldots$ are defined as follows:

$$U_\omega = \bigcap_i S_i \quad \text{and for each} \ i \in \mathbb{N} \ \ U_i = S_i \setminus S_{i+1}$$

Our reasoning shows that set S is *partitioned* into the sets $U_0, U_1, U_2, \ldots$ and U_ω.

- The sets V_ω and $V_0, V_1, V_2, \ldots$ are defined as follows:

$$V_\omega = \bigcap_i T_i \quad \text{and for each} \ i \in \mathbb{N} \ \ V_i = T_i \setminus T_{i+1}$$

Our reasoning shows that set T is *partitioned* into the sets $V_0, V_1, V_2, \ldots$ and V_ω.

Moreover, one sees that:

- $F^{(S \to T)}$ is a *bijection* between each U_i and V_{i+1}, while

- $F^{(T \to S)}$ is a *bijection* between each V_i and U_{i+1}.

Finally, since $V_\omega \subseteq F^{(S \to T)}(S)$, and

$$\left[F^{(S \to T)} \right]^{-1}(V_\omega) = \left[F^{(S \to T)} \right]^{-1} \left(\bigcap T_{i+1} \right) = \bigcap \left[F^{(S \to T)} \right]^{-1}(T_{i+1}) = \bigcap S_i = U_\omega$$

it follows that $F^{(S \to T)}$ maps U_ω *bijectively* onto V_ω.

Let us briefly "decode" the preceding symbol-heavy chain of equations. Recall first that $F^{(S \to T)}$ maps each set S_i *injectively* into T_{i+1}. This means that as we look at the *preimage* of the set $\bigcap T_{i+1}$ under $F^{(S \to T)}$, the various sets S_i do not get "mixed together"; they retain the separate identities that $F^{(S \to T)}$ endows them with because of the way it maps S into T. In other words,

$$\left[F^{(S \to T)} \right]^{-1} \left(\bigcap T_{i+1} \right) = \bigcap \left[F^{(S \to T)} \right]^{-1}(T_{i+1})$$

By similar calculation and reasoning, $F^{(T \to S)}$ maps V_ω bijectively onto U_ω.

The preceding identifications of bijections between portions of set S and portions of set T allow us to piece together the advertised bijection $F^{(S \leftrightarrow T)}$ between the complete sets S and T. In detail:

- For each $i \in \mathbb{N}$:

  – $F^{(S \leftrightarrow T)}$ maps each set U_{2i} bijectively onto V_{2i+1} in the same way that $F^{(S \to T)}$ does.

  – $F^{(S \leftrightarrow T)}$ maps each set U_{2i+1} bijectively onto V_{2i} in the same way that $\left[F^{(T \to S)} \right]^{-1}$ does.

- $F^{(S \leftrightarrow T)}$ maps V_ω bijectively onto U_ω in the same way that either $F^{(S \to T)}$ or $\left[F^{(T \to S)} \right]^{-1}$ does.

This sketch provides all of the salient ingredients of the proof in [107].   □

3.3.5 Sets, Strings, Functions: Important Connections

In his play, *Romeo and Juliet*, William Shakespeare pens a question whose significance goes far beyond its poetic origins: "What's in a name?" This section will hopefully get you thinking about that question in new ways!

Let us begin with a binary sequence—i.e., a sequence of bits

$$\beta = \beta_0 \beta_1 \beta_2 \cdots \beta_n(\cdots)$$

where each $\beta_i \in \{0,1\}$. The rather bizarre notation here, where the second set of (centered) dots is parenthesized, is intended to encompass both finite, length-n, sequences and infinite sequences. The question is:

What does sequence β denote (or, name)?

At least three respectable answers to this question should be considered.

1. With the most literal interpretation, β is just an uninterpreted sequence of bits.

2. With a little more imagination, one can interpret each bit β_i of β as information about the (index) integer i. For example, β can be viewed as specifying two mutually complementary sets of positive integers:

$$S_\beta = \{k \in \mathbb{N}^+ \mid \beta_k = 1\}$$
$$\overline{S}_\beta = \{k \in \mathbb{N}^+ \mid \beta_k = 0\}$$

When β has (finite) length n, then both S_β and $\overline{S}_\beta$ are viewed as subsets of the set $\{1, 2, \ldots, n\}$; when β is infinite, then both S_β and $\overline{S}_\beta$ are viewed as subsets of the set $\mathbb{N}^+$.

In this scenario, β is called the *characteristic sequence* of the set S_β.

The notion of characteristic sequence leads to a quite simple way of counting the number of distinct subsets that an n-element set has.

The number of length-n binary strings is determined in Proposition 11.1 and is then used in Proposition 11.3 to count the subsets of an n-element set.

3. Finally—*and we certainly do not mean to imply that our three alternatives exhaust the possibilities!*—each bit β_i of β can be interpreted as specifying the value at integer-argument i of a function f_β. When β has (finite) length n, then f_β is viewed as a function $f_\beta : \{1, 2, \ldots, n\} \to \{0,1\}$; when β is infinite, then f_β is viewed as a function $f_\beta : \mathbb{N} \to \{0,1\}$.

In this scenario, β is called the *characteristic vector* of the function f_β.

The notion of characteristic vector leads to a rather straightforward proof that there does not exist a bijection between the set $\mathbb{N}^+$ of positive integers and the set $\mathbb{F}$ of all functions from $\mathbb{N}^+$ to $\{0,1\}$; symbolically, $\mathbb{F} = \{f : \mathbb{N}^+ \to \{0,1\}\}$. See Proposition 10.7.

We now have a new way to think about sequences/strings, sets, and functions. Each way conveys insights and tools for reasoning about and analyzing these concepts that other "names" for the concepts do not. And, we now recognize conceptual ties among these three concepts that are not always intuitive. Our conceptual toolkit has been enriched!

3.4 Boolean Algebras

We remarked in Section 3.2.2 that there is an extensive repertoire of operations on sets. It is useful—and not difficult—to compile a list of laws that govern various sets of these operations.

- The *commutativity* of union and intersection is one example:

$$S \cup T = T \cup S$$
$$S \cap T = T \cap S$$

- The *distributivity* of either of union and intersection over the other provides another example:

$$R \cup (S \cap T) = (R \cup S) \cap (R \cup T) \tag{3.5}$$
$$R \cap (S \cup T) = (R \cap S) \cup (R \cap T) \tag{3.6}$$

Note that arithmetic (of numbers) has an analogue of Eq. (3.6)—with multiplication playing the role of intersection and addition playing the role of union—but it does not *have an analogue of Eq. (3.5).*

- The *idempotence* of complementation provides a third example:

$$\overline{\overline{S}} = S$$

Following the same general theme, there are special constants, $\emptyset$ and $U = \overline{\emptyset}$.

$$S \cap \overline{S} = S \cap \emptyset = \emptyset$$
$$S \cup \overline{S} = S \cup U = U \quad \text{(the "universal" set)}$$
$$S \cup \emptyset = S \cup S = S \cap S = S$$

The English logician/mathematician George Boole is historically credited with developing the system that we are describing here, with the goal of encapsulating a simple version of mathematical logic within an algebraic framework; see [22].[7] The system that Boole developed has been named *Boolean Algebra*, in his honor.

Rather than provide an extensive description of "abstract" Boolean Algebras in terms of their operations and axioms, we turn directly to the system of logic that was Boole's inspiration and that has since his day provided a mathematical foundation for areas of immeasurable importance to technology. (Much of an exposition of "abstract" Boolean Algebra would just repeat material from the "applied" areas.) We list citations to just three areas: the first citation is to Boole's original work; the other two are seminal studies by one of the greatest "technology-theorists" of the twentieth century, Claude E. Shannon.

---

[7] The cited source seems to be an expansion on Boole's original exposition of these thoughts, in *The Mathematical Analysis of Logic* (1847).

- the foundations of Mathematical Logic [22]
- the mathematical foundations of Switching-Circuit Theory [102]
- the mathematical foundations of Information Theory [103]

The genius of the mathematical system invented by Boole is manifest in the fact that it provides a starting point for formally understanding many seemingly disparate disciplines. Indeed, if one develops a level of understanding of Boole's algebraic framework for Propositional Logic, then one can transfer that understanding almost seamlessly to at least the preliminaries of Shannon's mathematical settings for Switching-Circuit Theory and Information Theory. We could use the verb "*translate*" here, rather than "transfer" because converting assertions within Boolean Algebra to kindred assertions within Switching-Circuit Theory often requires only a change in notation.

3.4.1 The Algebra of Propositional Logic

3.4.1.1 Logic as an algebra

A. Propositions: the objects of the algebra

As we noted earlier in this chapter, what distinguishes Propositional Logic from more general mathematical logic is the absence of quantifiers (THERE EXISTS, FOR ALL, etc.). The objects of Propositional Logic are *propositions*, i.e., fixed assertions, as exemplified by the following three sentences.

"The sky is pink."
"Elephants are good swimmers."
"Colorless green dreams sleep furiously."

Explanatory and Historical note.

The property of being a "proposition" is *syntactic*, rather than *semantic*. Our three sample propositions illustrate that the assertion made in a proposition need not be factually true—nor even sensible. Indeed, our third sample proposition was made famous by American linguist Noam Chomsky—in his seminal monograph *Syntactic Structures* (Mouton Press, 1957, *Janua Linguarum* series)—as he illustrated the independence of the notions "grammatical" and "meaningful".

The algebra that underlies Propositional Logic uses operations that are reminiscent of the set-theoretic operations to combine simple assertions into complex ones

if "the bananas are ripe" **and** "you are hungry" **then** "you should buy bananas"

either "the grass is green" **or** "the ocean is calm"

Two special propositions—the constants of the algebra—are denoted TRUE and FALSE. They are *intended* to represent factual truth and factual falsehood, but they are *defined* by the way they interact with each other and with other propositions. In order to specify these interactions, we have to specify the logic's *connectives*: the operations of the algebra.

B. Logical connectives: the algebra's operations

As we describe and define the basic connectives of the Propositional Logic, we point out their relationships to the Boolean set-related operations introduced in Section 3.2.2.

B.i The *unary* logical connective

- NOT: **negation** ($\neg$)   (Set-theoretic analogue: *complementation*).

  Two shorthand notations for "NOT P" have evolved:

  - the prefix-operator $\neg$, as in "$\neg P$"

  - the overline-operator, as in "$\overline{P}$"

  Whichever notation one uses, the defining properties of negation are encapsulated in the following equations.

  $$\left[\neg\text{TRUE} = \text{FALSE}\right] \quad \text{and} \quad \left[\neg\text{FALSE} = \text{TRUE}\right]$$

B.ii The *binary* logical connectives

- OR: **disjunction** ($\vee$)   (Set-theoretic analogue: *union*).

  The operation OR—which is also called *logical sum*—is usually denoted by the infix-operator $\vee$; the operation's defining properties are encapsulated as follows.

  $$[[P \vee Q] = \text{TRUE}] \quad \text{if, and only if,} \quad [P = \text{TRUE}] \text{ or } [Q = \text{TRUE}] \text{ or both.}$$

  Note that, as with union, logical OR is *inclusive:* The assertion
  $$[P \vee Q] \text{ is TRUE}$$
  is true when *both* propositions P and Q are true, as well as when only one of them is. Because such inclusivity does not always capture one's intended meaning, there is also an *exclusive* version of disjunction, as we see next.

- XOR: **XOR** ($\oplus$)   (Set-theoretic analogue: *disjoint union*).

  The operation *exclusive or* is a version of disjunction that does *not* allow both disjuncts to be true simultaneously. It is usually denoted by the infix-operator $\oplus$; the operation's defining properties are encapsulated as follows.

  $$[[P \oplus Q] = \text{TRUE}] \quad \text{if, and only if,} \quad [P = \text{TRUE}] \text{ or } [Q = \text{TRUE}] \text{ but } not \text{ both.}$$

  We emphasize the distinction between $\vee$ and $\oplus$, the (respectively) inclusive and exclusive versions of disjunction, by remarking that the assertion

$$[P \oplus Q] \text{ is TRUE}$$

is *false* when both propositions P and Q are true.

- AND: **conjunction** ($\wedge$)   (Set-theoretic analogue: *intersection*).

  The operation AND—which is also called *logical product*—is usually denoted by the infix-operator $\wedge$; the operation's defining properties are encapsulated as follows.

  $$[[P \wedge Q] = \text{TRUE}] \quad \text{if, and only if, } both \quad [P = \text{TRUE}] \text{ and } [Q = \text{TRUE}].$$

- IMPLIES: **logical implication** ($\Rightarrow$)   (Set-theoretic analogue: *subset*).

  The logical operation IMPLIES—which is often called *conditional*—is usually denoted by the infix-operator $\Rightarrow$; the operation's defining properties are encapsulated as follows.

  $$[[P \Rightarrow Q] = \text{TRUE}] \quad \text{if, and only if, } \quad [[\neg P] = \text{TRUE}] \text{ (inclusive) or } [Q = \text{TRUE}].$$

  We remark that the operation IMPLIES differs from the other logical operations that we have discussed in a way that the reader must always keep in mind. In contrast to NOT, OR, XOR, and AND, whose formal meanings pretty much coincide with their informal meanings, the formal version of implication carries with it connotations that we do not always associate with the word "implies" in the vernacular. One would not likely get universal agreement that the informal word "implies" satisfies the following properties that the formal word IMPLIES definitely does.

  – *If proposition P is false, then it implies* every *proposition.*

  – *If proposition Q is true, then it is implied by* every *proposition.*

- IFF: **logical equivalence** ($\equiv$)   (Set-theoretic analogue: *set equality*).

  The final logical operation that we shall discuss is known by many names, including *logical equivalence* and *biconditional*, as well as *if and only if* and its shorthand IFF. It is usually denoted via one of the following two infix-operators: $\equiv$ or $\Leftrightarrow$; the operation's defining properties are encapsulated as follows.

  $$[[P \equiv Q] = \text{TRUE}] \quad \text{if, and only if } [[P \Rightarrow Q] = \text{TRUE}] \quad \text{and} \quad [[Q \Rightarrow P] = \text{TRUE}].$$

  Of course, our warnings concerning the tension between "implies" and IMPLIES are inherited by IFF and its vernacular version.

The Propositional Logic is often called the *Propositional Calculus*, in acknowledgment of Boole's intention of developing an *algebra of logic*. In the system Boole created, *logical connectives* become *algebraic operations*. The "set-theoretic analogues" that we have specified in the course of introducing the logical connectives implicitly specify the laws that the Propositional operations obey.

As we shall discuss imminently, we never have to manipulate Propositional expressions algebraically in order to "do" logic—although we are free to engage in such manipulation if we choose to.

C. The goal of the game: *Theorems*

One learns as a student of either mathematics or mathematical logic that a *theorem* is a statement in the relevant logical system—in our case, the Propositional Calculus—such that one of the following is true:

- The statement is an *axiom* of the system.[8]

- The statement can be derived from the theorems of the system by repeated application of the system's rules of inference.

  This means that, by recursion, any theorem can be derived from the axioms of the system by repeated application of the system's rules of inference.

If this text were devoted to the topic of mathematical logic, then we would at this point develop the concepts that the preceding definition alludes to. However, for us "doers of mathematics", the Propositional Calculus is an avenue for introducing a variety of concepts within mathematics, not within mathematical logic. Therefore, in the next section, we invoke (without proof) the powerful *Completeness Theorem for the Propositional Calculus* (Theorem 3.3), which gives us access to a tremendous shortcut toward our goals.

3.4.1.2 Semantic completeness: Logic via Truth Tables

A. Truth, theoremhood, and completeness

Mathematical logic was invented to formalize the activity of mathematical reasoning. The overarching goal was to develop a set of rules for "playing the theorem-proving game" in a way that was guaranteed to uncover true facts about whatever domain one was studying. There are two complementary aspects to this goal:

1. *to ensure that every theorem is true*

   A logical system which achieves this goal is said to be *consistent*.

   The main avenue to consistency is to ensure that the system's "rules of inference" combine simple expressions into complex ones in a way that preserves the truth of the expressions. Of course, the algebra underlying the system plays a major role here.

2. *to ensure that every true expression is a theorem*

---

[8] The word "axiom" means "self-evident truth".

A logical system which achieves this goal is said to be *semantically complete*. (There are other forms of logical completeness, but semantical completeness is a major—and natural—one.)

Regrettably, semantical completeness is a goal that is achievable only for the simplest logical systems. Loosely speaking, any logical system that can correctly reason about elementary arithmetic propositions concerning integers, propositions such as

     "integer n is the sum of integers m and p"

or

     "integer n is the product of integers m and p"

cannot be semantically complete. (Chapter 5 contains a careful, but informal, discussion of incompleteness.)

In any semantically incomplete system of logic, there must be expressions that are *true but not provable*.

Historical note.

If the preceding discussion surprises you, then you are in good company. The publication that announced the incompleteness result, [53], turned the mathematical world on its ear when it appeared in 1931!

Happily for many mathematical and computational goals, the Propositional Calculus is sufficiently simple that it *is* semantically complete. We elaborate a bit on this fact now.

B. The completeness of Propositional Logic

The *semantic completeness* of the Propositional Calculus as a logical system is a consequence of the fact that we are able to view the expressions of the Calculus as Boolean functions, in the following way.

- All propositions, $P, Q, \ldots$, that appear in expressions are uninterpreted "names" of actual propositions.

- As we examine an expression in the Calculus, the only information we need about the propositions $P, Q, \ldots$, which appear in the expression is the array of truth-values for the propositions.

If we tabulate how the truth-values of propositions combine under the logical operators that interconnect them in the expression, then we remark immediately how the expressions can be viewed as *functions of binary tuples*, where the arity of the functions is the number of propositional variables. Using this viewpoint, the tables in Fig. 3.7 reproduce the definitions of the logical operators of Section 3.4.1.1.B, viewed as functions within the space of truth-values. Each propositional variable is instantiated with all of its possible truth-values, TRUE and FALSE—which we denote here, for convenience, by 1 and 0, respectively. Note how the truth-value entries

P	$\neg P$
0	1
1	0

P	Q	$P \vee Q$	$P \oplus Q$	$P \wedge Q$	$P \Rightarrow Q$	$P \equiv Q$
0	0	0	0	0	1	1
0	1	1	1	0	1	0
1	0	1	1	0	0	0
1	1	1	0	1	1	1

Fig. 3.7 Tabular presentations of the basic Propositional connectives

in the right-hand truth table of Fig. 3.7 reinforce the lessons of Section 3.4.1.1.B. This awareness is essential with the implication-related operations ($\Rightarrow$ and $\equiv$) in the light of our earlier discussion of the tension between their formal meanings (from the table) and the way they are used in "real-life" reasoning voiced in the vernacular.

Every Propositional expression is a binary function, so we can pass back and forth between logical and functional/operational terminology. This ability affords us very simple definitions of two important concepts that are somewhat harder to define in purely logical terms.

- Tautology.

  - *Mathematical formulation:*

    A Propositional expression is a *tautology* iff its corresponding function is the constant function $F(x) \equiv 1$.

  - *Logical formulation:*

    A Propositional expression is a *tautology* iff it evaluates to TRUE under every instantiation of truth-values for its Propositional variables.

- Satisfiable expression.

  - *Mathematical formulation:*

    A Propositional expression is *satisfiable* iff its corresponding function has 1 in its range; i.e., iff there is an argument x such that $F(x) = 1$.

  - *Logical formulation:*

    A Propositional expression is *satisfiable* iff there exists an instantiation of truth-values for its Propositional variables under which the expression evaluates to TRUE.

We are now ready to present the main result of this section.

Theorem 3.3 (The Completeness Theorem for Propositional Logic). *A Propositional expression $E(P, Q, \ldots, R)$ is a theorem of Propositional logic if, and only if, it is a tautology.*

Proving Theorem 3.3 is beyond the scope of this book. The interested reader can find a proof in numerous texts on logic. The proof in [94] might be particularly interesting to an aspiring mathematician because that source exists precisely to explain logic in terms that are accessible to a mathematician.

The next subsection presents illustrative applications of the Theorem.

C. The laws of Propositional Logic as theorems

The applications of Theorem 3.3 which we have chosen to discuss here are the *laws* of the logical algebra described in Section 3.4.1.

The law of double negation

This is a formal analogue of the homely adage that "a double negative is a positive."

Proposition 3.3 *For any proposition P,*

$$P \equiv \neg[\neg P]$$

Proof via truth table.

P	$\neg P$	$\neg[\neg P]$
0	1	0
1	0	1

(3.7)

Columns 1 and 3 of truth table (3.7) are identical. By Theorem 3.3, this fact verifies Proposition 3.3, the law of double negation. □

The law of contraposition

This is a very exciting example! Let us immediately convert this to a mathematical statement about the Boolean Algebra of Propositional Logic and prove the statement. We shall then contemplate the implications of this law for *logic* rather than for *mathematics*.

Proposition 3.4 *For any propositions P and Q,*

$$[[P \Rightarrow Q] \equiv [\neg Q \Rightarrow \neg P]]$$

Proof via truth table.

P	$\neg P$	Q	$\neg Q$	$P \Rightarrow Q$	$\neg Q \Rightarrow \neg P$
0	1	0	1	1	1
0	1	1	0	1	1
1	0	0	1	0	0
1	0	1	0	1	1

(3.8)

Columns 5 and 6 of truth table (3.8) are identical. By Theorem 3.3, this fact verifies Proposition 3.4, the law of contraposition. □

Now let us reconsider the whole concept of contraposition, including this law, in the light of *logic and reasoning*.

The *Law of Contraposition* states that the assertion

"Proposition P implies Proposition Q"

is *logically equivalent* to the assertion

"the negation of Proposition Q implies the negation of Proposition P".

Think about this! This is a version of Proof by Contradiction!

De Morgan's Laws:

Proposition 3.5 *For any propositions P and Q:*

- $[P \wedge Q] \equiv \neg[[\neg P] \vee [\neg Q]]$
- $[P \vee Q] \equiv \neg[[\neg P] \wedge [\neg Q]]$

Proof via truth table.

P	$\neg P$	Q	$\neg Q$	$[P \wedge Q]$	$[\neg P] \vee [\neg Q]$	$[P \vee Q]$	$[\neg P] \wedge [\neg Q]$
0	1	0	1	0	1	0	1
0	1	1	0	0	1	1	0
1	0	0	1	0	1	1	0
1	0	1	0	1	0	1	0

$$(3.9)$$

Columns 5 and 6 of truth table (3.9) are mutually complementary, as are columns 7 and 8. If we negate (i.e., complement) the entries of columns 6 and 8, then we can invoke Theorem 3.3 to verify Proposition 3.5, which encapsulates De Morgan's laws for Propositional Logic. □

The distributive laws for Propositional Logic

In numerical arithmetic, multiplication distributes over addition, but not conversely, so we have a single distributive law for arithmetic (see Section 5.2). In contrast, each of logical multiplication and logical addition distributes over the other, so we have *two* distributive laws for Propositional Logic.

- $P \vee [Q \wedge R] \equiv [P \vee Q] \wedge [P \vee R]$
- $P \wedge [Q \vee R] \equiv [P \wedge Q] \vee [P \wedge R]$

P	Q	R	$[P \vee Q]$	$[P \wedge Q]$	$[Q \wedge R]$	$[Q \vee R]$	$P \vee [Q \wedge R]$	$[P \vee Q] \wedge [P \vee R]$	$P \wedge [Q \vee R]$	$[P \wedge Q] \vee [P \wedge R]$
0	0	0	0	0	0	0	0	0	0	0
0	0	1	0	0	0	1	0	0	0	0
0	1	0	1	0	0	1	0	0	0	0
0	1	1	1	0	1	1	1	1	0	0
1	0	0	1	0	0	0	1	1	0	0
1	0	1	1	0	0	1	1	1	1	1
1	1	0	1	1	0	1	1	1	1	1
1	1	1	1	1	1	1	1	1	1	1

$$(3.10)$$

Columns 8 and 9 of truth table (3.10) are identical, as are columns 10 and 11. By Theorem 3.3, this fact verifies the distributive laws for Propositional Logic.

3.4.2 ⊕⊕ *A Purely Algebraic Setting for Completeness*

In the hope of whetting the appetite of at least some readers, we remark here that there is a rather sophisticated algebraic concept that can be used to derive Theorem 3.3 in a purely algebraic setting. We have noted that the Propositional Calculus is a Boolean Algebra. One can state more: The Propositional Calculus is a *free* Boolean Algebra.

The meaning of the qualifier *free* is easy to state but not so easy to understand. Here is a very informal try.

An algebra of type X—in our case, X is a Boolean Algebra—is *free* if it obeys no laws except those that it must obey to belong to the class of type-X algebras.

In more detail:

Any genre of algebra is defined as a set of objects accompanied by a set of basic operations, which obey certain "self-evident" axioms. In rough, but evocative, terms, the algebra is *free* if its operations do not relate to one another in any way that is not mandated by the axioms.

Here is an easy illustration of this decidedly *not-easy* concept. There is a (quite important) genre of algebra called a *semi-group*. A semi-group $\mathscr{S}$ is specified via a set (of objects) S, together with a binary operation on S, which we denote $\otimes$. For the (algebraic) structure $\mathscr{S} = \langle S, \otimes \rangle$ to be a semi-group, it must obey the following two laws.

1. $\mathscr{S}$ must be *closed* under operation $\otimes$; i.e., for all $s_1, s_2 \in S$, we must have

$$(s_1 \otimes s_2) \in S$$

2. The operation $\otimes$ must obey the *associative law;* i.e., for all $s_1, s_2, s_3 \in S$,

$$(s_1 \otimes s_2) \otimes s_3 = s_1 \otimes (s_2 \otimes s_3)$$

Now, it is easy to verify—we shall discuss this in Chapter 5—that the (algebraic) structure formed by the integers coupled with the operation of addition forms a semi-group (which the notation of Chapter 5 would denote $\langle \mathbb{Z}, + \rangle$). But this semi-group is *not* free, because integer addition obeys laws beyond just closure and associativity, namely:

1. Addition of integers is *commutative*.

2. Integer addition has an *identity*, namely 0.

3. Every integer $z \in \mathbb{Z}$ has an *additive inverse* (namely, $-z$).

So, it *is* meaningful—and consequential—that the Propositional Calculus is a *free* Boolean Algebra.

3.4.3 Satisfiability Problems and NP-Completeness

Satisfiability problems deal with propositional formulae that are populated by enti-ties that can assume the truth-values TRUE and FALSE. The *underlying* entities are *logical variables*, i.e., variables that range over the truth-values. The *actual* entities that appear in each formula are *logical literals*, i.e., **instances of logical variables** in either their *true* or *complemented* forms. In detail:

• In its *true* form, a literal evaluates to TRUE precisely when its associated variable does.

• In its *complemented* form, a literal evaluates to TRUE precisely when its associ-ated variable evaluates to FALSE.

The following simple examples should clarify these terms.
The formula: $(\neg x \vee y) \wedge (x \vee \neg y)$
The variables: x and y
The literals: x and y (true form); $\neg x$ and $\neg y$ (complemented form)

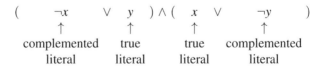

SAT: *The original NP-complete problem*
The general form of *The Satisfiability Problem for the Propositional Calculus* is de-noted SAT. The problem can be viewed in two formally distinct but algorithmically equivalent ways.

• In the *combinatorial formulation*:

– The Satisfiability Problem is specified by a set of *logical clauses*, each clause being a *set of* logical *variables*, i.e., variables that range over the truth-values TRUE and FALSE.

– The Satisfiability question is: Can one assign truth-values to all of the logical variables of the formula in such a way that every clause ends up with at least one TRUE literal?

- In the *formal logical formulation*:

  – The Satisfiability Problem is specified by a propositional formula Φ that is a *conjunction of disjuncts of logical literals*.

  In the lingo of the *cognoscenti* (the "in-crowd"), these expressions are said to be in *POS* form, shorthand for a *(logical) product of (logical) sums*.

  – The Satisfiability question is: Can one assign truth-values to all of the logical variables of formula Φ in such a way that every disjunct evaluates to TRUE?

The reader should be able to prove that these are just two views of the same logical problem.

Until 1971, POS expressions were largely viewed simply as the duals of the somewhat more common *SOP*-form expressions, shorthand for *sums of products*. A discovery in 1971 by Stephen A. Cook changed this view: Cook exposed, in [35], how POS-form expressions can be used to model a number of computational phenomena whose importance was only beginning to be understood at that time. These phenomena were *nondeterminism in computation* and *computational complexity*.

Nondeterminism can best be thought of in one of the following ways. The first, more concrete, way to think about a nondeterministic computing device is as a device that can "make guesses" at various junctures of a computation, rather than performing some prolonged search through plausible alternative fixed paths. The second, more idealistic, way to think about nondeterminism is by imagining that a computing device has the capability of spawning alternative universes that it can proceed through "in parallel".

Explanatory note.

We place the phrase "in parallel" within quotes because parallel computing is a concrete computational paradigm, while nondeterministic computing is a purely conceptual framework for thinking about the computational process.

Nondeterminism was invented during the 1950s, under a variety of names, as a conceptual framework for understand the computing power of *finite automata*. The concept's appearance in the seminal study [88] by Michael O. Rabin and Dana Scott established the standard way of formulating and discussing the topic.

By the early 1960s, computation theorists had recognized the importance of developing conceptual, mathematical tools for discussing, analyzing, and quantifying the *complexity* of performing computations. For close to a decade, models for studying such issues were proposed, discarded, improved, and embellished, but no consensus developed about an intellectually uniform approach to the topic. This situation largely changed toward the end of the 1960s. The issues were split into what one might term *concrete* questions, which focused on specific computational problems and specific computing platforms, and what one might term *abstract* questions, which tried to uncover foundational insights into what made—or appeared to make—certain types of computational problems *inherently* difficult. Substantial progress was made on the concrete front, but the abstract front proved quite recalcitrant until Cook's work at the end of the decade.

In 1971, Cook announced, in [35], a revolutionary foundation for the field of *computational complexity*, which built upon the conceptual frameworks[9] of the decades-old field known variously as *Computability Theory* and *Computation Theory*. Computability Theory enabled one to study questions about the *feasibility* of computing a function f,

Can one write a program that enables a digital computer to compute function f?

Cook's new theory enabled one to rigorously study a range of questions regarding the *efficiency* that could be achieved by a digital computer as it performs a (perforce, feasible) computation. One could now ask the following types of questions about the *complexity* of computations for functions f and g:

- Why is it easy (computationally) to prove that $f(x) = y$, but hard to compute $f(x)$?

- Why can you adapt any program that computes function f to a "similar" program that computes function g, even though these functions do not appear to have any relationship to one another?

- Is function f inherently harder to compute than function g?

Cook's ingenious adaptation created a new theory that was named (the theory of) NP-*completeness*, for reasons that are explained in sources such as [52]. In very informal, but hopefully evocative terms, NP-complete computational problems are problems whose solutions are *easy to check but hard to find*. (Of course, the adjectives "hard" and "easy" have technical meanings, but developing these meanings is beyond the scope of this text. These concepts are developed "gently" but systematically in texts such as [91].)

Tying up loose ends: The notion *easy to check* is formalized using the concept of computational nondeterminism which we just discussed. Moreover, the technical apparatus used to represent nondeterminism in Cook's formulation is manifest in the disjunctions that abound in POS-form expressions—and in the associated Satisfiability Problem SAT!

The reason we relate this story at this point in our study is that the problem SAT was the first "natural" problem that was shown by Cook to be NP-complete!

---

[9] These frameworks were the legacy of intellectual giants such as Kurt Gödel (in [53]) and Alan M. Turing (in [108]).

(The term "NP-complete" does not occur in [35]; there was not yet a name for this concept.) To round out the story, within a year, Richard M. Karp and colleagues had greatly lengthened the list of computational problems that Cook's new theory encompassed. The details of the theory of NP-completeness are beyond the scope of this text, but we do want the reader to recognize the following.

It remains mathematically *possible* that NP-complete problems can be solved "efficiently"—meaning, technically, "in time polynomial in the size of the description of the problem". (For the SAT problem, this size is measured in terms of the number of literals in the POS propositional formula.) However it is mathematically *certain* that if any one problem in this class has a polynomial-time solution, then *every* problem in this class does. This is because of the following intellectually exciting fact from [35], which is known as *Cook's Theorem*. Even without the technical details—which are available in sources such as [52, 91]—the excitement of this result should be clear.

Theorem 3.4 (Cook's Theorem). (a) *The Satisfiability Problem for the Propositional Calculus,* SAT, *is* NP*-complete.*
(b) *All* NP*-complete problems can be efficiently "translated" into one another.*

The detailed meaning of part (b) of Theorem 3.4 is the following. *Let* A *and* B *be* NP*-complete problems. There exists a polynomial-time computable function* $f_{A \to B}$ *that produces an instance of problem* B *when given an instance of problem* A.

Given the half-century that has passed since the work of Cook and Karp, and given the *practical* importance of many of the problems that are known to be NP-complete, it is *widely suspected* that no NP-complete problem can be solved via a polynomial-time algorithm.

There is a large, vibrant literature concerning approaches for lessening the negative computational impact of NP-completeness—by considering significant special cases of NP-complete problems, by developing approximate solutions to such problems, and by developing heuristics for such problems, whose behavior cannot be guaranteed but which seem to work well "in practice".

2SAT: SAT with two literals per clause

Once a framework for discussing the complexity of computations received general acceptance, researchers began to investigate the boundaries within the framework: How, and how much, could one "weaken" an NP-complete problem before one lost the problem's NP-completeness? One of the earliest examples of this activity occurred within the paper that actually invented the framework, namely, [35]! A careful analysis revealed the following strengthening of Theorem 3.4(a).

Theorem 3.5 (Cook's Theorem: stronger version). 3SAT, *the* SAT *problem with three literals per clause, is* NP*-complete.*

Of course, the natural question raised by Theorem 3.5 is: What is the complexity of the problem 2SAT: the variant of the SAT problem in which each clause has only two literals? This section announces the answer to this question but must defer the verification of the answer to a later chapter, where new mathematical material will be available. We include this computation-theoretic material in our "pure" mathematics text because it turns out that the answer reveals—in an elegant manner—an interesting, nonobvious intersection region of the fields of mathematical logic, computational complexity, algorithms, and graph theory. Indeed, we defer developing this answer until we introduce the basic notions concerning graphs, in Section 12.1.2.

Proposition 3.6 *The* 2SAT *problem can be solved in polynomial time.*

That is, given any instance Φ of 2SAT*, one can determine in time polynomial in the number of literals in Φ whether there exists a satisfying assignment of truth-values to the variables of Φ.*

The proof of Proposition 3.6 requires the basic connectivity-related notions underlying graphs. These are available in Section 12.1.1.3, and the proof follows immediately, in Section 12.1.2.

3.5 Exercises: Chapter 3

Throughout the text, we mark each exercise with 0 or 1 or 2 occurrences of the symbol $\oplus$, as a rough gauge of its level of challenge. The 0-$\oplus$ exercises should be accessible by just reviewing the text. We provide *hints* for the 1-$\oplus$ exercises; Appendix H provides *solutions* for the 2-$\oplus$ exercises. Additionally, we begin each exercise with a brief explanation of its anticipated value to the reader.

1. **What do power sets look like?**
   LESSON: Enhance understanding of power sets, subsets, supersets, etc.

   Focus on a set S and on its power set $\mathscr{P}(S)$.

   a. *Prove the following assertions.*

      Proposition 3.7 (a) *The biggest (i.e.,* most populous*) element of the power set $\mathscr{P}(S)$ is the set S.*
      (b) *The smallest (i.e.,* least populous*) element of $\mathscr{P}(S)$ is the empty set $\emptyset$.*

   b. Say that $\mathscr{P}(S)$ contains the subsets A and B of set S.
      Explain which of the following sets belong to $\mathscr{P}(S)$.
      i. $A \cup B$

      ii. $A \setminus B$

      iii. $\mathscr{P}(A)$

      iv. $A \times B$

      v. $S \setminus A$

2. **Justifying De Morgan's laws**
   LESSON: Analyzing operations on sets

   We justified one of De Morgan's laws in the text. *Justify the other one:*

   $$\overline{S \cup T} \;=\; \overline{S} \cap \overline{T}$$

3. **Formally defining set operations**
   LESSON: Analyzing operations on sets

   Define the following two set-operations by means of expressions that use only the operations UNION *and* SET DIFFERENCE. *Do this in two ways: (i) Algebraically, ending up with an equation of the form (3.1); (ii) Pictorially, using Venn diagrams*

   a. The operation INTERSECTION (denoted $S \cap T$) isolates all elements that are shared by sets S and T.

   b. The operation SYMMETRIC DIFFERENCE (denoted $S + T$) isolates all elements that belong to precisely one of S and T.

4. **A consequence of De Morgan's laws**
   LESSON: Analyzing operations on sets

   Prove the following assertion.

   Proposition 3.8 (a) *If the collection $\mathscr{S}$ of finite sets is closed under the operations of union and complementation, then it is closed under the operation of intersection.*
   (b) *If the collection $\mathscr{S}$ of finite sets is closed under the operations of intersection and complementation, then it is closed under the operation of union.*

5. **Weak vs. strong orders**
   LESSON: The negation of a weak order relation is a strong order relation, and vice versa.

   Prove the following assertion.

   Proposition 3.9 *Let ρ be a strong order relation on a set S, and let $\underline{\rho}$ be the weak version of ρ. Say that ρ is* total *in the sense that for distinct $s, t \in S$, either $s\rho t$ or $t\rho s$.*

   (a) $\overline{\underline{\rho}} = \rho$

       In other words, for all $s, t \in S$, $\big[[s\rho t] \;\; iff \;\; [s\overline{\underline{\rho}}t]\big]$.

   (b) *For all $s, t \in S$,* $\big[[s\underline{\rho}t] \;\; iff \;\; [t\overline{\rho}s]\big]$.

6. **Equivalence relations**
   LESSON: Relating equivalence relations and equality

a. *Prove Theorem 3.1:*
   The equality relation, $=$, on a set S is the *finest* equivalence relation on S, in
   the sense that $=$ refines every equivalence relation on S.

b. What is the *coarsest* equivalence relation on S, i.e., the equivalence relation
   which is refined by every other equivalence relation?

7. **Injections and surjections**
   LESSON: Recognizing properties of functions

   This problem concerns the following three functions from $\mathbb{N}^+$ to $\mathbb{N}^+$.

   $$f(x) = 2x$$
   $$g(x) = 2x + 1$$
   $$h(x) = \lfloor x/2 \rfloor$$

   a. *Prove that both $f(x)$ and $g(x)$ are* injections *from $\mathbb{N}^+$ to $\mathbb{N}^+$.*

   b. *Prove that $h(x)$ is a* surjection *from $\mathbb{N}^+$ to $\mathbb{N}^+$.*

   c. *Prove that the preceding assertions are as strong as possible, in the sense that:*
      i. *Neither $f(x)$ nor $g(x)$ is a* surjection *from $\mathbb{N}^+$ to $\mathbb{N}^+$.*

      ii. *$h(x)$ is not an* injection *from $\mathbb{N}^+$ to $\mathbb{N}^+$.*

   d. *Prove that $h(x)$ is a "post-inverse" to both $f(x)$ and $g(x)$, in the sense that
      applying h "undoes" an earlier application of either f or g. In detail, for all
      $x \in \mathbb{N}^+$:*
      $$h(f(x)) \;=\; h(g(x)) \;=\; x$$
      In other words, both $f \circ h$ and $g \circ h$ are the identity function on $\mathbb{N}^+$.

8. **More connections between strings and functions**
   LESSON: Exploit these connections in a way that leads you to discover the *factorial* operation (which arises in Chapter 5 and beyond)

   A permutation f of the set $S = \{1, 2, 3, 4\}$ can be specified by the length-4 string

   $$f(1)\, f(2)\, f(3)\, f(4)$$

   For instance, the identity function (which is clearly bijective, hence a permutation) is specified by the string

   $$1\,2\,3\,4 \quad \text{because} \quad f(1) = 1,\ f(2) = 2,\ f(3) = 3,\ f(4) = 4$$

   while the one-element right-cyclic-shift permutation is specified by the string

   $$4\,3\,2\,1 \quad \text{because} \quad f(1) = 4,\ f(2) = 3,\ f(3) = 2,\ f(4) = 1$$

   a. *List all permutations of S, specifying functions via length-4 strings.*

b. ⊕ *Craft an argument that predicts the number of permutations, based on the size of set S.*

Hint:

- As you create a new string of numbers, in how many ways can you choose *the first number? the second number? ...*

- Based on your answers for the first and second and third numbers of the new string, in how many ways can you choose *the first two numbers—i.e., the first* pair *of numbers? the next two numbers? ...*

c. *Strengthen your argument by listing all permutations of* $S' = \{1,2,3,4,5\}$.

Write small—there are a lot of permutations.

d. ⊕⊕ *Extrapolate from your argument to determine the number of permutations of the set* $S'' = \{1,2,3,\ldots,n\}$, *as a function of n.*

9. Theorems and non-theorems

LESSON: Practicing arguments within Propositional Logic

After consulting Fig. 3.7, prove the following assertions.

a. *The following Propositional expression,* $\mathbf{P}_1$, *is a tautology.*

$$\mathbf{P}_1 \equiv \left[\neg P \implies P\right]$$

Why does this mean that $\mathbf{P}_1$ *is a theorem of Propositional Logic?*

b. *The following Propositional expression,* $\mathbf{P}_2$, *is satisfiable.*

$$\mathbf{P}_2 \equiv \left[P \implies \neg P\right]$$

Why does this fact mean that the negation *of* $\mathbf{P}_2$ *is* not a theorem *of Propositional Logic?*

10. The equivalence of two formulations of SAT

LESSON: Enhance understanding of and facility with mathematical formulations

Prove that the text's two views of the problem SAT *are equivalent, i.e., that they specify the same mathematical/computational problem.*

- *Viewing a* **POS** *expression as a* set *of clauses*

- *Viewing a* **POS** *expression as a* disjunction *of clauses.*

Chapter 4
Numbers I: The Basics of Our Number System

*Die Mathematik ist die Königin der Wissenschaften und die
Zahlentheorie ist die Königin der Mathematik.*
(*Mathematics is the queen of the sciences and number theory
⟨often, "arithmetic"⟩ is the queen of mathematics.*)
Carl Friedrich Gauss
Quoted in *Gauss zum Gedächtnis* (1856) by
Wolfgang Sartorius von Waltershausen

4.1 Introducing the Three Chapters on Numbers

Many, perhaps most, of us take for granted the brilliant notations that have been developed for the many arithmetic constructs which we use daily. Our mathematical ancestors have bequeathed us notations that are not only perspicuous but also convenient for computing and for discovering and verifying new mathematical truths. Several chapters in this text are devoted to sharing the elements of this legacy with the reader. Three of these chapters focus on numbers themselves: the objects and how we represent them.

1. The current chapter focuses on elementary concepts and techniques of reasoning relating to the most familiar objects of mathematical discourse, namely, the four families of numbers that underlie all of quantitative mathematics.

2. Chapter 8 extends this introductory material, by introducing some more advanced topics concerning numbers. We look "inward" as we discuss important special classes of numbers which expose deep insights into the intrinsic nature of all numbers. And, we look "outward" as we reveal how numbers can "encode" structured data. In this latter regard, we suggest some nonobvious implications of being able to perform such encodings.

3. Chapter 10 looks at *numerals*, the strings that we use to manipulate and analyze numbers. Among a variety of representation-related topics, we discuss how much

© Springer Nature Switzerland AG 2020

A. L. Rosenberg, D. Trystram, *Understand Mathematics, Understand Computing*,
https://doi.org/10.1007/978-3-030-58376-7_4

one can learn about the intrinsics of various classes of numbers based only on the types of strings that we can use to name the numbers in the class.

Explanatory note.

Numbers and numerals embody what is certainly the most familiar instance of a very important dichotomy that pervades our intellectual lives: the distinction between objects and their names:

Numbers are intangible, abstract objects.
Numerals are the names we use to refer to and manipulate numbers.

This is a critically important distinction! You can "touch" a numeral: break it into pieces, combine two (or more) numerals via a large range of operations. When the representations we use as numerals are *operational*, then you can *compute* using them. Numbers are intangible abstractions or conceptualizations: you *reason* using numbers.

4.2 A Brief Biography of Our Number System

We begin our study of numbers and numerals with a short taxonomy of our number system. Although we assume that the reader is familiar with the most common classes of numbers, we do spend some time highlighting important features of each class, partly, at least, in the hope of heightening the reader's interest in this most basic object of mathematical discourse.

We present the four most common classes of numbers in what is almost certainly the chronological order of their discovery/invention.

Enrichment note.

Did humans *invent* these classes of numbers to fill specific needs, or did we just *discover* their preexisting selves as needs prompted us to search for them? The great German mathematician Leopold Kronecker, as cited in [112] (page 19), has asserted

Die ganzen Zahlen hat der liebe Gott gemacht, alles andere ist Menschenwerk.

[God made the integers, all else is the work of man].

[Dieu a créé les nombres entiers, tout le reste est invention humaine].

A pleasing narrative can be fabricated to account for our multi-class system of numbers. In the beginning, the story goes, we needed to count things (sheep, bottles of oil, weapons, . . .), and the positive *integers* were discovered to serve this need. As accounting practices matured, we needed to augment this class with the number zero (0) and with the negative integers. The former augmentation (of zero) allowed Merchant *A* to keep track of the inventory remaining after the last flask of wine is sold; the latter augmentation (of the negative numbers) allowed Merchant *A*'s banker to record *A*'s credit balance after taking a loan. The class of integers was now complete—although a variety of special classes of integers remained to be discovered, motivated by reasons ranging from the religious to the intellectual.

Moving on: As society matured, humans began to share materials that had to be subdivided—cloth, grain, etc.—rather than partitioned into discrete units. We needed to invent the *rational numbers* to deal with such materials. Happily for the mathematically inclined, the rational numbers could be developed in a manner that allowed one to view an integer as a special type of rational. This meant that our ancestors could build upon the systems they had developed to deal with integers, rather than having to scrap those systems and start a new.

Enrichment note.

The quest for extendible *frameworks rather than isolated unrelated frameworks is a hallmark of mathematical thinking.*

One cannot overemphasize the centrality of this maxim to the success of mathematics over the millennia.

We now enter the realm of "semi-recorded" history in the West: our legacy from the Babylonians, the Egyptians, the Greeks, and others. "Practical" mathematics was invented—and reinvented—to accommodate pursuits as varied as astronomy, commerce, navigation, and architecture. Our mathematical stalwarts, the integers and the rationals, were not adequate to deal with all of the measurements that we wanted to make, calculations that we wanted to do, structures that we wanted to design. So, we approximated and "fudged" and got pretty much where we wanted to get. The standard story at this point (at least in the West) is that the ancient Greeks began to try to systematize mathematical knowledge and practice. The Greek mathematician and geometer Euclid, and members of his school, verified—via one of the first *proofs* in recorded history—the uncomfortable fact that the lengths of portions of eminently buildable structures were not "measurable"—in modern lingo, the lengths *were not rational*. The poster child for this phenomenon was the *hypotenuse of the isosceles right triangle T with unit-length legs*. Thanks to the well-known theorem of the Greek mathematician Pythagoras, even schoolchildren nowadays know that the length of this eminently "buildable" (with straightedge and compass) line is $\sqrt{2}$. What Euclid discovered is: *There is—provably!—no way to find integers p and q whose ratio is $\sqrt{2}$, the length of T's hypotenuse.*[1]

⊕ **Digression: *The Pythagorean Theorem*.** The famed theorem of Pythagoras is widely known, at least informally. We pause now to provide a *formal* statement of this seminal result and to provide a proof of the special case that relates to our current discussion of the square root of 2. Quite aside from reviewing the Theorem's important content, this statement will provide the reader one more opportunity to ponder the "music" of mathematical discourse.

Let us be given a triangle T with vertices A, B, and C. Use the left-hand grey triangle in Fig. 4.1 as a model. Say that T is a *right triangle*, meaning that one of

---

[1] We present a version of Euclid's proof in Proposition 4.7.

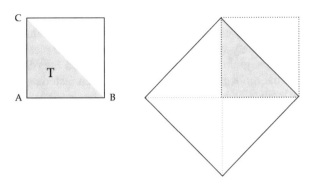

Fig. 4.1 A pictorial schematic proof of The Pythagorean Theorem for the special case of an isosceles right triangle

its angles is a *right angle*, i.e., that its measure is $90°$ (read: "90 degrees") or, equivalently, $\pi/2$ radians. In our example, T's right angle occurs at vertex A. Because T is a right triangle, the line from vertex A to vertex B and the line from vertex A to vertex C are called the *sides* (or the *legs*) of T, while the line from vertex B to vertex C is called the *hypotenuse* of T. Triangle T has a very special "shape": it is *isosceles*, meaning that its two sides /legs have the same length. The grey triangle in Fig. 4.1 is an isosceles right triangle.

Theorem 4.1 (The Pythagorean Theorem). *Let T be a right triangle whose two sides have respective lengths s_1 and s_2, and whose hypotenuse has length h. Then*

$$h^2 = s_1^2 + s_2^2.$$

Consequently, when T is an isosceles *right triangle, then $h^2 = 2s_1^2$.*

The special case of the Pythagorean Theorem that deals with isosceles right triangles admits the very perspicuous proof presented pictorially in Fig. 4.1. Follow along in the figure as we review this proof.

We focus first on the unit-side square S on the left of the figure. We partition S by its diagonal into two unit-side isosceles right triangles, one grey and one white. In this construction the diagonal of S is the (shared) hypotenuse of the two triangles. On the right-hand side of Fig. 4.1, we use our partitioned version of S to construct a new, bigger square, call it $\widehat{S}$, whose side-length is the hypotenuse-length of the grey triangle. The dotted lines in the figure tell us how big $\widehat{S}$ is (measured by its area).

- Square S is unit-sided, hence has *unit area*, i.e., area $= 1$.

- The grey triangle is (geometrically) half of S, hence has area $1/2$.

- Square $\widehat{S}$, being built from four copies of the grey triangle, has area $4 \cdot (1/2) = 2$.

Because the hypotenuse of the grey triangle is a side of an area-2 square, we have just proved the following special case of the Pythagorean Theorem. □

Proposition 4.1 *The hypotenuse of the unit-side isosceles right triangle has length* $\sqrt{2}$.

End of digression

As part of the movement toward formalizing mathematics, the Greek mathematician and polymath Archimedes, from Syracuse, was systematically observing that squares are better approximations to circles than triangles are; regular pentagons are better approximations than squares; regular hexagons are better than pentagons; and so on. Figure 4.2 illustrates this evidence. In fact, observed Archimedes, as the

Fig. 4.2 Octagons approximate circles much better than squares do

number of sides, n, in a regular polygon grows without bound (or, as we might say today, tends to infinity), each increase in n brings a regular polygon closer to being a circle.

In order to pursue their respective observations to their completions, both Euclid and Archimedes would have had to leave the world of the rationals and enter the world of the *real numbers* (so named by the French mathematician-philosopher René Descartes)—but this world did not yet exist! It would take roughly two thousand years from the days of Euclid and Archimedes before the real numbers were *formally* introduced to the world, by mathematical luminaries such as the early-nineteenth-century French mathematician-scientist Augustin-Louis Cauchy and the late-nineteenth-century German mathematician Richard Dedekind. It turned out to be much easier to recognize instances of non-rational real numbers—better known as *irrational* real numbers—than to formally delimit the entire family of such numbers. Once again, happily, one could develop the real numbers in a way that allowed one to view a rational number as a special type of real number.

During the millennia between the discoveries of Euclid, Archimedes, and their friends, and the full development of the real numbers, mathematics was enriched repeatedly by the discovery of new conceptual structures. One of these—polynomials and their roots—ultimately led to the final major subsystem of our number system.

In Section 3.3.4, we discussed the important notion of *function*. Polynomials are a practically important class of functions that are delimited by the operations needed to compute them. Specifically, an *n*-argument polynomial function—typically just called a *polynomial*—is a function $P(x_1, x_2, \ldots, x_n)$ whose values can be calculated using just the basic operations of arithmetic: addition/subtraction and multiplication/division.

Explanatory note.

We pair the operations in this way because addition and subtraction are *mutually inverse operations*, as are multiplication and division. This means that one can undo an operation (say, an addition) by performing its inverse operation (in this case, a subtraction). But *be careful:* one cannot undo multiplication by 0.

A *root* of a polynomial (function) $P(x_1, x_2, \ldots, x_n)$ is an argument $\langle r_1, r_2, \ldots, r_n \rangle$ that causes P to *vanish*, meaning that $P(r_1, r_2, \ldots, r_n) = 0$. In Fig. 4.3, we illustrate

	Polynomial $P(x)$	Root(s)
1	$x + 1$	$x = -1$
2	$x - 1$	$x = 1$
3	$x^2 + 2x + 1 = (x+1)^2$	$x = -1$
4	$x^2 - 2x + 1 = (x-1)^2$	$x = 1$
5	$x^2 - 1$	$x = 1$ and $x = -1$
6	$x^2 + 1$	*no real root*
7	$x^2 - 2$	$x = \sqrt{2}$ and $x = -\sqrt{2}$
8	$x^2 + 2$	*no real root*

Fig. 4.3 A sampler of univariate polynomials and their roots

a few sample *univariate*—i.e., *single-variable*—polynomials.

There are lessons, both major and minor, to be gleaned from the examples in Fig. 4.3.

- Entries 3 and 4 in the table illustrate that even simple polynomials can often be written in several different ways. These entries illustrate also that roots can occur with multiplicity: one can view the value $x = -1$ as causing the polynomial $x^2 + 2x + 1 = (x+1) \cdot (x+1)$ to vanish in two ways—(1) by setting the left-hand factor $x + 1$ to 0 and (2) by setting the right-hand factor $x + 1$ to 0.

- Entries 5 and 7 illustrate that polynomials can have multiple distinct roots.

- Perhaps most importantly, entries 6 and 8 provide explicit, simple polynomials—whose expressions involve only positive integers—that fail to have any real roots!

  The fact that the indicated polynomials have no real roots is immediate, because the square of a real number can never be negative. Hence, for instance, there is no real number c such that $c^2 = -1$, or, equivalently, $c^2 + 1 = 0$.

For both applied and purely intellectual reasons, there has always been considerable interest in developing techniques for finding the roots of polynomials. Indeed, much seminal mathematics was developed in the quest for such techniques;[2] we study the topic at length in Section 5.3.

Closely related to this interest in a polynomial's roots was the considerable discomfort within the mathematical and technical world at the fact that the then-current number system—built upon the integers, the rationals, and the reals—was inadequate to the important task of providing roots for every polynomial. The reaction to this deficiency was similar in kind to all earlier recognized deficiencies: a way was found to expand the number system! Centuries would pass before mathematics developed adequately to find the needed expansion. Once discovered, the expansion was based on the conception, in the seventeenth century, of a new *imaginary* number, so designated by Descartes.[3] This new number was named i (for "imaginary") and was defined to be a root of the polynomial $P_{-1}(x) = x^2 + 1$. The number i was evocatively often defined via the equation, $i = \sqrt{-1}$. By keeping our extended arithmetic consistent with our former arithmetic, $-i$ also became a root of $P_{-1}(x)$. When the imaginary number i was added to the real number system, and the combination was blended via the rules of arithmetic, the *complex number system* was born.

Thankfully, the imaginary number i was the only totally new concept that was needed to mend the observed deficiency in the real numbers. In formal terms, the complex numbers were shown to be *algebraically complete* in the sense expressed in the landmark *Fundamental Theorem of Algebra*.

Theorem 4.2 (The Fundamental Theorem of Algebra). *Every polynomial of degree n with complex coefficients has n roots over the complex numbers.*

The proof of the Theorem is beyond the scope of our introductory text, but the result is a notable milestone in our mathematical/technical culture which everyone should know of.

Our historical tour is now complete, so we can—finally—begin to get acquainted with the several components of our number system and the operations that bring each component to life in applications.

4.3 Integers: The "Whole" Numbers

The first class of numbers that most of us encounter comprises the *integers*, which are also referred to as the *whole numbers* or the *counting numbers*. Being used for elementary operations such as counting, the integers are almost certainly the first

---

[2] A giant in the development and transmission of this work was the ninth-century mathematician and astronomer Muhammad ibn Mūsā al-Khwārizmī. We discuss his seminal contributions at greater length in Chapter 5.

[3] The term "imaginary" is reputed to be a derogation of these numbers that flouted tradition.

numbers that our prehistoric ancestors would have used. This section is devoted to exploring some of the basic properties of the class of integers. We shall come to appreciate the sophistication lurking behind even this most primitive class in our number system. The details we provide in Section 8.1 regarding the building blocks of the integers, the *prime numbers*, will prepare the reader for a broad range of applications of the integers, including important *(computer-)security-related* applications. Our introduction to *pairing functions*, in Section 8.2, will open the door to many applications of the integers which build on the ordering properties of these numbers, coupled with tools for encoding highly structured data as integers.

4.3.1 The Basics of the Integers: The Number Line

We survey a number of the most important properties of the following three sets, which collectively comprise "the integers".

- The set $\mathbb{Z}$ comprises *all integers*—the positive and negative integers plus the special number zero (0).

- The set $\mathbb{N}$ comprises the *nonnegative integers*—the positive integers plus zero (0).

- The set $\mathbb{N}^+$ comprises the *positive integers*.

There is no universally accepted default when one refers to "the integers" without a qualifying adjective; therefore, we shall always be careful to indicate which set we are discussing at any moment—often by supplying the set-name: $\mathbb{Z}$ or $\mathbb{N}$ or $\mathbb{N}^+$.

4.3.1.1 Natural orderings of the integers

Several essential properties of the sets $\mathbb{Z}$, $\mathbb{N}$, and $\mathbb{N}^+$ are consequences of the integers' behaviors under their natural *order* relations:

- the two *less-than* relations:

  - the *strict* relation ($<$). We articulate "$a < b$" as "a is (strictly) less than b" or as "a is (strictly) smaller than b".

  - the *nonstrict* or *weak* relation ($\leq$). We articulate "$a \leq b$" as "a is less than or equal to b" or as "a is no larger than b".

- their *converses,* the *greater-than* relations:

  - the *strict* relation ($>$). We articulate "$a > b$" as "a is (strictly) larger than b".

  - the *nonstrict* or *weak* relation ($\geq$). We articulate "$a \geq b$" as "a is greater than or equal to b" or "a is no smaller than b".

One sometimes encounters *emphatic* versions of the strict relations: the notations $a \ll b$ and $a \gg b$ indicate that a is, respectively, *much smaller than* or *much larger than b*. Usually, the qualifier "much" is not quantified, but sometimes it is.

The reader will note throughout the text that order relations within a number system are among one's biggest friends when reasoning about numbers.

4.3.1.2 The order-related laws of the integers

A. Total order and the Trichotomy Laws
   The sets $\mathbb{Z}$, $\mathbb{N}$, and $\mathbb{N}^+$ are *totally ordered*, also termed *linearly ordered*.

   These facts are embodied in the *Trichotomy Laws for integers*.

   The Trichotomy Laws for integers.

   (i) *For each integer $a \in \mathbb{Z}$, precisely one of the following is true.*
       a equals 0: $(a = 0)$ a is *positive*: $(a > 0)$ a is *negative*: $(a < 0)$
   (ii) *For each integer $a \in \mathbb{N}$, precisely one of the following is true.*
       a equals 0: $(a = 0)$ a is *positive*: $(a > 0)$
   (iii) *For each integer $a \in \mathbb{N}^+$:*
       a is *positive*: $(a > 0)$

   Consequently:
   (i') $\mathbb{Z}$ can be visualized via the (two-way infinite) number line:

   $$\ldots, -3, -2, -1, 0, 1, 2, \ldots$$

   (ii') $\mathbb{N}$ can be visualized via the (one-way infinite) number line:

   $$0, 1, 2, 3, \ldots$$

   (iii') $\mathbb{N}^+$ can be visualized via the (one-way infinite) number line:

   $$1, 2, 3, \ldots$$

   The Trichotomy Laws can be expressed using arbitrary pairs of integers, rather than insisting that one of the integers be zero. For the set $\mathbb{Z}$, for instance, this version of the Laws takes the following form:
   For any integers $a, b \in \mathbb{Z}$, precisely one of the following is true.
       a equals b: $(a = b)$ a is less than b: $(a < b)$ a is greater than b: $(a > b)$

B. Well-ordering
   The sets $\mathbb{N}$ and $\mathbb{N}^+$ are *well-ordered*.

The Well-Ordering Principle for nonnegative and positive integers.
   Every subset of $\mathbb{N}$ or of $\mathbb{N}^+$ has a smallest element (under the ordering $<$).

We digress momentarily to point out a nontrivial consequence of the well-ordering of the positive integers. We have already seen this consequence stated as Proposition 2.2, which we now have the tools to prove.

Proposition 2.2 [COMPACT FORM]. *The set S of Proposition 2.2 contains* $\mathbb{N}^+$.

Proof (Proposition 2.2). Say, for the sake of contradiction, that the set S specified in the statement of the proposition does not contain every positive integer. By the Well-Ordering Principle, there must then be a *smallest* positive integer, call it k, that does not belong to S. Clearly, $k > 1$, because the proposition asserts that $1 \in S$. Therefore, the number $k - 1$ is a positive integer. Moreover, $k - 1$, being smaller than k, belongs to S.

But now, by definition of S, we see that $(k - 1) + 1 = k$ *does belong to S.*

Thus, the assumption that S does not contain every positive integer leads to a contradiction. We infer that Proposition 2.2 is true. $\square$

C. Discreteness

The set $\mathbb{Z}$ is *discrete*.

The discreteness of the integers.
    For every integer $a \in \mathbb{Z}$, there is no integer between a and $a + 1$; i.e., there is no $b \in \mathbb{Z}$ such that $a < b < a + 1$.

D. The law of "between-ness"

   The "between-ness" law for the set $\mathbb{Z}$:
For any integers $a, b \in \mathbb{Z}$, there are finitely many $c \in \mathbb{Z}$ such that $a < c < b$.

Any such c lies *between* a and b along the number line, whence the name of the law.

E. The cancellation laws

   There are two *cancellation laws* for the set $\mathbb{Z}$, one for the operation of addition and one for the operation of multiplication.

The cancellation law for addition.
    For any integers $a, b, c \in \mathbb{Z}$, if $a + c = b + c$, then $a = b$.
The cancellation law for multiplication.
    For any integers $a, b \in \mathbb{Z}$ and $c \in \mathbb{Z} \setminus \{0\}$, if $a \cdot c = b \cdot c$, then $a = b$.

The cancellation laws provide limited versions of the algebraic notion of mutually inverse arithmetic operations (Section 5.1.2).

4.3.2 Divisibility: Quotients, Remainders, Divisors

This section is devoted to studying the fundamental relation of *divisibility* between two integers. Let $m, n \in \mathbb{N}$ be nonnegative integers. We use any of the following notations to assert the existence of a positive integer q such that $n = q \cdot m$.

- *m divides n*

- *m is a divisor of n*

- *n is divisible by m*

- *m | n.*

We consider the possible *divisibility* relations between integers m and n. We begin by noting some general facts.

- *Every nonzero integer m divides* 0.

  This is because of the universal equations $m \cdot 0 = 0 \cdot m = 0$. The same equations verify that 0 *does not divide any integer.*

- 1 *divides every integer.*

  This is because of the universal equation $1 \cdot m = m$.

- *Every nonzero integer divides itself.*

  This is because of the universal equation $m \cdot 1 = m$.

Some nonzero integers have many distinct divisors, while some have very few. Consider, for illustration, the first twelve positive integers.

Number	Divisors
1	$\{1\}$
2	$\{1,2\}$
3	$\{1,3\}$
4	$\{1,2,4\}$
5	$\{1,5\}$
6	$\{1,2,3,6\}$
7	$\{1,7\}$
8	$\{1,2,4,8\}$
9	$\{1,3,9\}$
10	$\{1,2,5,10\}$
11	$\{1,11\}$
12	$\{1,2,3,4,6,12\}$

All nonzero integers (except for 1) have at least two divisors, 1 and themselves. The "sparsely divisible" integers that have only these two divisors are called *primes* or *prime integers* or *prime numbers*. We study these "building blocks of the positive integers" in more detail in Section 8.1. While there is, indeed, much of interest to discuss about prime numbers, the *composite* or *nonprime* integers are also quite interesting, particularly, when we focus of *pairs* of integers. The next section looks at the defining property of composite integers, namely, *divisibility*.

In order to better understand the fundamental concept of divisibility, we must broaden our perspective somewhat and consider the notion of *Euclidean division*, i.e., *division with remainders*. The notion of "perfect" divisibility that we have been discussing is the special case in which the remainder is 0. The next subsection studies Euclidean division; the remainder of this section investigates the ramifications of "perfect" divisibility.

4.3.2.1 Euclidean division

Divisibility is not always perfect: Given a pair of integers, neither needs be an integer multiple of the other. As we learned in elementary school, if an integer $m > 0$ does not "evenly" divide an integer n, then we are left with a "remainder" when we attempt to divide n by m. *Euclidean division*—so named for the Greek mathematician Euclid whose writings introduced the process in the West—is the process of producing, given integers $m > 0$ and n, an integer *quotient q* and an integer *remainder r* (where $0 \leq r < m$) such that $n = q \cdot m + r$. The process of Euclidean division always succeeds, in the following strong sense.

Theorem 4.3 (The Division Theorem). *Given any integers n and m > 0, there exists a unique pair of integers q and r, with $0 \leq r < m$, such that*

$$n = q \cdot m + r \tag{4.1}$$

Proof. We first prove that a result-pair $\langle q, r \rangle$, as described in the Proposition, exists for each argument-pair $\langle m, n \rangle$. Then we prove that the result-pair is unique.

There exists at least one result-pair. Given any argument-pair $\langle m, n \rangle$, let $N_{m,n}$ be the set of all integers of the form $(n - a \cdot m)$ for some integer a. Symbolically,

$$N_{m,n} \stackrel{\text{def}}{=} \{ (n - a \cdot m) \mid [a \in \mathbb{N}] \text{ and } [(n - a \cdot m) \in \mathbb{N}] \}$$

Each such set $N_{m,n}$ is a nonempty set of nonnegative integers: the nonemptiness follows because $n \in N_{m,n}$ (via the case $a = 0$). By the Well-Ordering Principle, $N_{m,n}$ contains a (perforce, unique) *smallest* element. Let us denote this smallest element by r, and let us denote by a_r the value of a that yields r; i.e.,

$$n - a_r \cdot m = r$$

To complete this section of the proof, we need to show that $r < m$. Say, for contradiction, that $r = m + r'$ for some $r' \in \mathbb{N}$. We then find that

$$n - (a_r + 1) \cdot m = r - m = r'$$

is an element of $N_{m,n}$ that is strictly smaller than r. This contradiction completes the first part of the proof.

2. There exists at most one argument-pair. We turn now to the issue of uniqueness. Say, for the sake of contradiction, that there exists an argument-pair $\langle m, n \rangle$ for which there exist distinct result-pairs $\langle q_1, r_1 \rangle$ and $\langle q_2, r_2 \rangle$. We therefore have

$$n = q_1 \cdot m + r_1 = q_2 \cdot m + r_2 \qquad (4.2)$$

where both r_1 and r_2 satisfy the inequalities $0 \le r_1, r_2 < m$. We consider two cases.

- **Case 1.** Assume first that $r_2 = r_1$. In this case, the equations of (4.2) tell us that

$$q_1 \cdot m = q_2 \cdot m$$

  The cancellation law for multiplication then tells us that $q_1 = q_2$. Therefore, the allegedly distinct result-pairs are, in fact, identical.

- **Case 2.** If $r_2 \ne r_1$, then say, with no loss of generality, that $r_2 > r_1$. In this case, the equations of (4.2) tell us that

$$(q_1 - q_2)m = r_2 - r_1$$

  Because the right-hand quantity is positive, so also must be the left-hand quantity; i.e., $q_1 > q_2$ because $r_2 > r_1$.

  On the one hand, the left-hand quantity, $(q_1 - q_2)m$, is no smaller than m. This is because q_1 and q_2 are integers and $q_1 > q_2$. On the other hand, the right-hand quantity, $r_2 - r_1$, is strictly smaller than m. This is because $r_1 \ge 0$ so that $r_2 - r_1$ is no larger than r_2.

Both of the relevant cases thus lead to contradictions, so we must conclude that no argument-pair gives rise to more than one result-pair.   □

4.3.2.2 Divisibility, divisors, GCDs

We begin to study the several important aspects of integer divisibility by considering a variety of simple, yet significant, consequences of an integer n's being divisible by an integer m. We leave the following applications of the basic definitions as exercises for the reader.

Proposition 4.2 *1. If m divides n, then m divides all integer multiples of n. That is: If m divides n, then m divides cn for all integers c.*

2. The relation "divides" is transitive.[4] *Specifically, if m divides n and n divides q for some integer q, then m divides q.*

3. The relation "divides" distributes over addition.[5] *That is, if m divides n and m divides $(n+q)$ for some integer q, then m divides q.*

4. For any integer $c \ne 0$,

$$[m \text{ divides } n \quad \text{if, and only if,} \quad cm \text{ divides } cn]$$

---

[4] See Section 3.3.2.

[5] We use the term "distributes" in the sense of the Distributive Law; see Section 5.2.

The following result follows from the preceding facts.

Proposition 4.3 *Given integers m, n, and q, if m divides both n and q, then m divides all linear combinations of n and q; i.e., m divides $(sn+tq)$ for all integers s and t.*

Proof. Because m divides both n and q, there exist integers k_1 and k_2 such that $k_1 \cdot m = n$ and $k_2 \cdot m = q$. By the distributive law, we therefore have:

$$(k_1 \cdot s \ + \ k_2 \cdot t)m \ = \ sn + tq$$

for any s and t.   □

Among the common divisors of integers n and q, a particularly significant one is their *greatest common divisor*, which is the *largest* integer that divides both n and q. We abbreviate "greatest common divisor" by GCD, and we write

$$m \ = \ \text{GCD}(n,q)$$

to identify an integer m as the GCD of n and q.

We are finally ready for our first major result about integer division and divisors.

Proposition 4.4 (Bézout's identity) *For positive integers n and q, $\text{GCD}(n,q)$ is the smallest positive linear combination of n and q.*

Stated alternatively, the following two statements both hold.

- *For any positive integers n and q, there exist integers s_0 and t_0, not necessarily positive, such that $s_0 n + t_0 q \ = \ \text{GCD}(n,q)$.*
- *For all integers s and t, not necessarily positive, such that $sn + tq > 0$, we have $sn + tq \ \geq \ s_0 n + t_0 q$.*

Proof. Let m denote the smallest positive linear combination of n and q. We prove in turn that $m \geq \text{GCD}(n,q)$ and $m \leq \text{GCD}(n,q)$.

1. By definition, $\text{GCD}(n,q)$ divides both n and q. By Proposition 4.3, then, for all integers s and t, $\text{GCD}(n,q)$ divides the linear combination, $sn + tq$, of n and q. In particular, $\text{GCD}(n,q)$ divides the *smallest* such combination, m.

   It follows, therefore, that $m \geq \text{GCD}(n,q)$.
2. First, notice that $m \leq n$. The expression $n = 1 \cdot n + 0 \cdot q$ is a particular linear combination of n and q, and m is the *smallest* positive such combination.

   Because $m \leq n$, Theorem 4.3 (the Division Theorem) assures us that there exists an integer r with $0 \leq r < m$ such that $n \ = \ km + r$. Further, because m can be written in the form $m = sn + tq$ for integers s and t, simple arithmetic manipulation expresses r in the form $r \ = \ (1 - ks)n + (-kt)q$, which is a linear combination *of n and q*. Because m is the *smallest positive* such combination and $0 \leq r < m$, it follows that $r = 0$. We conclude that m divides n.

   A symmetric argument, using exactly the same analysis, allows us to conclude that m divides q.

   We thus have $m \leq \text{GCD}(n,q)$, which completes the proof.   □

Bézout's identity has the following significant corollary.

Corollary 4.1 *Every linear combination of n and q is a multiple of* $\mathrm{GCD}(n,q)$, *and vice versa.*

Greatest common divisors are fundamental companions of pairs of integers, with manifold computational applications. How does one compute them? This question was addressed millennia ago by Euclid, who authored the following result which led to the GCD-computing algorithm that bears his name.

For positive integers m and n, let $\mathrm{REM}(m,n)$ denote the (integer!) remainder r in the Euclidean-division expression (4.1).

Proposition 4.5 *For any integers n and m > 0,*

$$\mathrm{GCD}(m,n) = \mathrm{GCD}(m, \mathrm{REM}(m,n))$$

Proof. For integers $x > 0$ and $y \geq 0$, we denote by $D(x,y)$ the set of common divisors of x and y, i.e., the set of integers that divide both x and y. We prove the proposition by showing, as follows, that the sets $D(m,n)$ and $D(m, \mathrm{REM}(m,n))$ actually contain precisely the same elements.

- $D(m,n) \subseteq D(m, \mathrm{REM}(m,n))$.

  Say that the integer d divides both m and n, i.e., that $d \in D(m,n)$. By Theorem 4.3, we know that $n = q \cdot m + \mathrm{REM}(m,n)$. By property 3 in Proposition 4.2, we must then have $d \mid \mathrm{REM}(m,n)$. This means that $d \in D(m, \mathrm{REM}(m,n))$.

- $D(m, \mathrm{REM}(m,n)) \subseteq D(m,n)$.

  Say that the integer d divides both m and $\mathrm{REM}(m,n)$, i.e., that $d \in D(m, \mathrm{REM}(m,n))$. By Proposition 4.3, d divides every linear combination of m and $\mathrm{REM}(m,n)$. In particular, d divides the specific combination $q \cdot m + 1 \cdot \mathrm{REM}(m,n) = n$. Thus, d divides n, so that $d \in D(m,n)$.

Since we thus have $D(m,n) = D(m, \mathrm{REM}(m,n))$, we know that the sets contain the same largest element:

$$\max\left(D(m,n)\right) = \max\left(D(m, \mathrm{REM}(m,n))\right)$$

The proposition follows. $\square$

4.4 The Rational Numbers

Each enrichment of our number system throughout history has been a response to a deficiency with the then-current system. The deficiency that instigated the introduction of the rational numbers was the fact that many integers do not divide certain other integers.

This situation led to practical problems when people began to share commodities that were physically divisible. With a bit of care, you can always cut a pizza into any desired number of slices—but mandating such an action is awkward if you lack the terminology to describe what you want to achieve.

The situation also led to an intellectual problem, when viewed from a modern perspective. The arithmetic operation *multiplication* was surely recognized not long after its slightly simpler sibling operation *addition*. In many ways, these two operations mimic one another. Both are *total bivariate functions* which take a pair of numbers and produce a number; both are *commutative*, in that the argument numbers can be presented in either order without changing the result:

$$(\forall a,b)\;\big[[a+b \;=\; b+a]\quad \text{and}\quad [a\cdot b \;=\; b\cdot a]\big]$$

and both are *associative*, in the sense asserted by the equations

$$(\forall a,b)\;\big[[a+(b+c) \;=\; (a+b)+c]\quad \text{and}\quad [a\cdot(b\cdot c) \;=\; (a\cdot b)\cdot c]\big]$$

If we restrict our focus to the *integers*, however, there is a glaring difference between addition and multiplication. That is, addition has a "partner operation", *subtraction*, which operates as an *inverse operation*:

$$(\forall a,b,c)\big[\text{if}\;\;[c=a+b]\;\;\text{then}\;\;[a=c-b]\big]$$

(We call $c-b$ the *difference* between c and b.) Within the context of the integers, multiplication has no such "partner". We respond to this imbalance by inventing a "partner" for multiplication, and we call it *division*, denoted $\div$. Now, division cannot completely mimic subtraction because of the technical problems that arise from the *multiplicative annihilation* properties of the integer 0:

$$(\forall a)\,[a\cdot 0 \;=\; 0\cdot a \;=\; 0]$$

There is no way to "undo" or "invert" the operation *multiply-by-0*, because that operation is not one-to-one. However, if we frame the operation of division carefully—specifically, by avoiding division by 0—then we can endow multiplication with the desired "partner":

$$(\forall a,b,c)\big[\text{if}\;\;[c=a\cdot b]\;\;\text{and if}\;\;[b\neq 0]\;\;\text{then}\;\;[a=c\div b]\big]$$

(We call $c\div b$ the *quotient of c by b*.) We are almost at the end of our journey. All we need is a way to speak about specific quotients. When integer b divides integer c, as when $c=12$ and $b=4$, it is natural to write $12\div 4 \;=\; 3$, but how should we denote the quotient $12\div 5$, which is not an integer? Enter the rational numbers!

4.4.1 The Rationals: Special Ordered Pairs of Integers

The set $\mathbb{Q}$ of *rational* numbers—often abbreviated as just "the rationals"—was invented to name the quotients referred to in the preceding paragraph. Formally:

$$\mathbb{Q} \overset{\text{def}}{=} \{0\} \cup \{p/q \mid p,q \in \mathbb{Z} \setminus \{0\}\}$$

Each element of $\mathbb{Q}$ is called a *rational* number; each *nonzero* rational number p/q is often called a *fraction*; some people reserve the word "fraction" for the case $q > p$, because the word seems to connote "less than the whole", but this does not seem to be a valuable distinction.

In analogy with our treatment of integers, we reserve the notation $\mathbb{Q}^+$ for the *positive* rationals.

An alternative, mathematically more advanced, way of defining the set $\mathbb{Q}$ is as *the smallest set of numbers that contains the integers and is closed under the operation of dividing any number by any nonzero number.* The word "*closed*" here means that, for every two numbers $p \in \mathbb{Q}$ and $q \in \mathbb{Q} \setminus \{0\}$, the quotient p/q belongs to $\mathbb{Q}$.

Numerous notations have been proposed for denoting rational numbers in terms of the integers they are "built from". Most of these notations continue our custom of employing the single symbol "0" for the number 0, but notations such as $0/q$ (where $q \neq 0$) are permissible when they arise as part of a calculation or an analysis. For the nonzero elements of $\mathbb{Q}$, we traditionally employ some notation for the operation of division and denote the quotient of p by q using one of the following:

$$p/q \quad \text{or} \quad \frac{p}{q} \quad \text{or} \quad p \div q \tag{4.3}$$

The integer p in any of the expressions of (4.3) is the *numerator* of the fraction; the integer q is the *denominator*.

4.4.2 Comparing the Rational and Integer Number Lines

There are many ways to compare the sets $\mathbb{Z}$ and $\mathbb{Q}$ that enhance our understanding of both sets. We craft a comparison that focuses on the similarities and differences in the two sets' number lines, using Section 4.3.1 as the reference for the integer number line.

As the first point in our comparison, we remark that every integer $n \in \mathbb{Z}$ can be encoded as a rational number. Specifically, we represent/encode the integer $n \in \mathbb{Z}$ by the rational p/q whose numerator is $p = n$ and whose denominator is $q = 1$. This encoding is so intuitive that most people would write "$n = n/1$" and ignore the fact that this is expressing an encoding rather than an equality. We know with hindsight that this intellectual shortcut can cause no problems, but it is important to be aware that we are using a shortcut, for (at least) two reasons.

1. We should contemplate *why* the encoding "can cause no problems." Answering this question will enhance our understanding of both $\mathbb{Z}$ and $\mathbb{Q}$. *What essential properties of rationals and integers does the proposed encoding preserve?* To get started, note that the encoding preserves the special characters of the numbers 0 and 1—because the following equations hold: $0/1 = 0$ and $1/1 = 1$.

2. There are intuitively similar situations wherein one's intuition turns out to be wrong! One such situation occupies Section 4.4.2.2, wherein we demonstrate that the sets $\mathbb{Z}$ and $\mathbb{Q}$ "have the same size", and the more advanced Section 10.4.2, wherein we show that the set of real numbers is (in a formal sense) "larger" than sets $\mathbb{Z}$ and $\mathbb{Q}$. (*Even the fact that we can discuss the relative "sizes" of infinite sets is interesting—and not obvious!*)

4.4.2.1 Comparing $\mathbb{Z}$ and $\mathbb{Q}$ via their number-line laws

The rational numbers share some, but not all, of the number-line laws of the integers, as enumerated in Section 4.3.1. We now adapt for $\mathbb{Q}$ that section's discussion of $\mathbb{Z}$'s number line.

The sets $\mathbb{Q}$ and $\mathbb{Q}^+$ are both *totally ordered*, in the manner expressed by the Trichotomy laws for rational numbers.

The Trichotomy Laws for the rational numbers
(i) *For each rational $a \in \mathbb{Q}$, precisely one of the following is true.*

$$a \text{ equals } 0\text{: } (a = 0) \qquad a \text{ is } positive\text{: } (a > 0) \qquad a \text{ is } negative\text{: } (a < 0)$$

(ii) *Every rational $a \in \mathbb{Q}^+$ is positive $(a > 0)$.*

The total ordering of $\mathbb{Q}$ is expressed as follows
(iii) *For any rationals $a, b \in \mathbb{Q}$, precisely one of the following is true.*

$$a = b \quad \text{or} \quad a < b \quad \text{or} \quad a > b$$

As with the integers, the rationals can be visualized via a (two-way infinite) number line. But the rational line is much harder to visualize, mainly because the rationals do *not* enjoy the well-ordering or discreteness or "between-ness" laws of the integers.

The set $\mathbb{Q}$ is not *well-ordered.*

For illustration: The set

$$S = \{a \in \mathbb{Q} \mid 0 < a \leq 1\}$$

has no smallest element. If you give me a rational $p \in S$ that you claim is the smallest element of the set, then I shall give you $p/2$ as a smaller one.

The set $\mathbb{Q}$ does not *obey the "Between" laws.*

In fact, $\mathbb{Q}$ violates the "Between" laws in a very strong way: *For any two unequal rationals, a and b > a, there are infinitely many rationals between a and b.*

One can specify such an infinite set for the pair a, b in a many different ways. Here is a simple such set for the case $b > 0$, call it $S_{a,b}$.

$$S_{a,b} = \left\{ \frac{ka+b}{k} \ \middle| \ k \in \mathbb{N}^+ \right\} \tag{4.4}$$

4.4.2.2 Comparing $\mathbb{Z}$ and $\mathbb{Q}$ via their cardinalities

Our final comparison between the rationals and the integers compares the relative "sizes" or *cardinalities* of $\mathbb{Z}$ and $\mathbb{Q}$. Informally,

> *Are there "more" rationals than integers?*

Consider the following facts.

- Every integer is a rational number, as attested to by the "encoding"

$$\text{Encode} \quad n \in \mathbb{Z} \quad \text{by the quotient} \quad \frac{n}{1} \in \mathbb{Q}. \tag{4.5}$$

- There are infinitely many non-integer rational numbers between every pair of adjacent integers, as attested to by every set $S_{n,n+1}$ as defined in Eq. (4.4).

Thus, the set $\mathbb{Z}$ of integers is a *proper* subset of the set $\mathbb{Q}$ of rationals: symbolically, $\mathbb{Z} \subset \mathbb{Q}$. To many, this subset relation provides an intuitively compelling argument that there are more rational numbers than integers.

For us—and for the general mathematical community—the preceding intuition provides a compelling argument only for the fact that reasoning about infinite sets demands subtlety and care. For us, only the formal setting of Section 8.2.3 allows us to reason cogently about the relative "sizes" of infinite sets. Within this rigorous setting, we now show that

> *the set $\mathbb{N}$ has the same cardinality as the set $\mathbb{Q}$.*

Hoping to whet the reader's appetite for further study, we note that the following result exposes some of the main intuitions from the proof of Proposition 8.7.

Proposition 4.6 $|\mathbb{Q}| = |\mathbb{N}|$.

Proof (Sketch). We leave the details of this proof as an exercise.

First, we note that the encoding f defined by

$$(\forall n \in \mathbb{N}) \left[f(n) = \frac{n}{1} \right]$$

provides an injection from $\mathbb{N}$ into $\mathbb{Q}$. This injection verifies that $|\mathbb{Q}| \geq |\mathbb{N}|$.

For the converse relation, we proceed in two steps.

1. Let the function g associate each rational $p/q \in \mathbb{Q}$ with the ordered pair $\langle a, b \rangle \in \mathbb{N} \times \mathbb{N}$ that is obtained by expressing p/q in *lowest terms*; that is,

   - $\dfrac{p}{q} = \dfrac{a}{b}$.

   - The rational $\dfrac{a}{b}$ is in *lowest terms*, in the sense that a and b share no non-unit common divisor.

   Clearly, g is an injection from $\mathbb{Q}$ into $\mathbb{N} \times \mathbb{N}$.

2. Let the function h be an injection from $\mathbb{N} \times \mathbb{N}$ into $\mathbb{N}$. Sample such injections can be found in the proof of Proposition 8.7.

Since the composition of injections is again an injection, the composite injection $g \circ h$ verifies that $|\mathbb{N}| \geq |\mathbb{Q}|$.

Combining the preceding derived inequalities completes the proof. □

4.5 The Real Numbers

4.5.1 Inventing the Real Numbers

Each subsequent augmentation of our system of numbers is more complicated than the last one: Many relatively sophisticated deficiencies in the system are not even visible until other, relatively simple deficiencies have been eliminated. The deficiency in the system of rational numbers which led to the system of real numbers actually dates to historical time, roughly two-and-a-half millennia ago. The ancient Egyptians were prodigious builders who mastered truly sophisticated mathematics in order to engineer their temples and pyramids. The ancient Greeks continued this engineering tradition, but they added to it a philosophical "soul". Both the earthly and ethereal sides of mathematics will be observed in this chapter.

Numbers were (literally) sacred objects to many (philosophically oriented) Greeks. This at least partially explains their concern with understanding *why* certain mathematical facts were true, in addition to knowing *that* they were true. Indeed, the Greek mathematicians of antiquity invented ways of thinking about mathematical phenomena that were quite "modern" from our perspective. One intellectual project in this spirit had to do with the way they designed constructions such as temples. They were attracted to geometric constructions that could be accomplished using only *straight-edges* (what we often call "rulers") *and compasses*. And—most relevant to our story—they preferred that the relative lengths of linear sections of their structures be *commensurable* in the following sense. *Integers* $x, y \in \mathbb{N}$ *are commensurable if there exist* $a, b \in \mathbb{N}$ *such that*

$$ax = by \quad \text{or, equivalently,} \quad x = \frac{b}{a}y.$$

As Greek philosophers contemplated their desire to employ commensurable pairs of integers in constructions, they discovered that this goal was not always possible— even in moderately simple constructions. The poster child of this assertion is perceptible in *the diagonal of the square with unit-length sides* or, equivalently, in *the hypotenuse of the isosceles right triangle with unit-length legs*. For both of these constructs, the unit lengths of the structure's sides or legs were accompanied by the inevitable *non-commensurability* of the length of the square's diagonal or the triangle's hypotenuse: In current terminology, the (shared) length of the diagonal and the hypotenuse is $\sqrt{2}$. The Greek mathematicians, as reported by Euclid,[6] proved, using current terminology, that $\sqrt{2}$ *is not rational*. (We rephrase Euclid's proof imminently, in Proposition 4.7.) The conclusion from this proof is that a number system based solely on the integers and rationals was inadequate for even elementary constructions. In response to this discovery, the philosophers/mathematicians augmented our number system by introducing *surds* or, as we more commonly term them, *radicals*. The augmentation thus begun culminated in what we know as the *real number system*. Since our intention in this introduction has been to justify the journey along that trajectory, we leave our historical digression and turn to our real focus, the set of *real numbers*.

4.5.2 Defining the Real Numbers via Their Representations

There are numerous ways to define the class $\mathbb{R}$ of *real numbers*. The definition which we have chosen has two advantages, given the goals for this text. Our definition: (1) has clear connections to the *computational aspects* of real numbers; (2) relies on *basic* mathematical concepts, hence can confidently appear at this early stage of the text. We define the class of real numbers in terms of their *operational* numerals, thereby laying the groundwork for Chapter 10's study of numerals. The definition we provide is completely correct and adequate, but its nuances will be appreciated by readers only incrementally, as they progress through Chapter 6 (especially Section 6.2.2) and Chapter 10.

Our definition is parameterized by an integer *base* $b > 1$: any such integer will work. Having fixed on a base b, a *real number* is any number that is specified by an *infinite* summation of the form

$$\sum_{i=0}^{n} \alpha_i \cdot b^i \;+\; \sum_{j=0}^{\infty} \beta_j \cdot b^{-j}$$

$$(4.6)$$

where: • $n \in \mathbb{N}$

• each $\alpha_i, \beta_j \in \{0, 1, \ldots, b-1\}$

---

[6] Euclid wrote extensively on this and related subjects, especially regarding geometry and what is currently known as number theory.

By prepending a *negative sign* or *minus sign* to a numeral or a number, one renders the thus-embellished entity as negative.

The perceptive reader will note that a proof is required to establish that every expression of the form (4.6) actually defines a specific number. Such a proof can be developed, but it would require mathematical tools that are beyond the scope of this text. For the practicalities of our study, we establish the following rule, which allows one to posit the membership in $\mathbb{R}$ of a particular subset of the real numbers, which we shall term *constructible*.

Constructibility Principle for real numbers. *Any length (of a line segment) that can be specified via a drawing made using only straightedge and compass is a real number.*

4.5.3 Not All Real Numbers Are Rational

We close this section by verifying that the real number $\sqrt{2}$ is not rational. We thereby conclude—via a specific example—that there exist real numbers that are not rational: In Section 10.4.2, we shall observe infinitely many such numbers. The current proof also provides the basis of a proof of the non-commensurability of the length of the hypotenuse of an isosceles right triangle—or, equivalently, of the (common) length of the diagonals of a square with the (common) length of its sides.

Proposition 4.7 **(a)** *The number $\sqrt{2} = 2^{1/2}$ is real; i.e., it belongs to $\mathbb{R}$.*
   (b) *The number $\sqrt{2}$ is not rational; i.e., it does not belong to $\mathbb{Q}$.*

Proof (Part (a)). The fact that $\sqrt{2}$ is a real number follows from the Constructibility Principle for real numbers, in the light of Proposition 4.1.   □-Part (a)

Enrichment note.
$\oplus$ *Digression: An infinite summation for $\sqrt{2}$.*
We have presented the Constructibility Principle for real numbers precisely because deriving, and reasoning about, infinite summations often requires mathematical tools that go beyond those that populate an introductory text. That said, one of our goals is to introduce the reader to the beauty in mathematics. To that end, we present, with no proofs, an infinite summation for $\sqrt{2}$. While this summation is not of the form (4.6), it is close enough to render plausible the fact that $\sqrt{2}$ can be represented by a summation that is of the form (4.6).

$$\sqrt{2} = 1 + \frac{3}{4} \cdot \left(\frac{1}{2}\right) + \frac{3}{8} \cdot \left(\frac{1\cdot 3}{2\cdot 4\cdot 6}\right) + \frac{3}{12} \cdot \left(\frac{1\cdot 3\cdot 5\cdot 7}{2\cdot 4\cdot 6\cdot 8\cdot 10}\right)$$
$$+ \frac{3}{16} \cdot \left(\frac{1\cdot 3\cdot 5\cdot 7\cdot 9\cdot 11}{2\cdot 4\cdot 6\cdot 8\cdot 10\cdot 12\cdot 14}\right) + \frac{3}{20} \cdot \left(\frac{1\cdot 3\cdot 5\cdot 7\cdot 9\cdot 11\cdot 13\cdot 15}{2\cdot 4\cdot 6\cdot 8\cdot 10\cdot 12\cdot 14\cdot 16\cdot 18}\right)$$
$$+ \cdots + \frac{3}{4k} \cdot \left(\frac{1\cdot 3\cdot 5\cdot 7\cdots (4k-5)}{2\cdot 4\cdot 6\cdot 8\cdot 10\cdots (4k-2)}\right) + \cdots$$

As we do for many results that we encounter in our mathematical journey, we provide multiple proofs, which build upon quite different mathematical insights. In this spirit, we now present two proofs for Proposition 4.7(b). In Section 4.5.3.1 we provide the classical proof of the result.[7] This proof invokes a simple provision of the Fundamental Theorem of Arithmetic (Theorem 8.1) to exploit the divisibility properties of integers. In Section 4.5.3.2, we provide a proof that builds on the Pythagorean Theorem (Theorem 4.1) to develop geometric insights. The reader may enjoy the more extensive exposition on $\sqrt{2}$ in [34].

4.5.3.1 A number-based proof that $\sqrt{2}$ is not rational: $\sqrt{2} \notin \mathbb{Q}$

Proof. Let us assume, for contradiction, that $\sqrt{2}$ is rational. By definition, then, $\sqrt{2}$ can be written as a quotient

$$\sqrt{2} = \frac{a}{b}$$

for positive integers a and b. In fact, we can also insist that a and b *share no common prime factor*—because if a and b did share the prime factor p, then we would have $a = p \cdot c$ and $b = p \cdot d$. In this case, we would also have

$$\sqrt{2} = \frac{a}{b} = \frac{p \cdot c}{p \cdot d} = \frac{c}{d}$$

by canceling the common factor p. We can eliminate all further common prime factors if necessary until, finally, we find a quotient for $\sqrt{2}$ whose numerator and denominator share no common prime factor. This must occur eventually because each elimination of a common factor leaves us with smaller integers, so the iterative elimination of common factors must terminate. Therefore, let us say, finally, that

$$\sqrt{2} = \frac{k}{\ell} \tag{4.7}$$

where k and ℓ share no common prime factor.

Let us square both expressions in Eq. (4.7) and multiply both sides of the resulting equation by ℓ^2. We thereby discover that

$$2\ell^2 = k^2 \tag{4.8}$$

This rewriting exposes the fact that k^2 is *even*, i.e., *divisible by 2*. We claim that the divisibility of k^2 by 2 betokens the same divisibility for k.

We could "bring in the howitzers" to verify this claim, via a forward reference to the *Fundamental Theorem of Arithmetic* (Theorem 8.1), but a simple calculation will suffice.

Say, for contradiction, that k^2 is even, but k is odd. Since every odd integer k can be written in the form $k = 2h + 1$, we observe that k^2 can be written in the form

---

[7] This proof is quite literally *classical*: it is based on the proof provided by Euclid in his *Elements*.

$$k^2 = (2h+1)^2 = 4h^2 + 4h + 1 = 4(h+1) + 1$$

This way of writing k^2 makes it clear that k^2 is *odd*, i.e., is *not* divisible by 2. This contradiction establishes the claim.

So we now know that $k = 2m$ for some positive integer m, which allows us to rewrite Eq. (4.8) in the form

$$2\ell^2 = k^2 = (2m)^2 = 4m^2. \tag{4.9}$$

Hence, we can divide the first and last quantities in Eq. (4.9) by 2, to discover that

$$\ell^2 = 2m^2.$$

If we now repeat the preceding indented argument, we discover that the integer ℓ must be even. This means that *both k and ℓ are even*. This fact contradicts our assumption that k and ℓ share no common prime divisor!

Since every step of our argument is ironclad, except for our assumption that $\sqrt{2}$ is rational, we conclude that that assumption is false! Part (b) of the Proposition is verified.   □

4.5.3.2 A geometric proof that $\sqrt{2}$ is not rational: $\sqrt{2} \notin \mathbb{Q}$

Proof. Our geometric proof is built around Fig. 4.4, which suggestively invokes

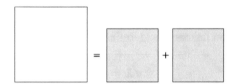

Fig. 4.4 *A geometric depiction of the Pythagorean Theorem and its underlying equation:* $a^2 = b^2 + b^2$

the Pythagorean Theorem. The figure displays three squares: The two small grey squares are identical, with common area A, while the large white square has double this area. By the Pythagorean Theorem, if the small squares have (common) side-lengths $b \in \mathbb{N}^+$, hence share area $A = b^2$ each, then the large square has side-lengths $a \stackrel{\text{def}}{=} \sqrt{2}b$, hence has area $a^2 = 2b^2$.

As in the classical proof of Section 4.5.3.1, we now assume that $\sqrt{2}$ is rational. Within the context of Fig. 4.4, this means that

$$\sqrt{2} = \frac{a}{b}$$

for $a, b \in \mathbb{N}^+$. Since all that we have said thus far holds for arbitrary a and b, we are free to insist, as before, that a and b do not share any common prime factor. Note additionally that because[8]

$$\sqrt{2} > 1.4$$

we know that $a > b$.

Explanatory note.

Of course, our demand that the numerator a and the denominator b do not share a common factor does not diminish the generality of our argument. This is because "a/b" is just one name for the depicted rational number, and choosing any specific name has no impact on the number itself.

Now that we have the suggestive "equation" presented in Fig. 4.4, we can manipulate the depicted squares. We embed both of the grey $b \times b$ squares of Fig. 4.4 into the white $a \times a$ square, in the overlapped manner depicted in Fig. 4.5: one grey

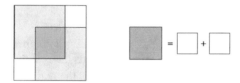

Fig. 4.5 *Constructing a smaller instance of the Pythagorean equation*

square is nestled into the northwestern corner of the white square, while the other is nestled into the southeastern corner. The overlapping of the grey squares under this embedding creates a new square—depicted in dark grey—in the center of the white square, while it leaves unoccupied two small squares, which remain white in the figure.

Now, let us get quantitative.

- On the one hand, the fact that the combined areas of the two grey squares equal the area of the white square guarantees that the area of the dark grey overlap-square is equal to the combined areas of the small unoccupied white squares.

- On the other hand, because the side-length of the large white square is a, while the (common) side-length of the grey squares is b, it follows that the side-lengths of the small white square is $a - b$, and the side-length of the dark grey overlap-square is $2b - a$. All of these side-lengths are positive because of the value of $\sqrt{2}$. That is: (*i*) $a > b$ because $\sqrt{2} > 1$; (*ii*) $2b > a$ because $\sqrt{2} < 2$.

The preceding facts allow us to label the squares of Fig. 4.4 differently than we did earlier—and thereby derive a different valid "equation". As we did at the beginning

---

[8] If this inequality is new to you, then just note that $(1.4)^2 = 1.96$, which is less than 2.

of this discussion, we again invoke the Pythagorean Theorem, but now we do so while focusing—see Fig. 4.5—on the dark grey overlap-square (which plays the role of the large square in Fig. 4.4) and the two small white squares (which play the role of the two small squares in Fig. 4.4). Whereas our original focus led to the putative rational value $\frac{a}{b}$ for $\sqrt{2}$, the new focus yields the putative rational value $\frac{2b-a}{a-b}$. We thus have

$$\sqrt{2} = \frac{a}{b} = \frac{2b-a}{a-b}$$

where $2b - a < a$ and $a - b < b$. In the light of the Fundamental Theorem of Arithmetic (Theorem 8.1), this new rational name for $\sqrt{2}$ contradicts our initial assumption that a/b was a fraction *in lowest terms*, i.e., in which a and b share no common factor. □

4.6 The Basics of the Complex Numbers

Let us denote by $\mathbb{C}$ the set of complex numbers. Each number $\kappa = a + bi$ in $\mathbb{C}$ has a *real part*—the part that *does not* involve the imaginary unit i—and an *imaginary part*—the part that *does* involve i. To be explicit: the real part of our number κ, is $\mathrm{Re}(\kappa) = a$; the *imaginary part* of our number κ is $\mathrm{Im}(\kappa) = b$. The notations $\mathrm{Re}(\kappa)$ and $\mathrm{Im}(\kappa)$ are common but not universal.

Based on the arithmetic laws that we have discussed thus far, plus the defining equation ($i^2 = -1$) of the imaginary unit i, we find that the *product* of two complex numbers, $a + bi \in \mathbb{C}$ and $c + di \in \mathbb{C}$ is the complex number

$$(a + bi) \cdot (c + di) = (ac - bd) + (ad + bc)i \qquad (4.10)$$

We note that a "direct" implementation of complex multiplication, i.e., one that implements Eq. (4.10) literally, requires *four* real multiplications—namely, $a \times c$, $b \times d$, $a \times d$, $b \times c$.

During the 1960s, people began to pay close attention to the differing costs associated with alternative ways of achieving computational results. They sought—and found—a number of procedures that replaced computations involving k real multiplications (a relatively expensive operation) and ℓ real additions (a relatively inexpensive operation) by computations that achieved the same result but used fewer multiplications and not too many more additions. Complex multiplication was one of the operations they studied. Here is the result.

Proposition 4.8 *One can compute the product of two complex numbers using* three *real multiplications rather than four.*

The proof of this result involves arithmetic manipulation that requires "thinking outside the box". We leave the proof as an exercise. The main message of the result is that we should never be lulled into assuming that the way an arithmetic expression *is* written is the way that it *has to be* written.

4.7 Exercises: Chapter 4

Throughout the text, we mark each exercise with 0 or 1 or 2 occurrences of the symbol $\oplus$, as a rough gauge of its level of challenge. The 0-$\oplus$ exercises should be accessible by just reviewing the text. We provide *hints* for the 1-$\oplus$ exercises; Appendix H provides *solutions* for the 2-$\oplus$ exercises. Additionally, we begin each exercise with a brief explanation of its anticipated value to the reader.

1. **Irrationality abounds!**
   LESSON: Extending an argument by understanding its basis.

   Extend Euclid's reasoning about $\sqrt{2}$ to prove the following assertion.

   Proposition 4.9 *For any prime $p > 1$, the number $\sqrt{p}$ is not rational, i.e., does not belong to $\mathbb{Q}$.*

2. **Appreciating the totality of algebra**
   Lesson: Even if a problem does not mention the inverses to addition and multiplication, you may benefit from paying attention to these operations.

   Show how to compute the product of two complex numbers using only three *real multiplications rather than the method that directly follows the definition (which uses four multiplications).*[9]

   > Of course, a full reckoning of the comparative costs of multiplying complex numbers using three real multiplications vs. four must consider that the three-multiplication procedure uses *three* real additions, whereas the four-multiplication procedure uses only *two* real additions. A complete comparison between the two procedures would require knowing the relative costs of multiplication and addition on the target computing platform.

3. **Basic properties of divisibility**
   LESSONS: Practice with basic proofs; enhance understanding of integers
   Prove the four assertions in Proposition 4.2.

4. **Observing an influence of parity**
   Lesson: Learn to recognize the influence of numerical properties such as parity whenever one deals with sequences of numbers.

   The notion of even-odd parity is so simple that one often overlooks the way it can influence the "shape" of computations. You will now observe such influence on

---

[9] Make a note of your solution technique. We shall encounter it again in Chapter 9.

the important area of *recurrences*. We shall study recurrent summations in depth in Chapter 9 and in Appendix D, but we begin this study with the simple topic of this exercise.

It is not uncommon, in mathematics, computing, and allied disciplines, to encounter sequences of integers that have recursive structures that are kindred to the following sequences:

- *A bilinear recurrence*

  Two base integers: a_0, a_1

  The recurrent remainder: $(\forall n \geq 1) \left[a_{n+1} = a_n + a_{n-1} \right]$ (4.11)

- *A trilinear recurrence*

  Three base integers: b_0, b_1, b_2

  The recurrent remainder: $(\forall n \geq 1) \left[b_{n+1} = b_n + b_{n-1} + b_{n-2} \right]$ (4.12)

Viewing these specifications inductively, formulas (4.11) and (4.12) generate two infinite sequence of positive integers:

$$a_0, a_1, a_2, a_3, a_4, \ldots \quad \text{and} \quad b_0, b_1, b_2, b_3, b_4, \ldots$$

We are interested in exposing some properties of these sequences that are consequences of the parities of their respective base integers.

On to the exercise:

Of course, if one chooses only even positive base integers, then the entire sequence A generated by formula (4.11) and the entire sequence B generated by formula (4.12) consist of even integers.

Answer the following questions for each of these sequences.

  a. *Is there a choice of positive base integers a_0, a_1 for which the entire sequence A consists of only odd integers?*

  b. *Is there a choice of positive base integers b_0, b_1, b_2 for which the entire sequence B consists of only odd integers?*

  c. *Is there a choice of positive base integers a_0, a_1 for which the entire sequence A alternates between even and odd integers?*

  d. *Is there a choice of positive base integers b_0, b_1, b_2 for which the entire sequence B alternates between even and odd integers?*

5. ⊕ **The rationals ($\mathbb{Q}$) and the integers ($\mathbb{N}$) are equinumerous**
   LESSON: Experience with a "multilayer" proof, which combines several elementary threads

   Provide a detailed *proof of Proposition 4.6. Informally, there are equally many integers as there are rationals.*

Chapter 5
Arithmetic: Putting Numbers to Work

*Dans les arithmétiques anciennes, un chapitre était consacré
à l'addition des toises, un autre à l'addition des sols, un
autre à l'addition des onces, etc. Seuls quelques grands
esprits en ces temps là parvenaient à l'abstraction de
l'addition d'objets quelconques. Maintenant on exige cette
abstraction d'enfants de huit ans.*

*(In the arithmetic of yore, one chapter was devoted to the
addition of lengths, another to the addition of areas, a
third to the addition of weights, etc. Only a few great spirits
in those days arrived at the abstraction of addition of
arbitrary items. Nowadays we expect this abstraction from
8-year-olds.)*

Raymond Quenaud (*Bords*)

Previous chapters have given us sets, numbers, and numerals, the simple objects that we use to count and measure and aggregate. The current chapter is devoted to expounding the rules of *arithmetic*, the system which enables us to manipulate these objects, and to develop complicated objects out of simple ones.

5.1 The Basic Arithmetic Operations

The basic tools of arithmetic consist of a small set of operations on numbers/numerals, together with two special integers which play important roles with respect to the operations. Since these entities, the operations and special numbers, are so tightly intertwined, we discuss them simultaneously.

The two special integers. The integers zero (0) and one (1) play special roles as the tools of arithmetic are applied to all the classes of numbers described in Chapter 4.

© Springer Nature Switzerland AG 2020
A. L. Rosenberg, D. Trystram, *Understand Mathematics, Understand Computing*,
https://doi.org/10.1007/978-3-030-58376-7_5

The operations of arithmetic. Arithmetic on the classes of numbers described in Chapter 4 is built upon a rather small repertoire of operations. When we say that an operation produces a number "of the same sort", we mean that it produces:

- an integer result from integer arguments;

- a rational (number) result from rational (number) arguments;

- a real (number) result from real (number) arguments;

- a complex (number) result from complex (number) arguments.

The fundamental operations on numbers are, of course, familiar to the reader. Our goal in discussing them is to stress the laws that govern the operations. Along the way, we also introduce a few operations that are less familiar but no less important.

5.1.1 Unary (Single-Argument) Operations

5.1.1.1 Negating and reciprocating numbers

i. The operation of negation:

- is a *total function* on the sets $\mathbb{Z}, \mathbb{Q}, \mathbb{R}, \mathbb{C}$. It replaces a number a by its *negative*, which is a number from the same class.

  The *negative* of a number a is denoted $-a$ and is usually articulated "minus a" or "negative a".

- is a *bijection* between the positive and negative subsets of $\mathbb{Z}$, of $\mathbb{Q}$, and of $\mathbb{R}$.

  The notions "positive" and "negative" do not apply in a straightforward manner to $\mathbb{C}$.

Zero (0) is the unique *fixed point* of the operation, meaning that 0 is the unique number a such that $a = -a$.

ii. The operation of reciprocation:

- is a *total function* on the sets $\mathbb{Q} \setminus \{0\}$, $\mathbb{R} \setminus \{0\}$, and $\mathbb{C} \setminus \{0\}$; it replaces each number a by its *reciprocal*, which is a number from the same class.

  The *reciprocal* of a number $a \neq 0$ is denoted $1/a$ or $\dfrac{1}{a}$; we employ whichever notation enhances legibility. The reciprocal of a is usually articulated "1 over a".

- is *undefined* on every integer a except for $a = 1$.

5.1.1.2 Floors, ceilings, magnitudes

i. The operations of taking floors and ceilings are total operations on the sets $\mathbb{N}, \mathbb{Z}, \mathbb{Q}$, and $\mathbb{R}$.

- The *floor* of a number a, also called *the integer part* of a, is denoted $\lfloor a \rfloor$. It is the largest integer which does not exceed a; i.e.,

$$\lfloor a \rfloor \overset{\text{def}}{=} \max_{b \in \mathbb{N}} \left[b \le a \right]$$

- The *ceiling* of a number a, denoted $\lceil a \rceil$, is the smallest integer which is not smaller than a:

$$\lceil a \rceil \overset{\text{def}}{=} \min_{b \in \mathbb{N}} \left[b \ge a \right]$$

The operations of taking floors and ceilings are two common ways to *round* rationals and reals to their "closest" integers.

ii. The operation of taking absolute values/magnitudes: Let a be a real number. The *absolute value* or *magnitude*, of a, denoted $|a|$, equals either a or $-a$, whichever is nonnegative. For a complex number a, the definition of $|a|$ is more complicated: it is a measure of a's "distance" from the "origin" complex number $0 + 0 \cdot i$. In detail:

$$|a| = \begin{cases} a & \text{if } \left[a \in \mathbb{R} \right] \text{ and } \left[a \ge 0 \right] \\ -a & \text{if } \left[a \in \mathbb{R} \right] \text{ and } \left[a < 0 \right] \\ \sqrt{b^2 + c^2} & \text{if } \left[a \in \mathbb{C} \setminus \mathbb{R} \right] \text{ and } \left[a = (b + ci) \right] \end{cases}$$

5.1.1.3 Factorials (of a nonnegative integer)

The *factorial* of a nonnegative integer $n \in \mathbb{N}$, which is commonly denoted $n!$, is the function defined via the following recursion.

$$\text{FACT}(n) = \begin{cases} 1 & \text{if } n = 0 \\ n \cdot \text{FACT}(n-1) & \text{if } n > 0 \end{cases} \tag{5.1}$$

We can "unwind" the recursion in Eq. (5.1) and, thereby, derive an iterative expression for $n!$. In what follows, we denote multiplication by the familiar symbol $\times$ rather than the centered dot ($\cdot$), to enhance legibility. We find that, for all $n \in \mathbb{N}$,

$$n! = \text{FACT}(n) = n \times (n-1) \times (n-2) \times \cdots \times 2 \times 1 \tag{5.2}$$

A three-step inductive argument validates this "unwinding":

1. If $n = 0$, then $\text{FACT}(n) = 1$, by definition (5.1).

2. Assume, for induction, that the expansion in Eq. (5.2) is valid for a given $k \in N$:

$$\text{FACT}(k) = k \times (k-1) \times (k-2) \times \cdots \times 2 \times 1$$

3. Then:

$$\text{FACT}(k+1) = (k+1) \cdot \text{FACT}(k) \qquad\qquad\qquad \text{by Eq. (5.1)}$$
$$= (k+1) \times k \times (k-1) \times (k-2) \times \cdots \times 2 \times 1 \quad \text{by induction}$$

5.1.2 Binary (Two-Argument) Operations

5.1.2.1 Addition and subtraction

The operation of *addition* is a *total function* which replaces any two numbers a and b by a number of the same sort. The resulting number is the *sum of a and b* and is denoted $a+b$.

The operation of *subtraction* is a *total function* on the sets $\mathbb{Z}, \mathbb{Q}, \mathbb{R}, \mathbb{C}$ which replaces any two numbers a and b by a number of the same sort. The resulting number is the *difference of a and b* and is denoted $a-b$. On the nonnegative subsets of the sets $\mathbb{Z}, \mathbb{Q}, \mathbb{R}, \mathbb{C}$—such as $\mathbb{N}$, which is the largest nonnegative subset of $\mathbb{Z}$—subtraction is a *partial function* which is defined only when $a \geq b$.

Subtraction can also be defined as follows. For any two numbers a and b, *the difference of a and b is the sum of a and the negation of b*:

$$a-b \ = \ a+(-b)$$

The special role of 0 under addition and subtraction. The number 0 is the *identity* under addition and subtraction; terminologically, 0 is an *additive identity*. This means that, for all numbers a,

$$a+0 \ = \ a-0 \ = \ a.$$

In Section 2.2.2.1, we illustrated the technique of Proof by Contradiction by proving the following basic result.

Proposition 5.1 *The number 0 is the unique additive identity.*

The special role of 1 under addition and subtraction. For any integer a, there is no integer between a and $a+1$ nor between $a-1$ and a. For this reason, on the sets $\mathbb{Z}$ and $\mathbb{N}$, one often singles out the following special cases of addition and subtraction, especially in reasoning about situations that are indexed by integers. Strangely, these operations have no universally accepted notations.

- The *successor* operation is a *total function* on both $\mathbb{N}$ and $\mathbb{Z}$ which replaces an integer a by the integer $a+1$.

- The *predecessor* operation is a *total function* on $\mathbb{Z}$ which replaces an integer a by the integer $a-1$. It is a *partial function* on $\mathbb{N}$ which is defined only when the argument a is positive (so that $a-1 \in \mathbb{N}$).

Functional inverses. When a function f admits an *inverse*—i.e., a function that "undoes" the result of applying f—knowledge of the inverse can help one verify and analyze algorithms that involve f. A number of arithmetic operations have some type of inverse. We discuss here only functions that are related to the operations of addition and subtraction. The reader should supply analogues of these definitions for functions that are based on other arithmetic operations.

- *Inverses of unary functions.* Unary function g is a *functional inverse* of unary function f if for all x in the domain of f,

$$g(f(x)) = x \tag{5.3}$$

Note the implicit assumption here that g is defined on every result $f(x)$ of f.

The operations of successor and predecessor are *mutually inverse operations* in the sense of Eq. (5.3). Symbolically, for all $a \in \mathbb{N}$:

$$\begin{aligned}
\text{PREDECESSOR}(\text{SUCCESSOR}(a)) &= \text{PREDECESSOR}(a+1) \\
&= (a+1) - 1 \\
&= a \\
\text{SUCCESSOR}(\text{PREDECESSOR}(a)) &= \text{SUCCESSOR}(a-1) \quad \text{for all } a \in \mathbb{N}^+ \\
&= (a-1) + 1 \\
&= a
\end{aligned}$$

- *Inverses of binary functions.* Binary function g is a *functional inverse* of binary function f if for all pairs (x,y) in the domain of f,

$$g(f(x,y),y) = x \tag{5.4}$$

Note the implicit assumption here that g is defined on every pair (x,y) whose first element is a result $f(x,y)$ of f.

Each of addition and subtraction is a functional inverse of the other, in the sense of Eq. (5.4). Symbolically, for all $a,b \in \mathbb{N}$:

$$\begin{aligned}
(a+b) - b &= a \\
(a-b) + b &= a \quad \text{whenever} \quad a \geq b
\end{aligned}$$

5.1.2.2 Multiplication and division

The operation of *multiplication* is a *total function* that replaces any two numbers a and b by a number of the same sort. The resulting number is the *product of a and b* and is denoted in three main ways: $a \cdot b$ or $a \times b$ or ab (the last one only when the absence of an explicit multiplication sign can cause no confusion). We shall usually favor the first notation (to save space), except when the second enhances legibility.

The operation of *division* is a *partial function* on all of our sets of numbers. Given two numbers a and b, the result of dividing a by b—*when that result is defined*—is the *quotient of a by b;* it is denoted by one of the following three notations:

$$a/b, \quad a \div b, \quad \frac{a}{b}$$

The *quotient of a by b* is defined precisely when *both*

(1) $b \neq 0$: one can never divide by 0
and
(2) there exists a number c such that $a = b \cdot c$

Assuming that condition (1) holds, *condition (2) always holds when a and b belong to $\mathbb{Q}$ or $\mathbb{R}$ or $\mathbb{C}$.*

Division can also be defined as follows. For any two numbers a and b, *the quotient of a and b is the product of a and the reciprocal of b* (assuming that the latter exists):

$$a/b = a \times (1/b)$$

Computing reciprocals of nonzero numbers in $\mathbb{Q}$ and $\mathbb{R}$ is standard high-school level fare; computing reciprocals of nonzero numbers in $\mathbb{C}$ requires a bit of calculational algebra which we do not cover. For completeness, we note that the reciprocal of the *nonzero* complex number $a + bi \in \mathbb{C}$ is the complex number $c + di$ where

$$c = \frac{a}{a^2 + b^2} \quad \text{and} \quad d = \frac{-b}{a^2 + b^2}$$

The special role of 1 *under multiplication and division.* The number 1 is the *identity* under the operations of multiplication and division; terminologically, 1 is a *multiplicative identity.* This means that, for all numbers a,

$$a \cdot 1 = a \cdot (1/1) = a$$

The reader has already shown—by proving Proposition 2.9—that 1 is the *unique* multiplicative identity.

The special role of 0 *under multiplication and division.* The number 0 is the *annihilator* under multiplication. This means that, for all numbers a

$$a \times 0 = 0$$

Using reasoning which parallels that of Propositions 5.1 and 2.9, the reader can prove that 0 is *the unique* multiplicative annihilator.

Proposition 5.2 *The number* 0 *is the unique multiplicative annihilator.*

The operations of multiplication and division are mutually inverse operations, in the sense of Eq. (5.4): each can be used to "undo" the other.

$$a = (a \times b) \div b = (a \div b) \times b$$

5.1.2.3 Exponentiation and taking logarithms

A conceptually powerful notational construct is the operation of *exponentiation*, i.e., *raising a number to a power*. For real numbers a and b, the bth *power* of a, denoted a^b, is defined by the system of equations

$$\text{(for all numbers } a, b, c) \quad a^b \cdot a^c = a^{b+c} \tag{5.5}$$

This deceptively simple definition has many consequences which we often take for granted.

- *Multiplication of exponentials is accomplished via addition within the exponents.*

- *For all numbers $a > 0$, the number $a^0 = 1$*

  This follows (via cancellation) from system (5.5) via the fact that

$$a^b \cdot a^0 = a^{b+0} = a^b = a^b \cdot 1$$

  Many texts provide an inductive definition of exponentiation whose base is the equation $a^0 = 1$. We see now that this base equation is actually a consequence of the inductive system (5.5) rather than an independent convention.

- *For all numbers $a > 0$, the number $a^{1/2}$ is the square root of a; i.e., $a^{1/2}$ is the (unique, via cancellation) number b such that $b^2 = a$. (Another common notation for the number $a^{1/2}$ is $\sqrt{a}$.)*

  This follows from system (5.5) via the fact that

$$a = a^1 = a^{(1/2)+(1/2)} = a^{1/2} \cdot a^{1/2} = \left(a^{1/2}\right)^2.$$

- *For all numbers $a > 0$ and b, the number a^{-b} is the multiplicative inverse of a^b, meaning that $a^b \times a^{-b} = 1$*

  This follows from system (5.5) via the fact that

$$a^b \cdot a^{-b} = a^{(b+(-b))} = a^0 = 1$$

When the power b is a positive integer, then definition (5.5) can be cast in the following attractive inductive form:

$$\text{for all numbers } a > 0 \qquad a^0 = 1$$

$$\text{for all numbers } a \text{ and integers } b \quad a^{b+1} = a \times a^b \tag{5.6}$$

We now have the background needed to manipulate and employ powers that are integral or fractional, positive, zero, or negative.

Explanatory note.

There are laws of arithmetic that deal with *imaginary*—hence, by extrapolation, also with *complex*—powers of numbers. These laws are traditionally beyond the scope of an introductory text on discrete mathematics, so we shall not develop this topic here.

That said, we will need to refer briefly to the topic a few times—when we cite the landmark Theorem 5.2, which exposes a wondrous relationship among the three fundamental constants e, π, and i—and when we briefly discuss the challenge of computing the roots of polynomials.

Just as addition has its inverse operation, subtraction, and multiplication has its inverse operation, division, the operation of exponentiation has its inverse operation, *taking logarithms*. Our main discussion of logarithms appears in Section 5.4; we just define the operation here.

For positive real numbers a and b, the operation of taking logarithms is defined by the following dual equation.

$$\log_b(b^a) = b^{\log_b a} = a$$

We shall note in Section 5.4 that most of the main properties of this operation can be inferred from properties of the operation of exponentiation.

Explanatory note.

This note points out one of the most beautiful patterns in mathematics.

We have remarked in several places about the importance of noticing patterns as we "do" mathematics. A pattern that leads to some beneficial insights has emerged in the current section. The pattern suggests that in certain senses:

- The operations of addition, multiplication, and exponentiation "behave" similarly to one another.

- The operations of subtraction, division, and taking logarithms "behave" similarly to one another.

- The operations addition and subtraction relate to one another
      in much the same way as
   the operations multiplication and division relate to one another
      and in much the same way as
   the operations exponentiation and taking logarithms relate to one another.

The pattern we have just exposed is summarized by the following suggestive figure.

$$
\begin{array}{ccc}
\text{addition} & \approx \text{multiplication} \approx & \text{exponentiation} \\
\updownarrow & \updownarrow & \updownarrow \\
\text{subtraction} \approx & \text{division} & \approx \text{taking logarithms}
\end{array}
$$

Of course, patterns are *de*scriptive rather than *pre*scriptive, so one must use them only as hints to be followed or as inspirations for further investigation, not as facts to be depended on.

5.1.2.4 Binomial coefficients and Pascal's triangle

We close our catalogue of arithmetic operations with a binary operation on the set $\mathbb{N} \times \mathbb{N}$[1] known as the *binomial coefficient*. The versatility of this operation—and the origin of the unexpected word "coefficient" in its name—will become clear throughout our text. We now lay the groundwork for the *many* applications of binomial coefficients.

Let n and $k \leq n$ be nonnegative integers (i.e., elements of $\mathbb{N}$). The *binomial coefficient* usually articulated as "*n choose k*" is a number m which is usually denoted by one of the following notations

$$
\binom{n}{k}, \quad C(n,k), \quad \Delta_{n,k}
$$

---

[1] In advanced contexts, one encounters binomial coefficients with arguments which are either nonintegral or nonpositive.

The number m denoted by these notations is defined by the following equation, which invokes the notation that we use most frequently in this text.

$$m = \binom{n}{k} \overset{\text{def}}{=} \frac{n!}{k!(n-k)!} = \frac{n(n-1)\cdots(n-k+1)}{\left(k(k-1)\cdots 1\right)\left((n-k)(n-k-1)\cdots 1\right)} \tag{5.7}$$

Many of the secrets of these wonderful numbers—including the fact that they are *integers*—can be deduced from the following results.

Proposition 5.3 *For all $n, k \in \mathbb{N}$ with $k \le n$:*
(a) *The symmetry rule:*

$$\binom{n}{k} = \binom{n}{n-k} \tag{5.8}$$

(b) *The addition rule:*

$$\binom{n}{k} + \binom{n}{k+1} = \binom{n+1}{k+1} \tag{5.9}$$

Proof. (**a**) We verify Eq. (5.8) by invoking the defining Eq. (5.7) plus the commutativity of multiplication (see Section 5.2):

$$\binom{n}{k} = \frac{n!}{k!(n-k)!}$$
$$= \frac{n!}{(n-k)!k!}$$
$$= \binom{n}{n-k}$$

(**b**) We verify Eq. (5.9) by adding the fractions exposed by Eq. (5.7):

$$\binom{n}{k} + \binom{n}{k+1} = \frac{n!}{k!(n-k)!} + \frac{n!}{(k+1)!(n-k-1)!}$$
$$= n! \cdot \frac{(k+1)+(n-k)}{(k+1)!(n-k)!}$$
$$= \frac{(n+1)!}{(k+1)!(n-k)!}$$
$$= \binom{n+1}{k+1}$$

We shall encounter a quite different verification of the addition rule in Chapter 11, by exploiting a fundamental combinatorial aspect of binomial coefficients, namely, the fact that $\binom{n}{k}$ is the number of ways of picking k objects out of a set of n objects. □

We shall have a lot to say about binomial coefficients throughout the text.

- In Section 5.3, binomial coefficients take center stage in the proof of Newton's *Binomial Theorem*; the name "binomial coefficient" emerges from this use of the operation.

- In Section 6.2.1, binomial coefficients play a prominent role in evaluating arithmetic summations.

- In Section 9.2, binomial coefficients provide an important example of *bilinear* recurrences and their computational importance.

- In Chapter 11, we see how binomial coefficients are indispensable when studying a great many topics related to *counting*; some examples:

  - What are the relative likelihoods of various five-card deals from an unbiased 52-card deck?

  - What is the likelihood of observing 15 HEADs and 25 TAILs in 40 flips of an unbiased coin?

  - What are the comparative operation-count costs of Merge-Sort and Quick-Sort when sorting n keys; cf. [36]?

One of the most common ways of presenting binomial coefficients is by means of *Pascal's Triangle*, an array of integers which is named in honor of the French polymath Blaise Pascal. The contents of this array are determined by the laws enunciated in Proposition 5.3. Since we shall devote ample attention in subsequent chapters to studying this array and its manifold applications, we merely present in Fig. 5.1 a "prefix" of this famed array, for the range of parameters $n, k \leq 9$. The *formation*

$\binom{n}{k}$	$k=0$	$k=1$	$k=2$	$k=3$	$k=4$	$k=5$	$k=6$	$k=7$	$k=8$	$k=9$	$\ldots$
$n=1$	1	1									$\ldots$
$n=2$	1	2	1								$\ldots$
$n=3$	1	3	3	1							$\ldots$
$n=4$	1	4	6	4	1						$\ldots$
$n=5$	1	5	10	10	5	1					$\ldots$
$n=6$	1	6	15	20	15	6	1				$\ldots$
$n=7$	1	7	21	35	35	21	7	1			$\ldots$
$n=8$	1	8	28	56	70	56	28	8	1		$\ldots$
$n=9$	1	9	36	84	126	126	84	36	9	1	$\ldots$
$\vdots$	$\vdots$	$\vdots$	$\vdots$	$\vdots$	$\vdots$	$\vdots$	$\vdots$	$\vdots$	$\vdots$	$\vdots$	$\ddots$

Fig. 5.1 A "prefix" of Pascal's Triangle, for $n, k \leq 9$

rule of Pascal's Triangle is that the array-entry at (row $n+1$, column $k+1$) is the sum of the array-entries at (row n, column k) and at (row n, column $k+1$).

> **Historical note.**
>
> The array in Fig. 5.1 is usually attributed to Blaise Pascal in the West, with a date in the seventeenth century. In fact, the array appears in *Siyuan Yujian* [118], a Chinese book from the early fourteenth century; the array is there attributed to Jia Xian in the eleventh century.

5.1.3 Rational Arithmetic: A Computational Exercise

In Section 4.4 we defined the rational numbers $\mathbb{Q}$ and reviewed why they were needed to compensate for the general absence of multiplicative inverses within the set $\mathbb{Z}$. But, our discussion did not reveal how to perform arithmetic within $\mathbb{Q}$. We make up for this shortcoming now. Of course, the reader will have encountered arithmetic on rationals long ago—but we are now reviewing the topic for two reasons: (1) We want to reinforce the systematic nature of the progression from the algebra of integers coupled with their arithmetic to the extended algebra of rationals coupled with their arithmetic. (2) We want to provide the reader with a set of valuable exercises to reinforce the mathematical thinking skills whose presentation is our main goal.

The basic arithmetic operations on the rational numbers obey rules that are adapted from the corresponding rules for integers. (This is why we were justified in referring to the algebra of rationals as an "*extension*" of the algebra of integers.). For all ratios p/q and r/s in $\mathbb{Q}$, these rules take the following form.

$$\text{Addition:} \qquad \frac{p}{q} + \frac{r}{s} = \frac{p \cdot s + r \cdot q}{q \cdot s}$$

$$\text{Subtraction:} \qquad \frac{p}{q} - \frac{r}{s} = \frac{p}{q} + \frac{(-r)}{s}$$

$$\text{Multiplication:} \quad \frac{p}{q} \cdot \frac{r}{s} = \frac{p \cdot r}{q \cdot s}$$

$$\text{Division:} \qquad \frac{p}{q} \div \frac{r}{s} = \frac{p}{q} \cdot \frac{s}{r}$$

An essential component of the culture of mathematics is to ensure that the preceding definitions do supply an *orderly* extension of the corresponding rules for integer arithmetic, as expounded in Section 5.2. We leave to the reader the valuable exercise of verifying, in particular, that rational arithmetic:

- works correctly when the argument rational numbers are, in fact, integers, i.e., when $q = s = 1$ in the preceding table;

- treats the number 0 appropriately, i.e., as an additive identity and a multiplicative annihilator;

- treats the number 1 appropriately, i.e., as a multiplicative identity;
- obeys the laws outlined in Section 5.2.

Verifying the distributivity of rational multiplication over rational addition is an especially valuable exercise because of the amount of manipulation that the verification demands.

5.2 The Laws of Arithmetic, with Applications

This section is devoted to enumerating the basic laws of arithmetic on the integers, rationals, and reals. Anyone seeking to "do" mathematics should be able to employ these laws cogently in rigorous argumentation.

5.2.1 The Commutative, Associative, and Distributive Laws

(i) The commutative law. For all numbers x and y:

$$\begin{array}{ll} \text{for addition:} & x+y \;=\; y+x \\ \text{for multiplication:} & x \times y \;=\; y \times x \end{array}$$

Commutativity of addition (resp., of multiplication) enables one to add (resp., to multiply) sequences of numbers in any order.

(ii) The associative law. For all numbers x, y, and z:

$$\begin{array}{ll} \text{for addition:} & (x+y)+z \;=\; x+(y+z) \\ \text{for multiplication:} & (x \times y) \times z \;=\; x \times (y \times z) \end{array}$$

Associativity of addition (resp., of multiplication) enables one to write sequences of additions (resp., of multiplications) without using parentheses for grouping.

(iii) The distributive law. For all numbers x, y, and z,

$$x \times (y+z) \;=\; (x \times y) + (x \times z) \tag{5.10}$$

One commonly articulates this law as, "*Multiplication distributes over addition.*"

One of the most common uses of the distributive law reads Eq. (5.10) "backwards", thereby deriving a formula for *factoring* structurally complicated expressions that involve both addition and multiplication.

Easily, addition does *not* distribute over multiplication; i.e., in general, $x+y \cdot z \neq (x+y) \cdot (x+z)$. Hence, when we see the expression

$$x+y \cdot z$$

we know that the multiplication is performed before the addition. This rule is commonly expressed via the mandate

Multiplication takes priority over addition.

This prioritization enables us to write the right-hand side of Eq. (5.10) without parentheses, as in

$$x \cdot (y + z) = x \cdot y + x \cdot z$$

By invoking the preceding laws multiple times, we can derive a recipe for multiplying complicated arithmetic expressions. We illustrate this via the "simplest" complicated expression, $(a+b) \cdot (c+d)$.

Proposition 5.4 *For all numbers a, b, c, d:*[2]

$$(a+b) \cdot (c+d) = a \cdot c + a \cdot d + b \cdot c + b \cdot d \qquad (5.11)$$

Proof. We perform a sequence of operations on the left-hand expression of Eq. (5.11), justifying each by an invocation of the commutative and associative laws.

$$
\begin{aligned}
(a+b) \cdot (c+d) &= (a+b) \cdot c + (a+b) \cdot d &&\text{distributive law} \\
&= c \cdot (a+b) + d \cdot (a+b) &&\text{commutativity of multiplication } (2\times) \\
&= c \cdot a + c \cdot b + d \cdot a + d \cdot b &&\text{distributive law } (2\times) \\
&= a \cdot c + b \cdot c + a \cdot d + b \cdot d &&\text{commutativity of multiplication } (4\times) \\
&= a \cdot c + a \cdot d + b \cdot c + b \cdot d &&\text{commutativity of addition}
\end{aligned}
$$

We thus finally derive the right-hand expression of Eq. (5.11). □

We close our short survey of the laws of arithmetic with the following important two-part law.

- *The law of inverses.*

  - Every number x has an *additive inverse*, i.e., a number y such that $x + y = 0$. This inverse is x's *negative*, $-x$.

  - Every *nonzero* number $x \neq 0$ has a *multiplicative inverse*, i.e., a number y such that $x \cdot y = 1$. This inverse is x's *reciprocal*, $1/x$.

5.3 Polynomials and Their Roots

Among the most basic functions are *polynomials*. A polynomial on the *k variables* $v_1, v_2, \ldots, v_k$ with *coefficients* in the set S is any function that can be written as a sum of *monomials*. A monomial on the *k* variables $v_1, v_2, \ldots, v_k$ whose coefficient is in the set S is a function that can be written in the following form:

---

[2] Because multiplication has priority over addition, the absence of parentheses in the right-hand side of Eq. (5.11) cannot introduce ambiguity.

$$a \times v_1^{c_1} \times v_2^{c_2} \times \cdots \times v_k^{c_k} \tag{5.12}$$

In this expression:

- The coefficient a belongs to the set S.

- The set S is typically a set of numbers, although it could in principle be a set of objects of another type, such as functions from a prespecified family.

  Exemplifying the typical case:

  - When $S = \mathbb{N}$, we have "a polynomial/monomial with *integer coefficients*".

  - When $S = \mathbb{Q}$, we have "a polynomial/monomial with *rational coefficients*".

- The exponents c_i of the variables are *nonnegative integers* that are called the *powers* to which the variables are raised.

For illustration, here is a specific *bivariate*—i.e., two-variable—polynomial with integer coefficients; we denote its two variables x_1 and x_2. (By convention a variable does not appear when its power or its coefficient is 0.)

$$3x_1^7 \cdot x_2^{19} + x_1^3 \cdot x_2 + 45x_2^9 + 7x_1^{11} \cdot x_2^5$$

Another way to look at polynomials is as the class of functions that are formed from *variables* and *numbers* using the basic algebraic operations: addition/subtraction and multiplication/division.

Polynomials on a single variable—the case $k = 1$—are said to be *univariate* (a Latinate form of "having a single variable"). Polynomials on two variables—the case $k = 2$—are said to be *bivariate* (a Latinate form of "having two variables"). These simplest polynomials are usually the only ones graced by Latinate nicknames. Our interest in this section will mostly be in univariate polynomials (Section 5.3.1); we also spend some time on one particular family of bivariate polynomials which has many important applications (Section 5.3.2); and we finally describe informally an advanced result on general multivariate polynomials which has truly foundational importance to the underpinnings of the activity of computing (Section 5.3.3).

A problem of particular interest when discussing a polynomial $P(x_1, x_2, \ldots, x_k)$ is to discover vectors of values of the variables $\{x_i\}$ that cause P to *vanish*, i.e., to evaluate to 0. Each such vector $\langle r_1, r_2, \ldots, r_k \rangle$ is called a *root* of P. As a very simple example, the roots of the polynomial

$$P(x, y) = x^2 - 2xy + y^2$$

are precisely the two-place vectors (i.e., ordered pairs) $\langle k, k \rangle$ whose first and second components are identical, $x = y$.

The problem of finding the roots of given polynomials has garnered tremendous attention for centuries, for both its practical applications and its theoretical implications. Historically, two of the major problems regarding roots of polynomials are:

- *Find all roots of a given polynomial.*

  Most of the studies of this problem are found in algorithmic settings—courses, books, software. Yet, there are many valuable mathematical lessons to be learned under the aegis of this problem. We discuss this problem for univariate polynomials in Section 5.3.1.

- *Determine whether a given multivariate polynomial has any integer roots.*

  This problem is often found under the name *Diophantine analysis*, so named in honor of the Greek mathematician Diophantus of Alexandria, who has been called the *father of algebra* for his seminal studies of the process of equation-solving. By its very nature—namely, seeking a "YES"/"NO" answer, rather than an actual root-finding procedure—this problem has been studied largely within theoretical or mathematical settings. While most work on this subject uses advanced techniques and hence is beyond the scope of the current text, there is one result in this domain whose underlying message is so profound that we at least tell its "story", in Section 5.3.3.

Our concern with the roots of polynomials, as well as with the functions themselves, forces us to expand our horizon to consider a new classification of numbers and functions. This new classification goes by the name *algebraic*; it encompasses number and functions that arise in the solutions of equations of the form $P(x) = 0$, where P is a polynomial. This expanded view mandates that we be prepared to discuss and analyze "polynomials" whose variables have fractional exponents, i.e., whose variables are raised to fractional powers.

5.3.1 Univariate Polynomials and Their Roots

We focus throughout this section on univariate polynomials with complex coefficients, i.e., coefficients from $\mathbb{C}$. We are concerned with using mathematics to elucidate *the structure* of the roots of a given univariate polynomial; we leave to our algorithmic colleagues the practicalities of actually computing the roots. We discuss two topics which have strong lessons for the endeavor of doing mathematics. In Section 5.3.1.1, we discuss univariate polynomials of arbitrary (maximum) degree d. We derive a number of results about the d complex roots that every degree-d polynomial has; we highlight the *Fundamental Theorem of Algebra*, the storied result that verifies the existence of these d roots. In Section 5.3.1.2, we focus on the quest for "simple" formulas for roots, in terms of algebraic operations and *radicals*—i.e., surds such as square roots, cube roots, etc. (Of course, "radicals" are just another framework for discussing fractional powers.) We will discover that such formulas are readily calculated for degree-2 (*quadratic*) polynomials, arduously calculated for degree-3 (*cubic*) polynomials, computable only in principle for degree-4 (*quartic*) polynomials—and generally *nonexistent* for polynomials of degree 5 (*quintic*) and higher.

5.3.1.1 The d roots of a degree-d polynomial

When we discussed the history of our number system, in Section 4.2, we remarked that each step in the progression from the integers ($\mathbb{Z}$) to the rational numbers ($\mathbb{Q}$), to the real numbers ($\mathbb{R}$), and finally to the complex numbers ($\mathbb{C}$) was motivated by a perceived deficiency in the number system up to that point. We then commented that the complex numbers were the culmination of this process, in that they were *algebraically complete*. We promised at that time to clarify the meaning of the term "algebraically complete"—and the time has come to do that. The "algebraic completeness" of the complex numbers resides in the fact that:

Polynomials with coefficients in $\mathbb{C}$ can always be solved, via roots in $\mathbb{C}$.

The common way of stating this *practically and intellectually powerful* property is via the storied theorem known as *The Fundamental Theorem of Algebra*.

Theorem 5.1 (The Fundamental Theorem of Algebra). *Every degree-d univariate polynomial with complex coefficients has d complex roots.*

Explanatory note.

The tally of roots in Theorem 5.1 counts multiple roots according to their multiplicities. It does *not* promise d *distinct* roots.

For instance, the polynomial

$$P(x) = x^2 - 2x + 1 = (x-1)^2$$

does, indeed, have *two* roots, but these roots *are equal*.

Despite its rather simple form, Theorem 5.1 resisted formal proof for literally *centuries*, resisting the proof attempts of mathematical luminaries such as Fermat, Euler, Lagrange, and Laplace. The theorem was finally proved in the early nineteenth century by the amateur(!) French mathematician Jean-Robert Argand. Argand publicized his proof in 1806 [9], but the proof became well known only when it appeared in the famous *Cours d'analyse* [28] of Augustin-Louis Cauchy.

A few cultural/historical comments are in order.

- The many failed attempts at proving Theorem 5.1 should not be viewed as casting shadows over the luster of any of the greats who failed. In most of the cases, the failures pointed out new mathematics that had yet to be developed. Such is the trajectory of mathematics and the sciences!

- Despite the word "fundamental" in the name of Theorem 5.1, the result is no longer viewed as *the* fundamental theorem of algebra. The field of algebra has grown in a variety of directions in the past two centuries, so the *Theory of Equations*, which was largely coextensive with the field of algebra into the nineteenth century, is presently just one branch of the field.

- Despite the word "algebra" in the name of Theorem 5.1, there does not yet exist a purely *algebraic* proof of the Theorem: Some input from other areas of mathematics enters every known proof in some essential way.

- The complete march toward Theorem 5.1 took literally millennia. Among the luminaries who made major contributions along the way were the following pioneers in the Theory of (Solving) Equations:

  - *Diophantus of Alexandria*. He authored a series of books known collectively as *Arithmetica*, which expounded the basics of a theory of solving equations. These books earned Diophantus the name "father of algebra".

  - *Muhammad ibn Mūsā al-Khwārizmī*. Better known as *al-Khwārizmī*, this Persian mathematician and scholar of the ninth century lent his name to the modern term *algorithm*. His extensive writings, especially the book [2], introduced into Europe both the Hindu-Arabic numerals which we all use in everyday discourse and the elements of what was known of the field of algebra in the ninth century—particularly the solution of equations.

  > **Historical/cultural note.**
  >
  > The important role of the Middle East in the development of mathematics is testified to eloquently by the origin of the word "algebra". According to the *Oxford English Dictionary*, this word comes from the Arabic "*al-jabr*", which literally means "reunion of broken parts"—an allusion to the manipulation of terms in algebraic computations.

  - *René Descartes*. The seventeenth-century mathematician and philosopher Descartes is credited with establishing the algebraic notation that we use to this day. Readers should not minimize the importance of notation until they have tried to perform arithmetic using Roman numerals! At a more abstract level, Descartes's invention of *Analytical Geometry* enabled the use of geometric concepts and techniques in algebra. Both of these contributions were of incalculable value in the history leading to Theorem 5.1.

The earliest proofs of Theorem 5.1 were not *constructive*—they did not provide a roadmap for actually finding the roots of a given polynomial. There currently do exist proofs of the Theorem that are constructive, in the following sense. Given a degree-d polynomial $P(x)$, these proofs determine a disk in two-dimensional space which contains all d roots of $P(x)$. How might such a disk emerge from a mathematical argument? Here is a *very simple* illustration which focuses on a *very special* type of polynomial.

Let us be given a degree-d polynomial with real coefficients:

$$Q(x) = a_d x^d + a_{d-1} x^{d-1} + a_{d-2} x^{d-2} + \cdots + a_1 x + a_0$$

Since we are interested only in *the roots* of $Q(x)$, we lose no generality by focusing only on polynomials that are *monic*, i.e., that have leading coefficient $a_d = 1$. When the polynomial $Q(x)$ whose roots we are seeking is not monic—i.e., when $a_d \neq 1$—then we replace $Q(x)$ by the monic polynomial

$$P_Q(x) = x^d + \frac{a_{d-1}}{a_d}x^{d-1} + \frac{a_{d-2}}{a_d}x^{d-2} + \cdots + \frac{a_1}{a_d}x + \frac{a_0}{a_d}$$

which has the same roots as $Q(x)$. (*The reader should verify this assertion!*)

When the polynomial

$$P(x) = x^d + a_{d-1}x^{d-1} + a_{d-2}x^{d-2} + \cdots + a_1 x + a_0$$

whose roots we are seeking *is* monic, we can rewrite $P(x)$ in the following form:

$$P(x) = x^d \cdot \left(1 + \frac{a_{d-1}}{x} + \frac{a_{d-2}}{x^2} + \cdots + \frac{a_1}{x^{d-1}} + \frac{a_0}{x^d} \right)$$

The benefit of this rewriting is that it exposes the following inequality.

$$P(x) > 0 \quad \text{whenever} \quad x > (|a_{d-1}| + |a_{d-2}| + \cdots + |a_1| + |a_0|) \qquad (5.13)$$

And this inequality implies that all real roots of $P(x)$, if there are any, lie within the region

$$x \leq (|a_{d-1}| + |a_{d-2}| + \cdots + |a_1| + |a_0|)$$

This observation can be used to develop an efficient algorithm for finding all of P's roots—but this is outside the scope of this text.

This is just one very simple example, but it does illustrate that one can sometimes delimit bounded regions within which all of $P(x)$'s roots must lie. Numerous texts (and research papers) deal with the general problem of computing the roots of polynomials; see, e.g., [81].

5.3.1.2 Solving polynomials by radicals

The problem of *solving* arbitrary degree-d polynomials—i.e., of discovering their d roots, as promised by Theorem 5.1—is computationally very complex, even for moderately low degrees. (This assertion can be made mathematically precise, but the required notions are beyond the scope of this text.) For univariate polynomials of low degree, there do exist computationally feasible root-finding algorithms. Indeed, for polynomials of *very* low degree—specifically, degrees 2, 3, and 4—there actually exist "simple" *formulas* that specify the polynomial's roots. The word "simple" is used in a technical sense here: it refers to a formula that can be constructed using the following algebraic operations: adding/subtracting two quantities, multiplying/dividing two quantities, and raising a quantity to a rational power. Because the last of these operations is often expressed by using a *radical sign*, rather than an exponent—as when we write $\sqrt{x}$ for $x^{1/2}$—these formulas are often referred to as *solutions by radicals*.

The remainder of this section is devoted to deriving the *quadratic* formula—the one that specifies the roots of any degree-2 *(quadratic)* polynomial—and the *cubic* formula—the one that specifies the roots of any degree-3 *(cubic)* polynomial. We shall observe that the cubic formula is so onerous calculationally that it is seldom actually written out; it is instead specified *algorithmically*. The *quartic* formula— the one that specifies the roots of any degree-4 *(quartic)* polynomial—is so complex that it is virtually never written out. The courageous reader can attack the quartic formula as an exercise, using the conceptual techniques which we derive here.

It is useless to try to solve polynomials of degrees > 4 by radicals. In the early nineteenth century, a (mathematically brilliant, for sure!) French teenager, Evariste Galois, proved that *one cannot solve the degree-5 (quintic) polynomial by radicals, no matter how much abstruse computation one is willing to do!* Galois achieved this result by developing a mathematical theory which has since been named for him. Tragically for the world of mathematics, Galois was killed in a duel just two years after announcing his theory via a memoir submitted to the Paris Academy of Sciences.

On to solving low-degree polynomials:

A. Solving quadratic polynomials by radicals

We now derive the *quadratic formula*, which solves an arbitrary quadratic polynomial with real coefficients:

$$P(x) = ax^2 + bx + c \quad \text{where } b, c \in \mathbb{R};\ a \in \mathbb{R} \setminus \{0\} \tag{5.14}$$

While the formula and its derivation are specialized to the structure of quadratic polynomials, several aspects of the derivation can be extrapolated to polynomials of higher degree. The formula that we derive is announced in the following proposition.

Proposition 5.5 *The two roots, x_1 and x_2, of the generic quadratic polynomial (5.14) are:*

$$x_1 = \frac{-b + \sqrt{b^2 - 4ac}}{2a}$$

$$\tag{5.15}$$

$$x_2 = \frac{-b - \sqrt{b^2 - 4ac}}{2a}$$

Proof. We find the roots of $P(x)$ by solving the polynomial equation $P(x) = 0$. We simplify our task by dividing both sides of this equation by a; easily, this does not impact the two solutions we seek, since it merely replaces $P(x)$ by a monic polynomial that shares the same roots. We thereby reduce the root-finding problem to the solution of the equation

$$x^2 + \frac{b}{a}x = -\frac{c}{a} \tag{5.16}$$

The technique of *completing the square* gives us an easy path toward solving this equation. This technique involves adding to both sides of the equation a variable-free expression E that turns the left-hand expression into a perfect square. In the case of Eq. (5.16), the expression

$$E = \frac{b^2}{4a^2}$$

does the job, because

$$x^2 + \frac{b}{a}x + E = x^2 + \frac{b}{a}x + \frac{b^2}{4a^2} = \left(x + \frac{b}{2a}\right)^2$$

We have thereby converted Eq. (5.16) to the equation

$$\left(x + \frac{b}{2a}\right)^2 = \frac{b^2 - 4ac}{4a^2}. \tag{5.17}$$

Elementary calculation on Eq. (5.17) identifies $P(x)$'s two roots as the values x_1 and x_2 specified in Eq. (5.15). □

We close our discussion of quadratic polynomials with a few comments that may inspire some readers to read further about roots of quadratic polynomials.

1. Using a common shorthand, the expressions for x_1 and x_2 in Eq. (5.15) are often abbreviated within a single expression, by using the operator $\pm$, which is a shorthand for "plus-or-minus". The quadratic formula can then be written in the following familiar form:

$$x = \frac{1}{2a}\left(-b \pm \sqrt{b^2 - 4ac}\right)$$

2. The reader can verify that our proof essentially proceeds by replacing $P(x)$'s variable x with the variable

$$u = x + \frac{b}{2a}$$

This replacement streamlines the process of completing the square and finding the solutions x_1 and x_2. We presented a more elementary version of the proof so that the reader could watch the solution process proceed step by step. As we turn now to the clerically more complicated solution of the cubic polynomial, we get around some of the calculational complexity by employing the variable-substitution stratagem.

3. When we look carefully at the roots x_1 and x_2 in Eq. (5.15), we note that there are three genres of solutions. These can be discriminated in the graphical form of $P(x)$, as illustrated in Fig. 5.2. (Note that the *horizontal* displacement of $P(x)$'s plot is irrelevant; it is only the *vertical* displacement which affects the nature of $P(x)$'s roots.)

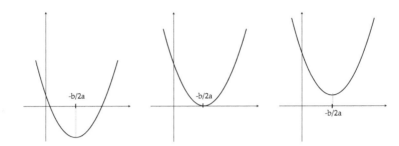

Fig. 5.2 $P(x)$ has: [left] two distinct real roots; [center] two equal real roots; [right] no real root (both roots are complex)

- The solutions x_1 and x_2 are distinct real numbers.

  This corresponds to the left-hand plot in the figure: $P(x)$ crosses the (real) X-axis at two distinct points.

- The solutions x_1 and x_2 are identical real numbers.

  This corresponds to the central plot in the figure: $P(x)$ crosses/touches the (real) X-axis at precisely one point.

- The solutions x_1 and x_2 are distinct *non*-real numbers.

  This corresponds to the right-hand plot in the figure: $P(x)$ never touches the (real) X-axis.

B. Solving cubic polynomials by radicals

We derive a formula for the roots of the general cubic polynomial with real coefficients:

$$P(x) \,=\, ax^3 + bx^2 + cx + d \quad \text{where } b,c,d \in \mathbb{R};\ a \in \mathbb{R} \setminus \{0\} \qquad (5.18)$$

Although the so-called *cubic formula*, which we derive now, is daunting in form, its proof and derivation should be quite accessible because they follow the same flow as the clerically much simpler proof and derivation of the quadratic formula. Because the cubic formula is so complex in form, we present the derivation step by step *before* we present the formula.

Proof. (Derivation of the cubic formula)

Step 1. Convert $P(x)$ to a *monic* cubic polynomial $P^{(1)}(x)$ which shares $P(x)$'s roots.

We accomplish this via a change of coefficients that rewrites $P(x)$ as the monic cubic polynomial

$$P^{(1)}(x) = x^3 + Bx^2 + Cx + D \tag{5.19}$$

The required change of coefficients is specified implicitly as follows

$$B = \frac{b}{a}; \quad C = \frac{c}{a}; \quad D = \frac{d}{a}$$

Step 2. Convert $P^{(1)}$ to a *reduced form* cubic polynomial $P^{(2)}$ which shares P's roots. By "reduced", we mean that $P^{(2)}$ has no quadratic term—i.e., no term in which the variable is raised to the power 2.

We accomplish this transformation as follows. We make the following change of variable in Eq. (5.19).

$$y = x + \frac{B}{3} \tag{5.20}$$

We thereby convert $P^{(1)}(x)$, a polynomial in variable x, into the following polynomial in variable y:

$$
\begin{aligned}
P^{(2)}(y) &= \left(y - \frac{B}{3}\right)^3 + B\left(y - \frac{B}{3}\right)^2 + C\left(y - \frac{B}{3}\right) + D \\
&= y^3 + \left(\frac{B^2}{3} - \frac{2B^2}{3} + C\right)y + \left(\frac{2B^3}{27} - \frac{BC}{3} + D\right)
\end{aligned} \tag{5.21}
$$

For simplicity, we rewrite $P^{(2)}(y)$, which clearly is in reduced form, as

$$P^{(2)}(y) = y^3 + Ey + F \tag{5.22}$$

where

$$E = -\frac{B^2}{3} + C$$

$$F = \frac{2B^3}{27} - \frac{BC}{3} + D$$

Step 3. Convert $P^{(2)}(y)$ to its *associated* quadratic polynomial.

We accomplish this step by applying a transformation that is attributed to the sixteenth-century French mathematician François Viète (often referred to by his Latinized name, Franciscus Vieta); see [56]. Vieta's transformation converts $P^{(2)}(y)$ into a quadratic polynomial by means of the following variable-substitution in Eq. (5.22):

$$y = z - \frac{E}{3z} \tag{5.23}$$

We thereby obtain (after calculations involving several cancellations) an expression

$$P^{(3)}(z) = \left(z - \frac{E}{3z}\right)^3 + E\left(z - \frac{E}{3z}\right) + F$$

$$= z^3 - \frac{E^3}{27z^3} + F \tag{5.24}$$

Clearly, $P^{(3)}(z)$ is not a polynomial in z, because of the term in which variable z appears in the denominator. But it is a valuable stepping stone because the function $P^{(3)}(z)$ vanishes—i.e., $P^{(3)}(z) = 0$—precisely when the following polynomial (in the "variable" z^3) vanishes.

$$P^{(4)}(z^3) = (z^3)^2 + (z^3)F - \frac{E^3}{27}$$

We wrote both instances of z^3 in the expression for $P^{(4)}(z^3)$ within parentheses to facilitate the view of $P^{(4)}(z^3)$ as a (quadratic) polynomial in the "variable" z^3. The quadratic formula (5.15) provides us with two roots for $P^{(4)}$, which we express as the following two values for z^3 (abbreviated via the shorthand operator $\pm$).

$$(z^3) = -\frac{F}{2} \pm \sqrt{\frac{F^2}{4} + \frac{E^3}{27}} \tag{5.25}$$

We can now derive all solutions for the variable z in Eq. (5.25) via back-substitution in transformation (5.23). But completing this derivation requires a bit of background.

Theorem 5.1 assures us that the polynomial $P^{(2)}(z)$ has three roots. In order to compute these roots, we invoke a truly remarkable result that is known as *Euler's Formula*, in honor of its discoverer, the much-traveled eighteenth-century mathematician Leonhard Euler. This result/formula exposes a fundamental relationship among:

- the imaginary unit i

- the ratio of the circumference of a circle to its radius, $\pi = 3.141592653\cdots$

- the base of natural logarithms, Euler's constant $e = 2.718281828\cdots$

Theorem 5.2 (Euler's formula).

$$e^{i\pi} = -1$$

Back to Eq. (5.25): Theorem 5.1 tells us that within the complex number system $\mathbb{C}$, the cubic polynomial

$$u^3 - 1$$

has three distinct roots. These numbers are known as the *primitive third roots of unity* and are traditionally denoted ω^0, ω^1, and ω^2. Using Theorem 5.2, we can provide explicit values for these numbers:

$$\omega^0 = 1; \quad \omega^1 = e^{2i\pi/3}; \quad \omega^2 = e^{4i\pi/3}$$

Aside: There are analogues of these numbers for any parameter n, not just for $n = 3$. For general n, the *primitive nth roots of unity* comprise the set

$$\{e^{2ki\pi/n} \mid k = 0, 1, \ldots n - 1\}$$

When we unite the abbreviated double equation (5.25) for z^3 with Euler's formula (Theorem 5.2), we discover *six* solutions for the variable z, namely,

$$z_1 = \omega^0 \cdot \left(-\frac{F}{2} + \sqrt{\frac{F^2}{4} + \frac{E^3}{27}}\right)^{1/3} \quad z_2 = \omega^0 \cdot \left(-\frac{F}{2} - \sqrt{\frac{F^2}{4} + \frac{E^3}{27}}\right)^{1/3}$$

$$z_3 = \omega^1 \cdot \left(-\frac{F}{2} + \sqrt{\frac{F^2}{4} + \frac{E^3}{27}}\right)^{1/3} \quad z_4 = \omega^1 \cdot \left(-\frac{F}{2} - \sqrt{\frac{F^2}{4} + \frac{E^3}{27}}\right)^{1/3} \quad (5.26)$$

$$z_5 = \omega^2 \cdot \left(-\frac{F}{2} + \sqrt{\frac{F^2}{4} + \frac{E^3}{27}}\right)^{1/3} \quad z_6 = \omega^2 \cdot \left(-\frac{F}{2} - \sqrt{\frac{F^2}{4} + \frac{E^3}{27}}\right)^{1/3}$$

The algorithmically interesting portion of the process of solving cubics by radicals is now complete. The remainder of the process consists of "reversing" the two transformations, (5.23) and (5.20), which have taken us from the problem of solving a polynomial in x to the problem of solving a polynomial in z. The calculations that embody this reverse transformation are onerous when solved symbolically, so we make do with some exercises in which the reader will solve numerical instances. The most interesting and noteworthy feature of these exercises will be the observation of "collapsing" of intermediate expressions, whose impact is to leave us with only *three* solutions for x—i.e., the number promised by Theorem 5.1—rather than the six solutions that the array (5.26) of z-values would lead us to expect.

While the promise of a visually appealing cubic analogue of the quadratic formula (5.15) is appealing, the actual cubic formula is so complex visually that it offers no important insights. The curious reader can find renditions of the formula on the web. $\square$

5.3.2 Bivariate Polynomials: The Binomial Theorem

A polynomial P is *bivariate* if each of its summand monomials has the form

$$a \cdot x^b \cdot y^c$$

In this expression: x and y are the monomial's two variables (*two* because the polynomial is *bivariate*); a is the monomial's coefficient; b and c are the respective powers of variables x and y in this monomial.

Probably the simplest bivariate polynomials are the ones in the following family.

$$\text{For } n \in \mathbb{N}^+, \quad P_n(x,y) \stackrel{\text{def}}{=} (x+y)^n \tag{5.27}$$

There are lessons to be learned from studying the structure of the polynomials in this family, so let us begin to expand the polynomials using the arithmetic techniques we have learned earlier.

$$
\begin{aligned}
P_1(x,y) &= (x+y)^1 = x+y \\
P_2(x,y) &= (x+y)^2 = (x+y)\cdot(x+y) \\
&= x^2 + 2xy + y^2 \\
P_3(x,y) &= (x+y)^3 = (x+y)\cdot(x^2 + 2xy + y^2) \\
&= (x^3 + 2x^2y + xy^2) + (x^2y + 2xy^2 + y^3) \\
&= x^3 + 3x^2y + 3xy^2 + y^3
\end{aligned}
$$

Let us take a moment to review what we are observing. We have remarked before that doing mathematics can sometimes involve a wonderfully exciting—and quite sophisticated—pattern-matching game. So, let us pattern-match!

1. The sequence of coefficients of the expanded $P_1(x,y)$ is $\langle 1,1 \rangle$.

2. The sequence of coefficients of the expanded $P_2(x,y)$ is $\langle 1,2,1 \rangle$.

3. The sequence of coefficients of the expanded $P_3(x,y)$ is $\langle 1,3,3,1 \rangle$.

There is a familiar pattern emerging here. Can you spot it? Where have we seen a pattern of tuples that begins in the same manner? As a rather broad hint, look at Fig. 5.1! Could the coefficients of each P_n possibly be the successive binomial coefficients:

$$\binom{n}{0}, \ \binom{n}{1}, \ \dots, \ \binom{n}{n-1}, \ \binom{n}{n}$$

Let us use induction to explore this possibility by expanding a generic P_n with symbolic "dummy" coefficients and see what this says about P_{n+1}. To this end, let $a_{n,n-r}$ denote the coefficient of $x^{n-r}y^r$ in the expansion of $P_n(x,y)$. Using our symbolic coefficients $a_{n,k}$, we have:

$$
\begin{aligned}
P_n(x,y) = x^n &+ a_{n,n-1}x^{n-1}y + \cdots + \\
a_{n,n-r}x^{n-r}y^r &+ a_{n,n-r-1}x^{n-r-1}y^{r+1} + a_{n,n-r-2}x^{n-r-2}y^{r+2} \\
+\cdots &+ a_{n,1}xy^{n-1} + y^n
\end{aligned}
$$

Continuing with this symbolic evaluation, we have:

$$
\begin{aligned}
x\cdot P_n(x,y) = x^{n+1} &+ a_{n,n-1}x^n y + \cdots + \\
a_{n,n-r}x^{n-r+1}y^r &+ a_{n,n-r-1}x^{n-r}y^{r+1} + a_{n,n-r-2}x^{n-r-1}y^{r+2} \\
+\cdots &+ a_{n,1}x^2 y^{n-1} + xy^n
\end{aligned} \tag{5.28}
$$

and

$$y \cdot P_n(x,y) = x^n y + a_{n,n-1}x^{n-1}y^2 + \cdots +$$
$$a_{n,n-r}x^{n-r}y^{r+1} + a_{n,n-r-1}x^{n-r-1}y^{r+2} + a_{n,n-r-2}x^{n-r-2}y^{r+3}$$
$$+ \cdots + a_{n,1}x^2 y^n + y^{n+1} \tag{5.29}$$

Because

$$P_{n+1}(x+y) = (x+y) \cdot P_n(x,y) = x \cdot P_n(x,y) + y \cdot P_n(x,y)$$

the symbolic coefficient $a_{n+1,n-r+1}$ of $x^{n-r+1}y^r$ in $P_{n+1}(x+y)$ is the sum of the following symbolic coefficients in $P_n(x,y)$:

- the coefficient $a_{n,n-r}$ of $x^{n-r}y^r$
- the coefficient $a_{n,n-r+1}$ of $x^{n-r+1}y^{r-1}$

By induction, then, for all $n, r \in \mathbb{N}$ with $r \leq n$, we have

$$a_{n,r} + a_{n,r+1} = a_{n+1,r+1}$$

Combining this equation with the observed initial conditions yields

$$a_{1,0} = a_{1,1} = 1$$

We thereby see that each coefficient $a_{n,r}$ is actually the binomial coefficient $\binom{n}{r}$.

The preceding observation is attributed to the renowned English mathematician/physicist Isaac Newton and is enshrined in Newton's famous *Binomial Theorem*. In fact, the calculations preceding the observation constitute a proof of this seminal result.

Theorem 5.3 (The Binomial Theorem). *For all $n \in \mathbb{N}$,*

$$(x+y)^n = \sum_{i=0}^{n} \binom{n}{i} x^{n-i}y^i \tag{5.30}$$

5.3.3 ⊕ Integer Roots of Polynomials: Hilbert's Tenth Problem

This section is devoted to a true milestone in the history of mathematics, of logic, and of computing. The mathematical details necessary to fully describe the analyses that lead to this blockbuster theorem go way beyond the scope of an introductory text. However, this result has cultural lessons that make even its story valuable. An important source of this value is that the reader who understands even the outlines of the theorem will be receptive to the broad implications of the material in Section 10.4.2, Section 8.2, and Appendix B. This material supplies essential mathematical underpinnings for the theorem whose tale we are about to embark on.

In 1900, the eminent, influential German mathematician David Hilbert set forth a list of 23 problems to serve as a "bucket list"[3] for the world mathematics community at the dawn of the twentieth century.[4] Of Hilbert's 23 problems, the tenth stands out within the world of computing. Stated in modern terminology, with a modern perspective, the Problem can be stated as follows.

Hilbert's Tenth Problem   *Develop an algorithm that will decide—via a* YES *or* NO *answer—whether a given multivariate polynomial with integer coefficients has any integer roots.*

The Problem attracted the attention of many of the best mathematical minds of the twentieth century. Building on the work of American mathematicians Martin Davis and Julia Robinson, the Problem was finally resolved by the Russian then-graduate student Yuri Matiyasevich in the mid-1970s.

Historical note.

This is certainly not the only instance of a young person making a world-class contribution to mathematics! Look, for instance, back at our reference to Galois! The message is clear: Study hard, and aim high!

Theorem 5.4 (Hilbert's Tenth Problem, Resolved). *There is no algorithm that, when presented with an arbitrary multivariate polynomial P with integer coefficients, will correctly determine whether P has any integer roots.*

The long history leading to Matiyasevich's proof of Theorem 5.4 is documented in [41, 42, 43, 80]. The first three of these sources give accessible, informal descriptions of the exciting journey from Hilbert to Matiyasevich.

The mathematics needed even to understand Hilbert's Tenth Problem and its resolution are beyond an introductory text. However, a few words are in order about the exciting way in which Theorem 5.4 fundamentally changed the course of twentieth-century mathematics by turning the tables on all of the mathematics that had been done to that point in history.

We recall from Section 2.1.1 that one of the main movements that took root in nineteenth-century mathematics had as its goal to codify the notion of "rigorous proof". Toward this end, one of Hilbert's goals in his to-do list of 1900—especially with regard to the tenth problem on his list—was to have mathematicians rigorously prove that the notions of "Truth" and "Provability" coincided. Hilbert's immediate challenge was to find such a proof at least for "elementary" areas of mathematics.

---

[3] The "bucket" is not yet empty. A number of Hilbert's original 23 problems are yet to be solved.

[4] Hilbert's list was published in 1902 [58]; it was translated into English by Mary Frances Winston Newson, the first American woman to receive a doctorate from the University of Göttingen (which was probably the world's premier university for mathematical research in the late 19th and early 20th centuries).

This hope was, in fact, the first of Hilbert's dreams to be dashed: In the early 1930s, Austrian logician-mathematician Kurt Gödel proved the first of his celebrated *Incompleteness Theorems* [53]. Stated *very(!)* informally, this first theorem asserts that in any "reasonable" mathematical system, there will always be *true* statements that cannot be proved.

Explanatory note.

Two crucial words remain unspecified in our story, namely, "elementary" and "reasonable". Both refer to what is, intuitively, the simplest domain in formal mathematics—the study of the basic properties of positive integers. Specifically, both words point to a mathematical system which is capable of dealing with elementary assertions about triples, x, y, z, of positive integers, specifically, the following assertions:

- One given integer, say z, is the sum of the other two ($z = x + y$).

- One given integer, again say z, is the product of the other two ($z = x \times y$).

Building on Gödel's insights, English mathematician Alan Turing soon thereafter proved—again, stated *very(!)* informally—that no matter how much "smarts" we build into a digital computer using digital logic, there will always be functions—even (0-1)-valued functions—that the computer cannot compute [108]. Theorem 5.4 can be reworded to assert that the integer-root-finding behavior called for in Hilbert's Tenth Problem is one of Turing's *uncomputable* functions![5]

5.4 Exponential and Logarithmic Functions

This section introduces the fundamentals of the mathematically and computationally important functions that emerge from the operations of exponentiation and taking logarithms. These functions are mutually inverse, in the sense of Eqs. (5.3) and (5.4).

5.4.1 Basic Definitions

A function f is *exponential* if there is a positive number b such that, for all x,

$$f(x) = b^x \tag{5.31}$$

The number b is the *base* of function $f(x)$. The basic arithmetic properties of exponential functions are easily derived from the recurrent system of equations (5.5).

---

[5] The connections between Gödel's work and Turing's, and the manner in which both authors' work connects to the work of German logician Georg Cantor on infinite sets of numbers [25, 26], are described in [91] using consistent terminology to discuss all three authors' work.

The important message is that multiplication "at the ground level" is addition at the "exponent level":

$$b^x \times b^y = b^{x+y}$$

Given an integer $b > 1$ (mnemonic for "base"), the *base-b logarithm* of a real number $a > 0$ is denoted $\log_b a$ and defined by the equation

$$a = b^{\log_b a} \qquad (5.32)$$

Logarithms are partial functions: $\log_b a$ is not defined for non-positive values of a.

Two bases are so prominent in the contexts of computation theory and information theory that we commonly invoke one of two special notations for logarithms to those bases. These bases are $b = 2$ and $b = e = 2.71828\cdots$; the latter is called the base of *natural* logarithms. The special notations associated with these bases are:

- For base-2 logarithms, we often elide the subscript 2 and write $\log a$ for $\log_2 a$.

- For base-e logarithms, we not only elide the subscript e, we actually employ the specialized notation $\ln a$ for $\log_e a$.

We leave to the reader the easy verification, from Eq. (5.32), that the *base-b logarithmic function* $f(x) = \log_b x$ is the functional inverse of the base-b exponential function.

5.4.2 Learning from Logarithms (and Exponentials)

Definition (5.32) exposes and—even more importantly—explains many facts about logarithms which we often take for granted.

Proposition 5.6 *For any base $b > 1$, for all numbers $x > 0$, $y > 0$,*

$$\log_b(x \cdot y) = \log_b x + \log_b y$$

Proof. Definition (5.32) tells us that $x = b^{\log_b x}$ and $y = b^{\log_b y}$. Therefore,

$$x \cdot y = b^{\log_b x} \cdot b^{\log_b y} = b^{\log_b x + \log_b y}$$

by the laws of powers. Taking base-b logarithms of the first and last terms in the chain yields the claimed equation. $\square$

Historical note.

Proposition 5.6 provides the mathematical underpinnings of the analogue computing device known as a *slide rule*. Without going into details, a slide rule is built from two straight sticks that are both marked off according to a logarithmic scale. The sticks are attached in a way that allows each to slide along the other (whence the name of the device); see Fig. 5.3. The important property of this device is illustrated in the figure: When one positions the origin (number 1) of stick A at the position of stick B that corresponds to number a, then the number on stick B that corresponds to the number b on stick A is $a \times b$. In short, the logarithmic scale on the device's sticks enables it to translate linear motion (the sliding of the sticks relative to one another) into numerical multiplication.

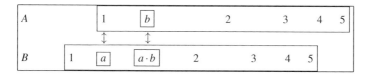

Fig. 5.3 A schematic view of a slide rule performing a multiplication. If one slides Stick A to position a of Stick B, then position b of Stick A appears directly above position $a \cdot b$ of Stick B. Thereby, the addition of distances on the rule's logarithmic scale is used to compute products

Many people believe that the following important property of logarithms is a *convention* rather than a consequence of the basic definitions:

The logarithm of 1 *to any base is* 0.

We correct this misapprehension now.

Proposition 5.7 *For any base $b > 1$,*

$$\log_b 1 = 0$$

Proof. We note the following chain of equalities.

$$b^{\log_b x} = b^{\log_b(x \cdot 1)} = b^{(\log_b x) + (\log_b 1)} = b^{\log_b x} \cdot b^{\log_b 1}$$

Hence, $b^{\log_b 1} = 1$. If $\log_b 1$ did not equal 0, then $b^{\log_b 1}$ would differ from 1. $\square$

Similar exploitation of the basic definitions yields the following fundamental properties of logarithms (to any base).

Proposition 5.8 *For all bases $b > 1$ and all numbers x, y,*

$$x^{\log_b y} = y^{\log_b x}$$

Proof. We invoke Eq. (5.32) twice to remark that

$$\left[x^{\log_b y} = b^{(\log_b x)\cdot(\log_b y)}\right] \quad \text{and} \quad \left[y^{\log_b x} = b^{(\log_b y)\cdot(\log_b x)}\right]$$

The commutativity of multiplication completes the verification. □

Proposition 5.9 *For any base $b > 1$,*

$$\log_b(1/x) = -\log_b x$$

Proof. By the product law for logarithms,

$$\log_b x + \log_b(1/x) = \log_b(x \cdot (1/x)) = \log_b 1$$

An invocation of Proposition 5.7 completes the proof. □

Proposition 5.10 *For any bases $a, b > 1$,*

$$\log_b x = (\log_b a) \cdot (\log_a x) \tag{5.33}$$

Proof. We begin by noting that, by definition,

$$x = b^{\log_b x} = a^{\log_a x} \tag{5.34}$$

We now take the base-b logarithm of the second and third expressions in Eq. (5.34) and then invoke the product law for logarithms. The second expression tells us that

$$\log_b\left(b^{\log_b x}\right) = \log_b x \tag{5.35}$$

The third expression tells us that

$$\log_b\left(a^{\log_a x}\right) = (\log_b a) \cdot (\log_a x) \tag{5.36}$$

We know from the double equation (5.34) that the righthand expressions in Eqs. (5.35) and (5.36) are equal. Eq. (5.33) follows by transitivity. □

By setting $x = b$ in Eq. (5.33), we discover the following marvelous equation.

Proposition 5.11 *For any integers $a, b > 1$,*

$$(\log_b a) \cdot (\log_a b) = 1 \quad \text{or, equivalently,} \quad \log_b a = \frac{1}{\log_a b} \tag{5.37}$$

5.5 ⊕ **Pointers to Specialized Topics**

Emerging technologies prolifically give rise to novel specialties that grab the imagination of the computer-literate public. Since many such specialties build upon mathematical concepts and tools, they afford the designer of a mathematics course an opportunity to add enrichment to a syllabus that is dominated by the "classics". During the first decades of the twenty-first century, application areas such as *robotics* or *data science* or *data mining* or *computer security* or *energy-awareness* can all benefit from supplements to a standard mathematics syllabus—of course, at levels consistent with the students' preparation. We now flesh out this suggestion by discussing two supplements which, broadly construed, could fit under the mantle of *Arithmetic*.

5.5.1 Norms and Metrics for Tuple-Spaces

By the 1950s, computers had become sophisticated enough to handle structured data, and computer users had become sophisticated enough to think in terms of structured models. Thus, from the 1960s and 1970s onward to the present, one sees *tuple-spaces* used to algorithmic advantage in application areas as varied as *databases* [32], *parallel computing* [21, 104], and *robotics* [79, 92]. While the specific concepts drawn from the study of tuple-spaces vary a bit from one application area to another, certain concepts recur—and among these, the *norms* and *metrics* that allow us to talk about notions such as "distance" play a very important role. Of course, notions of "distance" relevant to databases may differ from those used in, e.g., robotic path planning, but in very many applications it is important to know whether tuple t_1 is "closer to" tuple t_2 than to tuple t_3.

It is an often-frustrating phenomenon that new application areas all too frequently rename concepts and tools that they inherit from their predecessors. So how shall we name the norms and metrics that we inject into our mathematics curriculum? Probably the most scholarly course is to employ the names that honor the early-twentieth-century French mathematician and function theorist Henri Léon Lebesgue, whose *L*-measures form the classical taxonomy of norms and distances in tuple-spaces.

We focus on *two-dimensional* tuple-spaces, although everything we say generalizes in a straightforward manner to arbitrary finite-integer dimensionalities. Also, we restrict attention to the three *L*-measures that are the most common in application areas such as informatics and robotics. Focus on arbitrary 2-tuples of numbers $z_1 = \langle x_1, y_1 \rangle$ and $z_2 = \langle x_2, y_2 \rangle$.

The L^1 measure.

- The L^1-*norm* of z_1 is the sum of the magnitudes of its coordinates, i.e.,

$$L^1(z_1) = |x_1| + |y_1|$$

- The L^1-*distance* between z_1 and z_2 is the sum of the magnitudes of the differences between the coordinates of the two pairs, i.e.,

$$L^1(z_1, z_2) = |x_1 - x_2| + |y_1 - y_2|$$

The L^1-distance is also called *Manhattan distance* or *rook distance*, because a path between z_1 and z_2, viewed as points in two-dimensional space, in which adjacent points are unit L^1-distance apart, follows a rectilinear grid-like pattern such as one observes in a map of Manhattan or in the path taken by a rook in chess. From a less picturesque point of view, one observes that each point in two-dimensional space is unit L^1-distance from four other points, one each in the four NEWS (north, east, west, south) compass directions. If one maps out the points in two-dimensional space that are at successive L^1-distances from a "center" point z, then one observes the L^1-*disc* of Fig. 5.4(left), which generalizes in higher dimensions to the L^1-*sphere*.

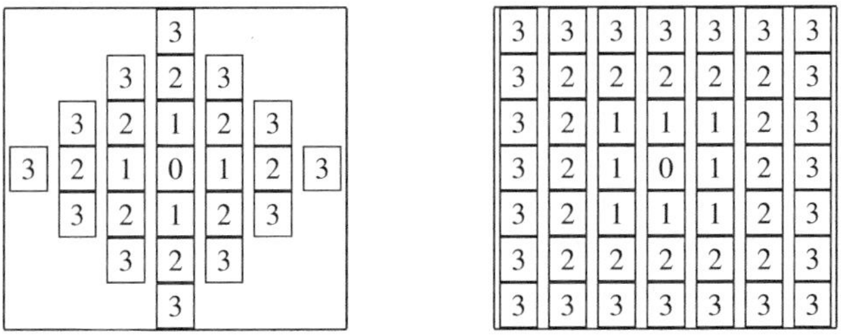

Fig. 5.4 The radius-3 L^1-disc (left) and L^∞-disc (right), with annotated L^1-distances and L^∞-distances, respectively, from the "center" point

The L^∞ measure.

- The L^∞-*norm* of z_1 is the larger of the magnitudes of its coordinates, i.e.,

$$L^\infty(z_1) = \max(|x_1|, |y_1|)$$

- The L^∞-*distance* between z_1 and z_2 is the larger of the magnitudes of the differences between the coordinates of the two pairs, i.e.,

$$L^\infty(z_1, z_2) = \max(|x_1 - x_2|, |y_1 - y_2|)$$

L^∞-distance is also called *king's-move distance*, because each step in a path between z_1 and z_2, in which adjacent points are unit L^∞-distance apart follows a pattern such as one observes in a path taken by a king in chess. From a less picturesque point of view, one observes that each point in two-dimensional space is unit L^∞-distance from eight other points, one each in the eight compass directions, N, NE, E, SE,

S, SW, W, NW. If one maps out the points in two-dimensional space that are at successive L^∞-distances from a "center" point z, then one observes the L^∞-*disc* of Fig. 5.4(right), which generalizes in higher dimensions to the L^∞-*sphere*.

As one observes in Fig. 5.4, the structures of the L^1-disc and the L^∞-disc are, respectively, dominated by diagonals and squares; see the "equipotential" lines in the figure. One can exploit this fact when crafting algorithms that are governed by these norms. Indeed, we invoke the structures of these discs as we develop the "pairing functions" that we showcase in Section 8.2.2.

We have not (yet) mentioned what is likely the most familiar L-measure, namely the L^2 (*Euclidean*) measure which is named for Euclid in honor of his celebrated treatise on geometry, *The Elements*. For completeness, we now define this measure, but we do not develop the subject further because it is not frequently used when dealing with the discrete mathematical structures that are the primary foci of this text.

The L^2 measure.

- The L^2-*norm* of z_1 is the Euclidean distance of the point from the *origin* of two-dimensional space, which is the point $\langle 0,0 \rangle$:

$$L^2(z_1) = \sqrt{x_1^2 + y_1^2}$$

- The L^2-*distance* between z_1 and z_2 is the Euclidean distance between the points in 2-dimensional space, i.e.,

$$L^2(z_1, z_2) = \sqrt{(x_1 - x_2)^2 + (y_1 - y_2)^2}$$

5.5.2 Edit-Distance: Measuring Closeness in String Spaces

From the earliest days of digital computing, the challenge of employing computers as "editorial assistants" was viewed as a prime domain within which to achieve a significant practical payoff. An early problem within this domain was to determine of "how different" two strings, x and y, composed of letters from an alphabet A, are from one another. The quotation marks in the preceding sentence acknowledge that it is actually not obvious how to measure the "distance" between x and y in a way that matters. The intersecting needs of three separate developing communities finally supplied the definition that has generally been accepted. Roughly contemporaneously, during the decade from 1955–1965:

- Computational linguists attempted to enlist computers in the processing of natural language. The goal of automatic translation between Russian and English became the "holy grail" of this community, at least in the USA and USSR. Reminiscences by Anthony G. Oettinger, a pioneer in this effort, appear in [61].

• As the quickly growing power of computers became appreciated, more sophisticated process-specification languages began to emerge. From the very early days of this development, programming languages became more "natural"-language-like. (There was a split among computer-language designers: Some interpreted "natural" as in the phrase "natural language": They wanted to program in [an approximation to] English or French or Russian or .... Others interpreted "natural" as a synonym for "problem-specific". They wanted scientists to be able to feed computers systems of equations, while engineers might feed some textual encoding of a circuit design.) One of the earliest entrants in this arena was the business-oriented language COBOL, which was inspired by early work of the computing giant Grace Murray Hopper and developed under the leadership of Jean E. Sammet; see [97]. A second early entrant was the string-processing language COMIT developed under the leadership of Victor Yngve, with an eye toward applications such as language processing; see [116]. A champion by any metric was the ever-evolving FORTRAN language, which was developed in the mid-1950s by a small team at IBM, headed by John Backus; see [10]. FORTRAN was the first programming language which aimed to enable scientists to specify the problems they wished to compute in a manner that was at least reminiscent of the mathematical notation they would use to communicate with one another.

• As it became clear that FORTRAN was a step toward satisfying the programming needs of an enormous market, a veritable army of systems programmers began to develop increasingly sophisticated processors for its genre of programming languages. The era of "smart compilers" was dawning.

A common feature in the preceding developments was that people were typing more as they used computers. Inevitably, therefore, they were making more typing mistakes. A very specific computing challenge arose, to make much more concrete the desire to understand how close a string x was to a string y. Could one develop an algorithm that would rewrite one of these strings to the other while "editing", or rewriting, as few symbols as possible? While such an algorithm would not completely solve the "mistyping" problem—consider, e.g., that the strings
    "SORTH" and "NOUTH"
are both "edit-distance 1" from both
    "SOUTH" and "NORTH"
hence cannot be "corrected" automatically—being able to correct a mistyped string to its edit-nearest legitimate string would probably be very useful in practice. Happily, although this algorithmic problem was more difficult than many had imagined, it did admit the efficient, elegant solution that appears in [111].

5.6 Exercises: Chapter 5

Throughout the text, we mark each exercise with 0 or 1 or 2 occurrences of the symbol $\oplus$, as a rough gauge of its level of challenge. The 0-$\oplus$ exercises should be accessible by just reviewing the text. We provide *hints* for the 1-$\oplus$ exercises; Appendix H provides *solutions* for the 2-$\oplus$ exercises. Additionally, we begin each exercise with a brief explanation of its anticipated value to the reader.

1. **Additive and multiplicative identities are unique**
   LESSON: Enhance familiarity with basic arithmetic argumentation
   Prove the following assertions about 0 and 1.

   Proposition 5.12 *The numbers 0 and 1 play unique roles:*

   a. *0 is the unique additive identity:* $(\forall n)\, [n+0 = 0+n = n]$

   b. *1 is the unique multiplicative identity:* $(\forall n)\, [n \cdot 1 = 1 \cdot n = n]$

   c. *0 is the unique multiplicative annihilator:* $(\forall n)\, [n \cdot 0 = 0 \cdot n = n]$

2. **Additive and multiplicative inverses are unique**
   LESSON: Enhance familiarity with basic arithmetic argumentation
   Prove the following assertions.

   Proposition 5.13 *Additive and multiplicative inverses are unique:*

   a. *For each number x in* $\mathbb{Z}$, *or in* $\mathbb{Q}$, *or in* $\mathbb{R}$, *there is a unique number y in the same class such that* $x + y = 0$.

   b. *For each* nonzero *number x in* $\mathbb{Q}$ *or in* $\mathbb{R}$, *there is a unique number y in the same class such that* $x \cdot y = 1$.

   The class $\mathbb{Z}$ is (intentionally) missing from Proposition 5.13.b: integers other than ± 1 do not have integral multiplicative inverses. We shall take some of the sting out of this lack in Section 8.4 by means of Proposition 8.14, which shows that integers have a rudimentary version of multiplicative inverse.

3. **Verifying the critical properties of rational arithmetic**
   LESSONS: Rational arithmetic works the way that it should.
   Prove the following properties of rational arithmetic ("RA" for short):

   - *RA works correctly when the argument rational numbers are integers;*

   - *RA treats zero (0) appropriately, as both an additive identity and a multiplicative annihilator;*

   - *RA treats the number 1 appropriately, i.e., as a multiplicative identity;*

- *RA obeys the laws outlined in Section 5.2.*

  Verifying the distributivity of rational multiplication over rational addition is an especially valuable exercise because of the amount of manipulation that the verification demands.

4. **Roots of polynomials**
   LESSON: Practice with roots and polynomial manipulation
   Prove the following assertions, which employ the notation of the chapter.

   a. Not surprisingly, the roots of a polynomial $P(x)$ tell us something about $P(x)$'s algebraic structure.

      Proposition 5.14 *If the number r is a root of the polynomial $P(x)$, then $(x - r)$ divides $P(x)$; i.e., there exists a polynomial $Q(x)$ such that*

      $$P(x) = (x - r) \times Q(x)$$

   b. The following result can often simplify the calculations needed to find a polynomial's roots.

      Proposition 5.15 *The polynomial*

      $$P_1(x) = a_d x^d + a_{d-1} x^{d-1} + a_{d-2} x^{d-2} + \cdots + a_1 x + a_0$$

      and its monic "sibling"

      $$P_2(x) = x^d + \frac{a_{d-1}}{a_d} x^{d-1} + \frac{a_{d-2}}{a_d} x^{d-2} + \cdots + \frac{a_1}{a_d} x + \frac{a_0}{a_d}$$

      have the same roots.

   c. Certain properties of polynomial $P_1(x)$, such as its dominant growth rate, are easier to analyze when one writes $P_1(x)$ in the following form.

      $$P_3(x) = x^d \cdot \left(a_d + \frac{a_{d-1}}{x} + \frac{a_{d-2}}{x^2} + \cdots + \frac{a_1}{x^{d-1}} + \frac{a_0}{x^d} \right)$$

      Prove that $P_3(x) = P_1(x)$.

   d. Prove that the inequality (5.13) holds for any monic polynomial $P(x)$ with real coefficients.

5. **Logarithmic functions and exponential functions**
   LESSON: Taking logarithms and exponentiating are mutually inverse operations
   Prove the following assertion.

   Proposition 5.16 *The base-b logarithmic function*

   $$f(x) = \log_b x$$

and the base-b exponential function

$$g(x) = b^x$$

are functional inverses of each other.

6. **An identity for binomial coefficients**
   LESSON: Experience with algebraic manipulation

   Prove the following.

   Proposition 5.17 *For all n and all $k \in \{1, \ldots, n\}$*

   $$k \times \binom{n}{k} = n \times \binom{n-1}{k-1}$$

Chapter 6
Summation: A Complex Whole from Simple Parts

The whole is greater than the sum of its parts
Aristotle (*Metaphysics*, Book VIII)

6.1 The Many Facets of Summation

The operation of *summation*—adding up aggregations of numbers—is of fundamental importance in the world of digital computing. While we humans are able to deal handily with abstractions such as "smoothness" and "continuity", we must employ sophisticated *discretizations* of these concepts in order to enlist the aid of digital computers in dealing with such abstractions. Summations provide a very useful discretization of *continuous*, or "smooth", phenomena that are typically dealt with the aid of the (differential and integral) calculus.

This chapter is dedicated to exploring how to employ summations as a computational tool. We deal throughout with *series*, i.e., (possibly infinite) sequences of numbers

$$a_1, a_2, a_3, \ldots$$

whose sum

$$a_1 + a_2 + a_3 + \cdots \tag{6.1}$$

is the target of interest. Striving to live up to the quotation that heads this chapter, we emphasize the methodology that enables us to reason about and manipulate series, rather than just techniques for deriving their sums.

© Springer Nature Switzerland AG 2020
A. L. Rosenberg, D. Trystram, *Understand Mathematics, Understand Computing*,
https://doi.org/10.1007/978-3-030-58376-7_6

Explanatory note.

Of course, when we deal with *infinite* series, wherein there are infinitely many numbers a_i, we must address the question of whether the sum (6.1) exists as a finite number. For some infinite series the sum *does* exist: We note, for example, the well-known sum

$$1 + \frac{1}{2} + \frac{1}{4} + \frac{1}{8} + \frac{1}{16} + \cdots + \frac{1}{2^k} + \frac{1}{2^{k+1}} + \cdots = 2 \qquad (6.2)$$

(We shall encounter this summation in several guises during our journey through this text.) Infinite series such as in Eq. (6.2) are said to *converge*.

But sometimes an infinite series does *not* have a finite sum. This is true, for instance, of the well-known *harmonic* series

$$1 + \frac{1}{2} + \frac{1}{3} + \frac{1}{4} + \frac{1}{5} + \cdots + \frac{1}{k} + \frac{1}{k+1} + \cdots \qquad (6.3)$$

As more and more terms are added to this summation, the accumulated sum eventually exceeds every number. Such an infinite series is said to *diverge*.

Summations such as in Eq. (6.3) diverge because their initial partial sums—i.e., the sequence $1, 1 + \frac{1}{2}, 1 + \frac{1}{2} + \frac{1}{3}, \ldots$, for Eq. (6.3)—grow without bound. There are other infinite series whose behavior is harder to describe. One such example is the series

$$1 - 1 + 1 - \cdots + 1 - 1 + - \cdots$$

which we discuss briefly in the exercises of this chapter and in Chapter 7.

Summations such as in Eqs. (6.2) and (6.3) illustrate some of the complexities of dealing with infinite entities. Most obviously, as we just remarked, when dealing with summations that are infinite, some of them have finite sums while others do not. Even more subtle, the series that *do* have (finite) sums illustrate the unintuitive fact that sometimes finite objects or entities—such as the integer 2 in Eq. (6.2)—have infinite "names": In this example, the series is an infinite "name" of integer 2. Lots to think about!

Enrichment note.

The conceptual dissonance of dealing with *infinite objects* that have *finite values*—as exemplified by the series in Eq. (6.2)—has been recognized in various forms for more than 25 centuries. Several charming and familiar examples appear in the paradoxes attributed to Zeno of Elea.

In his *Paradox of Achilles and the Tortoise*, for instance, Zeno appears at first glance to prove that all motion is illusory! In this story, the slow-of-foot Tortoise (T) tries to convince the speedy Achilles (A) of the futility of trying to win any race in which A gives T even the most minute head start. As long as T is ahead of A, argues T, every time A traverses half the distance between the competitors, T will respond by moving a bit further ahead. Thereby, T will always be a positive distance ahead of A, so that A can *never* catch T.

A similar "argument" demonstrates that an arrow shot at you by an adversary can never reach you, as long as you continually move away from the archer. *DO NOT TRY THIS AT HOME!*

The notion of *infinitesimals*, as invented by Isaac Newton and Gottfried Wilhelm Leibniz, explains the fallacy of assertions such as the Tortoise's. This notion, which plays a huge role in modern mathematics, underlying such foundational concepts as *limits* and *continuity* (of functions), was not well understood until just a few hundred years ago.

The general topic of the convergence or divergence of infinite series is beyond the scope of this text, but we shall observe several examples of each concept throughout this chapter. It is a fascinating subject for further study.

Toward the end of guiding the reader through the forest of abstractions and operations and techniques associated with summations, we categorize the targets of our discussions in three ways.

1. We study a number of *fundamental summations* which have intrinsic interest.

   Examples of this topic category include *arithmetic summations*, *geometric summations*, and *mathematically "smooth"* summations, including sums of positive and negative powers of integers. Here is a sampler of six summations that appear later in this chapter.

   1. $1 + 2 + 3 + 4 + 5 + \quad \cdots + k + (k+1) + \cdots + n$
   2. $1 + 2 + 4 + 8 + 16 + \quad \cdots + 2^k + 2^{k+1} + \cdots + 2^n$
   3. $1 + 4 + 9 + 16 + 25 + \cdots + k^2 + (k+1)^2 + \cdots + n^2$
   4. $1 + \frac{1}{2} + \frac{1}{4} + \frac{1}{16} + \frac{1}{32} + \cdots + \frac{1}{2^k} + \frac{1}{2^{k+1}} + \cdots$
   5. $1 + \frac{1}{2} + \frac{1}{3} + \frac{1}{4} + \frac{1}{5} + \quad \cdots + \frac{1}{k} + \frac{1}{k+1} + \cdots$
   6. $1 + \frac{1}{4} + \frac{1}{9} + \frac{1}{16} + \frac{1}{25} + \cdots + \frac{1}{k^2} + \frac{1}{(k+1)^2} + \cdots$

2. We study a variety of *fundamental techniques* for evaluating summations.

We include specialized techniques that work for specific classes of summations, as well as more general techniques that work in a broad range of situations.

Examples of such techniques include, e.g., estimating summations by integrating functions related to the summation; grouping/replicating terms within a summation; verifying "guessed" sums via induction.

3. We study a variety of *fundamental representations* of the elements being summed. We observe that being able to study the same phenomenon in a variety of seemingly unrelated ways often gives one unexpected mathematical understanding of, and operational control over, the phenomenon.

Examples of such representations include, among others, representations of numbers by: numerals in a positional number system; slices of pie; tokens arranged in stylized ways; basic geometrical structures, including the unit-width rectangles of so-called Riemann sums.

In conclusion, we treat each topic in multiple ways, as long as each new way supplies new intuition and teaches a new lesson.

To illustrate the power of summation methodology, consider the following modernized version of the *legend of Sissa ibn Dahir*. Sissa, goes the legend, has invented a marvelous game that is played on a chessboard, i.e., an 8×8 array of unit-size squares. An entrepreneur proposes to buy the rights of this marvelous game from Sissa. The entrepreneur offers Sissa a one-time payment of *one million million (i.e.,* $10^{12} = 1,000,000,000,000$) *euros* in return for all rights to the new game[1]. As a counteroffer, Sissa asked the entrepreneur instead for all of the money amassed in the following way. Sissa requested that the entrepreneur proceed row by row along a chessboard, placing money in the board's squares, according to the following regimen. The entrepreneur should place 1 euro in the first square, 2 euros in the second square, $4 (= 2 \times 2)$ euros in the third square, $8 (= 4 \times 2)$ euros in the fourth square, and so on, doubling the number of euros at each step of the procedure—so the last square would contain 2^{63} euros. Figure 6.1 illustrates the growth of the pile of euros during the first few steps of the procedure.

Fig. 6.1 The money in the first seven squares of the chessboard based on Sissa's counteroffer

Has Sissa made a good bargain?

By the end of this chapter, you will be able to determine in minutes that under the procedure that amasses money on the chessboard, Sissa would receive $2^{64} - 1$

---

[1] The original legend was about grains of rice; this is what is called inflation ...

euros—which is more than 10^{20} euros! Sissa would, therefore, amass *much* more money via his counteroffer than the mere 10^{12} euros that the entrepreneur offered! *A good bargain, indeed!*

6.2 Summing Structured Series

6.2.1 Arithmetic Summations and Series

6.2.1.1 General development

We define *arithmetic sequences* and learn how to calculate their sums, *arithmetic series.*

An n-term *arithmetic sequence*:
$$a, \; a+b, \; a+2b, \; a+3b, \; \ldots, a+(n-1)b$$

The corresponding *arithmetic series*:
$$\begin{aligned} & a+(a+b)+(a+2b)+(a+3b)+\cdots+(a+(n-1)b) \\ = \; & an+b\cdot(1+2+\cdots+n-1) \end{aligned}$$

(6.4)

The common inter-element difference b in the sequence and the series in Eq. (6.4) is the *period* of the sequence and the series.

The message from the factorization of the arithmetic series in Eq. (6.4) is that we can calculate the sum of the series by determining the sum of the first $n-1$ positive integers. In the next subsection, we use this fact as an opportunity to introduce important notation.

6.2.1.2 Example #1: Summing the first n integers

Our first goal in this section is to sum the first n positive integers:

$$1 + 2 + \cdots + n$$

that is, to find a *closed-form expression* for the sum. In somewhat informal terms, we say that an expression of the type

$$f(n) \stackrel{\text{def}}{=} \sum_{i=1}^{n} i \tag{6.5}$$

is in *closed form* if it provides a prescription for evaluating the sum using a *fixed-length* sequence of arithmetic operations (e.g., addition/subtraction, multiplication/division, exponentiation/taking logarithms). The notion "closed form" contrasts with the recipe implicit in the notation (6.5), which takes $n-1$ additions to evaluate—because its length depends on input n, the indicated computation does not have *fixed* length.

The sum $f(n)$ of the special summation (6.5) is commonly denoted Δ_n. Within this chapter, we usually prefer the notation $S_1(n)$ for this sum, because it exposes this summation as one instance of a related family of such summations that will occupy us throughout this chapter; we denote the sums of these summations by the notation $S_c(n)$ for various values of parameter c.

The remainder of this section develops multiple ways to derive the following *closed-form* expression for $\Delta_n = S_1(n)$.

Proposition 6.1 *For all $n \in \mathbb{N}$,*

$$S_1(n) \;=\; \sum_{i=1}^{n} i \;=\; \frac{1}{2}n(n+1) \tag{6.6}$$

Proof. **A textual proof.** We begin with a *constructive, textual* proof[2] of summation (6.6) which employs an approach that was, famously, used in a school exercise by the eminent German mathematician Carl Friedrich Gauss—as a preteenager. This approach proceeds in two steps.

Write $S_1(n)$ "forwards": $\qquad \sum_{i=1}^{n} i = 1 + \quad 2 \quad + \cdots + (n-1) + n$

$\qquad\qquad\qquad\qquad\qquad\qquad\qquad\qquad\qquad\qquad\qquad\qquad\qquad\qquad\qquad$ (6.7)

Write $S_1(n)$ "in reverse": $\qquad \sum_{i=n}^{1} i = n + (n-1) + \cdots + \quad 2 \quad + 1$

Now add the two representations of $S_1(n)$ in Eq. (6.7) *column by column*. Because each of the n column-sums equals $n+1$, we find that $2S_1(n) = n(n+1)$, which we easily rewrite in the form (6.6)—after we multiply both sides of the equation by 2. $\square$

Explanatory note.

Let us step back from the specific result in Proposition 6.1 and concentrate on the textual proof. What Gauss noticed about the problem of computing $S_1(n)$ is that when the sum is doubly written as in Eq. (6.7), the column-sums are all the same. This phenomenon of finding *invariants* is a "pattern" of the form referred to in Chapter 2 as we discussed how to "do mathematics". We observe how the pattern can be exploited to determine the sum of any arithmetic series. *What seemed to be a "trick" turns out to be an insightful instance of pattern-matching!* We soon see how to employ the pattern for other, related, ends.

Not everyone thinks the same way—even within the context of mathematics. It is, therefore, very important for the reader to recognize that even the simplest mathematical facts can be proved and analyzed in a broad variety of ways. We illustrate this assertion by developing several more proofs of Proposition 6.1.

---

[2] The proof is *constructive:* it actually derives an explicit answer. This contrasts with, say, the inductive validation of the sum $S_1(n)$ in Section 6.3.2.2.C, which just verifies a "guessed" answer.

Proof. **A "pictorial", graphical proof.** Now we shall look at the problem of summing $S_1(n)$ by estimating the area of a simple (in the *good sense* of the word) geometric figure. In this worldview, integers are represented by concatenating *unit-side* (i.e., 1×1) *squares*, as in Fig. 6.2.

Our summation process proceeds in the three steps illustrated in Figs. 6.2-6.3.

1. We begin, in Fig. 6.2, by depicting the problem of calculating $S_1(n)$ as the problem of determining the area of a surface built from unit-side squares.

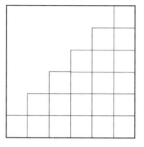

Fig. 6.2 Representing the first n integers using unit squares; $n = 6$ in this example.

2. Next, we illustrate in Fig. 6.3(Left) that the area of the (light grey) lower-right triangle of the $n \times n$ square is one-half the area of the entire $n \times n$ square.

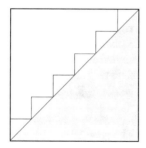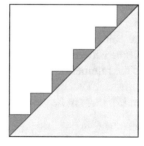

Fig. 6.3 (Left) The area of the lower-right triangle (light grey) is one-half that of the entire $n \times n$ square. (Right) The area of the (dark) triangles upon the upper diagonal of the $n \times n$ square is $\frac{1}{2}n$

3. Finally, we indicate in Fig. 6.3(Right) that the cumulative area of the small (dark grey) triangles which sit on the upper diagonal of the $n \times n$ square is $\frac{1}{2}n$. This reckoning notes that there are n triangles, each of area equal to one-half the area of a unit-side square.

We thereby reckon the area of the surface depicting $S_1(n)$ as

One-half the area of the $n \times n$ *square, i.e.,* $\frac{1}{2}n^2$
    plus
n times the area of one-half a unit-side square, i.e., $\frac{1}{2}n$

This reasoning thus derives the value of $S_1(n)$.   □

We present one final, combinatorial, proof of Proposition 6.1.

Proof. **A combinatorial proof.** The following argument is *combinatorial* in that it achieves its goal by *counting* instances of the first n integers, laid out in a line.

Place (tokens that represent) the integers 0 to n along a line. For each integer i, count how many integers $j > i$ lie to its right. We see that in general, there is a *block* of $n - i$ integers that lie to the right of integer i. In detail: the block of integers lying to the right of $i = 0$ contains n values of j; the block to the right of $i = 1$ contains $n - 1$ values of j, and so on, as suggested in Fig. 6.4.

All integers ≤ 4:	0 1 2 3 4
integers to the right of 0:	1 2 3 4
integers to the right of 1:	2 3 4
integers to the right of 2:	3 4
integers to the right of 3:	4

Fig. 6.4 A two-dimensional (triangular) depiction of the right-lying integer-instances

On the one hand, we see that the total number, j, of right-lying integers equals $n + (n - 1) + \cdots + 1 = S_1(n)$.

On the other hand, every instance of a right-lying integer can be identified uniquely by the pair of nonnegative integers, i (the instance's block) and $j > i$ (the instance's position-within-block). The total number of right-lying integer-instances corresponds to the number of ways one can select two objects (here, integers) from among $n + 1$ objects.[3] This number is the *binomial coefficient* , whose definition we specialize from Eq. (5.7) in Section 5.1.2.4.

$$\Delta_n = \binom{n+1}{2} \stackrel{\text{def}}{=} \frac{1}{2}n(n+1)$$

We have arrived at two distinct—but, of course, equal—expressions for $S_1(n)$.   □

Explanatory note.

We have exposed the sums Δ_n as special binomial coefficients, namely, those whose "bottom" parameter is 2. The many ways of viewing the underlying summation in terms of *triangles*—as in Figs. 6.2–6.4—have led to the naming of these special binomial coefficients as *triangular numbers*.

---

[3] We study such counting techniques in depth in Section 11.1.

> **Cultural note.**
>
> Our combinatorial derivation of summation (6.6) illustrates one of the most important roles of mathematical abstraction. There is no obvious intuition to explain the relationship between the activity of summing n consecutive integers and the activity of extracting two objects out of a set of n objects. Yet, our combinatorial derivation exposes an intimate connection between the two.

Now that we know—*and understand*—how to derive the value of $S_1(n)$, we can finally evaluate our original series (6.4).

Proposition 6.2 *The arithmetic series (6.4) has the sum*

$$a + (a+b) + (a+2b) + (a+3b) + \cdots + (a+(n-1)b) \;=\; an + b \cdot \Delta_n \qquad (6.8)$$

6.2.1.3 Example #2: Squares as sums of odd integers

In this section, we build on Proposition 6.1 to craft multiple constructive proofs of the fact that each perfect square n^2 is the sum of the first n odd integers, $1, 3, 5, \ldots, 2n-1$. All of these proofs complement the "guess-and-verify" inductive proof of this result in Section 2.2.1.1 (Proposition 2.4).

Proposition 6.3 *For all $n \in \mathbb{N}^+$,*

$$\sum_{i=1}^{n} (2i-1) \;=\; 1 + 3 + 5 + \cdots + (2n-1) \;=\; n^2 \qquad (6.9)$$

That, is, the nth perfect square is the sum of the first n odd integers.

Before we present our proofs of this result, we want to stress that the notation for odd integers in the summation within Eq. (6.9) is completely general: *Every positive odd integer can be written in the form $2i-1$ for some positive integer i.*

Our first two proofs of Proposition 6.3 note that the result is a corollary of both Proposition 6.1 and Proposition 6.2.

Proof. **A proof by calculation.** By direct calculation, we find that

$$
\begin{aligned}
\sum_{i=1}^{n} (2i-1) &= 2\sum_{i=1}^{n} i \;-\; n \\
&= 2\Delta_n \;-\; n \qquad \text{(by Proposition 6.1)} \\
&= (n^2 + n) - n \\
&= n^2
\end{aligned}
$$

We thus have a proof by algebraic manipulation. □

Proof. **A proof by formula instantiation.** Because summation (6.9) is an arithmetic series with parameters $a = 1$ and $b = 2$, we know from Proposition 6.2 that the summation evaluates to

$$(1 \cdot n) + 2\Delta_{n-1} = n + n^2 - n = n^2$$

We thereby have a proof by applying our formula for arithmetic sums.   □

Our next proof builds on the stratagem of *finding invariants* that we exploited in the textual proof of Proposition 6.1.

Proof. **A textual proof.** We adapt Gauss's "trick" to this summation ; i.e., we write the current summation both forwards and backwards, and then add the two summations term by term. Let us denote the target summation $\sum_{k=1}^{n} (2k - 1)$ by $S(n)$. We record $S(n)$ both forwards and backwards:

"Forwards": $S(n) = \quad 1 \quad + \quad 3 \quad + \cdots + (2n - 3) + (2n - 1)$

$$(6.10)$$

"Backwards": $S(n) = (2n - 1) + (2n - 3) + \cdots + \quad 3 \quad + \quad 1$

Now, we add these two representations of $S(n)$ *column by column*. Because each of the n column-sums equals $2n$, we find that

$$2S(n) = 2 \sum_{i=1}^{n} (2i - 1) = 2n^2 \qquad (6.11)$$

We thus derive the sum (6.9) when we halve (i.e., divide by 2) the three equated quantities in Eq. (6.11).   □

Proof. **A "pictorial" proof.** We now build up to a proof that is almost purely pictorial, with only a bit of reasoning turning the pictures into a narrative. The only "sophisticated" knowledge this proof requires is that

$$(n + 1)^2 = n^2 + 2n + 1 \qquad (6.12)$$

This well-known equation is verified by symbolically *squaring* the expression $n + 1$:

$$(n + 1) \times (n + 1) = n \cdot (n + 1) + (n + 1) = n^2 + n + n + 1$$

Enrichment note.

Eq. (6.12) is the simplest instance of the *restricted Binomial Theorem*, which appears later in this chapter, as Theorem 6.1.

Our pictorial proof begins by representing the generic positive integer n as a horizontal sequence of n tokens, i.e., darkened circles. The problem of summing the

Fig. 6.5 Representing the first n odd integers using tokens. In this illustration, $n = 5$

first n odd integers then begins with a picture such as appears in Fig. 6.5, for the illustrative case $n = 5$.

Starting with this picture, we take each row of $2i - 1$ tokens and "fold" it at its midpoint so that it becomes a reversed letter "L". The row of $2i - 1$ tokens becomes an "L" (see Fig. 6.6) whose horizontal portion (at the bottom of the reversed "L") is a row of i tokens and whose vertical portion (at the right of the reversed "L") is a column of i tokens (the "hinge" token resides in both portions).

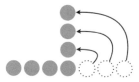

Fig. 6.6 Folding a single row into a reversed letter "L"

Once we have folded every row of tokens into a reversed "L", we nest the different-size occurrences of "L" into one another, in the manner depicted in Fig 6.7. Clearly, this nesting produces an $n \times n$ square array of (perforce, n^2) tokens. □

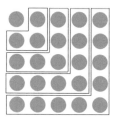

Fig. 6.7 The final picture organized as an $n \times n$ square array of tokens

Proof. **Another "pictorial" proof.** The reader who enjoyed the preceding "pictorial" proof may be amused by the challenge of completing the kindred slightly more complex proof illustrated by Fig. 6.8. In the figure we have illustrated four copies

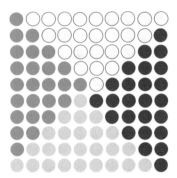

Fig. 6.8 Four copies of $S(n)$ represented as a triangle of tokens. The triangles are arranged to yield a $2n \times 2n$ square of tokens

of the triangle of tokens that depicts summation $S(n)$ (which sums the first n odd numbers). We have arranged the triangles in a way that produces a $2n \times 2n$ square of tokens. In the arrangement, each side of the square is a line of $1 + 2n - 1 = 2n$ tokens. This construction illustrates that $4S(n) = (2n)^2 = 4n^2$. $\square$

Explanatory note.

Full disclosure: Figure 6.8 provides the basis for a proof of Proposition 6.3, but it does not provide a complete proof: We must somehow verify that the construction in the figure is completely general, i.e., that the depicted emergence of the $2n \times 2n$ square of tokens from four copies of the Δ_n-triangle is not an artifact of the depicted case $n = 5$. This verification is not difficult.

What is notable about Fig. 6.8, even with this caveat, is how it truly facilitates the *discovery* of the Proposition.

Of course, it should not be surprising that pictorial reasoning is useful in highlighting the kinds of patterns that lead to discoveries as we *do* mathematics. It should also come as no surprise that nothing comes for free. As this example indicates: Even when pictorial reasoning leads us to intriguing discoveries, it does not obviate the invocation of additional modalities of reasoning for rigorously verifying the facts that the patterns reveal.

A preview of coming attractions. In Section 6.3.2.1, we develop the underpinnings of techniques that incrementally compute:

- the sums of the first n consecutive integers (the summations $S_1(n)$),

- the sums of the first n squares of the integers (the summations $S_2(n)$),

- the sums of the first n cubes of the integers (the summations $S_3(n)$),

and so on. We have thus begun to establish a base for a summation technique that is *inductive*, i.e., that computes each summation $S_c(n)$ from the lower-index summations: $S_1(n), S_2(n), \ldots, S_{c-1}(n)$. Our inductive technique awaits a bit more background.

6.2.2 Geometric Sums and Series

6.2.2.1 Overview and main results

We define geometric sequences and series and learn how to calculate their sums via the following generic examples.

An n-term geometric sequence:

$$a, \ ab, \ ab^2, \ \ldots, ab^{n-1} \tag{6.13}$$

The corresponding geometric summation:

$$S_{a,b}(n) \overset{\text{def}}{=} \sum_{i=0}^{n-1} ab^i = a + ab + ab^2 + \cdots + ab^{n-1} \tag{6.14}$$

$$= a \cdot (1 + b + b^2 + \cdots + b^{n-1})$$

The associated geometric *(infinite) series* (used only when $b < 1$):

$$S_{a,b}^{(\infty)} \overset{\text{def}}{=} \sum_{i=0}^{\infty} ab^i = a + ab + ab^2 + \cdots \tag{6.15}$$

Two related facts are clear from these definitions:
We can evaluate the summation (6.14) by evaluating just the sub-summation

$$S_b(n) \overset{\text{def}}{=} \sum_{i=0}^{n-1} b^i = 1 + b + b^2 + \cdots + b^{n-1} \tag{6.16}$$

We can evaluate the series (6.15) by evaluating just the sub-series

$$S_b^{(\infty)} \overset{\text{def}}{=} \sum_{i=0}^{\infty} b_i = 1 + b + b^2 + \cdots \tag{6.17}$$

This section's major results are embodied in the following proposition.

Proposition 6.4 *Let $S_b(n)$ be a geometric summation, as defined in Eq. (6.16).*
(a) *When $b > 1$, $S_b(n)$ evaluates to the following sum.*

$$S_b^{(b>1)}(n) = \frac{b^n - 1}{b - 1} \tag{6.18}$$

(b) *When $b < 1$, $S_b(n)$ evaluates to the following sum.*

$$S_b^{(b<1)}(n) = \frac{1 - b^n}{1 - b} \tag{6.19}$$

(c) *In the uninteresting,* degenerate *case $b = 1$*

$$S_b^{(b=1)}(n) = 1 + 1 + \cdots + 1 \ (n \ times) = n$$

Explanatory notes.

(1) Of course, the expressions (6.18) for $S_b^{(b>1)}(n)$ and (6.19) for $S_b^{(b<1)}(n)$ are algebraically equivalent as functions of b and n. But they will be read differently:

$$\left[S_2^{(2)}(n) = 2^n - 1 \right] \quad \text{while} \quad \left[S_2^{(1/2)}(n) = \frac{1 - \left(\frac{1}{2}\right)^n}{1 - \left(\frac{1}{2}\right)} = 2 - \left(\frac{1}{2}\right)^{n-1} \right]$$

(2) The *New Oxford Dictionary* defines the word "degenerate" as follows:

lacking some property, order, or distinctness of structure previously or usually present.

Mathematics relating to or denoting an example of a particular type of equation, curve, or other entity that is equivalent to a simpler type, often occurring when a variable or parameter is set to zero.

In our case, the structures of the terms in summations (6.18) and (6.19) play a critical role in determining the sums of the summations. In the case $b = 1$, all terms lack structure: they "degenerate" to value 1.

The infinite case of Eq. (6.17) can be dealt with as a corollary of Proposition 6.4(b), by letting n grow without bound and observing that the resulting sequence of values converges.

Proposition 6.5 *When $b < 1$, the* infinite *series $S_b^{(\infty)}$ converges to the following sum.*

$$S_b^{(\infty)} = \sum_{i=0}^{\infty} b^i = 1 + b + b^2 + \cdots = \frac{1}{1 - b}$$

6.2.2.2 Techniques for summing geometric series

We now develop several proofs of Propositions 6.4 and 6.5, each providing a new insight on the summation process.

Proof. **A proof via textual replication.** Toward the end of developing our first method for summing $S_b(n)$, we note that we can rewrite the summation in two ways that are *(textually) recurrent*.

Recurrent subexpressions constitute a mathematically exploitable *pattern* of the form described in Chapter 2. We now describe how such patterns can be exploited to find explicit sums for geometric summations and series.

Both of the recurrent expressions for $S_b(n)$ have the form

$$S_b(n) = \alpha \cdot S_b(n) + \beta(n) \tag{6.20}$$

where α is a constant and $\beta(n)$ is a function of n; both α and $\beta(n)$ may depend on the parameter b. We provide two recurrent expressions for $S_b(n)$, one of which is more interesting when $b > 1$, the other when $b < 1$.

$$S_b(n) \overset{\text{def}}{=} 1 + b + b^2 + \cdots + b^{n-1}$$

$$= b \cdot S_b(n) + (1 - b^n) \tag{6.21}$$

$$= \frac{1}{b} \cdot S_b(n) + \frac{b^n - 1}{b} \tag{6.22}$$

The significance of a recurrent expression of the type (6.20) is that it exposes an explicit value for $S_b(n)$:

$$S_b(n) = \frac{\beta(n)}{1 - \alpha} \tag{6.23}$$

We now combine the generic value (6.23) of $S_b(n)$ with the specialized recurrent expressions in Eqs. (6.21) and (6.22) to derive two explicit solutions for $S_b(n)$.

1. The first solution is most useful and perspicuous when $b > 1$. In this case:

$$\left(1 - \frac{1}{b}\right) S_b^{(b>1)}(n) = b^{n-1} - \frac{1}{b}$$

which is easily rearranged to the equivalent and more perspicuous form of expression (6.18).

2. The second solution is most useful and perspicuous when $b < 1$. In this case:

$$(1 - b) S_b^{(b<1)}(n) = 1 - b^n$$

which is easily rearranged to the equivalent and more perspicuous form in expression (6.19).

This completes our proof by replication. $\square$

Note that both $S_b^{(b>1)}(n)$ and $S_b^{(b<1)}(n)$ have simple *approximate* values which are useful in "back-of-the-envelope" calculations: For very large values of n, we have

$$S_b^{(b>1)}(n) \approx \frac{b^n}{b-1} \quad \text{and} \quad S_b^{(b<1)}(n) \approx \frac{1}{1-b} \tag{6.24}$$

The expression for $S_b^{(b<1)}(n)$ in Eq. (6.24) is actually a rewording of Proposition 6.5.

Proof. **A "pictorial" summation of** $S_{1/2}^{(\infty)}$. Figure 6.9 pictorially depicts a process whose analysis provides a rigorous proof of Proposition 6.5 for the case $b = 1/2$, i.e., a rigorous argument that the series $S_{1/2}^{(\infty)} = \sum_{i=0}^{\infty} 2^{-i}$ sums to 2. In this evalua-

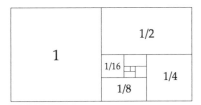

Fig. 6.9 Arranging successive rectangles to evaluate $S_{1/2}^{(\infty)}$

tion of $S_{1/2}^{(\infty)}$, we measure fractional quantities by the portion of a unit-side rectangle that they fill. Thus (follow in the figure): the initial term of $S_{1/2}^{(\infty)}$, namely, 1, is represented by the unit square that is labeled "1" in the figure. The next term of the series, namely, $1/2$, is represented by the rectangle labeled "$1/2$" in the figure. And so on, with successively smaller rectangles. Because we design each rectangle to have half the area of its predecessor, the sequence of rectangles represents successively smaller inverse powers of 2. As the process proceeds, we observe increasingly more of the right-hand unit-side square being filled. In fact, one can verify, via an elementary induction, that *every* point in the right-hand unit-side square eventually gets covered by some small rectangle (as n tends to ∞), thereby establishing that the infinite series $S_{1/2}^{(\infty)}$ does, indeed, sum to 2.

The preceding procedure is difficult, but not impossible, to adapt to values of $b < 1$ other than $1/2$. The sequence Fig. 6.10–Fig. 6.12 suggests how to achieve such an adaptation for any value of b with $0 \le b < 1$, via an appropriate cascade of shrinking squares. The unit-side square in Fig. 6.10 begins the construction of the cascade. The two squares in Fig. 6.11 illustrate the second step in constructing the cascade; the suggestive cascade depicted in Fig. 6.12 illustrates what the final cascade looks like: The cumulative length of the bases of its abutting rectangles is the value of the infinite series $S_b^{(\infty)}$; cf. Eq. (6.15). The proof that the base of the cascade in the figure yields the desired sum is completed by a version of the theorem

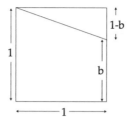

Fig. 6.10 Initial state: the unit square and the base b

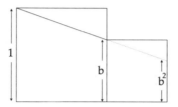

Fig. 6.11 Beginning to develop the geometric series via cascading shrinking squares

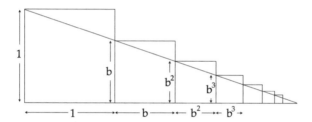

Fig. 6.12 The complete process for computing a geometric series by using a cascade of shrinking squares

about similar triangles which is associated with Thales of Miletus and mentioned in Euclid's *Elements*. □

Explanatory/historical note.

Thales of Miletus is credited with a classical theorem about similar right triangles. The following figure depicts right triangle T and its (shaded) sub-triangle T' whose right angle is distance 1 from T's.

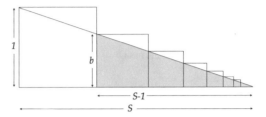

Proposition 6.6 (The Theorem of Thales) *Let us be given a unit-height right triangle T whose base has length S. The ratio $(S-1)/S$ equals the height b of the sub-triangle T' whose right angle is at distance 1 from T's.*

We use this proportionality of the side-ratios of the big and small triangles to evaluate the summation $\sum_{i=0}^{\infty} b^i$ for $0 \le b < 1$.

Proof. **Another "pictorial" summation of** $S_{1/2}^{(\infty)}$. The pictorial derivation of the sum $S_{1/2}^{(\infty)}$ can be accomplished using geometric shapes other than squares. We now present a natural derivation of the sum $S_{1/2}^{(\infty)}$ by vigorously slicing a pie.

The process of pie-slicing works most naturally with the modified series

$$\overline{S}_{1/2}^{(\infty)} \;=\; \sum_{i=1}^{\infty} 2^{-i} \;=\; S_{1/2}^{(\infty)} - 1$$

which omits the initial summand 1 from $S_{1/2}^{(\infty)}$. Of course, $\overline{S}_{1/2}^{(\infty)}$ sums to 1, because $S_{1/2}^{(\infty)}$ sums to 2.

The pie-slicing evaluation of $\overline{S}_{1/2}^{(\infty)}$ is depicted in Fig. 6.13. In the figure, the inverse powers of 2 are represented by appropriate fractions of a unit-diameter disk (the pie). The evaluation begins with this disk before it is sliced: this represents the number 1, which we eventually show to be the sum $\overline{S}_{1/2}^{(\infty)}$. We slice the disk in half using the depicted diameter and label one of the resulting half-disks "1/2". Next, we slice the other half-disk in half using a radius of the unit disk and label one of the quarter-disks "1/4". We continue in this manner *ad infinitum*. The analysis that yields the sum $\overline{S}_{1/2}^{(\infty)}$ amounts to a proof that every point in the unit-diameter disk eventually resides in a slice that is not further sliced. Details are left to the reader.
□

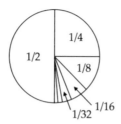

Fig. 6.13 Computing the sum of $1/2^i$ by slicing a unit disk

6.2.2.3 Extended geometric series and their sums

We now build on our ability to evaluate geometric summations of the forms (6.18) and (6.19) to evaluate summations that we term *extended* geometric summations (not a standard term), i.e., summations of the form

$$S_b^{(c)}(n) \overset{\text{def}}{=} \sum_{i=1}^{n} i^c b^i$$

where c is an arbitrary fixed positive integer, and b is an arbitrary fixed real number.

We restrict attention in this section to summations $S_b^{(c)}(n)$ that satisfy the joint inequalities $c \neq 0$ and $b \neq 1$.

- We have already adequately studied the case $c = 0$, which characterizes "ordinary" geometric summations.

- The degenerate case $b = 1$ removes the "geometric growth" of the sequence underlying the summation. We study various aspects of this *summation-of-fixed-powers* case in Section 6.3, with special treatment of summations of fixed powers of consecutive integers in Section 6.3.2.

The summation method that we now develop for evaluating all other summations $S_b^{(c)}(n)$—i.e., those with $b \neq 1$ and $c \neq 0$—has the following notable properties.

- The method is *inductive in parameter c*.

  Specifically, we will express our sum for $S_b^{(c)}(n)$ in terms of sums for summations $S_b^{(c-1)}(n)$, $S_b^{(c-2)}(n)$, ..., $S_b^{(1)}(n)$, and $S_b^{(0)}(n) = S_b(n)$.

- For each fixed value of c, the method is *inductive in the argument n*.

- The method relies on the recurrent-subexpression strategy which we used effectively in Section 6.2.2.

A. The summation $S_b^{(1)}(n) = \sum_{i=1}^{n} ib^i$

We illustrate our strategy in detail for the case $c = 1$ and sketch only briefly how it deals with larger values of c. Elementary algebraic manipulations which are suggested by the analysis of the case $c = 1$ should thereby allow the reader to deal with any value $c > 1$.

Proposition 6.7 *For all bases $b > 1$,*

$$S_b^{(1)}(n) = \sum_{i=1}^{n} ib^i = \frac{(b-1)n-1}{(b-1)^2} \cdot b^{n+1} + \frac{b}{(b-1)^2} \tag{6.25}$$

Proof. **Deriving a sum via algebraic manipulation.** We begin to develop our strategy by writing the natural expression for

$$S_b^{(1)}(n) = b + 2b^2 + 3b^3 + \cdots + nb^n$$

in two different ways. First, we isolate the summation's last term:

$$S_b^{(1)}(n+1) = S_b^{(1)}(n) + (n+1)b^{n+1} \tag{6.26}$$

Then we isolate the left-hand side of expression (6.26):

$$S_b^{(1)}(n+1) = b + \sum_{i=2}^{n+1} ib^i$$

$$= b + \sum_{i=1}^{n} (i+1)b^{i+1}$$

$$= b + b \cdot \sum_{i=1}^{n} (i+1)b^i$$

$$= b + b \cdot \left(\sum_{i=1}^{n} ib^i + \sum_{i=1}^{n} b^i \right)$$

$$= b \cdot \left(S_b^{(1)}(n) + S_b^{(0)}(n) \right) + b$$

$$= b \cdot \left(S_b^{(1)}(n) + \frac{b^{n+1}-1}{b-1} - 1 \right) + b$$

$$= b \cdot S_b^{(1)}(n) + b \cdot \frac{b^{n+1}-1}{b-1} \tag{6.27}$$

Combining expressions (6.26) and (6.27) for $S_b^{(1)}(n+1)$, we finally find that

$$(b-1) \cdot S_b^{(1)}(n) = (n+1) \cdot b^{n+1} - b \cdot \frac{b^{n+1}-1}{b-1}$$

$$= \left(n - \frac{1}{b-1}\right) \cdot b^{n+1} + \frac{b}{b-1} \qquad (6.28)$$

We now use standard algebraic manipulations to derive expression (6.25) from Eq. (6.28). ☐

Proof. **Solving the case** $b = 2$ **using subsum rearrangement.** We evaluate the sum

$$S_2^{(1)}(n) = \sum_{i=1}^{n} i2^i$$

in an especially interesting way, by rearranging the sub-summations of the target summation.

Explanatory note.

The reader should pay careful attention to this technique. It shows that one can sometimes decompose a cumbersome expression for a summation into a readily manipulated one.

Underlying our evaluation of $S_2^{(1)}(n)$ is the fact that we can rewrite the summation as a *double* summation:

$$S_2^{(1)}(n) = \sum_{i=1}^{n} \sum_{k=1}^{i} 2^i \qquad (6.29)$$

By suitably applying the laws of arithmetic that appear in Section 5.2—specifically, the distributive, associative, and commutative laws—we can perform the required double summation in a different order than that specified in Eq. (6.29). In fact, we can exchange the indices of summation in a manner that enables us to compute $S_2^{(1)}(n)$ in the order implied by the following expression:

$$S_2^{(1)}(n) = \sum_{k=1}^{n} \sum_{i=k}^{n} 2^i$$

This process is depicted in Fig. 6.14. The indicated summation is much easier to perform in this order, because its core consists of instances of the "ordinary" geometric summation $\sum_{i=k}^{n} 2^i$ (Proposition 6.4). Expanding these instances, we find finally that

$$S_2^{(1)}(n) = \sum_{k=1}^{n} \left(2^{n+1} - 1 - \sum_{i=0}^{k-1} 2^i\right)$$
$$= \sum_{k=1}^{n} \left(2^{n+1} - 2^k\right)$$
$$= n \cdot 2^{n+1} - (2^{n+1} - 1) + 1$$
$$= (n-1) \cdot 2^{n+1} + 2$$

This completes the proof. ☐

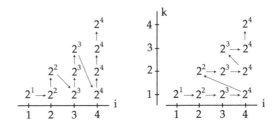

Fig. 6.14 Illustrating the case $n = 4$ of the exchange of indices in the summation. The original summation is drawn on the left. In the right-hand drawing, each term $i \cdot 2^i$ is reproduced i times within column i. We then express the summation on the right by introducing a new index k and scanning the terms of the summation row by row (on the right). The arrows represent the flow of the scan

We remark that the process of obtaining the original summation can also be seen in the figure, by scanning the elements of the summation along diagonals, as depicted in Fig. 6.15. Each of the n diagonals contains exactly the difference between the complete geometric summation and the partial summation that is truncated at the kth term.

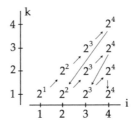

Fig. 6.15 The successive diagonal patterns correspond to the single summation obtained after exchanging the two initial sums

B. The summations $S_b^{(c)}(n) = \sum_{i=1}^{n} i^c b^i$

We now develop a strategy which adapts the evaluation of summation $S_b^{(1)}(n)$ in the proof of Proposition 6.7 to an evaluation of the general extended geometric summation

$$S_b^{(c)}(n) = \sum_{i=1}^{n} i^c b^i = b + 2^c b^2 + 3^c b^3 + \cdots + n^c b^n$$

The strategy is *recursive*, in that it computes a value for $S_b^{(c)}(n)$ from values for $S_b^{(c-1)}(n), S_b^{(c-2)}(n), \ldots, S_b^{(0)}(n)$. It proceeds in three steps.

Step 1. As in the case $c = 1$, we write summation $S_b^{(c)}(n)$ in two ways. The expression that embodies the first way isolates the summation's first term:

$$S_b^{(c)}(n+1) = b + \sum_{i=1}^{n} (i+1)^c b^{i+1}$$

The expression that embodies the second way isolates the summation's last term:

$$S_b^{(c)}(n+1) = S_b^{(c)}(n) + (n+1)^c b^{n+1}$$

By combining these expressions, we find that

$$S_b^{(c)}(n) = b \cdot \left(1 - (n+1)^c b^n + \sum_{i=1}^{n} (i+1)^c b^i \right) \tag{6.30}$$

Step 2. We next invoke the Restricted Binomial Theorem (Theorem 6.1) to see that

$$(i+1)^c = i^c + c \cdot i^{c-1} + \binom{c}{2} \cdot i^{c-2} + \cdots + \binom{c}{k} \cdot i^{c-k} + \cdots + 1$$

We use this expansion of $(i+1)^c$, together with multiple applications of the laws of arithmetic from Section 5.2 to verify that

$$\sum_{i=1}^{n} (i+1)^c b^i = S_b^{(c)}(n) + c \cdot S_b^{(c-1)}(n) + \binom{c}{2} \cdot S_b^{(c-2)}(n) + \cdots$$

$$\cdots + \binom{c}{k} \cdot S_b^{(c-k)}(n) + \cdots + S_b^{(0)}(n) \tag{6.31}$$

Step 3. We finally combine Eqs. (6.30) and (6.31) to discover that

$$(b-1) \cdot S_b^{(c)}(n) = (n+1)^c b^n - c \cdot S_b^{(c-1)}(n) - \binom{c}{2} \cdot S_b^{(c-2)}(n) - \cdots$$

$$\cdots - \binom{c}{k} \cdot S_b^{(c-k)}(n) - \cdots - S_b^{(0)}(n) - 1 \tag{6.32}$$

We now have the promised method of evaluating the summation $S_b^{(c)}(n)$ associated with the fixed power c in terms of the sums of kindred summations associated with smaller fixed powers.

Each incremental step in our summation technique is an elementary application of an idea that we have seen previously. Careful pattern-matching—primarily always focusing on the "top level" of the target summation, i.e., on the largest current values of parameters b and c—has enabled us to achieve a decidedly nonelementary goal via a sequence of elementary steps.

6.3 On Summing "Smooth" Series

6.3.1 Approximate Sums via Integration

This section develops and illustrates a powerful strategy for obtaining nontrivial upper and lower bounds on the sums of summations and series, by finding continuous *envelopes* that bound the summations both above and below. The areas under the enveloping continuous functions—which we can calculate via integration—provide the desired bounds on the summations.

The stratagem follows a three-step procedure. Focus on a summation

$$S = a_1 + a_2 + \cdots + a_n$$

whose sum we want to determine. We have specified S as a *finite* summation to simplify exposition; our stratagem often works with infinite summations also, as we shall see via specific examples.

Step 1. Represent the summands of S as a series of n abutting unit-width rectangles of respective heights $a_1, a_2, \ldots, a_n$.

Step 2. Construct a continuous curve $\overline{C}(x)$ that passes through the corners of the unit-width rectangles specified by the summation, in such a way that the rectangles lie completely within the area under $\overline{C}(x)$.

Because the aggregate areas of the abutting rectangles lie completely under curve $\overline{C}(x)$, the area under the curve—obtained by integrating $\overline{C}(x)$ between appropriate limits—yields an *upper bound* on the value of the summation of interest.

Step 3. Construct a continuous curve $\underline{C}(x)$ that passes through the corners of the unit-width rectangles specified by the summation, in such a way that the area under $\underline{C}(x)$ lies completely within the rectangles.

Because the area under the curve $\underline{C}(x)$ lies completely within the abutting rectangles, the area under the curve—obtained by integrating $\underline{C}(x)$ within appropriate limits—yields a *lower bound* on the summation of interest.

We now describe two specific summations, to illustrate the stratagem in action.

1. Figure 6.16 illustrates our stratagem applied to the seven-term version of the summation $S_2(n) = \sum_{i=1}^{n} i^2$, namely,

$$S_2(7) = 1 + 4 + 9 + 16 + 25 + 36 + 49$$

The figure represents the terms of $S_2(7)$ by means of seven abutting unit-width rectangles whose respective heights are given by the terms. The sum of $S_2(7)$ is, then, given by the aggregate area of the abutting rectangles. If we were to extend the figure rightward—which increases n and thereby extends the summation by encompassing more addends—then the next two rectangles would have respective heights 64 and 81.

Our stratagem mandates embellishing the rectangles in Fig. 6.16 by two continuous curves—labeled $\overline{C}$ and $\underline{C}$ in the figure—that connect the upper corners of the rectangles. Curve $\overline{C}$ completely "covers" the rectangles; therefore, the area $A(\overline{C})$

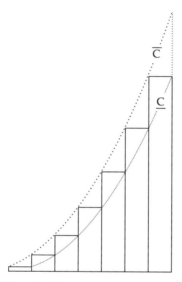

Fig. 6.16 The summation $S_2(n) = \sum_{i=1}^{n} i^2$ represented as the aggregate area of a sequence of unit-width shaded rectangles. The summation is *bounded from above* by the area under the continuous curve $\overline{C}(x)$ that connects the upper left-hand corners of the rectangles; this curve is labeled $\overline{C}$ in the figure. In detail, curve $\overline{C}$ begins at the upper left-hand corner of the leftmost rectangle; it continues to what would be the upper left-hand corner of the rightmost rectangle. The area under curve $\overline{C}$ is $A(\overline{C}(x)) = \int_0^n (x+1)^2 dx$. The summation is *bounded from below* by the area under the continuous curve $\underline{C}(x)$ that connects the upper right-hand corners of the rectangles; this curve is labeled $\underline{C}$ in the figure. In detail, curve $\underline{C}$ begins at the upper right-hand corner of the leftmost rectangle; it continues to the upper right-hand corner of the rightmost rectangle. The area under curve $\underline{C}$ is $A(\underline{C}(x)) = \int_1^n x^2 dx$.

under the curve is an *upper bound* on the aggregate area of the rectangles:

$$A(\overline{C}(x)) = \int_0^n (x+1)^2 dx$$

Curve $\underline{C}$ lies completely within the rectangles; therefore, the area $A(\underline{C})$ under the curve is a *lower bound* on the aggregate area of the rectangles:

$$A(\underline{C}(x)) = \int_1^n x^2 dx$$

2. Figure 6.17 illustrates our stratagem as it applies to the ten-term version of the *harmonic* summation

$$S^{(H)}(10) = \sum_{i=1}^{10} i^{-1} = \sum_{i=1}^{10} 1/i = 1 + \frac{1}{2} + \frac{1}{3} + \frac{1}{4} + \frac{1}{5} + \frac{1}{6} + \frac{1}{7} + \frac{1}{8} + \frac{1}{9} + \frac{1}{10}$$

The figure represents these terms via ten abutting unit-width rectangles whose respective heights are given by the terms. The sum of $S^{(H)}(10)$ is, then, the aggregate area of the abutting rectangles. If we were to extend the figure rightward (which increases n and thereby extends the summation to encompass more addends), then the next two rectangles would have respective heights $1/11$ and $1/12$.

Our stratagem mandates embellishing the rectangles in Fig. 6.17 by two continuous curves—labeled $\overline{C}$ and $\underline{C}$ in the figure—that connect the rectangles' upper corners. The left-hand rectangle *plus* curve $\overline{C}$ completely covers the rectangles;

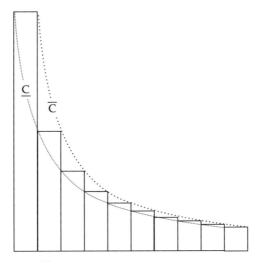

Fig. 6.17 The summation $S^{(H)}(n) = \sum_{i=1}^{n} 1/i$ represented as the aggregate area of a sequence of unit-width rectangles. The summation is *bounded from above* by the area of the leftmost rectangle *plus* the area $A(\overline{C}(x))$ under the continuous curve $\overline{C}$ that connects the upper righthand corners of the rectangles. In detail, this curve begins at the upper right-hand corner of the leftmost rectangle; it continues to the upper right-hand corner of the rightmost rectangle. (The figure's resolution makes the right-hand end of the curve invisible.) The upper-bounding area is $A(\overline{C}(x)) = 1 + \int_1^n 1/x dx$. The summation is *bounded from below* by the area $A(\underline{C}(x))$ under the continuous curve $\underline{C}$ that connects the upper left-hand corners of the rectangles. In detail, curve $\underline{C}$ begins at the upper left-hand corner of the leftmost rectangle and continues to the upper left-hand corner of the rightmost rectangle. The lower-bounding area is $A(\underline{C}(x)) = \int_0^{n-1} 1/(x+1) dx$.

therefore, their combined areas provide an *upper bound* on the aggregate area of the rectangles. Curve $\underline{C}$ lies completely within the rectangles; therefore, the area $A(\underline{C}(x))$ under the curve is a *lower bound* on the aggregate area of the rectangles.

Explanatory note.

The curves labeled $\overline{C}$ in Figs. 6.16 and 6.17 are instances of the stratagem's mandated continuous curve $\overline{C}(x)$. *Note that curve $\overline{C}$ in Fig. 6.17 must be considered to contain the top of the leftmost rectangle in the figure.*

The curves labeled $\underline{C}$ in Figs. 6.16 and 6.17 are instances of the stratagem's mandated continuous curve $\underline{C}(x)$.

In the next subsection, we apply our stratagem to summations of fixed powers of successive integers, i.e., summations of the form

$$S_c(n) \stackrel{\text{def}}{=} \sum_{i=1}^{n} i^c$$

for various (classes of) values of the fixed power c.

6.3.2 Sums of Fixed Powers of Consecutive Integers: $\sum i^c$

We derive bounds on the summations $S_c(n)$ which are rather good for large n. In some cases, the bounds are very good, sometimes even exact for all n.

6.3.2.1 $S_c(n)$ for general *nonnegative* real cth powers

We begin to illustrate the technique of bounding summations via integrals by focusing on summations of the form

$$S_c(n) \stackrel{\text{def}}{=} \sum_{i=1}^{n} i^c$$

for arbitrary positive numbers c. The reader can garner intuition for the upcoming bounds from the general shapes of the continuous curves that rest upon the rectangles in Fig. 6.16. The bounds that we describe now can be specified more elegantly using the formalism which awaits us in Section 7.1.

We obtain our bounds on each summation $S_c(n)$ by evaluating the integrals that yield the area $\overline{C}_c(n)$ under the continuous curve $\overline{C}$ in (the cth-power analogue of) Fig. 6.16 and the area $\underline{C}_c(n)$ under the continuous curve $\underline{C}$ in the figure. We thereby find that, for all sufficiently large n:

- the former bound is given by

$$\overline{C}_c(n) = \int_0^n (x+1)^c dx = \frac{1}{c+1}(n+1)^{c+1} + \overline{\kappa} \tag{6.33}$$

- the latter bound is given by

$$\underline{C}_c(n) = \int_1^n x^c dx = \frac{1}{c+1}n^{c+1} + \underline{\kappa} \tag{6.34}$$

The terms $\overline{\kappa}$ and $\underline{\kappa}$ are real constants which do not depend on n—although they could depend on c.

We combine the bounds (6.33) and (6.34) into a two-sided bound on $S_c(n)$.

- We use the quantity $\underline{C}_c(n)$ from Eq. (6.34) as our lower bound on $S_c(n)$.

- To derive our upper bound $S_c(n)$:

  – We begin with the quantity $\overline{C}_c(n)$ from Eq. (6.33)

  $$\frac{1}{c+1}(n+1)^{c+1} + \overline{\kappa}$$

  – We multiply this expression by the quantity $n^{c+1}/n^{c+1} = 1$. This operation clearly does not change the value of the expression, although it does change its form into

  $$\frac{1}{c+1} \cdot n^{c+1} \cdot \left(\frac{n+1}{n}\right)^{c+1} + \overline{\kappa}$$

  This expression can be simplified by algebraic manipulation into the form

  $$\frac{1}{c+1} \cdot n^{c+1} \cdot \left(1 + \frac{1}{n}\right)^{c+1} + \overline{\kappa} \tag{6.35}$$

  – Next, we remark that there exists a constant κ which depends on $\overline{\kappa}$ and c, *but not on n*, such that expression (6.35) is no larger than (i.e., is less than or equal to) the expression

  $$\frac{1}{c+1} \cdot n^{c+1} \cdot \left(1 + \frac{1}{n}\right)^{n+1} + \kappa \tag{6.36}$$

  This inequality follows from the fact that $1 + 1/n$ always exceeds 1, so increasing the exponent in expression (6.35) from $c+1$ to $n+1$ can only increase the value of the expression.

  – As our final discrete step, we note that when $n \geq c$,

$$\frac{1}{c+1} \cdot n^{c+1} \cdot \left(1 + \frac{1}{n}\right)^{n+1} + \kappa \leq \frac{1}{c+1} \cdot \frac{c+1}{c} \cdot n^{c+1} \cdot \left(1 + \frac{1}{n}\right)^{n} + \kappa$$

$$= \frac{1}{c} \cdot n^{c+1} \cdot \left(1 + \frac{1}{n}\right)^{n} + \kappa \qquad (6.37)$$

– Finally, we invoke a somewhat surprising, elegant result from the calculus.

Lemma 6.1. *For all positive n,* $(1 + 1/n)^{n} \leq e$, *where* $e = 2.71828\cdots$ *is Euler's constant.*

Lemma 6.1 tells us that expression (6.37) no larger than (i.e., is less than or equal to) the expression

$$\frac{e}{c} \cdot n^{c+1} + \kappa \qquad (6.38)$$

We finally have our two-sided bound on $S_c(n)$:

Proposition 6.8 *There exist constants* $\underline{\kappa}$ *and* κ *which do not depend on n—but which could depend on c—such that, for all sufficiently large n:*

$$\frac{1}{c+1} \cdot n^{c+1} + \underline{\kappa} \leq S_c(n) \leq \frac{e}{c} \cdot n^{c+1} + \kappa \qquad (6.39)$$

The main message here is that when one views each $S_c(n)$ as a function of n:

The behavior of $S_c(n)$ *is dominated by* n^{c+1} *as n grows without bound.*

6.3.2.2 Nonnegative integer cth powers

When c is a nonnegative integer, we can strengthen Proposition 6.8.

Proposition 6.9 *There exist constants* $\underline{\kappa}$ *and* κ *which do not depend on n—although they could depend on c—such that, for all sufficiently large n:*

$$\frac{1}{c+1} n^{c+1} + \underline{\kappa} \leq S_c(n) \leq \frac{1}{c+1} n^{c+1} + \kappa n^{c} \qquad (6.40)$$

The upper bound in the inequality-pair (6.40) follows from a restricted version of Newton's celebrated Binomial Theorem.

A. A better bound via the Binomial Theorem

When c is a positive integer, the following special case of Newton's *Binomial Theorem*[4] affords us a much more detailed upper bound on the sum of $S_c(n)$.

---

[4] The general form of the Binomial Theorem expands the polynomial $(x+y)^{k}$ rather than $(x+1)^{k}$; see Section 5.3.2.

Theorem 6.1 (The Restricted Binomial Theorem). *For all positive integers k,*

$$(x+1)^k = \sum_{i=0}^{k} \binom{k}{i} x^{k-i} \tag{6.41}$$

We improve our upper bound on $S_c(n)$ by paralleling the reasoning that led us to the relation (6.33). Our improved bound emerges also by evaluating the integral that yields the area $\overline{C}_c(n)$ under the continuous curve that passes through the upper left-hand corners of the unit-width rectangles specified by the summation $S_c(n)$. For sufficiently large n:

$$\overline{C}_c(n) = \int_0^n (x+1)^c dx = \int_0^n \left(\sum_{i=0}^{c} \binom{c}{i} x^{c-i} \right) dx$$

$$= \sum_{i=0}^{c} \left(\int_0^n \binom{c}{i} x^{c-i} \right) dx$$

$$= \sum_{i=0}^{c} \frac{1}{c-i+1} \binom{c}{i} n^{c-i+1} \tag{6.42}$$

This is a proper upper bound because the region defined by this curve totally contains the region covered by the rectangles.

We now have access to the upper bound of (6.40), hence to Proposition 6.40, by exploiting the fact that the right-hand expression in Eq. (6.41) has $k+1$ terms. Therefore, for sufficiently large n:

$$\sum_{i=0}^{c} \frac{1}{c-i+1} \binom{c}{i} n^{c-i+1} = \frac{1}{c+1} n^{c+1} + \sum_{i=1}^{c} \frac{1}{c-i+1} \binom{c}{i} n^{c-i+1}$$

$$\leq \frac{1}{c+1} n^{c+1} + c \cdot \binom{c}{\lceil c/2 \rceil} n^c \tag{6.43}$$

We obtain the bound of inequality (6.43) via two majorizations. First, we remark that $1/(c-i+1) \leq 1$ when $1 \leq i \leq c$. Then we determine that the binomial coefficient $\binom{c}{i}$ achieves its maximum when $i = \lceil c/2 \rceil$. The latter fact is discovered via the following reasoning.

- Binomial coefficients are symmetric; i.e.,

$$\binom{c}{i} = \binom{c}{c-i}$$

- Binomial coefficients increase "from either end"; i.e.,

$$\binom{c}{i} < \binom{c}{i+1}$$

when $i < c/2$

The calculations that detail the preceding inequalities are left to the reader.

We have, of course, not changed the dominant behavior of $S_c(n)$ as n grows without bound, but we have taken a significant step toward developing explicit expressions for the sums of the summations $S_c(n)$ when c is a positive integer.

B. Using *undetermined coefficients* to refine sums

We now introduce the *Method of Undetermined Coefficients* and illustrate its use in deriving explicit expressions for the sums $S_c(n)$ when c is a positive integer. Our development builds on the intuition (garnered from the bounds in Eq. (6.43)) that

$$S_c(n) = \frac{1}{c+1}n^{c+1} + a_c^{(c)}n^c + a_{c-1}^{(c)}n^{c-1} + \cdots + a_2^{(c)}n^2 + a_1^{(c)}n + a_0^{(c)}$$

for some nonnegative numbers $a_c^{(c)}, \ldots, a_0^{(c)}$. To begin, we know that $a_0^{(c)} = 0$, because $S_c(0) = 0$.

Explanatory note.

Because our starting point is a conjecture based only on intuition, we will have to verify the explicit expressions that we derive. We do this after deriving the expressions.

The Method becomes computationally cumbersome for large values of c, as it involves solving c linear equations in c unknowns—a problem that is outside the scope of the text. Therefore, all we do here is introduce the Method by illustrating its use with the first few positive integer values of c.

The case $c = 1$. We begin with the sum $S_1(n)$, whose value we already know. Reasoning from the case $c = 1$ of Eq. (6.43), we intuit that

$$S_1(n) = \frac{1}{2}n^2 + a_1^{(1)}n$$

for some positive *undetermined coefficient* $a_1^{(1)}$. We can discover the value of the single unknown, $a_1^{(1)}$ by evaluating $S_1(n)$ at a single value for the variable n. Since *any* value of n will work—because we are solving *one linear equation in one unknown*—we use the *smallest* one, $n = 1$, to simplify our calculation. Because $S_1(1) = 1$, we have

$$S_1(1) = 1 = \frac{1}{2} + a_1^{(1)}$$

Therefore, $a_1^{(1)} = 1/2$, which gives us yet one more derivation of the value

$$S_1(n) = \frac{1}{2}\left(n^2 + n\right) = \frac{n(n+1)}{2}$$

The case $c = 2$. We derive an explicit expression for $S_2(n) \overset{\text{def}}{=} 1 + 4 + \cdots + n^2$.

Proposition 6.10 For all $n \in \mathbb{N}$,

$$S_2(n) \overset{\text{def}}{=} \sum_{i=1}^{n} i^2 = \frac{1}{3}n^3 + \frac{1}{2}n^2 + \frac{1}{6}n \tag{6.44}$$

$S_2(n)$ is often expressed in a more aesthetic form:

$$S_2(n) = \frac{1}{6}n(n+1)(2n+1) = \frac{2n+1}{3} \cdot \binom{n}{2}$$

Proof. Reasoning from the case $c = 2$ of Eq. (6.43), we propose the conjecture:

$$S_2(n) = \sum_{i=0}^{n} i^2 = \frac{1}{3}n^3 + a_2^{(2)}n^2 + a_1^{(2)}n \tag{6.45}$$

for some positive *undetermined coefficients* $a_2^{(2)}$ and $a_1^{(2)}$. We thereby express $S_2(n)$ as a polynomial in two unknowns, $a_2^{(2)}$ and $a_1^{(2)}$. We determine values for the unknowns by instantiating the polynomial with (any) two values of n; to simplify calculations, we select the smallest two values of n, namely, $n = 1, 2$. These instantiations of the polynomial leave us with the following pair of linear equations.

$$n = 1 : \sum_{i=0}^{1} i^2 = 1 = 1/3 + a_2^{(2)} + a_1^{(2)}$$

$$n = 2 : \sum_{i=0}^{2} i^2 = 5 = 8/3 + 4a_2^{(2)} + 2a_1^{(2)}$$

By elementary arithmetic, these equations simplify to yield the pair

$$a_2^{(2)} + a_1^{(2)} = 2/3$$

$$2a_2^{(2)} + a_1^{(2)} = 7/6$$

These equations reveal that

$$2/3 - a_2^{(2)} = 7/6 - 2a_2^{(2)}$$

so that

$$a_2^{(2)} = 1/2$$

By back-substitution, then,

$$a_1^{(2)} = 1/6$$

We have, thus, derived Eq. (6.44). □

Extending the preceding examples via more (calculational) work but no new (mathematical) ideas, one can derive explicit expressions for the sum $S_c(n)$, for any positive integer c.

C. A pictorial solution for $c = 2$

We now provide a pictorial, rather Fubini-esque,[5] derivation of the sum of the first n perfect squares; cf. Eq. (6.44). The summation process begins with each perfect square k^2 depicted as a $k \times k$ square array of copies of the unit-side square. We then progressively rearrange these arrays in a way that sums the squares.

We pictorially sum the first four perfect squares, 1^2, 2^2, 3^2, and 4^2, to illustrate the process. The depiction of these four perfect squares appears at the top of each of Figs. 6.19–6.22. In each of the four figures, we have darkened certain component unit-side squares. These darkened squares are the ones that will play a lead role in each successive rearrangement in our pictorial summation process.

1. Figure 6.18 provides a schematic "scaffolding" for the process: The figure depicts

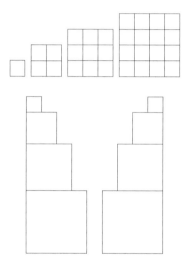

Fig. 6.18 The initial configuration for pictorially computing the sum of squares: the "scaffolding"

two mirrored copies of the perfect squares 1^2, 2^2, 3^2, and 4^2 piled in decreasing order of size from bottom to top. (To emphasize the pattern embodied by our procedure: the process of summing the first *five* perfect squares, 1^2, 2^2, 3^2, 4^2, 5^2 would begin by just placing the 5×5 square array beneath the 4×4 array in Fig. 6.18.) The two piles are positioned in such a manner that there is:

---

[5] See Section 2.3.2.2.

- a unit-width height-$(k = 4)$ "alley" between the $k \times k$ squares at the bottom of the "scaffolding"

- a width-3 height-$(k - 1)$ "alley" between the $(k - 1) \times (k - 1)$ squares at the next lowest level of the "scaffolding"

- a width-5 height-$(k - 2)$ "alley" between the $(k - 2) \times (k - 2)$ squares at the third lowest level of the "scaffolding"

    $\vdots$

- a width-$(2k - 1)$ unit-height "alley" between the two unit-side squares at the top level of the "scaffolding"

2. Figure 6.19 depicts the first rearrangement step. The $(k = 4)$ lower left unit-side

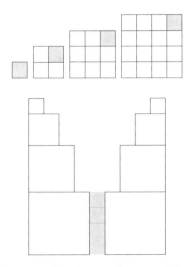

Fig. 6.19 Step 1 for pictorially computing the sum of squares: filling the unit-width "alley" with the lower-left-hand unit-side squares

squares are copied from their initial locations and moved to fill the height-$(k = 4)$ "alley" in the "scaffolding".

3. Figure 6.20 depicts the next step in the process. From each perfect square $k > 1$, we copy the 3-square "L-shape" that sits atop the lower left-hand square. We use the resulting length-3 "bars" to fill in the width-3 "alley" at the next lowest level in the "scaffolding".

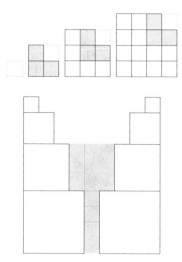

Fig. 6.20 Step 2 for pictorially computing the sum of squares: filling the width-3 "alley" with the width-3 "bars" produced by flattening the length-3 L's

---

Explanatory note.

The reader should recognize the kinship of the just-described process of taking L-shapes and flattening them into "bars" with an analogous process in our pictorial proof of Proposition 6.3. In that proof, we represented each odd integer $2k+1$ as a length-$(2k+1)$ "bar", and we bent the "bar" into a reversed-L whose horizontal and vertical sub-"bars" each had length $k+1$. In the current process, we are performing the exact inverse process—flattening instead of bending.

---

4 and 5. Figures 6.21 and 6.22 peel off successive-size L-shaped arrays from the adequately large initial square arrays, flatten these L's into "bars", and use these "bars" to fill in the next largest "alley" in the scaffolding.

We leave the detailed validation of the described process to the reader.

D. ⊕ Semi-pictorial solutions for $c = 3$

We described in Section 2.3.2.2 how the Italian mathematician Guido Fubini was able to make notable mathematical progress by rearranging the representations of a variety of structured mathematical objects; see [51]. Within the context of the current chapter, we apply such rearrangements to achieve exact values for some special subsums of $S_3(n)$, i.e., special sums of the cubes of integers.

The proof we develop here establishes relationships among three sets: the odd numbers $\{2i-1\}$, the perfect cubes $\{i^3\}$, and the triangular numbers $\{\Delta_i\}$. Specif-

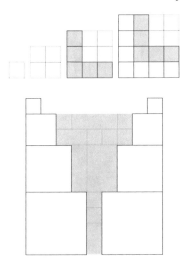

Fig. 6.21 Step 3 for pictorially computing the sum of squares: filling the width-5 "alley" with the width-5 "bars" produced by flattening the length-5 L-shapes

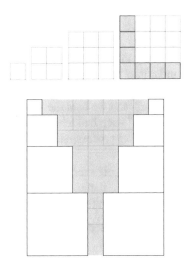

Fig. 6.22 Final step for pictorially computing the sum of squares: filling the width-7 "alley" with the width-7 "bars" produced by flattening the length-7 L-shapes

ically, we show that the sum of all odd numbers up to the triangular number Δ_k is equal to each of the following quantities:

- the square of the triangular number Δ_k
- the sum of perfect cubes up to k.

We provide pictorial reasoning that allows one to observe the equality between the preceding quantities. We state the result formally as follows

Proposition 6.11 *For all $k \in \mathbb{N}^+$,*

$$\Delta_k^2 \;=\; \sum_{i=1}^{\Delta_k} (2i-1) \;=\; \sum_{i=1}^{k} i^3 \tag{6.46}$$

Proof. The left-hand equation in (6.46) follows from Proposition 6.3: the first Δ_k odd numbers sum to Δ_k^2. We focus, therefore, only on the right-hand equation.

We take the odd integers in order and arrange them into groups whose successive sizes increase by 1 at each step, as in the following triangular table (6.47).

$$
\begin{array}{ll}
\text{group 1 (size 1):} & 1, \\
\text{group 2 (size 2):} & 3, \;\; 5, \\
\text{group 3 (size 3):} & 7, \;\; 9, \;\; 11, \\
\text{group 4 (size 4):} & 13, \;\; 15, \;\; 17, \;\; 19 \\
\;\;\;\; \vdots & \;\;\;\; \vdots
\end{array}
\tag{6.47}
$$

Group 1 comprises just the smallest odd number, 1; group 2 comprises the next two odd numbers, 3, 5; group 3 comprises the three odd numbers 7, 9, and 11; and so on. Let us observe some of the features of Table (6.47).

We observe first that—at least within the illustrated portion of the table—the i elements of the ith group add up to i^3:

$$
\begin{array}{lll}
\text{group 1 (size 1):} & 1, & : \text{sum} = 1^3 \\
\text{group 2 (size 2):} & 3, \;\; 5, & : \text{sum} = 2^3 \\
\text{group 3 (size 3):} & 7, \;\; 9, \;\; 11, & : \text{sum} = 3^3 \\
\text{group 4 (size 4):} & 13, \;\; 15, \;\; 17, \;\; 19 & : \text{sum} = 4^3
\end{array}
$$

We observe next that, by construction, the ith group/row of odd integers in the table consists of the i consecutive odd numbers beginning with the $(\Delta_{i-1}+1)$th odd number, $2\Delta_{i-1}+1$. Since consecutive odd numbers differ by 2, this means that the ith group (for $i > 1$) comprises the following i odd integers:

$$2\Delta_{i-1}+1, \; 2\Delta_{i-1}+3, \; 2\Delta_{i-1}+5, \; \ldots, \; 2\Delta_{i-1}+(2i-1)$$

Therefore, invoking Proposition 6.3, the *sum* of the i integers in group i, call it σ_i, equals

$$
\begin{aligned}
\sigma_i &= 2i\Delta_{i-1} + \big(1+3+\cdots+(2i-1)\big) \\
&= 2i\Delta_{i-1} + (\text{the sum of the first } i \text{ odd numbers}) \\
&= 2i\Delta_{i-1} + i^2
\end{aligned}
$$

By direct calculation, then,

$$\sigma_i \;=\; 2i \cdot \frac{i(i-1)}{2} + i^2 \;=\; (i^3 - i^2) + i^2 \;=\; i^3$$

The proof is now completed by concatenating the rows of Table (6.47) and observing the pattern that emerges:

$$(1) + (3+5) + (7+9+11) + \cdots \;=\; 1^3 + 2^3 + 3^3 + \cdots$$

Note the "Fubini-esque" arrangements and groupings throughout this argument. (To understand this reference, see Fubini's role in Section 2.3.2.2.) $\square$

We now present a *pictorial* proof of a portion of Proposition 6.11, namely, the relation between sums of perfect cubes and squares of triangular numbers. This illustration provides a *non-textual* way to understand this result, and—at least as importantly—it provides a fertile setting for seeking other facts of this type.

Proposition 6.12 *For all $k \in \mathbb{N}^+$,*

$$1^3 + 2^3 + \cdots + k^3 \;=\; \Delta_k^2$$

Proof. We develop an induction that reflects the structure of Table (6.47).

Base cases. The base of our induction is the first case of the claimed result:

$$1^3 \;=\; 1 \;=\; \Delta_1^2$$

While this first (and obvious) case is enough for the induction, it does not tell us much about the structure of the problem. Therefore, we consider also the next step, namely, the case $k = 2$:

$$1^3 + 2^3 \;=\; 9 \;=\; \Delta_2^2$$

We depict in Fig. 6.23 how to obtain a "larger" base case for our induction by pictorially summing the numbers from the first two groups in Table (6.47). Observe that

Fig. 6.23 (Left) Pictorially depicting the first two groups from Table (6.47) as stacked unit-side squares: the set $\{1\}$ of group 1 and the set $\{3,5\}$ of group 2. (Right) Illustrating how to form a 3×3 square by pictorially summing the numbers 1, 3, and 5. The clear portions of the square come from group 2; the shaded small square at the top comes from group 1

we can fit the shapes from the left side of the figure together to form the $\Delta_2 \times \Delta_2$ square.

Inductive hypothesis. Say for the sake of the induction that the target equality holds for all $i < k$; i.e.,

$$1^3 + 2^3 + \cdots + i^3 \;=\; \Delta_i^2 \;=\; \left(\frac{i+1}{2}\right)^2$$

If we go one step further, to incorporate group 3, i.e., the set $\{7, 9, 11\}$, into our pictorial summation process, then we discover that mimicking the process of Fig. 6.23 is a bit more complicated for this case. A bit of analysis exposes that the more complicated manipulation required to form the $\Delta_3 \times \Delta_3$ square is a consequence of the odd cardinality of the group-3 set. Accordingly, we must extend our induction in slightly different ways for the cases of odd and even k.

Inductive extension for odd k. In order to extend the inductive hypothesis in this case, we need to prove that

$$\Delta_k^2 \;=\; \Delta_{k-1}^2 + k^3$$

We begin to garner intuition for this extension by comparing the quantities Δ_k^2 and $1 + 2^3 + \cdots + k^3$.

Moving to the pictorial domain, we write k^3 as $k \times k^2$, and we distribute $k \times k$ square blocks around the $\Delta_{k-1} \times \Delta_{k-1}$ square, as shown in Fig. 6.24 for the case $k = 3$. Because k is odd, the small squares pack perfectly—since $(k - 1)$ is even,

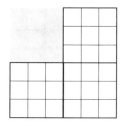

Fig. 6.24 Extending the inductive summation-by-manipulation for the case of odd k: The manipulation of small squares produces the next bigger perfect square. (The case $k = 3$ is depicted.)

hence divisible by 2. The depicted case depicts pictorially the definition of triangular numbers: $k \cdot \frac{1}{2}(k - 1) \;=\; \Delta_{k-1}$.

Inductive extension for even k. The basic reasoning here mirrors that for odd k, with one small, but major difference. Now, as we assemble small squares around the large square, two subsquares overlap, as depicted in Fig. 6.25. We must manipulate the overlapped region in the manner depicted in Fig. 6.26 in order to get a tight packing around the large square. Happily, when there is a small overlapping square region, there is also an identically shaped empty square region, as suggested by these two figures. We provide a bit more detail. Because $(k - 2)$ is even, the like-configured square blocks can be allocated to two sides of the initial $\Delta_{k-1} \times \Delta_{k-1}$ square—namely, its right side and its bottom—in the manner depicted in Fig. 6.25. One can discern in the figure the overlap we described: it has the shape of a square

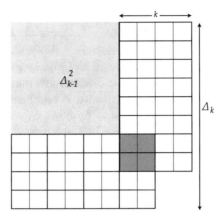

Fig. 6.25 The construction for even k produces an overlapped area represented in dark grey. (The figure depicts the case $k = 4$.)

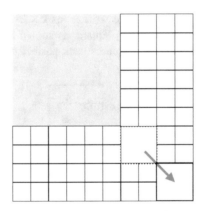

Fig. 6.26 How to obtain a large ($\Delta_k \times \Delta_k$) square by sliding the overlapped small square region into the like-sized small square empty region

that measures $\frac{1}{2}(\Delta_k - \Delta_{k-1})$ on a side. One also sees in the figure an empty square in the extreme bottom right of the composite $\Delta_k \times \Delta_k$ square, which matches the overlapped square identically. This situation is the pictorial version of the equation

$$\Delta_k^2 - \Delta_{k-1}^2 \;=\; \frac{1}{4}k^2\left((k+1)^2 - (k-1)^2\right) \;=\; k^3$$

We have thus extended the inductive hypothesis for both odd and even k, whence the result. $\square$

6.3.2.3 $S_c(n)$ for general *negative* cth powers

We focus finally on summations of the form

$$S_c(n) \stackrel{\text{def}}{=} \sum_{i=1}^{n} i^c$$

for arbitrary *negative* numbers c. The reader can garner intuition for the upcoming bounds from the general shape of the rectangles and continuous curves in Fig. 6.17.

Proposition 6.13 *For summations $S_c(n)$ with fixed negative powers $c < 0$,*

$$\underline{C}_c(n) \leq S_c(n) \leq \overline{C}_c(n) \tag{6.48}$$

where

$$\underline{C}_c(n) = \int_0^{n-1} (x+1)^c dx \quad and \quad \overline{C}_c(n) = 1 + \int_1^n x^c dx$$

Proof. We obtain our bounds on the sum of $S_c(n)$ by evaluating the integrals that yield the areas under the continuous curves $\overline{C}_c(n)$ and $\underline{C}_c(n)$ in the instantiation of Fig. 6.17 for $S_c(n)$. □

When $c \neq -1$,[6] we can provide more detail, by using reasoning similar to that underlying the bounds (6.39) that hold for positive values of c.

A. Negative powers $-1 < c < 0$

In this case, we obtain essentially the same bounds as in the case of nonnegative c.

Proposition 6.14 *The sum of any summation $S_c(n)$ having fixed negative powers in the range $-1 < c < 0$ satisfies the following inequalities.*

$$\frac{1}{c+1} n^{c+1} - \frac{1}{c+1} \leq S_c(n) \leq \frac{1}{c+1} n^{c+1} + \left(1 - \frac{1}{c+1}\right) \tag{6.49}$$

The infinite version of summation $S_c(n)$, i.e., the series $S_c^{(\infty)} \stackrel{\text{def}}{=} \sum_{i=0}^{\infty} i^c$, diverges.

We thus observe that $S_c(n)$ has the same growth *pattern* with increasing n as it does when c is positive, but that $S_c(n)$'s growth *rate* is slower because of the damping effect of the negative c in the exponent. This damped growth notwithstanding, the infinite series $S_c^{(\infty)}$ diverges because n^{c+1}, which is the variable portion of the lower bound on $S_c(n)$, grows without bound as n grows without bound.

---

[6] We need to avoid the case $c = -1$ so that we do not attempt to divide by 0.

B. Negative powers $c < -1$

When c is "very negative", specifically, when $c < -1$, the infinite version of $S_c(n)$, call it $S_c^{(\infty)}$, is a *convergent* infinite series. In this case, n^c *shrinks* as n grows, so an analysis mirroring the one that leads to Eq. (6.49) provides the following sum for $S_c^{(\infty)}$.

Proposition 6.15 *When the fixed negative power c is smaller than -1, then the infinite version, $S_c^{(\infty)}$, of $S_c(n)$, converges, with the following sum.*

$$S_c^{(\infty)} \leq \frac{c}{c+1} \tag{6.50}$$

Proof. We see as in Eq. (6.49) that, for $c < -1$, as n grows without bound, $S_c(n)$ is majorized by $\dfrac{c}{c+1}$. □

C. Negative powers $c = -1$: the *harmonic* series

The singular case defined by the value $c = -1$ defines the important *harmonic series*,

$$S^{(H)} = \sum_{i=1}^{\infty} \frac{1}{i}$$

and its finite prefixes that comprise the *harmonic summation*

$$S^{(H)}(n) = \sum_{i=1}^{n} \frac{1}{i}$$

   (i) The asymptotic behavior of $S^{(H)}(n)$. It has been known since the time of the well-traveled Swiss mathematician Leonhard Euler that $S^{(H)}$ and $S^{(H)}(n)$ are closely related to the *natural* or *Napierian*[7] logarithm that $S^{(H)}$ and $S^{(H)}(n)$ are closely Euler's constant $e = 2.718281828\ldots$.

Proposition 6.16 *The behavior of the harmonic summation $S^{(H)}(n)$ as a function of n is given by*

$$S^{(H)}(n) \approx \ln n$$

It follows, in particular, that the harmonic series $S^{(H)}$ diverges.

---

[7] The natural logarithm, i.e., the logarithm to the base e, is commonly referred to as the *Napierian logarithm*, in honor of the Scottish polymath John Napier.

> **Enrichment note.**
>
> The adjective "harmonic" calls to mind a number of concepts associated with *music*, such as "harmonics" and "harmony". The association between our series and these musical concepts is not a coincidence. The name of the harmonic series derives from the concept of *overtones*, or *harmonics*, in music. When one observes a vibrating string, one finds that the wavelengths of its overtones, as fractions of the string's fundamental wavelength, are the terms of the *harmonic sequence*, namely, $\frac{1}{2}, \frac{1}{3}, \frac{1}{4}, \ldots$.

(ii) Bounds on the asymptotic behavior of $S^{(H)}(n)$. Figure 6.27 depicts the harmonic summation $S^{(H)}(n)$ as the area of abutting unit-width rectangles of respective heights (from left to right) of $1, 1/2, 1/3, \ldots, 1/n$.

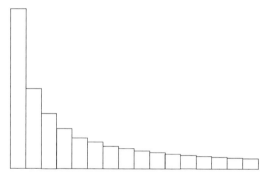

Fig. 6.27 The harmonic summation $S^{(H)}(n)$ represented by the area of abutting unit-width rectangles whose (decreasing) heights form the sequence of consecutive integer reciprocals

We now reorganize the sequence of $S^{(H)}(n)$'s summands in a way that helps us better understand the behavior of $S^{(H)}(n)$ as a function of n.

- We group $S^{(H)}(n)$'s summands into sub-summations whose sizes are the consecutive powers of 2:

The first $(2^0 = 1)$ reciprocal: $\qquad\qquad A_0 = 1$

the next $(2^1 = 2)$ consecutive reciprocals: $A_1 = \dfrac{1}{2} + \dfrac{1}{3}$

the next $(2^2 = 4)$ consecutive reciprocals: $A_2 = \dfrac{1}{4} + \dfrac{1}{5} + \dfrac{1}{6} + \dfrac{1}{7}$

$\qquad\vdots \qquad\qquad\qquad\qquad\qquad\qquad\qquad \vdots$

the next 2^i consecutive reciprocals: $\qquad A_i = \dfrac{1}{2^i} + \dfrac{1}{2^i + 1} + \cdots + \dfrac{1}{2^{i+1} - 1}$

$\qquad\vdots \qquad\qquad\qquad\qquad\qquad\qquad\qquad \vdots$

- We derive constant upper and lower bounds for each subsum. Focus on the 2^i consecutive reciprocals of the generic subsum A_i. The largest of these reciprocals is $1/2^i$; the smallest is $1/(2^{i+1} - 1)$. We therefore note that

$$\frac{1}{2} < \frac{2^i}{2^{i+1} - 1} < 2^i \cdot A_i < \frac{2^i}{2^i} = 1$$

To summarize: each subsum A_i lies between $1/2$ and 1.

We can interpret our bounds on the subsums A_i in the light of Fig. 6.27. Let us proceed left to right along the abutting rectangles in the figure, and let us recall, from Section 5.4.1, the definition of "logarithm to the base b". As we double the number of rectangles that we have traversed, consider the increase in the aggregate area of the thus-far traversed rectangles:

1. *This increase is no larger than* 1.
   This means that $S^{(H)}(n)$ grows *no faster than* $\log_2 n$.

2. *This increase is greater than* $1/2$.
   This means that $S^{(H)}(n)$ grows *faster than* $\log_4 n$.

Of course, these observations are consistent with the verified actual natural-logarithmic growth rate of $S^{(H)}(n)$, because $2 < e < 4$.

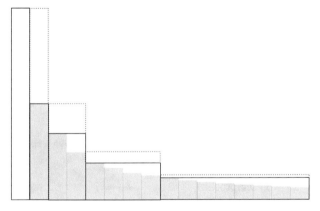

Fig. 6.28 Upper-bounding A_i by larger unit-size rectangles in the harmonic sum

6.4 Exercises: Chapter 6

Throughout the text, we mark each exercise with 0 or 1 or 2 occurrences of the symbol $\oplus$, as a rough gauge of its level of challenge. The 0-$\oplus$ exercises should

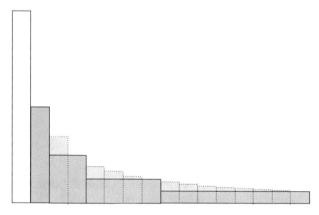

Fig. 6.29 A pictorial lower bound for the harmonic summation

be accessible by just reviewing the text. We provide *hints* for the 1-⊕ exercises; Appendix H provides *solutions* for the 2-⊕ exercises. Additionally, we begin each exercise with a brief explanation of its anticipated value to the reader.

1. **A "pictorial" summation of the first *n* integers**
   LESSON: A practice solo ride toward a simple "pictorial" argument

   Flesh out the use of Fig. 6.8 and its following text to produce a rigorous derivation of the formula for the sum of the first n integers.

2. **A "pictorial" summation of $S_{1/2}^{(\infty)}$**
   LESSON: Another practice solo ride toward a simple "pictorial" argument

   Derive the sum of $S_{1/2}^{(\infty)}$ by developing the pie-slicing argument of Fig. 6.13 and its accompanying text.

3. **Solving summation $S_b^{(2)}(n)$ algebraically**
   LESSON: A practice solo ride which expands a complicated inductive argument by one step

   Build on the proof of Proposition 6.7 to solve summation $S_b^{(2)}(n)$.

4. ⊕ **Evaluating $S_1(n) = \sum_{i=1}^{n} i$, using the fact that $S_2(n) = \sum_{i=1}^{n} i^2$**
   LESSON: Practice in manipulating mathematical expressions

   Say that someone gives you a *machine* that can compute the summation of the first *n* perfect squares, given *n* as an input:

   $$S_2(n) = 1 + 4 + 9 + \cdots + n^2$$

   Show how to use this machine in order to compute the sum of the first n integers.

   Hint. Try to think creatively. Think of how one can (mathematically) relate the first *n* integers to their squares.

5. **Evaluating $S_2(n) = \sum_{k=1}^{n} k^2$, given knowledge that $S_2(n) = \Theta(n^3)$**

   LESSON: Practice with solving systems of n linear equations in n unknowns, applied to the *method of undetermined coefficients*

   Say that you know—perhaps from reading Section 6.3.1—that the sum of the first n perfect squares is a *cubic* polynomial in n.

   Use the method of undetermined coefficients to derive the exact formula (2.2) for the sum.

6. ⊕ **Evaluating a geometric summation pictorially**

   LESSON: Experience using pictorial representations

   In Section 6.2.2.2, we used Thales's theorem about similarity in triangles (Proposition 6.6) to sum the simple infinite geometric series $\sum_{i=0}^{\infty} b^i$. It turns out that a modest modification of that evaluation strategy allows us also to evaluate the truncated versions of that series, namely, the summations

   $$S^{(b)}(n) = \sum_{i=0}^{n} b^i$$

   Use the following figure to evaluate $S^{(b)}(n)$.

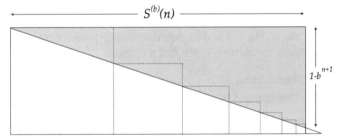

7. ⊕ **A direct proof that the harmonic series diverges**

   LESSON: Practice with manipulating summations and series

   There are many ways to prove that the harmonic series

   $$S^{(H)} = \sum_{k=1}^{\infty} \frac{1}{k}$$

   diverges. In Proposition 6.16, we developed a method that provided additional information: not only did it prove that $S^{(H)}$ diverges—i.e., tends to infinity as one looks at more and more terms grows—it also exposed the rate of this divergence as logarithmic in the number of terms one looks at. If one wants only to prove divergence, then the following more direct argument will suffice.

   Develop a proof that the harmonic series diverges by

   - *partitioning $S^{(H)}$'s terms into groups whose sizes are successive powers of 2*

   - *developing an argument based on the sums within the groups.*

For clarification: The partitioning step operates as follows:

$$S^{(H)} = 1 + \tfrac{1}{2} + \tfrac{1}{3} + \tfrac{1}{4} + \tfrac{1}{5} + \tfrac{1}{6} + \tfrac{1}{7} + \tfrac{1}{8} + \tfrac{1}{9} + \tfrac{1}{10} + \tfrac{1}{11} + \tfrac{1}{12} + \tfrac{1}{13} + \tfrac{1}{14} + \tfrac{1}{15} + \tfrac{1}{16} + \cdots$$

$$= 1 + \tfrac{1}{2} + \left(\tfrac{1}{3} + \tfrac{1}{4}\right) + \left(\tfrac{1}{5} + \tfrac{1}{6} + \tfrac{1}{7} + \tfrac{1}{8}\right) + \left(\tfrac{1}{9} + \tfrac{1}{10} + \tfrac{1}{11} + \tfrac{1}{12} + \tfrac{1}{13} + \tfrac{1}{14} + \tfrac{1}{15} + \tfrac{1}{16}\right) + \cdots$$

Propose another analysis by gathering the terms by powers of 3.

8. ⊕ **Summations, and summations of summations**
   LESSON: Practice manipulating summations; gain familiarity with basic sums
   By definition:

$$Id_n \stackrel{\text{def}}{=} \sum_{k=1}^{n} 1 = 1 + 1 + \cdots + 1 = n$$

We have proved in multiple ways:

$$\Delta_n \stackrel{\text{def}}{=} \sum_{k=1}^{n} k = 1 + 2 + 3 + \cdots + n = \frac{1}{2} Id_n \cdot (n+1)$$

This exercise picks up where these two examples leave off.

a. *Prove that*[8]

$$\widehat{\Theta}_n \stackrel{\text{def}}{=} \sum_{k=1}^{n} \Delta_k = \Delta_1 + \Delta_2 + \cdots + \Delta_n = \frac{1}{3} \Delta_n \cdot (n+2)$$

b. *Prove the following identities involving* Δ_n. *For all* $n \in \mathbb{N}^+$:
   i. $\Delta_n + \Delta_{n-1} = n^2$

   This is a straightforward exercise in algebraic manipulation. Can you *also* find an "interesting" proof of this identity, say one that employs a "pictorial" representation of Δ_n?

   ii. $\Delta_n^2 - \Delta_{n-1}^2 = n^3$
   Hint: Try "playing" with simple manipulations of the equation

$$\Delta_n = n + \Delta_{n-1}$$

   which defines Δ_n.

c. ⊕ *Derive a closed-form expression for the sum*

$$\widehat{\Theta}_n + \widehat{\Theta}_{n-1}$$

   Hint: Recall that the sums $\widehat{\Theta}_n$ are defined as sums of Δ_k, which, in turn, are defined by closed-form expressions. Lots to "play" with!

---

[8] The sum of the the first n triangular numbers, $\sum_{k=1}^{n} \Delta_k$, is often denoted Θ_n. We employ instead the nonstandard name $\widehat{\Theta}_n$ to ensure that we do not confuse the newcomer who has recently learned about the role of the letter Θ in asymptotic notation.

Chapter 7
The Vertigo of Infinity: Handling the Very Large and the Infinite

One Two Three . . . Infinity
George Gamow (title of 1947 book)

An immensely powerful attribute of the way mathematics allows us to deal with quantity is our ability—in principle, at least—to use identical reasoning and tools of analysis to manipulate all finite objects: We do not need—to cite just two examples—a hierarchy of arithmetics to deal with real numbers of various sizes nor a variety of logical calculi to deal with propositions having various numbers of satisfying assignments of truth values. This situation changes, however, when we deal with objects—say, sets or numbers—whose sizes can change *dynamically* or that are composed of *infinitely many* distinguishable objects. In order to reason rigorously about quantities that can grow—or, equivalently, can shrink—without bound, we need new conceptual machinery. This is true whether or not the growth can continue forever! Also, we need yet other new conceptual tools to deal with objects that are actually infinite.

This chapter is devoted to developing the needed tools and to heightening readers' awareness of the care that they must take when reasoning in the domain of this chapter's title: the *very large* and the *infinite*.

7.1 Asymptotics

Asymptotics can be viewed as a language and a system of reasoning that allow one to talk in a *qualitative* voice about *quantitative* topics. We thereby generalize to arbitrary growth functions terms such as "linear", "quadratic", "exponential", and "logarithmic".

Such a language and system are indispensable if one needs to reason about computational topics over a range of situations, such as a range ("all existing"?) of com-

© Springer Nature Switzerland AG 2020
A. L. Rosenberg, D. Trystram, *Understand Mathematics, Understand Computing*,
https://doi.org/10.1007/978-3-030-58376-7_7

puter architectures and software systems. As two simple examples: (1) Carry-ripple adders perform additions in a number of steps that is linear in the lengths n of the summands (measured in number of bits)—no matter what these lengths are. (2) Comparison-based sorting algorithms can sort lists of n keys in a number of steps proportional to $n \log n$, but no faster—where the base of the logarithm depends on the characteristics of the computing platform and the set of keys being sorted. More precise versions of the preceding statements require specification of the number n and other details, possibly down to the clock speed of the host computer's circuitry.

The need for the material in this section can be discerned (almost) daily, as the news media report—and all too often *misreport*—about dynamic quantities: the word "exponential" is very often used when "fast" or "large" is what is actually meant. It is not true (usually) that "Country A's GDP" or "The X-virus epidemic" is growing *exponentially*. Country A's strategic planning, in the first example, and the government's public health department, in the second example, cannot proceed rationally without trustworthy quantitative information.

7.1.1 The Language of Asymptotics

The language of asymptotics has its origins in the subfield of mathematics known as *Number Theory*, in the late nineteenth century. The basic version of the language— which handles almost all situations one encounters when studying the discrete mathematics germane to computational phenomena—builds on the terminology we discuss here. There is an advanced companion to the following notions, which builds on nondiscrete concepts such as limits; this advanced material will be beyond the likely needs of most students of computing—excepting, of course, specialists in specific, advanced subject areas. The basics of the language of asymptotics build on three primitive notations and notions. We need only the rudiments of this material within this text, so we refer the reader to standard texts on algorithm design and analysis (e.g., [36]) or number theory (e.g., [85]), to go beyond the basics of the following ideas.

The following notation actually has several variants in both symbology and articulation. We elaborate on this situation only for our first notation, the "big-O" (articulated "big O"). The reader should be aware that analogous complications accompany our other two notations, the "big-Ω" (articulated "big Omega") and the "big-Θ" (articulated "big Theta").

- *The big-O notation.* The assertion $f(x) = O(f_1(x))$ says, intuitively, that the function f *grows no faster than* the function f_1. It is, thus, the asymptotic analogue of "less than or equal to".

  Formally, the assertion
  $$f(x) = O(f_1(x))$$
  means
  $$(\exists c > 0)(\exists x_1)(\forall x > x_1)[f(x) \leq c \cdot f_1(x)]$$

Explanatory notes.

1. *Variables and unknowns.*
Our formal definitions of big-O, big-Ω, and big-Θ, employ the letters x, x_1, x_2, and x_3. It is important to note the following crucial distinction.

- x is a *variable* which ranges over the real numbers.

- x_1, x_2, and x_3 are *unknowns* which are selected from the same set of numbers.

  As explained in Chapter 2, this means that x_1, x_2, and x_3 are three specific real numbers—we just do not know which ones!

2. *When "equals" is not "equals".*
The "equals sign" in the notation $f(x) = O(f_1(x))$ does *not* mean "equals".

In fact, the assertion "$f(x) = O(f_1(x))$" actually means that function f *belongs to the class of functions that* eventually *grow no faster than function* f_1 *does.*
Consequently:

- The assertion "$f(x) = O(f_1(x))$" is *articulated*

    "$f(x)$ *is* $O(f_1(x))$".

  Never substitute "equals" for "is"

- Many people acknowledge the "belonging to a class" facet of the definition by writing

    "$f(x) \in O(f_1(x))$"

  rather than

    "$f(x) = O(f_1(x))$"

An esoteric aside.
It appears strange that precisely two of the three asymptotic letters, namely, Ω and Θ, come from the Greek alphabet, while the third, letter is the Latin O. To correct this apparent imbalance, many scholars insist that the letter O in "big-O" is actually the Greek letter Omicron, *not* the Latin letter O.

- *The big-Ω notation.* The assertion $f(x) = \Omega(f_2(x))$ says, intuitively, that the function f *grows at least as fast as* function f_2. It is, thus, the asymptotic analogue of "greater than or equal to".

Formally, the assertion

$$f(x) = \Omega(f_2(x))$$

means

$$(\exists c > 0)(\exists x_2)(\forall x > x_2)[f(x) \geq c \cdot f_2(x)]$$

- *The big-Θ notation.* The assertion $f(x) = \Theta(f_3(x))$ says, intuitively, that the function f *grows at the same rate as* function f_3. It is, thus, the asymptotic analogue of "equal to".

Formally, the assertion

$$f(x) = \Theta(f_3(x))$$

means

$$(\exists c > 0)(\exists c' > 0)(\exists x_3)(\forall x > x_3)[c \cdot f_3(x) \leq f(x) \leq c' \cdot f_3(x)] \tag{7.1}$$

It is not hard to verify—in fact, we pose this as an exercise—that the defining expression (7.1) is logically equivalent to either of the following expressions.

$$[f(x) = O(f_3(x))] \text{ and } [f_3(x) = O(f(x))] \tag{7.2}$$

$$[f(x) = \Omega(f_3(x))] \text{ and } [f_3(x) = \Omega(f(x))] \tag{7.3}$$

One renders the preceding intuitive explanations precise by pointing out that the three specified relations, i.e., the big-O, the big-Ω, and the big-Θ:

1. take hold *eventually*, i.e., only for large arguments to the functions f, f_1, f_2, f_3;

2. hold only up to an unspecified constant of proportionality.

The plots in Fig. 7.1 illustrate the definitions of big-O and big-Ω graphically.

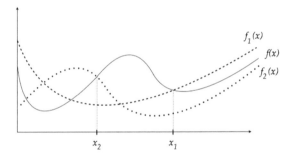

Fig. 7.1 Plots of three functions: f (solid curve), f_1 (dotted curve that originates on the left, above f), and f_2 (dotted curve that originates on the left, below f). Assume that the three illustrated trajectories continue in the depicted relationships for large values of x—i.e., for all $x > x_1$, the plot for f lies strictly below that for f_1 and strictly above that for f_2 for all $x > x_2$. With this assumption, the figure illustrates the relations: $f(x) = O(f_1(x))$ and $f(x) = \Omega(f_2(x))$

7.1.2 The "Uncertainties" in Asymptotic Relationships

The formal definitions of all three of our asymptotic relationships are bracketed by two important quantifiers, of the forms:

$$(\exists c > 0) \quad \text{and} \quad (\forall x > x_k).$$

The former, *uncertain-size* quantifier ($\exists c > 0$) asserts that asymptotic notions describe functional behavior "in the large". Thus, in common with more common qualitative descriptors of quantitative growth such as "linear", "quadratic", "cubic", "quartic", "exponential", "logarithmic", etc., asymptotic relationships give no information about constants of proportionality. *We are not saying that constant factors do not matter!* Rather, we are saying that *we sometimes want to discuss growth patterns* **in the large** *rather than in detail.*

The latter, *uncertain-time* quantifier ($\forall x > x_k$) asserts that asymptotic relationships between functions are promised to hold only "eventually", i.e., "for sufficiently large values of the argument x". Therefore: *Asymptotic notions cannot be employed to discuss or analyze quantities that can never grow beyond a fixed finite value.* The fact that all instances of a quantity throughout history have been below some value N—which is fixed, unknown, and possibly unimaginably large—is immaterial, as long as it is conceivable that an instance larger than N could appear at some time in the future.

These two quantifiers in particular distinguish claims of asymptotic relationship from the more familiar definite inequalities such as

$$[f(x) \leq g(x)] \quad \text{or} \quad [f(x) \geq 7 \cdot g(x)]$$

In fact, it is often easier to think about our three asymptotic bounding assertions as establishing *envelopes* for $f(x)$—cf. Fig. 7.1:

- $f(x) = O(g(x))$.
  If one draws the graphs of the functions $f(x)$ and $c \cdot g(x)$, then as one traces the graphs with increasing values of x, one eventually reaches a point x_k beyond which the graph of $f(x)$ never enters the territory *above* the graph of $c \cdot g(x)$.

- $f(x) = \Omega(g(x))$.
  This situation is the up-down mirror image of the preceding one: just replace the highlighted "*above*" with "*below.*"

- $f(x) = \Theta(g(x))$.
  This relation provides a *two-sided* envelope: Beyond x_k, the graph of $f(x)$ never enters the territory *above* the graph of $c \cdot g(x)$ and never enters the territory *below* the graph of $c' \cdot g(x)$.

In addition to allowing one to make familiar growth-rate comparisons such as "$n^{14} = O(n^{15})$" and "$1.001^n = \Omega(n^{1000})$", we can now also make assertions such as "$\sin x = \Theta(1)$", which are much clumsier to explain in words.

⊕ **Beyond the big letters.** There are "little"-letter analogues of the preceding "big"-letter asymptotic relations and notations: o, ω, and θ. The definitions and domains of application of these notations build on the notion of *limit*, which is fundamental in continuous mathematics—e.g., in the differential and integral calculus—but rare in discrete mathematics. We therefore do not include this topic in this text, instead referring the reader to any good text on the calculus.

7.1.3 Inescapable Complications

The story we have told thus far is more or less standard fare in courses on discrete mathematics and algorithms. Two complications to the story are covered less consistently, despite the fact that, lacking them, one cannot perform cogent asymptotic reasoning about modern computing environments—including environments that support parallel and distributed computing. Both complications involve the notion of *uniformity*.

1. *Multiple functions.* Say that we have four functions, f, g, h, k, and we know that

$$f(x) = O(g(x)) \quad \text{and} \quad h(x) = O(k(x))$$

It is compellingly suggestive that

$$f(x) + h(x) = O(g(x) + k(x))$$

— but is it true?

In short, the answer is *YES*—but verifying the answer requires a bit of subtlety. The challenge is that, absent hitherto-undisclosed information, the proportionality constants $c = c_{f,g}$ and $N = N_{f,g}$ that witness the big-O relationship between functions f and g have no connection with the constants, $c = c_{h,k}$ and $N = N_{h,k}$ that witness the analogous relationship between functions h and k. Therefore, in order to verify the posited relationship between functions $f + h$ and $g + k$, one must find witnessing constants $c = c_{f+h,g+k}$ and $N = N_{f+h,g+k}$. Of course, this task requires only elementary reasoning—*but it must be done*!

Explanatory note.

Notations such as "$c = c_{f,g}$" and "$N = N_{f,g}$" seem to be oxymoronic. Either c (or N) is a constant or it isn't! The explanation is at once simple and subtle.

When we discuss the relationships between fixed functions, f and g, then even though these functions are fixed throughout the discussion, the discussion may include implicit or explicit references to any number of characteristics of f and g. This means that, in some sense, a discussion about the relationships between f and g occurs in a logically different world than does a discussion about the relationships between two other fixed functions, h and k. And, we, as authors, must make it totally clear to you, our readers, which world we are in as we make any specific assertion. The notational mechanism that we use to stay "synchronized" with you resides in assertions such as "$c = c_{f,g}$" and "$N = N_{f,g}$". These assertions announce that, while both c and N are constants, the specific—fixed(!)—values of these constants are chosen with knowledge that the constants are being used to discuss the fixed functions f and g.

> **Explanatory note (cont'd).**
>
> The preceding paragraph stresses that, in order to compare functions $f + h$ and $g + k$, we must operate within the world defined by functions $f + h$ and $g + k$. Of course, this new world—and its associated constants—will be related to the world defined by f and g and the world defined by h and k, but it will almost certainly be distinct from those worlds.

2. *Multivariate functions.* Finally, we discuss the scenario that almost automatically accompanies the transition from *sequential, single-agent* computing to *parallel and/or distributed and/or multi-agent* computing. One must account for a system's many resources: computing, memory/storage, and communication, etc. One must assess the relationships among the costs of operating the system: time, energy, memory usage, etc. One must account for possible tradeoffs among costs and possible cost-altering emulations of one subsystem by another. Within such scenarios, every assertion of an asymptotic relationship of the form

$$f(\mathbf{x};\mathbf{y}) = O(g(\mathbf{x};\mathbf{y}))$$

must explicitly specify the following information:

- which variables represent resources that can grow without bound;
- among such unbounded variables, which participate in the posited asymptotic relation;
- for each participating unbounded variable x, what are the constants c_x and N_x that witness the posited asymptotic relationship(s).

Asymptotic reasoning becomes much more complicated in settings that demand multivariate functions and measures. Modern computing paradigms, such as those involving parallel and distributed activity by multiple agents, therefore present much greater challenges than do the simple settings we discuss here. However, the benefits of being able to cogently reason qualitatively about the quantitative aspects of computing increase at least commensurately!

7.2 Reasoning About Infinity

7.2.1 Coping with infinity

A recurring challenge when one "does" mathematics is dealing with *infinity*. Infinite objects, such as sets and summations, behave rather differently from the more familiar finite objects that we encounter in our daily lives. Two examples will suffice.

1. There are "equally many" integers within the set $\mathbb{N}$ of *all* integers as there are in

- $\mathbb{N}$'s proper *subset* that comprises just the *odd* integers.

- $\mathbb{N}$'s proper *superset* that comprises ordered pairs of integers.

Does this mean that "infinite is infinite", i.e., that one can match up the elements of any two infinite sets.

Decidedly NOT!

Alas! We have to await Section 10.4 before we make the acquaintance of infinite sets that are "bigger" than $\mathbb{N}$. We shall appreciate this introduction all the more after Section 8.2, where we shall encounter a lot of infinite sets that "should be bigger" than $\mathbb{N}$—but are not!

2. We have seen in Chapter 6 that there exist summations of infinitely many positive numbers whose sum is finite (in the sense that the series *converges*). The series

$$1 + \frac{1}{2} + \frac{1}{2^2} + \cdots + \frac{1}{2^k} + \cdots = 2$$

provides an example.

However, there exist other infinite summations whose "sum is infinite" (in the sense that the series *diverges*). The *harmonic* series

$$1 + \frac{1}{2} + \frac{1}{3} + \cdots + \frac{1}{k} + \cdots$$

provides an example.

There are clearly nonobvious concepts within the world of the infinite, which distinguish those infinite objects that behave more or less as we would expect from those infinite objects that don't.

The world of the infinite is even more subtle than the preceding examples suggest. This world hosts many *paradoxes*—situations that, in the words of the *New Oxford Dictionary*, "combine contradictory features or qualities".[1]

The remainder of this section is devoted to discussing and explaining the sometimes-confusing but always-fascinating world of the infinite.

7.2.2 The "Point at Infinity"

In many situations, the difficulties encountered when dealing with infinite objects result from the conceptual fiction that there is, in fact, a "point at infinity"—i.e.,

---

[1] In detail, the *New Oxford Dictionary* defines a "paradox" as
"a statement or proposition that, despite sound (or apparently sound) reasoning from acceptable premises, leads to a conclusion that seems senseless, logically unacceptable, or self-contradictory".

that one can treat infinity as just another number. In many mathematical environments, this fiction is an aid to reasoning which can be handled with totally rigor—but often only when accompanied by rather sophisticated mathematical machinery. Two familiar examples of such mathematical machinery are the notions of *limit* and *continuity* (of a function). A more advanced example of such machinery is the *Riemann sphere*, an invention of the nineteenth-century German mathematician Bernhard Riemann, which allows one to reason about the infinite two-dimensional plane by "conformally" wrapping the plane into a sphere whose "south pole" represents the zero-point (i.e., the origin) of the plane and whose "north pole" represents the "point at infinity." There are many other, less familiar examples of such mathematical machinery, including advanced topics such as *types* in the domain of mathematical logic.

Of course, we have already had some success in dealing with a quite sophisticated genre of infinite object: In Chapter 6, we summed and manipulated a broad variety of infinite summations. We are close to having the tools necessary to deal with a variety of other infinite objects; to cite just two:

1. In Section 8.2, we shall demonstrate how to establish one-to-one correspondences between certain infinite sets, including, notably, any two of: the set $\mathbb{N}$ of integers, the set $\mathbb{Q}$ of rationals, and the set $\mathbb{N} \times \mathbb{N}$ of ordered pairs of integers.

2. In Section 10.4, we shall demonstrate that one *cannot* establish a one-to-one correspondence between the set $\mathbb{N}$ of integers and the set $\mathbb{R}$ of real numbers.

The lesson of the preceding paragraphs is that there is no need to avoid dealing with infinity and its related notions—as long as one has the mathematical machinery necessary to: (*a*) define all needed notions unambiguously; (*b*) obtain well-defined results from all required operations and manipulations; (*c*) reason cogently about all the concepts and processes one employs. That said, the scenarios described in the coming two subsections warn us to treat all aspects of infinity with care and respect. The subsections point out two challenges often encountered when one reasons about the infinite. Both challenges leave us with some form of *paradox*.

7.2.2.1 Underspecified problems

The paradoxes we present in this section require refined ground rules for their resolution. The underlying problems *seem* to be totally specified—until one tries to develop their solutions.

A. Summing ambiguous infinite summations

The first question we tackle was the subject of much concern as the topic of infinite summations emerged from its infancy in the eighteenth century. The overriding question is, What can one learn from an infinite series that does not have a unique sum? Much valuable work has been done on this question, most of which is beyond the scope of this text. But the conundrum presented by such summations is valuable to consider.

For our first source of puzzlement, we invoke the following infinite summation.

$$S = 1 - 1 + 1 - 1 + 1 - 1 + \cdots$$

There are many conflicting, but well-reasoned, answers to questions about S:

- *Does summation S have a finite sum?*

- If so: *Is this sum positive or negative?*

Here are four plausible answers to these questions; you may find others. The fourth answer, in particular, merits contemplation.

1. YES: $S = 0$

   This response is justified by the following association of terms in summation S.

$$S = \underbrace{(1 - 1)}_{0} + \underbrace{(1 - 1)}_{0} + \underbrace{(1 - 1)}_{0} + \underbrace{(1 - 1)}_{0} + \ \cdots$$

2. YES: $S = 1$

   This response is justified by the following association of terms in summation S.

$$S = 1 \ \underbrace{(- 1 + 1)}_{0} \underbrace{(- 1 + 1)}_{0} \underbrace{(- 1 + 1)}_{0} \ + \ \cdots$$

3. YES: $S = 1/2$

   This response is justified by the fact that S satisfies the equation $S = 1 - S$.

4. THERE IS NO VALID ANSWER.

   This response is justified by the fact that the problem statement does not specify how to associate the terms of S. As the first three answers suggest, "mischievously" playing with the parenthesization of summation S enables us to arrive at many "plausible" sums.

The mysteries that arise from contemplating summation S are not exhausted by playing with parentheses. Consider, for instance, the infinite summation that consists of all integers, with alternating signs:

$$S' = 1 - 2 + 3 - 4 + 5 - \cdots + \cdots$$

If we associate corresponding summands in summations S and S', then we find that

$$S' = S - S'$$

If, then, we accept the valuation $S = 1/2$, then we would find that

$$S' = 1/4$$

Another defensible analysis of S' would say that S' has no finite valuation because of the association of terms

$$S' = (1 - 2) + (3 - 4) + (5 - 6) + \cdots$$
$$= (-1) + (-1) + (-1) + \cdots$$

One can add numerous examples to the preceding two to illustrate how very differently summations having infinitely many terms behave from summations having finitely many terms.

The conclusion we offer is: *Beware: Infinity is not a number!*

B. The Ross-Littlewood paradox

The following story is known as the *Ross-Littlewood paradox*, after its creators. A version of the story appeared first in John Littlewood's enlightening and entertaining book *Littlewood's Miscellany* [77]; the story was amplified to its present form by Sheldon M. Ross [93].

Let us imagine a system that contains

- a *really big* bin (in fact, one whose capacity grows as the story progresses)

- an unbounded sequence of ordered balls, labeled 1, 2, ...

- a *very* (read: infinitely) precise clock.

The system is watched over by a *Keeper*. We observe the *Keeper* executing the following process.

Step 1. At time *midnight minus 1 minute*, the *Keeper* places the first ten balls in the sequence (i.e., balls #1, ..., #10) into the bin, and *immediately* removes the first ball (ball #1).

Step 2. At time *midnight minus 1/2 minute*, the *Keeper* places the next ten balls in the sequence (i.e., balls #11, ..., #20) into the bin, and *immediately* removes the second ball (ball #2).

The *Keeper* repeats this process endlessly, at midnight minus 1/4 minute (putting balls #21, ..., #30 into the bin and removing ball #3), then at midnight minus 1/8 minute (putting balls #31, ..., #40 into the bin and removing ball #4), and so on.

Question: *How many balls are present in the bin at midnight?*
(Note that "infinity" now measures the number of steps executed in the process.)

As in paragraph A, there are several plausible answers to this question. We provide just three.

1. THERE ARE INFINITELY MANY BALLS.

This response is justified by the following reasoning. Each step of the process inserts 10 balls into the bin but removes only one ball. Hence, the population of the bin grows by nine balls after each step of the process—and it never decreases! It follows, therefore, that after infinitely many steps, the bin's population must be infinite.

2. THE BIN IS EMPTY—THERE ARE ZERO BALLS!

This response is justified by the following reasoning. Every ball is eventually removed from the bin at some (finite) step of the process. Specifically, ball #n is removed at step n, i.e., at time midnight minus $60/2^{n-1}$ seconds.

3. THERE IS NO VALID ANSWER.

This response is justified by the following reasoning. There is no "moment at infinity" that is actually ever encountered during the process. That "moment" is just a conceptual idealization.

In other words, infinity is not a number!

C. Zeno's paradox: Achilles and the tortoise

In his celebrated *Paradox of Achilles and the Tortoise*, Zeno of Elea presented a problem whose solution had to await the seventeenth century.

In Zeno's story, the slow-footed Tortoise (T) tries to convince the speedy Achilles (A) of the futility of trying to win any race in which A gives T even the smallest head start. As long as T is ahead of A, says T, every time A traverses half the distance between himself and T, T will respond by moving a bit further ahead. Thereby, T will always be a positive distance ahead of A, so that A can *never* catch T.

> **Explanatory note.**
>
> At first glance, Zeno's story seems to call into question the physical reality of all motion. In fact, the resolution of the apparent paradox resides in the notion of *infinitesimals*—quantities that dynamically grow smaller than any finite number. While infinitesimals are familiar today to anyone who has studied subjects such as the differential calculus, the notion of infinitesimal actually dates back only a few hundred years, to the seventeenth century.

> **Historical notes.**
>
> The notion of infinitesimal has aspects that are of interest beyond the "mere" technical.
>
> The philosophically inclined reader might be interested in the essay "The Two Infinities"[a] by the French mathematician-philosopher (or philosopher-mathematician?) Blaise Pascal, whose work we revisit several times in this text, including, notably, in Section 9.2.1.
>
> The historically inclined reader will certainly be interested in the drama that underlies the discovery/invention[b] of infinitesimals. The question of earliest discovery is one of the great real-life mysteries of all time. Who invented/discovered infinitesimals? The parties to this dispute were the German mathematician Gottfried Leibniz [76] and the English polymath (Sir) Isaac Newton [84]. The cases favoring each of these great men contain enough merit to guarantee that the dispute will likely never be settled. We therefore list Leibniz and Newton alphabetically and give a coarse dating of the seventeenth century for the discovery. This real-life mystery is as full of intrigue and suspense as any that one encounters in fiction.
>
> ───────────
>
> [a] The essay appears in Pascal's *Pensées*.
>
> [b] Were infinitesimals *discovered* or *invented*? As you ponder this choice, recall the words of Leopold Kronecker that end our *MANIFESTO*.

D. Hilbert's hotel paradox

While the final story of this section does not describe an actual paradox, it is often referred to by that term. Terminology aside, this story definitely points out a fundamental difference between the real world of finite capacities and an idealized world that is not so encumbered.

Imagine that you are running a hotel that has an infinite number of rooms which are labeled by the (entire set of) positive integers: there is a room #1, a room #2, a room #3, and so on. Say that on a particular evening, every room of the hotel is occupied by a guest—and then a new guest arrives!

In a desire to accommodate the newcomer, you initiate the following procedure, which was first proposed by the eminent German mathematician David Hilbert.

By means of a broadcast message to all current guests, you move each guest who currently occupies room #k into room #$k+1$. Of course, this total shift empties room #1, hence renders it available for housing the newcomer—so you assign this room to the newly arrived guest. The world is quiet once more!

Of course, this humorous story has its roots in a fundamental distinction between the world of finite-capacity hotels that we live in and the idealized infinite-capacity hotel proposed in Hilbert's story. In a word, every finite set of integers—think of the integers as the room numbers in a finite-capacity hotel—*has a largest number*, while an infinite set of positive integers *does not have a largest number*.

This paradox can be extended in a way that makes it even more perplexing: The story's hotelier can accommodate even *an infinite number* of new guests! Consider again that every room of the hotel is occupied, and say that infinitely many new guests arrive! (Pity the poor desk clerk!) The hotelier can accommodate all of the new arrivals by moving each guest who currently occupies room #*k* into room #2*k*.

The story of Hilbert's hotel provides yet another demonstration that ∞ is not a number![2] Interestingly, ∞ behaves like a number in many ways—but it violates many of the rules of arithmetic that govern actual numbers. For example, mathematical logicians who are concerned with *orders of infinity* will often employ expressions such as

$$(2 + \infty) \quad \text{or} \quad (4 \times \infty) \quad \text{or even} \quad (\infty + \infty)$$

as though ∞ actually were a number—but they never forget that ∞ behaves in nonstandard ways, such as being an "absorber" under the operations of addition and multiplication; e.g.,

$$(2 + \infty = \infty) \quad \text{and} \quad (\infty + \infty = \infty)$$

7.2.2.2 Foundational paradoxes

The *foundational* paradoxes that we present now can be resolved only via the development of new, sophisticated, mathematical machinery.

A. Gödel's paradox: Self-reference in language

The paradox attributed to logician/philosopher/mathematician Kurt Gödel is described most perspicuously by means of a simple utterance, which we shall call "Sentence *S*", and an accompanying query.

Sentence *S*: *The sentence you are reading at this moment is false.*

The query: *Is Sentence S true, or not?*

Let us analyze the two options that the question affords us.

1. *If Sentence S is true*, then one must accept its assertion that *Sentence S is false.*

2. *If Sentence S is false*, then one must reject its assertion that Sentence *S* is false. In other words, one must conclude that *Sentence S is true.*

Sentence *S* cannot be either true or false. This flies in the face of conventional two-valued logic's demand that every assertion is *precisely one of* true *or* false!

---

[2] The *lemniscate* curve ∞ is the traditional symbol for (the point at) "infinity".

What does this mean? Has Gödel "fooled us" with a strange linguistic construction?
—Regrettably, NO!

In the early 1930s, Gödel turned the mathematical world on its head with his *rigorous* demonstration that, speaking informally:

> Any language that is self-referential—*i.e., that can refer to its own sentences as objects of discourse*—must contain a sentence such as Sentence S, which is neither true nor false. [53]

The shocking implication of Gödel's work is that in any sufficiently sophisticated language *L*, the notions *true* and *false* do *not* totally partition (into two pieces) the set of legitimate utterances in language *L*. The simplicity of Sentence *S* and encodings thereof—see, e.g., [91]—can be used to show that the following classes of languages, and their kin, are "sufficiently sophisticated":

- Natural languages (Swahili, German, Urdu, etc.)

- General-purpose programming languages (assembly language, C, Java, LISP, Python, etc.)

- Quantified mathematical languages—i.e., languages that contain logical quantifiers such as FOR ALL ($\forall$), THERE EXISTS ($\exists$), etc.

Of course, the world was spinning and circling the sun before Gödel's earthshaking proof, and it is still spinning and circling after the proof. However, we are just aware now that we must be more careful in our use of language! For instance, we must take care to employ pre-validated transformations in our compilers and pre-justified "small steps" in our schedulers: We cannot be deceived that our "noble intentions" as we crafted our systems are sufficient safeguards from the dangers that result from unanticipated "encodings" within our system.

B. Russell's paradox: The "anti-universal" set

As we remarked in Chapter 3, the notion of set is perhaps the most basic one in mathematics. One of the fundamental features of sets is that their elements are not governed by any *a priori* restrictions. Most specifically for our discussion here, *a set can have sets as elements*. Indeed, there is no inherent reason why a set cannot contain *itself* as an element! At first blush, the possibility for "self-membership" in sets seems to be a rather innocuous, albeit unexpected, freedom. However, the twentieth-century English philosopher/logician Bertrand Russell pointed out in [95, 113] that, when the population of sets that we focus on contains infinite sets, the capacity for self-membership can be (intellectually) hazardous.

The hazard that Russell foresaw resided in the fact that, absent any restrictions on the way that one can form sets, one could employ *any* proposition $\mathbf{P}(x)$ to define the set

$$\{x \mid \mathbf{P}(x)\}$$

One could thereby specify many different "benign" sets such as

Set	**P**(x)
Even integers:	$[x \in \mathbb{N}]$ and $(\exists y \in \mathbb{N})[x = 2y]$
Perfect squares:	$[x \in \mathbb{N}]$ and $(\exists y \in \mathbb{N})[x = y \cdot y]$
Satisfiable POS propositions:	$[x$ is an n-variable POS formula$]$
	and $(\exists \, n$-place bit vector $y)[x(y) = \text{TRUE}]$
Infinite set:	$[x$ is a set$]$ and $[x$ is infinite$]$

But one could also specify the *anti-universe* set A:

In text: A is the set of sets that *do not* contain themselves (as elements)
In symbols: $A = \{x \mid [x$ is a set$]$ and $[x \notin x]\}$

The paradox posed by the existence of the set A within our universe of discourse becomes clear when we ponder the following question.

Question. *Is the set A a member of itself?*

Let us analyze the two options that the question affords us.

1. *If set A is a member of A*, then by definition, A *is not* a member of A.

   This is because A is the set of all sets that *do not* contain themselves as members.

2. *If set A is not a member of A*, then by definition, A is a member of A.

We once again find ourselves in a conundrum: A belongs to A if, and only if, A does not belong to A!

There have been many attempts over the years to resolve the dilemma inherent in the preceding analysis. Many have striven for logical edifices that declare the question "*Is the set A a member of itself?*" somehow illegitimate. One option that appeals to many is to assign each sentence within a language L a *type* (say, a positive integer)—and to adjudge a sentence of L to be *legitimate* only if it refers only to sentences of *lower* type-number.

The stratagem of typing utterances within a language L *disables* self-reference within L, hence defines away both Russell's paradox and Gödel's paradox. The stratagem of typing helps also to cope with problems that arise in programming languages which allow recursion. Typing is, thus, a powerful mechanism for imposing helpful structure in languages. But, the advantages of typing come only at the expense of adding an obtrusive layer of formalism to language(s) and their associated systems.

7.3 Exercises: Chapter 7

Throughout the text, we mark each exercise with 0 or 1 or 2 occurrences of the symbol $\oplus$, as a rough gauge of its level of challenge. The 0-$\oplus$ exercises should be accessible by just reviewing the text. We provide *hints* for the 1-⊕ exercises; Appendix H provides *solutions* for the 2-⊕ exercises. Additionally, we begin each exercise with a brief explanation of its anticipated value to the reader.

1. **Three equivalent definitions of big-Θ**
   LESSON: Experience arguing about asymptotic relations

   Prove that definitions (7.1), (7.2), and (7.3) provide three logically equivalent definitions of the assertion

   $$f(x) = \Theta(f_3(x))$$

2. **An inequality that we used repeatedly in Section 6.3.2.1**
   LESSON: Appreciating asymptotic language by having to reason without it.

   a. *Prove the following asymptotic assertion.*

   Proposition 7.1 *Let $c > 0$ be any fixed integer. For all $n \in \mathbb{N}^+$,*

   $$(n+1)^c = \Theta(n^c)$$

   b. If we could not use asymptotic notation, then we would employ the following wording.

   Proposition 7.2 *There exist fixed positive constants $\underline{\kappa}$ and $\overline{\kappa}$ such that, for all sufficiently large $n \in \mathbb{N}^+$,*

   $$\underline{\kappa} n^c \leq (n+1)^c \leq \overline{\kappa} n^c$$

   Prove that the assertions in Propositions 7.1 and 7.2 are equivalent.

3. **The transmission of asymptotics when functions are combined**
   LESSONS: Enhance understanding of how asymptotic parameters work
   Prove the following result.

   Proposition 7.3 *Focus on four "primary" real functions, f_1, f_2, f_3, and f_4, and two "derived" functions*

   $$f_5(x) \stackrel{def}{=} f_1(x) + f_2(x) \quad and \quad f_6(x) \stackrel{def}{=} f_3(x) + f_4(x)$$

   If $f_1(x) = O(f_2(x))$ and $f_3(x) = O(f_4(x))$, then $f_5(x) = O(f_6(x))$.

4. **Polynomials' degrees and their asymptotic growth rates**
   LESSON: Practice with asymptotic arguments

   Let us be given polynomials with positive coefficients: $P(x)$ of degree a and $Q(x)$ of degree $b > a$. The numbers a and b need not be integers.

   Prove that there exists a constant $X_{P,Q}$—i.e., a constant that depends on the degrees and coefficients of polynomials P and Q—such that

$$(\forall\, x > X_{P,Q})\ [P(x) < Q(x)]$$

The preceding formulation can be rephrased in the following two ways:
(1) Polynomial Q *eventually majorizes* polynomial P.
(2) Polynomial Q *grows asymptotically faster than* polynomial P.

5. **Exponentials grow asymptotically faster than polynomials**
   LESSON: Practice with asymptotic arguments

   Let us be given a degree-b polynomial Q with positive coefficients, together with an arbitrary real number $c > 1$.

   Prove that there exists a constant $Y_{c,Q}$—i.e., a constant that depends on the constant c and on the degree and coefficients of polynomial Q—such that

$$(\forall\, x > Y_{c,Q})\ [c^x > Q(x)]$$

Chapter 8
Numbers II: Building the Integers and Building with the Integers

Great oaks from little acorns grow.

Proverb

Chapter 4 developed the basic concepts that underlie objects of mathematical discourse that we traffic in daily: numbers and the numerals that we use to manipulate them. The current chapter builds on those basics with the help of the material in the intervening chapters, which have given us advanced tools for discussing and manipulating numbers and aggregations of numbers. We focus here on three advanced subjects. In Section 8.1, we develop a number of important topics concerning the *prime numbers*, a set of integers that can aptly be termed the *building blocks of the integers*. In Section 8.2, we focus on the important topic of *pairing functions*. These functions allow us, in both theory and practice, to mathematically and computationally treat tuples of numbers—as well as many other aggregates—with the same ease as we treat ordinary numbers. One particularly important contribution of pairing functions is their endowing tuples and other aggregates of numbers with a natural *total order*. Finally, in Section 8.3, we establish the elements of *finite number systems*. We use such systems every day—often without even being aware of doing so—as we tell time and measure angles: It is important to understand the ways in which such systems mirror our more familiar infinite number systems and the ways in which they do not.

8.1 Prime Numbers: Building Blocks of the Integers

We now single out a subclass of the positive integers whose mathematical importance has been recognized for millennia but which have found important new applications (e.g., within the domain of computer security) as recently as within the past several decades. This subclass is defined by its *divisibility* characteristics.

© Springer Nature Switzerland AG 2020
A. L. Rosenberg, D. Trystram, *Understand Mathematics, Understand Computing*,
https://doi.org/10.1007/978-3-030-58376-7_8

We all know that every positive integer n is divisible by 1 and by n. The class of integers that we focus on now consists of those $n > 1$ in $\mathbb{N}^+$ that have no other positive divisors. These are the *prime numbers*.

An integer $p > 1$ is *prime* if its *only* positive integer divisors are 1 (which divides every integer) and p itself (which always divides itself).

Explanatory note.

We often use the shorthand assertion, "p is a prime" (or even the simpler "p is prime") instead of the longer, but equivalent, "p is a prime integer".

8.1.1 The Fundamental Theorem of Arithmetic

8.1.1.1 Statement and proof

A very consequential way to classify a positive integer $n > 1$ is to list the primes that divide it, coupling each such prime p with its *multiplicity*, i.e., the number of times that p divides n. To elaborate: Let $p_1, p_2, \ldots, p_r$ be all of the distinct primes that divide a given interger $n > 1$, and let each p_i divide n with multiplicity m_i. The *prime factorization* of n is the product $p_1^{m_1} \times p_2^{m_2} \times \cdots \times p_r^{m_r}$. Note that this product satisfies the equation

$$n = p_1^{m_1} \times p_2^{m_2} \times \cdots \times p_r^{m_r} \tag{8.1}$$

It is traditional to write the factorization of an integer n in *canonical form*, i.e., with the primes $p_1, p_2, \ldots, p_r$ that divide n listed *in increasing order*, i.e., so that $p_1 < p_2 < \cdots < p_r$.

The following classical theorem asserts that a positive integer n is totally characterized by its canonical prime factorization. This theorem has been known for millennia and has been honored with the title *The Fundamental Theorem of Arithmetic*. We state the Theorem in two equivalent ways, which suggest somewhat different ways of thinking about the result.

Theorem 8.1 (The Fundamental Theorem of Arithmetic).
(Traditional formulation.) *The canonical prime factorization of every positive integer is unique.*

(Alternative formulation.) *Let $n \in \mathbb{N}^+$ be a positive integer, and let $\widehat{P}_n$ denote the ordered sequence of all prime numbers that are no larger than n:*

$$\widehat{P}_n = \langle p_1, p_2, \ldots, p_{r-1}, p_r \rangle$$
where: $p_1 = 2$
       each $p_i < p_{i+1}$
       $p_r \leq n.$

There exists a unique sequence of nonnegative integers, $\langle a_1, a_2, \ldots, a_r \rangle$, such that

$$n = \prod_{i=1}^{r} p_i^{a_i} = p_1^{a_1} \times p_2^{a_2} \times \cdots \times p_{r-1}^{a_{r-1}} \times p_r^{a_r}$$

A simple, yet important, corollary of Theorem 8.1 is the following result, whose proof we leave to the reader.

Proposition 8.1 *Every integer $n > 1$ is divisible by at least one prime number.*

Proving the Fundamental Theorem of Arithmetic. The proof of Theorem 8.1 is actually rather elementary, providing that one approaches it gradually. It employs a lot of important techniques and concepts involved in "doing mathematics" which we have discussed in the eponymous Chapter 2.

We begin with a purely technical result.

Proposition 8.2 *Let p be a prime, and let m be any positive integer that is not divisible by p. There exist integers a, b, not necessarily positive, such that*

$$ap + bm = 1$$

Proof. This result is a special case of Proposition 4.4. To see this, note that, for any prime p and integer m that is not divisible by p, $\text{GCD}(p, m) = 1$. $\quad\square$

Proposition 8.3 *If the prime p divides a composite number $m \times n$, then either p divides m, or p divides n, or both.*[1]

Proof. Let p, m, and n be as asserted, and say that p does not divide m. By Proposition 8.2, there exist integers a, b, not necessarily positive, such that

$$ap + bm = 1$$

Let us multiply both sides of this equation by n. After some manipulation—specifically, applying the distributive law—we find that

$$apn + bmn = n$$

Now, p divides the expression to the left of the equal sign: p divides p by definition, and p divides $m \times n$ by assumption. It follows that p must divide the expression to the right of the equal sign—which is the integer n. $\quad\square$

We are finally ready to develop the proof of the Fundamental Theorem.

Proof (The Fundamental Theorem of Arithmetic). Our dominant tool for proving Theorem 8.1 will be *proof by contradiction* (see Section 2.2.2). We assume, for the sake of contradiction, that there is a positive integer n that has two distinct canonical prime factorizations.

---

[1] The closing phase "or both" signals our use of the *inclusive* or.

Our argument will be a trifle simpler if we employ the *alternative* form of the Theorem. To this end, let

$$p_1 < p_2 < \cdots < p_{r-1} < p_r$$

denote, in increasing order, the set of all primes that do not exceed n; i.e., every $p_i \leq n$.

Under this formulation of the Theorem, the fact that n has two distinct canonical prime factorizations manifests itself in the assumption that there exist *two* distinct sequences of *nonnegative* integers,

$$\langle a_1, a_2, \ldots, a_r \rangle \quad \text{and} \quad \langle b_1, b_2, \ldots, b_r \rangle$$

such that n is expressible by—i.e., is equal to—both of the following products of the primes p_1, p_2, ..., p_{r-1}, p_r.

$$p_1^{a_1} \times p_2^{a_2} \times \cdots \times p_{r-1}^{a_{r-1}} \times p_r^{a_r} \tag{8.2}$$

$$p_1^{b_1} \times p_2^{b_2} \times \cdots \times p_{r-1}^{b_{r-1}} \times p_r^{b_r} \tag{8.3}$$

Let us now "cancel" from the products in Eqs. (8.2) and (8.3) the longest common prefix. Because the two products are, by hypothesis, distinct, at least one of them will not be reduced to 1 by this cancellation. We are, therefore, left with residual products of the forms

$$p_i^{a_i - c_i} \cdot X \tag{8.4}$$

$$p_i^{b_i - c_i} \cdot Y \tag{8.5}$$

where:

- Precisely one of $a_i - c_i$ and $b_i - c_i$ equals 0.

  To see this, begin canceling instances of prime p_1, one by one, from the two products. If one string of p_1's runs out before the other, say after c_1 cancellations, then the pair $(a_1 - c_1, b_1 - c_1)$ witnesses our claim. Otherwise, if both strings of p_1's run out simultaneously—because $a_1 = b_1$—then we start canceling instances of prime p_2, one by one, from the two residual products. If these strings do not run out simultaneously, then we have a witness for our claim; if they do run out simultaneously, then we turn our attention to p_3. And so on ...

  Because the two products in Eqs. (8.4) and (8.5) are distinct, we must eventually reach a prime p_i for which $a_i \neq b_i$. At that point,

  $$c_i = \max(a_i, b_i) - \min(a_i, b_i)$$

  cancellations of instances of p_i will lead us to the situation we have described.

  Say, with no loss of generality (because we have no *a priori* way to distinguish the products), that $b_i - c_i = 0$ while $a_i - c_i \neq 0$.

- Products X and Y are composed only of primes that are strictly bigger than p_i.

Note that
$$p_i^{a_i-c_i} \cdot X = p_i^{b_i-c_i} \cdot Y = Y$$
because these products result from cancelling the same prefix from the equal products (8.2) and (8.3), and because $b_i - c_i = 0$ so that $p_i^{b_i-c_i} = 1$.

We have finally reached the point of contradiction.

On the one hand, p_i *must* divide the product Y, because it divides the product $p_i^{a_i-c_i} \cdot X$ which equals Y.

On the other hand, p_i *cannot* divide the product Y, because every prime factor of Y is bigger than p_i. We now invoke Proposition 8.3 to infer that p_i can divide Y only if it divides one of Y's prime factors. Since all of Y's prime factors are larger than p_i, such divisibility would require one prime (namely p_i) to divide a bigger prime—which is impossible by defintion.

We conclude that one of the products (8.2) and (8.3) cannot exist, so the theorem must hold. $\square$

8.1.1.2 A "prime" corollary: There are infinitely many primes

The main result of this section builds upon Theorem 8.1 in a crucial way.

Proposition 8.4 *There are infinitely many prime numbers.*

We provide two proofs of this seminal result. The first, classical, proof is traditionally attributed to (our friend, by now) Euclid. The second, mathematically more interesting, proof is the work of Euler, who is quickly becoming another friend.

Proof (Euclid's classical proof). We know that the first several primes are

$$(p_1 = 2), \ (p_2 = 3), \ (p_3 = 5), \ (p_4 = 7), \ (p_5 = 11), \ldots$$

How far does this sequence extend? Does it ever end?

Let us assume, for the sake of contradiction, that there are only finitely many primes (so that our sequence ends). Say, in particular, that the following r-element sequence of integers enumerates all (and only) primes, in order of magnitude:

$$\textbf{Prime-Numbers} = \langle p_1, \ p_2, \ \ldots, \ p_r \rangle$$
$$\text{where} \qquad p_1 < p_2 < \cdots < p_{r-1} < p_r$$

We verify the *falseness* of the alleged completeness of the sequence **Prime-Numbers** by analyzing the positive integer

$$n^\star = 1 + \prod_{i=1}^{r} p_i = 1 + (p_1 \times p_2 \times \cdots \times p_r).$$

In fact, we claim that $n^\star$ is a prime that is not in the sequence **Prime-Numbers**. We begin to verify our claim by making three crucial observations.

1. *The number $n^\star$ is not divisible by any prime in the sequence* **Prime-Numbers**.

    To see this, note that for each p_k in the sequence,

    $$n^\star/p_k \;=\; \frac{1}{p_k} + \prod_{i \neq k} p_i.$$

    Because $p_k \geq 2$, we see that $n^\star/p_k$ obeys the inequalities

    $$\prod_{i \neq k} p_i \;<\; n^\star/p_k \;<\; 1 + \prod_{i \neq k} p_i.$$

    The discreteness of the set $\mathbb{Z}$—see Section 4.3.1—implies that $n^\star/p_k$ is not an integer, because it lies strictly between two adjacent integers.

2. Because of Observation 1, if the sequence **Prime-Numbers** actually did contain *all* of the prime numbers, then we would have to conclude that *the number $n^\star$ is not divisible by any prime number*.

3. Finally, we invoke Proposition 8.1, which asserts that *every integer $m > 1$ is divisible by (at least one) prime number*.

The preceding chain of assertions leads to a mutual inconsistency. On the one hand, the integer $n^\star > 1$ has no prime-integer divisor. On the other hand, no such integer can fail to have a prime-integer divisor!

Let us analyze how we arrived at this uncomfortable place.

- At the front end of this string of assertions we have the assumption that there are only finitely many prime numbers. We have (as yet) no substantiation for this assertion.

- At the back end of this string of assertions, we have a contradiction of Proposition 8.1, which we have *proved*.

- In between these two assertions we have a sequence of assertions, each of which follows from its predecessors via irrefutable rules of inference.

It follows that the *only* brick in this edifice that could be faulty—i.e., the only assertion that could be false—is the initial assumption, which states that there are only finitely many prime numbers. *We therefore conclude that this vulnerable assumption is false!*

In other words: *There are infinitely many prime numbers.*   □

Proof (Euler's modern proof). We now present Euler's proof of the infinitude of primes, which postdates Euclid's by more than two millennia. Being an elegant tapestry crafted from several quite distinct pieces of mathematics, this proof is a wonderful example of the kinds of unexpected connections that appear within mathematics once you learn where to look.

The idea underlying this proof is to establish that a given expression involving products of primes is not bounded.

Let $\widehat{P}$ denote the *set* of all prime numbers. (We highlight the word "set" to distinguish $\widehat{P}$, which certainly exists, from Euclid's sequence **Prime-Numbers**, which we now know does not exist.)

Consider the product:

$$\prod_{p \in \widehat{P}} \frac{p}{p-1} = \prod_{p \in \widehat{P}} \frac{1}{1-1/p} \tag{8.6}$$

We know from Section 6.2.2 that $p/(p-1)$ is the sum of the geometric series of all nonnegative integer powers of $1/p$:

$$\frac{1}{1-1/p} = \sum_{k \in \mathbb{N}} \frac{1}{p^k} = \sum_{k=0}^{\infty} \frac{1}{p^k} \tag{8.7}$$

Combining equations (8.6) and (8.7), we see that

$$\prod_{p \in \widehat{P}} \frac{p}{p-1} = \prod_{p \in \widehat{P}} \left(\sum_{k=0}^{\infty} \frac{1}{p^k} \right) \tag{8.8}$$

We can now expand the product of sums on the right-hand side of Eq. (8.8). After applying all of the relevant laws of arithmetic—the commutativity and associativity of both addition and multiplication, and the distributivity of multiplication over addition—we can rewrite the product of sums as the following summation

$$\prod_{p \in \widehat{P}} \frac{p}{p-1} = \sum \frac{1}{p_{i_1}^{a_1} p_{i_2}^{a_2} \cdots p_{i_k}^{a_k}} \tag{8.9}$$

where the summation is over all choices of

- a positive integer k
- a k-element sequence of primes $2 \leq p_{i_1} < p_{i_2} < \cdots < p_{i_k}$
- a k-element sequence of positive integers $a_1, a_2, \ldots, a_k$

But, now, Theorem 8.1 (the Fundamental Theorem of Arithmetic) tells us that the summation in Eq. (8.9) includes the reciprocal of every positive integer as a summand exactly once. This is because every such integer is completely characterized by its (unique) decomposition as a product of powers of primes. This means, though, that the product (8.6) has the same value as the harmonic series $S^{(H)}$:

$$\prod_{p \in \widehat{P}} \frac{1}{1-1/p} = \sum_{n \in \mathbb{N}^+} \frac{1}{n} = S^{(H)}$$

which, by Proposition 6.16, is infinite!

Since any product of finitely many numbers is finite, the only way that product (8.6) could be infinite is if there were infinitely many primes. We conclude that this must be the case.   □

8.1.1.3 Number sieves: Eratosthenes and Euler

The third-century BCE Greek polymath Eratosthenes of Cyrene is widely credited with a conceptual algorithm—*the sieve*—which exhibits the prime factors of integers in a perspicuous manner. We define the *sieve of Eratosthenes* and suggest some of its applications.

We formulate the sieve as a regimen for labeling integers with their prime factors. We use the label λ_p to identify integers which are multiples of prime p.

We begin with the linear array of positive integers:

$$1|2|3|4|5|6|7|8|9|10|11|12|13 \ldots$$

We next label the multiples of 2 (i.e., the even numbers):

$$
\begin{array}{c|c|c|c|c|c|c|c|c|c|c|c|c}
1 & 2 & 3 & 4 & 5 & 6 & 7 & 8 & 9 & 10 & 11 & 12 & 13 \\
 & \lambda_2 & & \lambda_2 & & \lambda_2 & & \lambda_2 & & \lambda_2 & & \lambda_2 &
\end{array}\cdots
$$

We then label the multiples of 3:

$$
\begin{array}{c|c|c|c|c|c|c|c|c|c|c|c|c}
1 & 2 & 3 & 4 & 5 & 6 & 7 & 8 & 9 & 10 & 11 & 12 & 13 \\
 & \lambda_2 & & \lambda_2 & & \lambda_2 & & \lambda_2 & & \lambda_2 & & \lambda_2 & \\
 & & \lambda_3 & & & \lambda_3 & & & \lambda_3 & & & \lambda_3 &
\end{array}\cdots
$$

We then label the multiples of 5:

$$
\begin{array}{c|c|c|c|c|c|c|c|c|c|c|c|c}
1 & 2 & 3 & 4 & 5 & 6 & 7 & 8 & 9 & 10 & 11 & 12 & 13 \\
 & \lambda_2 & & \lambda_2 & & \lambda_2 & & \lambda_2 & & \lambda_2 & & \lambda_2 & \\
 & & \lambda_3 & & & \lambda_3 & & & \lambda_3 & & & \lambda_3 & \\
 & & & & \lambda_5 & & & & & \lambda_5 & & &
\end{array}\cdots
$$

The sieve affords one a wonderful perspective for perceiving the patterns that drive mathematical insights. Here are a few results that you will be asked to prove as exercises which are suggested by studying the sieve.

1. *Every integer $n > 1$ is divisible by at least one prime number.*

2. *Every sequence $(k+1), (k+2), \ldots, (k+p)$ of p consecutive integers contains a multiple of p.*

3. *Every product of four consecutive integers, $(k+1) \cdot (k+2) \cdot (k+3) \cdot (k+4)$ is divisible by* FACT(4).

Our respected friend Leonhard Euler developed an alternative version of the sieve, which is more convenient for studying certain topics involving integer divisibility. In his version of the sieve, each prime and all of its multiples are removed from the sieve as soon as their existence is acknowledged. Our earlier construction of Eratosthenes's version of the sieve now becomes:

Initial array:	1	2	3	4	5	6	7	8	9	10	11	12	13	⋯
after prime 2 is processed:	1		3		5		7		9		11		13	⋯
after primes 2, 3 are processed:	1				5		7				11		13	⋯
after primes 2, 3, 5 are processed:	1						7				11		13	⋯

We shall revisit both versions of the sieve in the Exercises for this chapter.

8.1.1.4 Applying the Fundamental Theorem in *encryption*

One of the most important applications of Theorem 8.1 is its supplying a basis for designing mechanisms that *encrypt* data. While the details of both encryption and the use of prime numbers to achieve encryption are beyond the scope of this text, we now provide a "peek" into that area by means of the following result concerning *encodings* of sequences of positive integers as single integers!

Explanatory note.

There is a crucial difference between *encoding* and *encryption*, despite the words' often being confused in the vernacular.

Encodings seek representations of objects which achieve some benefit, such as efficient computation or compactness. An example might be the conversion of Roman numerals to conventional positional numerals to enable efficient computation of arithmetic operations.

Encryption usually has some notion of *secrecy* attached. An example might be some key-based cipher which is intended to limit unauthorized access to some information.

We illustrate (and achieve) the sought encodings as follows. Consider the (infinite) ordered sequence of *all primes:*

$$(p_1 = 2), (p_2 = 3), (p_3 = 5), \ldots$$

Let

$$\bar{s} = \langle m_1, m_2, \ldots, m_k \rangle \tag{8.10}$$

be an arbitrary sequence of positive integers. Then Theorem 8.1 assures us that the (single) positive integer

$$\iota(\bar{s}) = p_1^{m_1} \times p_2^{m_2} \times \cdots \times p_k^{m_k}$$

is a (uniquely decodable) integer-representation of sequence $\bar{s}$.

We return to the idea of encoding-via-integers in Section 8.2, using a quite different approach.

8.1.1.5 ⊕⊕ The "density" of the prime numbers

Whether or not one ascribes any spiritual significance to the prime numbers, as the early Greeks did, it can be fun to look for patterns in the sequence **P** of primes; we know by Proposition 8.4 that the sequence **P** is infinite. As we begin to list integers looking for primes, we are impressed by how "densely" they start out: There are six primes in the first 13 integers—the "unboxed" integers in the following sequence

$$2, 3, \boxed{4}, 5, \boxed{6}, 7, \boxed{8}, \boxed{9}, \boxed{10}, 11, \boxed{12}, 13$$

However, even as soon as the sequence

$$23, \boxed{24}, \boxed{25}, \boxed{26}, \boxed{27}, \boxed{28}, 29, \boxed{30}, 31, \boxed{32}, \boxed{33}, \boxed{34}, \boxed{35}, \boxed{36}, 37$$

we encounter only four primes in a run of 15 integers. Which of the preceding sequences is more representative of the true density of the primes among the integers? This question goes back to antiquity. Even after it became clear that "over the long enough haul", the primes were actually rather sparse among the integers, it took the labors of a string of mathematical luminaries until the end of the nineteenth century to determine exactly how sparse. It was finally ascertained only in 1896 that, asymptotically, roughly the fraction $1/\ln n$ of the integers in the interval

$$\{1, 2, 3, \ldots, n\}$$

are prime. The landmark *Prime Number Theorem*, which establishes this density result, was first proved independently and almost simultaneously (in 1896) by the French mathematician Jacques Hadamard [54] and the Belgian mathematician Charles Jean de la Vallée-Poussin [45].

In order to state the *Prime Number Theorem* precisely, we need to add a non-discrete asymptotic notion to our purely discrete treatment of asymptotics in Section 7.1. Let us be given real-valued functions $f(n)$ and $g(n)$, both defined on the positive integers: symbolically,

$$f : \mathbb{N}^+ \to \mathbb{R} \quad \text{and} \quad g : \mathbb{N}^+ \longrightarrow \mathbb{R}$$

We write

$$f(n) \ \sim \ g(n)$$

precisely when the following holds. If one uniformly approximates function f on each argument $n \in \mathbb{N}^+$ by function g on the same argument, then the *relative error* of the approximation, i.e., the ratio

$$\frac{|f(n) - g(n)|}{f(n)}$$

approaches 0 as n increases without bound. We can finally state the theorem.

Theorem 8.2 (The Prime Number Theorem). *The nth prime number, p_n, satisfies the asymptotic "equation"*

$$p_n \sim n \cdot \ln n$$

8.1.2 Fermat's Little Theorem: $a^p \equiv a \bmod p$

A measure of the greatness of the seventeenth-century French mathematician Pierre de Fermat is that the following fundamental number-theoretic result is called "Fermat's *little* theorem". Aside from its exposing an important basic property of prime numbers, the theorem provides the basis for a valuable algorithm that tests whether a given integer is a prime.

Theorem 8.3 (Fermat's Little Theorem). *Let a be any integer, and let p be any prime.*

(Formulation 1): *The number $a^p - a$ is divisible by p.*

(Formulation 2): $a^p \equiv a \bmod p$.

We provide two proofs for this result, each providing a rather different insight on the result.

8.1.2.1 A proof that uses "necklaces"

Proof. Our first proof reduces the theorem to the problem of counting a special set of strings.

Letting a and p be as in the theorem, focus on the set $S(\mathscr{A}, p)$ of all length-p words/strings over an a-symbol alphabet/set $\mathscr{A} = \{\alpha_1, \alpha_2, \ldots, \alpha_a\}$. For instance, when $p = 3$ and $\mathscr{A} = \{0, 1\}$ (so that $a = 2$), the set $S(\mathscr{A}, p)$ consists of the words:

$$000, \ 001, \ 010, \ 011, \ 100, \ 101, \ 110, \ 111$$

We begin with some basic definitions and observations.

- The number of words in $S(\mathscr{A}, p)$ is a^p.

  A straightforward induction on p validates this numeration: (1) There are a words of length 1. (2) If there are n words of length ℓ, then there are $a \times n$ words of length $\ell + 1$: simply append each of $\mathscr{A}$'s a symbols to each length-ℓ word.

  We prove this simple result a few times throughout the book—see Propositions 9.2 and 11.1—to emphasize the different intuitions that lead to the result.

- Let us view a circular shift of the words in $S(\mathscr{A}, p)$ as a function that associates a given word $w \in S(\mathscr{A}, p)$ with another word $w' \in S(\mathscr{A}, p)$.

  From this vantage point, a *(one-place) circular shift* c of a word $w \in S(\mathscr{A}, p)$ is accomplished by placing the last symbol of w into the first position and shifting all other symbols one position rightward. For illustration, if $w = \alpha_1 \alpha_2 \cdots \alpha_p$, then

$$c(\alpha_1 \alpha_2 \cdots \alpha_p) = \alpha_p \alpha_1 \cdots \alpha_{p-1}$$

- By iterating the shift c on a length-p word w at most $p - 1$ times, we obtain the *necklace* $\mathscr{N}(w)$, which is the sequence

$$\mathscr{N}(w) = w, c(w), c(c(w)), \ldots, c(c(\cdots c(w) \cdots))$$

  in which *one further shift would replicate word w.*

- The *period* of the necklace $\mathscr{N}(w)$ is the number of words in the sequence.

  Note that *the period of $\mathscr{N}(w)$ can never exceed $p - 1$*—because a previously seen word must recur by the time the length-p word w has been shifted p times.

Consider now a word w that is a *replicate* of a (perforce, shorter) word u, in the sense that $w = uu \cdots u$. Say that $u^\star$ is the shortest word of which w is a replicate and that $u^\star$ has length m. Then:

- *The length m of string $u^\star$ divides p.*

  This is obvious from our ability to write w in the indicated form.

- *The period of $\mathscr{N}(w)$ is $m - 1$.*

  This holds because, by the time one has shifted w m times, one has transferred a copy of $u^\star$ from the end of w to the beginning—hence, one has recreated w.

- In our special situation—where w has prime length—*the only candidates for the shortest replicated word $u^\star$ have length 1 or p.*

  This holds because m divides the prime p.

Summing up, one of the following two situations must hold.

Possibility #1: *The word w has the form $w = \alpha\alpha \cdots \alpha$ for some symbol $\alpha \in \mathscr{A}$.*
   This can occur in a distinct ways because of $\mathscr{A}$'s cardinality..

Possibility #2: *The word w is not a replicate of any shorter word.*

   Because p is a prime, Possibility #2 must hold for every one of the $a^p - a$ words over $\mathscr{A}$ that contain at least two distinct symbols. Because the period of any necklace $\mathscr{N}(w)$ for a word that contains at least two distinct symbols is exactly $p - 1$, the lengths of such necklaces must be exactly p.

   This means that the $a^p - a$ words that each contain at least two distinct symbols partition $S(\mathscr{A}, p)$ into disjoint sets of size p. This partition is possible only if p divides $a^p - a$. □

Fig. 8.1 The three necklaces composed of the same symbol ($a = 3$)

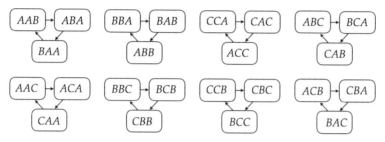

Fig. 8.2 Eight groups of necklaces of size $p = 3$ (for $a = 3$)

The ideas underlying this multi-step proof are sketched in Figs. 8.1 and 8.2, which jointly illustrate all necklaces $\mathcal{N}(w)$ for ($p = 3$)-letter words over the alphabet $\mathcal{A} = \{A, B, C\}$. Figure 8.1 depicts the necklaces for words that use only a single letter; Fig. 8.2 depicts the necklaces for words that use at least two distinct letters.

8.1.2.2 A proof that uses the Binomial Theorem

Proof. Our next proof is built upon Formulation 2 of the Proposition. We focus on a fixed prime p and argue by induction on the size a of alphabet $\mathcal{A}$ that $a^p \equiv a \bmod p$.

Base of the induction. The base case $a = 1$ is straightforward because $1^p = 1$

Inductive hypothesis. Assume for induction that $a^p \equiv a \bmod p$ for all alphabet sizes not exceeding the integer b., i.e., all $a \leq b$.

Extending the induction. By invoking the restricted form of the Binomial Theorem, we observe—see Eq. 6.41—that

$$(b+1)^p = (b^p + 1) + \sum_{i=1}^{p-1} \binom{p}{i} b^{p-i} \tag{8.11}$$

Pondering this equation, we make two important observations.

1. We learn from the development in Section 9.2.1 that p divides all "internal" binomial coefficients, i.e., all coefficients $\binom{p}{i}$ with $0 < i < p$. This means that there exists an integer n_1 such that

$$\sum_{i=1}^{p-1} \binom{p}{i} b^{p-i} = p \cdot n_1 \tag{8.12}$$

Explanatory note.

One can observe the just-exposed divisibility property of "internal" binomial coefficients by looking at Pascal's Triangle; see the rows corresponding to primes—i.e., rows $n = 2$, $n = 3$, and $n = 5$—in the triangle of Fig. 8.3. (Recall that rows are indexed beginning with $n = 0$.)

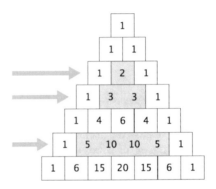

Fig. 8.3 The rows of Pascal's Triangle that correspond to $n = 0, n = 1, \ldots, n = 6$. The "internal" entries of the rows that correspond to prime numbers—in this case, $n = 2$, $n = 3$, and $n = 5$—are divisible by that number

2. By the inductive hypothesis, p divides $b^p - b$, which means that there exists an integer n_2 such that

$$b^p = b + p \cdot n_2 \tag{8.13}$$

When we combine relations (8.12) and (8.13), and we use them to rewrite Eq. (8.11), we find that

$$(b+1)^p = (b+1) + p \cdot (n_1 + n_2)$$

This means that p divides $(b + 1)^p - (b + 1)$, which extends the induction and completes the proof. □

8.1.3 ⊕ *Perfect Numbers and Mersenne Primes*

This short section is dedicated to two related topics whose intrinsic charm has garnered attention for more than three millennia from mathematicians who study numbers and their properties. We place the section here because of the central role of a particular class of prime numbers in the following story.

8.1.3.1 Definitions

Returning to the ancient Greeks' mystical affinity for special classes of integers, we term a positive integer $n \in \mathbb{N}^+$ *perfect* if n equals the sum of its proper divisors.[2] In preparation for our brief look at perfect numbers, let us denote by $\mathbf{D}(n)$ the set of *positive integer divisors of the integer $n \in \mathbb{N}^+$.*

It is not intuitively obvious that perfect numbers even exist, but only a short search is needed until one discovers the paired equations

$$\mathbf{D}(6) = \{1, 2, 3\}$$
$$6 = 1 + 2 + 3$$

which expose $n = 6$ as a perfect number. A few more minutes will lead one to an analogous pair of equations for $n = 28$:

$$\mathbf{D}(28) = \{1, 2, 4, 7, 14\}$$
$$28 = 1 + 2 + 4 + 7 + 14$$

It will take the curious reader only a bit longer to discover the paired equations

$$\mathbf{D}(496) = \{1, 2, 4, 8, 16, 31, 62, 124, 248\}$$
$$496 = 1 + 2 + 4 + 8 + 16 + 31 + 62 + 124 + 248$$

The preceding three pairs of equations demonstrate that there do exist perfect numbers. You have hopefully absorbed enough of our "How to be a mathematician" lore by this point to anticipate the following "natural" (to a mathematician, at least) questions.

Question #1. *Are there* infinitely many *perfect numbers?*
Answer. We will shortly derive the answer—YES.

Question #2. *Are there any* odd *perfect numbers?*
Answer. No one knows—as of 2020.

We pause to introduce the second of this section's topics of interest. A prime number p is a *Mersenne prime*—so named for the seventeenth-century French monk Marin Mersenne—if it has the form $p = 2^q - 1$ for some integer q.

It is obvious that not all primes are Mersenne primes—consider, e.g., 5 and 13. But it is equally obvious that some primes *are* Mersenne primes; to wit,

---

[2] Euclid refers to perfect numbers in his *Elements*, using adjectives such as "perfect" and "ideal".

The number $p = 3$ is a Mersenne prime, because $3 = 2^2 - 1$

The number $p = 7$ is a Mersenne prime, because $7 = 2^3 - 1$ (8.14)

The number $p = 31$ is a Mersenne prime, because $31 = 2^5 - 1$

Despite the ease with which we began a list of Mersenne primes, we "run out of steam" rather quickly: Only about 50 Mersenne primes are known! Indeed, it is not known—as of 2019—whether there exist infinitely many Mersenne primes! It is true, though, that *largest known prime*—as of 2019—*is a Mersenne prime:* $2^{77,232,917} - 1$. We close our very brief focus on Mersenne primes as an isolated topic—i.e., unrelated to perfect numbers—with the following result, which limits one's search for Mersenne primes to expressions with prime powers of 2.

Proposition 8.5 *The integer $2^q - 1$ is prime only if the integer q is prime.*

Proof. Say, for contradiction, that there is a composite number $q = m \cdot n$, with both $m, n > 1$, such that $2^q - 1$ is a prime. We invoke identity (6.18) from Proposition 6.4 to show that $2^q - 1$ is, in fact, *not* prime. We note, by direct calculation, that

$$2^{m \cdot n} - 1 = (2^m - 1) \cdot \frac{2^{m \cdot n} - 1}{2^m - 1}$$

$$= (2^m - 1) \cdot \sum_{i=0}^{n-1} (2^m)^i$$

$$= (2^m - 1) \cdot \left(1 + (2^m) + (2^m)^2 + \cdots + (2^m)^{n-1}\right) \quad (8.15)$$

The number $r = 2^q - 1$ is, thus the product of two numbers, both strictly greater than 1 and less than r. By definition, then, r is not a prime.   □

8.1.3.2 Using Mersenne primes to generate perfect numbers

We meld the two subjects of the preceding subsection to derive the basis of a procedure that uses Mersenne primes to generate perfect numbers.

Let us revisit the three sample perfect numbers presented at the beginning of Section 8.1.3.1. As the following table illustrates, all three numbers share the form $2^{p-1} \cdot (2^p - 1)$, where both p and $2^p - 1$ are primes.

$$n = \ \ 6 \ = \ \ 2 \cdot 3 \ \ = 2^1 \cdot (2^2 - 1) \ \ \text{so that} \ \ p = 2 \ \ \text{and} \ \ 2^p - 1 = 3$$
$$n = \ \ 28 \ = \ \ 4 \cdot 7 \ \ = 2^2 \cdot (2^3 - 1) \ \ \text{so that} \ \ p = 3 \ \ \text{and} \ \ 2^p - 1 = 7$$
$$n = 496 = 16 \cdot 31 = 2^4 \cdot (2^5 - 1) \ \ \text{so that} \ \ p = 5 \ \ \text{and} \ \ 2^p - 1 = 31$$

It was no accident that the three perfect numbers we found have intimate formal relationships with the first three primes. In fact, the result we present now establishes that *every* Mersenne prime has such a formal relationship with a perfect number—thereby providing us the promised path toward generating perfect numbers.[3]

---

[3] This result—of course with no mention of Mersenne—appears in Euclid's *Principiae IX-36*.

Proposition 8.6 *For every Mersenne prime* $2^p - 1$, *the following integer is perfect:*

$$2^{p-1} \times (2^p - 1) = \binom{2^p}{2} \tag{8.16}$$

Proof. Focus on the instance of expression (8.16) associated with the Mersenne prime $2^p - 1$. With the aid of Theorem 8.1 (the Fundamental Theorem of Arithmetic), we enumerate the factors of $2^{p-1} \cdot (2^p - 1)$. Our list consists of two groups:

1. the p powers of 2: $\{2^0 = 1, 2^1 = 2, \ldots, 2^{p-1}\}$

2. all products: $(2^p - 1) \times \big(\text{the } p \text{ powers of } 2 : \{2^0, 2^1, \ldots, 2^{p-1}\}\big)$.

Summing all of these factors leads to the expression

$$\sum_{i=0}^{p-1} 2^i + (2^p - 1) \cdot \sum_{i=0}^{p-1} 2^i = 2^p \cdot \sum_{i=0}^{p-1} 2^i = 2^p \cdot (2^p - 1) \tag{8.17}$$

We derive the ultimate expression in Eq. (8.17) from the penultimate expression by invoking Proposition 6.4 for the case $b = 2$.

We thus see that the factors of $n = 2^p \cdot (2^p - 1)$ sum to n, so that n is perfect. $\square$

8.2 Pairing Functions: Bringing Linear Order to Tuple Spaces

Paraphrasing an often-used quip by the late stand-up comic Rodney Dangerfield, integers "don't get no respect"! Superficially, it appears that integers are useful for counting things but for little else. The Fundamental Theorem of Arithmetic (Theorem 8.1) hints at the potential importance of the prime numbers, but it does little to inspire respect for the non-prime integers. Once one augments the integers with a bit of structure, *then* one can begin to represent interesting situations.

As we shall see shortly, we can accomplish our goals by focusing on very simple integer-based structures, namely, *ordered pairs of integers*—i.e., the sets $\mathbb{Z} \times \mathbb{Z}$, $\mathbb{N} \times \mathbb{N}$, and $\mathbb{N}^+ \times \mathbb{N}^+$. Using $\mathbb{Z} \times \mathbb{Z}$ as the exemplar of these three sets of ordered pairs of integers:

- Each *element* of $\mathbb{Z} \times \mathbb{Z}$ has the form $\langle a_1, a_2 \rangle$, where a_1 and a_2 belong to $\mathbb{Z}$.

- The *semantics* of this set allow one

  - to *aggregate* any two elements of $\mathbb{Z}$, call them b_1, b_2, to create the ordered pair $\langle b_1, b_2 \rangle \in \mathbb{Z} \times \mathbb{Z}$;

  - to *select*, given any pair $\langle a_1, a_2 \rangle \in \mathbb{Z} \times \mathbb{Z}$, the two elements $a_1 \overset{\text{def}}{=} \text{first}(\langle a_1, a_2 \rangle)$ and $a_2 \overset{\text{def}}{=} \text{second}(\langle a_1, a_2 \rangle)$.

8.2.1 Encoding Complex Structures via Ordered Pairs

One can use ordered pairs to represent—or, *encode*—myriad complex structures. Here is a brief sampler, which the reader can readily add to:

(i) *(Ordered) tuples of integers.* Focus on the set of k-tuples of integers, for any integer $k > 1$. One way to encode this set using ordered pairs is by recursion: the base case $k = 2$ consists of the *ordered pairs* themselves. For any $k > 2$, encode the k-tuple $\langle a_1, a_2, \ldots, a_k \rangle$ as the ordered pair whose *first* is the ordered $(k-1)$-tuple $\langle a_1, a_2, \ldots, a_{k-1} \rangle$ and whose *second* is the integer a_k:

$$\langle a_1, a_2, \ldots, a_k \rangle \text{ is encoded as } \langle \langle a_1, a_2, \ldots, a_{k-1} \rangle, a_k \rangle$$

(ii) *Strings of integers.* One way to encode the string of integers $a_1 a_2 \cdots a_n$ using ordered pairs is as follows.

$$a_1 a_2 \cdots a_n \text{ is encoded as } \langle a_1, \langle a_2, \langle a_3, \ldots, \langle a_{n-2}, \langle a_{n-1}, a_n \rangle \rangle \cdots \rangle \rangle \rangle$$

The following example should make the aggregation clear, without any possibly confusing dots:

$$a_1 a_2 a_3 a_4 \text{ is encoded as } \langle a_1, \langle a_2, \langle a_3, a_4 \rangle \rangle \rangle$$

(iii) *Binary trees.*[4] One can use ordered pairs to encode any binary tree by an appropriate "parenthesization" of the sequence of the tree's leaves. We provide just two illustrations, involving binary trees with leaves $a_1, a_2, a_3, a_4, a_5, a_6, a_7, a_8$.

The *complete binary tree* with these leaves would be encoded via the following aggregation of ordered pairs,

$$\langle \langle \langle a_1, a_2 \rangle, \langle a_3, a_4 \rangle \rangle, \langle \langle a_5, a_6 \rangle, \langle a_7, a_8 \rangle \rangle \rangle$$

The *comb-structured binary tree* with the same leaves would be encoded via the following aggregation of ordered pairs,

$$\langle a_1, \langle a_2, \langle a_3, \langle a_4, \langle a_5, \langle a_6, \langle a_7, a_8 \rangle \rangle \rangle \rangle \rangle \rangle \rangle$$

Explanatory note.

It should be no surprise that a single character string, such as $\langle a_1, \langle a_2, \langle a_3, a_4 \rangle \rangle \rangle$, can be used to represent many distinct, but *isomorphic* (literally, "same-shaped"), objects—for instance a length-4 string and a four-leaf comb-structured binary tree. Indeed, one of the biggest strengths of mathematics is its ability to expose such structural similarities.

---

[4] Trees, and binary trees, have become part of the vernacular, so we assume that the reader has an intuitive grasp of the notion. We refer the reader seeking a formal treatment of trees and related notions to Chapter 12.

The rather lengthy preamble to this section has been our attempt to enhance the reputation of the integers—all three sets, $\mathbb{Z}$, $\mathbb{N}$, and $\mathbb{N}^+$—in the eyes of the reader. With that process hopefully begun, we turn now to a demonstration via multiple examples that what we have accomplished has largely been an exercise in form rather than essence. Specifically, we show that one can easily and efficiently encode complicated structures such as the ones we have been describing as positive integers!

What does it mean to *encode* one class of entities, A, as another class, B? Our definition is a rather strict mathematical one. We insist that there exists a *bijection* $F_{A,B}$ that maps A *one-to-one onto* B (cf. Section 3.3.4). In other words, when presented with an element $a \in A$, the function $F_{A,B}$ "produces" a unique element $b \in B$. And conversely, when presented with an element $b \in B$, the function $F_{A,B}^{-1}$ "produces" a unique element $a \in A$.

8.2.2 The Diagonal Pairing Function: Encoding $\mathbb{N}^+ \times \mathbb{N}^+$ as $\mathbb{N}^+$

The remainder of this section—and its companion in the Appendix, Chapter B—are devoted to developing a few *easily computed* bijections between the set $\mathbb{N}^+ \times \mathbb{N}^+$ and the set $\mathbb{N}^+$. These easily computed *encodings* of ordered pairs of integers illustrate that—at least within the world of integers and entities encodable as integers, such as tuples, strings, and binary trees of integers—there is no intrinsic new representational power inherent in tuple-spaces:

Ordered pairs of integers play a special role in our study of encodings of structured sets of integers: They form the fundamental puzzle from whose solution all other aggregations will follow. It is appropriate, therefore, that a special name has been associated with bijections between $\mathbb{N}^+ \times \mathbb{N}^+$ and $\mathbb{N}^+$. These special bijections are called *pairing functions*.

One of the most valuable by-products of encodings via pairing functions is that such encodings provide a *linear/total ordering* of the set being encoded. We noted in Section 4.3.1.1 that "order within a number system is among one's biggest friends when reasoning about the numbers within the system." The orderings provided by pairing functions are particularly valuable when the structured sets being encoded as integers do not have their own "intrinsic" or "natural" orderings. Included in this category are structures such as tuples, strings, and trees.

Explanatory note.

Of course some structured sets do have natural, native linear orders: consider, as one such, strings under lexicographic ordering. Even for such sets, we often benefit from having alternative orderings as we design and analyze algorithms on the sets.

We focus on $\mathbb{N}^+$ as the avatar of *integer*, rather than on $\mathbb{Z}$ or $\mathbb{N}$, primarily for definiteness. We could easily rewrite this section with a focus on bijections between

$\mathbb{Z} \times \mathbb{Z}$ and $\mathbb{Z}$ or on bijections between $\mathbb{N} \times \mathbb{N}$ and $\mathbb{N}$ or, for that matter, between $\mathbb{Z} \times \mathbb{N}$ and $\mathbb{N}^+$.

We now describe our first explicit pairing function—the *diagonal* pairing function $\mathscr{D}(x, y)$. This is probably one of the oldest pairing functions, given that it was known as early as 1821. In Appendix B, we provide both a *"shell-based"* methodology for crafting pairing functions and three explicit siblings of the diagonal pairing function $\mathscr{D}$. Two of the siblings of $\mathscr{D}$ are constructed via the shell-based methodology. Two of the siblings of $\mathscr{D}$ were invented for their usefulness in certain computational domains; the third sibling was invented to uncover an inherent limit on the "compactness" of pairing functions.

Pairing functions first appeared in the literature early in the nineteenth century. Perhaps the simplest and "prettiest" such function (since it is a *polynomial*) appears, pictorially, in an 1821 work by the great French mathematician Augustin-Louis Cauchy [28]. This *diagonal* pairing function was formally specified a half-century later by the German logician Georg Cantor, whose studies [25, 26] revolutionized how we think about *infinite* sets.

$$\mathscr{D}(x, y) = \binom{x + y - 1}{2} + (1 - y) \tag{8.18}$$

($\mathscr{D}$ of course has a twin that exchanges the roles of x and y.) $\mathscr{D}$'s mapping of $\mathbb{N}^+ \times \mathbb{N}^+$ onto $\mathbb{N}^+$ is depicted in Fig. 8.4; it exposes that we can view $\mathscr{D}$'s mapping of $\mathbb{N}^+ \times \mathbb{N}^+$ as a two-step conceptual process:

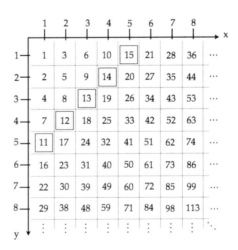

Fig. 8.4 The diagonal pairing function $\mathscr{D}$. The shell $x + y = 6$ is highlighted

1. partitioning $\mathbb{N}^+ \times \mathbb{N}^+$ into "diagonal shells".

For each index $k \in \mathbb{N}^+$, shell #k is the set

$$\text{Shell } \mathscr{K} \;=\; \{\langle x, y \rangle \mid x + y = k\}$$

The partitioning is an integral part of the specification of $\mathscr{D}$, as witnessed by the following subexpression in Eq. (8.18).

$$\frac{1}{2}(x+y)\cdot(x+y-1) \;=\; \binom{x+y-1}{2}$$

2. "climbing up" these shells in order.

The climbing is specified by the subexpression "$+1 - y$" in Eq. (8.18).

Explanatory note.

Keep in mind that we have just described a *conceptual* process specified by the definition of $\mathscr{D}$: $\mathscr{D}$ is a function, *not* an algorithm. "Running" this infinite process on a computing device would take forever.

An understanding of $\mathscr{D}$'s structure leads to a broadly applicable strategy for inductively constructing a broad range of "shell-based" pairing functions. We develop this strategy in Appendix B, accompanied by a rather simple inductive verification of the bijectiveness of functions created using the strategy. One finds a computationally more satisfying proof of $\mathscr{D}$'s bijectiveness in [40]—made more satisfying by its explicit recipe for computing $\mathscr{D}$'s inverse. The material in [40] builds in an essential way on $\mathscr{D}$'s structure.

⊕ **Esoterica for enrichment**

Topic-specific references.

R. Fueter, G. Pólya (1923): Rationale Abzählung der Gitterpunkte. *Vierteljschr. Naturforsch. Ges. Zürich 58*, 380–386.

J.S. Lew, A.L. Rosenberg (1978a): Polynomial indexing of integer lattices, I. *J. Number Th. 10*, 192–214.

J.S. Lew, A.L. Rosenberg (1978b): Polynomial indexing of integer lattices, II. *J. Number Th. 10*, 215–243.

The fact that the diagonal pairing function $\mathcal{D}$ is a *polynomial* in x and y raises the natural (to a mathematician!) question of whether there exist any *other* polynomial pairing functions. This is a quite advanced topic; indeed, even as a research problem, the question remains largely open. However, a few nontrivial pieces of an answer are known.

1. There is no *quadratic* polynomial pairing function other than $\mathcal{D}$ (and its twin) [Fueter, Pólya, 1923], [Lew, Rosenberg, 1978a].

2. No *cubic* or *quartic* polynomial is a pairing function [Lew, Rosenberg, 1978b].

3. The development in [Lew, Rosenberg, 1978b] excludes large families of higher-degree polynomials from being pairing functions; e.g., a *super-quadratic* polynomial—i.e., of degree > 2—whose coefficients are all positive cannot be a pairing function (because it cannot be surjective).

8.2.3 $\mathbb{N} \times \mathbb{N}$ *Is* **Not** *"Larger Than"* $\mathbb{N}$

8.2.3.1 Comparing infinite sets via cardinalities

We remarked earlier (and will do so again) that one must be very careful when reasoning about infinite sets. They can behave in ways that seem quite contradictory to our experience with finite sets. One of the most dramatic instances of this fact is encountered when we ask whether the set $\mathbb{N}^+ \times \mathbb{N}^+$ is "larger" than the set $\mathbb{N}^+$. We need to put the word "larger" in quotes because we do not know (yet) what the word means in the setting of infinite sets.

Finite beings that we are, it is quite comfortable to many of us to believe that "Infinite is infinite"—i.e., that once one makes the leap from *finite* to *infinite*, further refinement is either impossible or fruitless. Fortunately, the nineteenth-century German mathematician and logician Georg Cantor was not among that "many of us". Among his seminal contributions, Cantor demonstrated, in a way that is at once mathematically tractable and intuitively plausible, that there exist infinitely many *orders of infinity* and that the leap from one order of infinity to the next higher one is as consequential mathematically as is the leap from finite to infinite. Much of this material is beyond the scope of this text, but two specific results will be proved in detail.

1. *The rationals reside at the same order of infinity as do the integers.*

   This result occupies much of the remainder of this section.

2. *The reals reside at a higher order of infinity than do the rationals.*

   This result occupies much of Section 10.4.2. (Section 10.4 exhibits other residents of $\mathbb{R}$'s order of infinity.)

Cantor began his work on infinity by seeking a mathematically sound way to frame the assertion that one infinite set is—or is not—"larger" than another. The framework he developed occupies his groundbreaking study of the relative "sizes" of infinite sets [25, 26]. We now build on Cantor's formulation as we begin to study the relative "sizes" of the integers and the rationals.

We take our lead from finite sets. Is there a way to frame the notion "bigger" that works equally well for both finite and infinite sets? We begin to contemplate this question by focusing on a set A of apples and a set O of oranges, together with the challenge of determining which set is bigger.

If sets A and O are both finite, then we can just *count* the number of apples in A, call it a, and the number of oranges in O, call it o, and then compare the sizes of the (nonnegative) integers a and o. The Trichotomy Laws for integers (Section 4.3.1.2.A) guarantee that we shall be able to settle the question.

But we cannot count the elements in an infinite set, so this approach fails us when we have access to infinitely much fruit! So we need another approach.

Here is an approach that works for finite sets and that promises to extend to infinite sets. Let us assume that we can "prove"—we shall explain the word shortly—the following.

For every apple that we extract from set A *for the first time*, we can extract an orange from set O *for the first time*. It will then follow (at least in the finite case), that

There are at least as many oranges as apples!

This is really promising, because there is another way to describe the fruit-matching process that readily extends to infinite sets.

There is an injection,[5] call it f, from set A to set O.

In more formal terms: *Every time you pull the apple α from set A, I pull the orange $f(\alpha)$ from O.*

Inspired by this formulation using injections—and by Cantor's work—we craft the following definition.

Given sets A and O (finite or infinite), we say

---

[5] Recall from Section 3.3.4 that "*injection*" is synonymous with "*one-to-one function*".

Set O is at least as big as set A, denoted $|O| \geq |A|$
precisely when there is an injection from O to A.

By extension, we say that
Sets O and A have the same cardinality, denoted $|O| = |A|$
precisely when there is an injection from O to A and an injection from A to O.

Finally, back to numbers!

There has always been special interest in comparing the cardinalities of specific infinite sets with the cardinality of the integers. This interest has led to the following pair of adjectives.

- A (finite or infinite) set S is *countable* if $|S| \leq |\mathbb{N}|$.

- An infinite set S is *uncountable* if $|S| \not\leq |\mathbb{N}|$.

8.2.3.2 Comparing $\mathbb{N}$ and $\mathbb{N} \times \mathbb{N}$ via cardinalities

An obvious first candidate whose cardinality we might compare with that of the integers $\mathbb{N}$ (or $\mathbb{Z}$, or $\mathbb{N}^+$) is the corresponding set of *ordered pairs*, $\mathbb{N} \times \mathbb{N}$ (or $\mathbb{Z} \times \mathbb{Z}$, or $\mathbb{N}^+ \times \mathbb{N}^+$). Cantor announced in the 1870s that
Pairing does not increase cardinality in infinite sets.
In other words,
Pairing does not increase a set's order of infinity.
We prove this for the set $\mathbb{N}$. The reader can easily adapt our argument to $\mathbb{Z}$ or $\mathbb{N}^+$.

Proposition 8.7 *The set $\mathbb{N} \times \mathbb{N}$ is countable; i.e., $|\mathbb{N} \times \mathbb{N}| = |\mathbb{N}|$.*

Proof. We prove the following propositions in turn.

(a) *There exists an injection from $\mathbb{N}$ to $\mathbb{N} \times \mathbb{N}$. Therefore, $|\mathbb{N}| \leq |\mathbb{N} \times \mathbb{N}|$. Informally, $\mathbb{N} \times \mathbb{N}$ is "at least as big" as $\mathbb{N}$.*

Sub-proposition (a) is witnessed by any of the numerous injections from $\mathbb{N}$ into $\mathbb{N} \times \mathbb{N}$, including the following function f_1.

$$\text{For each } n \in \mathbb{N}, \quad f_1(n) = \langle n, n \rangle$$

(b) *There exists an injection from $\mathbb{N} \times \mathbb{N}$ to $\mathbb{N}$. Therefore, $|\mathbb{N} \times \mathbb{N}| \leq |\mathbb{N}|$. Informally, $\mathbb{N}$ is "at least as big" as $\mathbb{N} \times \mathbb{N}$.*

We establish sub-proposition (b) by defining an injection from $\mathbb{N} \times \mathbb{N}$ into $\mathbb{N}$. We employ a function that is inspired by the Fundamental Theorem of Arithmetic (Theorem 8.1). Specifically, the Theorem assures us that the function

$$f_2(p,q) \overset{\text{def}}{=} 2^p 3^q$$

is an *injection* from $\mathbb{N} \times \mathbb{N}$ into $\mathbb{N}$.

Sub-propositions (a) and (b) combine to prove the result. □

The rather elementary Proposition 8.7 Illustrates a slight awkwardness that is inherent in two-step proofs of equinumeracy of infinite sets A and B. Once one proves the relation "$|A| \leq |B|$", one must "reset one's thinking" and prove the relation "$|B| \leq |A|$". The underlying intuition is often lost in the midst of changing tools. Now, we know from the remarkable theorem of Schröder and Bernstein (Theorem 3.2) that *every* two-step proof can be replaced by a single-step proof: There is *always* a bijection whenever there exist paired injections! But the proof of Theorem 3.2 offers no hope that any bijection that affords one a single-step proof will expose the intuition that underlies the equinumeracy of sets A and B. The *pairing functions* of Section 8.2 provide a *nicely structured* way to design bijections—at least between certain number- and numeral-related sets: e.g., between $\mathbb{N}$ and $\mathbb{N} \times \mathbb{N}$, between $\mathbb{Z}$ and $\mathbb{Z} \times \mathbb{Z}$, and between[6] $\{0,1\}^{\star}$ and $\{0,1\}^{\star} \times \{0,1\}^{\star}$. Proposition 8.7 should, therefore, enhance our appreciation of the Schröder-Bernstein Theorem and of pairing functions.

8.3 Finite Number Systems

The sets that underlie the number systems that we use in most daily tasks—namely $\mathbb{N}, \mathbb{Z}, \mathbb{Q}$, and $\mathbb{R}$—are infinite: we can always find a number in each set that is bigger than all the numbers we have seen thus far. Indeed, the last three of these sets are "two-way" infinite: we can also always find a number in each set that is smaller than all the numbers we have seen thus far. We mention only two examples of finite number systems.

- The *clocks* that we use to indicate daily time are calibrated into a fixed finite number of major subdivisions, *hours*. We endow our days with 24 hours and, depending on circumstances, have our clocks measure each day's time via repeating cycles of either 12 or 24 hours. Once a clock's limit (of 12 or 24 hours) has been reached, it begins its numeration once again—with no memory of the past.

- We typically orient all manner of location specification relative to a fixed reference point in terms of *angles*. There are two coexisting, competing systems for such measurement. One system subdivides the "circle" around the reference point into 360 *degrees;* the other subdivides the "circle" into 2π *radians*. For our purposes, the main interesting point is that both of these systems are *cyclically repetitive*. Once we have circled the reference point by 360 degrees (or, equivalently, by 2π radians), then we measure further circumnavigation starting over at 0 degrees/radians.

This section is dedicated to integer-based *finite* number systems that were invented to describe and measure cyclically repetitive situations such as the two just described.

---

[6] Recall that $\{0,1\}^{\star}$ denotes the set of all binary strings.

8.3.1 Congruences on Nonnegative Integers

For each positive integer $q \in \mathbb{N}^+$, we denote by $\mathbb{N}_q$ the q-element "prefix" of the set $\mathbb{N}$ of nonnegative integers:

$$\mathbb{N}_q \overset{\text{def}}{=} \{0, 1, \ldots, q-1\}$$

For nonnegative integers $m, n \in \mathbb{N}$ and positive integer $q \in \mathbb{N}^+$, we say that *m is congruent to n modulo q*, denoted

$$m \equiv n \bmod q$$

precisely when q divides $|m-n|$. We call q the *modulus* of the congruence (relation).

Proposition 8.8 *The relation [congruence modulo a positive integer] is an equivalence relation on the set $\mathbb{N}$ of nonnegative integers.*

Proof. We verify in turn the three defining properties of an equivalence relation (see Section 3.3.3). Focus on nonnegative integers m, n, and r and on an arbitrary positive integer modulus q.

1. Congruence modulo $q \in \mathbb{N}^+$ is a *symmetric* relation on $\mathbb{N}$.

   Verification. Because $|m-n| = |n-m|$, the assertions

   $$[q \text{ divides } |n-m|]$$

   and

   $$[q \text{ divides } |m-n|]$$

   must hold simultaneously: Either both assertions are true or neither is.

2. Congruence modulo $q \in \mathbb{N}^+$ is a *reflexive* relation on $\mathbb{N}$.

   Verification. We always have $[m \equiv m \bmod q]$ because every positive integer divides $m - m = 0$.

3. Congruence modulo $q \in \mathbb{N}^+$ is a *transitive* relation on $\mathbb{N}$.

   Verification. Say that $[m \equiv n \bmod q]$ and $[n \equiv r \bmod q]$. Consider the arithmetic that one must perform in order to manipulate these two congruences into a proof that $m \equiv r \bmod q$. The required arithmetic breaks down into cases defined by the relative sizes of m, n, and r. We supply the details for the case $m > n > r$ and leave the other cases as exercises.

   Note first that the two assumed congruences can be rewritten as assertions of divisibility, namely,

   $$[q \text{ divides } |m-n|] \quad \text{and} \quad [q \text{ divides } |n-r|]$$

   Therefore, in the chosen case $[m > n > r]$, the two congruences imply that there exist integers c_1 and c_2 such that:

a. $c_1 q = m - n$, which implies that $n = m - c_1 q$

b. $c_2 q = n - r$, which implies that $n = r + c_2 q$

The two congruences thus imply that $m - r = (c_1 + c_2)q$, which means that $[m \equiv r \bmod q]$. In other words, the relation $[\equiv \bmod q]$ is transitive.

The preceding three properties define an equivalence relation, hence, jointly verify the proposition. ☐

8.3.2 Finite Number Systems via Modular Arithmetic

Once we embellish the sets $\mathbb{N}_q$ with arithmetic operations—namely, the "big four" of addition, subtraction, multiplication, and division—we see why we are able to use the resulting finite systems in the same way as their infinite counterparts, $\mathbb{N}$, $\mathbb{Z}$, and $\mathbb{Q}$. The coming subsections are devoted to showing that:

- *For every $q \in \mathbb{N}^+$, the set $\mathbb{N}_q$ can "mimic" $\mathbb{N}$ and $\mathbb{Z}$ with respect to addition, subtraction, and multiplication.*

- *When q is a prime number—and only then—$\mathbb{N}_q$ can, additionally, "mimic" $\mathbb{Q}$ with respect to division.*

8.3.2.1 Sums, differences, and products exist within $\mathbb{N}_q$

Every set $\mathbb{N}_q$ is, when embellished with the operations addition, subtraction, and multiplication, *closed* under these operations, in the sense spelled out in the following result.

Proposition 8.9 *For every integer $q \in \mathbb{N}^+$ and all $m, n \in \mathbb{N}_q$, the sum $[m + n \bmod q]$ and the difference $[m - n \bmod q]$ and the product $[m \cdot n \bmod q]$ exist within $\mathbb{N}_q$.*

Proof. For the operations of addition and multiplication, the result is true by definition of congruence modulo q: Since the sum $m + n$ and the product $m \cdot n$ exist within $\mathbb{N}^+$, their "reductions" modulo q exist within $\mathbb{N}_q$. For the case of subtraction, we augment the preceding sentence with the following equation. For all $r \in \mathbb{Z}$

$$q - r \equiv -r \bmod q$$

One verifies this equation by noting the following chain of equalities and congruences (parentheses added to enhance legibility)

$$(q - r) - (-r) = (q - r) + r = q \equiv 0 \bmod q$$

In all cases, therefore, the result of the operation remains in the set $\mathbb{N}_q$. ☐

We cannot generally include division among the operations listed in Proposition 8.9. For instance, the following table shows that the equation

$$2x \equiv 1 \bmod 6$$

is not solvable for all $x \in \mathbb{N}_6 \setminus \{0\}$.

x	$2 \cdot x \bmod 6$
1	2
2	4
3	0
4	2
5	4

The next subsection identifies the modulus 6's non-primality as the culprit in this example. In fact, the reader can easily show that $\mathbb{N}_q$ *is never closed under the operation of division when the modulus q is composite, i.e., nonprime.*

8.3.2.2 Quotients exist within $\mathbb{N}_p$ for every prime p

This section focuses on congruences modulo a prime number. We begin with our main result: *For every prime number p, every nonzero $n \in \mathbb{N}_p$ has a multiplicative inverse*, i.e., an element $m \in \mathbb{N}_p$ such that $m \cdot n \equiv 1 \bmod p$. Of course, the existence of multiplicative inverses allows one to *divide* any number in $\mathbb{N}_p$ by any nonzero number in $\mathbb{N}_p$.

Proposition 8.10 *For every prime number p, every nonzero number $n \in \mathbb{N}_p$ has a multiplicative inverse within $\mathbb{N}_p$.*

Proof. Our proof combines applications of the Fundamental Theorem of Arithmetic (Theorem 8.1) and the Pigeonhole Principle (see Section 2.2.3), alongside a proof by contradiction. It thereby exercises several proof techniques from the mathematician's kitbag.

Lemma 8.1. *Let p be prime and n a nonzero number in $\mathbb{N}_p$. There do not exist nonzero numbers m_1 and m_2 in $\mathbb{N}_p$ with $m_1 \not\equiv m_2 \bmod p$ such that $m_1 \cdot n \equiv m_2 \cdot n \bmod p$.*

Proof. (Lemma 8.1) Assume for contradiction that there *do* exist m_1 and m_2 as described in the lemma such that $m_1 \cdot n \equiv m_2 \cdot n \bmod p$. Say, with no loss of generality, that $m_1 > m_2$ within the set $\mathbb{N}$. We must then have

$$p \text{ divides } (m_1 - m_2) \cdot n \qquad (8.19)$$

The fact that p is a prime number ensures—by Proposition 8.3—that p divides at least one of the integers n or $(m_1 - m_2)$. Because both of these integers belong to

$\mathbb{N}_p$, hence lie strictly between 0 and $p-1$ (within the infinite set $\mathbb{N}$), the divisibility posited in expression (8.19) is impossible! The lemma follows. $\square$

Lemma 8.1 guarantees that all of the following $p-1$ elements of $\mathbb{N}_p$ are nonzero and distinct:

$$1 \cdot n, \quad 2 \cdot n, \ldots, \quad (p-1) \cdot n$$

Because $\mathbb{N}_p$ has precisely $p-1$ nonzero elements, these $p-1$ multiples of n must exhaust these elements. In other words, some multiple of n, say $c \cdot n$, must equal 1. This means that the number $c \in \mathbb{N}_p$ is n's multiplicative inverse within $\mathbb{N}_p$. $\square$

Of course, once we have multiplicative inverses, we have the operation of division and, consequently, arbitrary quotients and fractions. Of course, fractions within finite number systems such as $\mathbb{N}_p$ are going to look strange to our eyes, as the following example indicates.

What does the number $7/4$ look like within $\mathbb{N}_5$? We develop the answer in steps.

Explanatory note.

We are able to proceed in the following manner because the relations $[\equiv \mod q]$ are *congruences*, i.e., equivalence relations whose class structures are consistent with the algebraic structure of the arithmetic systems exemplified by $\mathbb{Z}$, $\mathbb{Q}$, $\mathbb{R}$, and $\mathbb{N}_p$ under the four classical arithmetic operations. A full treatment of this topic is beyond the scope of this text.

1. The numbers $4, 7 \in \mathbb{N}$ correspond, respectively, to the numbers $4, 2 \in \mathbb{N}_5$.
   Verification: $4 \equiv 4 \mod 5$, and $7 \equiv 2 \mod 5$.

2. The multiplicative inverse of 4 within $\mathbb{N}_5$ is 4.
   Verification: $4 \cdot 4 = 16 \equiv 1 \mod 5$.

3. Therefore, we have the following "translation" of the quotient $7/4$:
   Within $\mathbb{N}$:
   the product of $7 \in \mathbb{N}$ by the multiplicative inverse of $4 \in \mathbb{N}$
   Within $\mathbb{N}_5$:
   the product of $2 \in \mathbb{N}_5$ by the multiplicative inverse of $4 \in \mathbb{N}_5$, which is 4

4. the product of $2 \in \mathbb{N}_5$ by $4 \in \mathbb{N}_5$ is 3.
   Verification: $2 \cdot 4 = 8 \equiv 3 \mod 5$.

We thus have the unintuitive fact that the rational number $7/4$ corresponds to the number $3 \in \mathbb{N}_5$.

Of course, we do not often perform arbitrary arithmetic within the finite number systems $\mathbb{N}_p$, so we do not often struggle with the unfamiliar results of this subsection. That said, we do sometimes intermix "ordinary" numeration with "modular" numeration, as when we coordinate talk about elapsed time (measured in the "ordinary" way) with wall-clock time (which is a "modular" system). So, in summation, it is worth the effort to understand this seldom-used material. Plus, it can be amusing to announce at a party that you can "prove" that $1.75 = 3$.

8.4 Exercises: Chapter 8

Throughout the text, we mark each exercise with 0 or 1 or 2 occurrences of the symbol $\oplus$, as a rough gauge of its level of challenge. The 0-$\oplus$ exercises should be accessible by just reviewing the text. We provide *hints* for the 1-$\oplus$ exercises; Appendix H provides *solutions* for the 2-$\oplus$ exercises. Additionally, we begin each exercise with a brief explanation of its anticipated value to the reader.

1. **Organizing the integers into an $\infty \times \infty$ array**
   LESSONS: Enhance understanding of pairing functions; understand feasible progressions along rows and columns

   a. Consider the two-way infinite array A of positive integers whose kth row is the sequence of odd integers, each times the $(k-1)$th power of 2, namely, 2^{k-1}. For illustration:
   The first row of A begins
      $1, 3, 5, 7, \ldots$
   The second row of A begins
      $2, 6, 10, 14, \ldots$
   The third row of A begins
      $4, 12, 20, 28, \ldots$
   and so on.

   Proposition 8.11 *Prove that every $n \in \mathbb{N}^+$ appears in array A precisely once. Infer that there exists a pairing function $f : \mathbb{N} \times \mathbb{N} \leftrightarrow \mathbb{N}$ which maps each row of $\mathbb{N} \times \mathbb{N}$ onto an arithmetic sequence.*

   b. Proposition 8.11 goes about as far as one can in employing arithmetic sequences to construct pairing functions. The arguments needed to prove the following result employ a version of the Pigeonhole Principle.

   Proposition 8.12 **(a)** *Prove that there does not exist a pairing function $f : \mathbb{N} \times \mathbb{N} \leftrightarrow \mathbb{N}$ which maps every row of $\mathbb{N} \times \mathbb{N}$ onto an arithmetic sequence in such a way that all rows have the same period.*

   (b) *Prove that there does not exist a pairing function $f : \mathbb{N} \times \mathbb{N} \leftrightarrow \mathbb{N}$ which maps every row of $\mathbb{N} \times \mathbb{N}$ and every column of $\mathbb{N} \times \mathbb{N}$ onto an arithmetic sequence—even if each row and each column gets a distinct period.*

2. $\oplus$ **Discovering fractal-like structure in Pascal's Triangle**
   LESSON: Practice with nonelementary mathematical manipulations

   Modular arithmetic can do wondrous, unexpected things with regular structures. If one takes a system whose structure is governed by arithmetic and substitutes *modular* arithmetic for ordinary arithmetic, then the system's structure is somehow going to be "folded into itself". We illustrate this fact within a rather complex framework in Appendix Section F.2; the "folded" structures there are trees.

We illustrate the fact within a rather elementary framework in this exercise; the "folded" structures here arise from Pascal's Triangle.

Take Pascal's Triangle, choose a positive integer m, and replace each of the triangle's entries—which are, of course, the binomial coefficients—by that entry modulo m. As illustrated in Fig. 8.5, a wondrous transformation happens. The

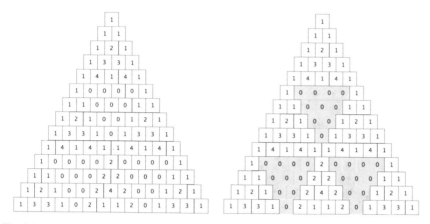

Fig. 8.5 Pascal's Triangle modulo 5: (left) the "raw" version; (right) with the reproducible pattern highlighted

first m levels of the triangle get replicated endlessly, with an inverted $(m-1)$-level triangle whose entries are all 0. (In the figure, the inverted triangle of 0s is depicted in grey.) The original triangle has been transformed into a fractal-like repetitive structure whose pattern of repetitions is dictated by the parameter m.

Prove that the described transformation occurs.

The $\oplus$ rating that we have assigned to this problem reflects the challenge of figuring out how the various parameters that determine the triangle get transformed by the "folding".

3. **A nexus between perfect numbers and triangular numbers**
   LESSON: Practice with non elementary mathematical concepts

   Proposition 8.6 (Section 8.1.3.2) tells us that, for every Mersenne prime $2^p - 1$, the integer is $2^{p-1} \times (2^p - 1)$ is a perfect number. Let us denote that number by PN_p. A small calculation verifies that both

   $$PN_2 = 6 \quad \text{and} \quad PN_3 = 28$$

   are triangular numbers, in addition to being perfect; that is, $6 = \Delta_3$ and $28 = \Delta_7$. This is not a coincidence!

   Prove the following result.

Proposition 8.13 *For every Mersenne prime* $2^p - 1$, *we have*

$$PN_p = \Delta_{2^p - 1}$$

Hint: Explore the defining expression $\Delta_n = \frac{1}{2}n(n+1)$ for arguments $n = 2^{p-1}$.

4. **Divisibility among integers: via the Fundamental Theorem of Arithmetic**
   LESSON: Understand the vestiges of divisibility that do exist within the integers

   a. **Cancellation in integer multiplication**
      LESSON: An exercise in applying the Fundamental Theorem of Arithmetic

      We noted after Proposition 5.13.b that integers lack multiplicative inverses. The notion of *cancellation* that we describe now can be viewed as a rudimentary form of multiplicative inverses for integers.

      Prove the following result.

      Proposition 8.14 *Let m, n, and p be three integers that exceed* 1 *(i.e., $m > 1$, $n > 1$, and $p > 1$). If $m \times n = m \times p$, then $n = p$.*

      The Fundamental Theorem of Arithmetic yields a particularly charming proof of Proposition 8.14.

   b. **The sieve of Eratosthenes and its implications**
      LESSONS: Understand many properties of the integers via the "sieve"

      Prove the following results which are hinted at by the sieve.

      i. **Proposition 8.15** *Every integer $n > 1$ is divisible by at least one prime number.*

      ii. **Proposition 8.16** *Every sequence $(k+1)$, $(k+2)$, ..., $(k+p)$ of p integers contains a multiple of p.*

      iii. ⊕ The following result calls for "second-order" insights from the sieve.
          Proposition 8.17 *Every product of four consecutive integers,*

          $$(k+1) \times (k+2) \times (k+3) \times (k+4)$$

      is divisible by FACT(4).

      Hint: Look at the *multiplicity* with which a given prime divides a given number, particularly within the context of the spacings of the divisibilities— e.g., every even number that is $2\times$ (an odd number) is followed by an even number that is $2\times$ (an even number).

      iv. ⊕⊕
          Proposition 8.18 *For every integer n, every product of n consecutive integers,*
          $$(k+1) \times (k+2) \times \cdots \times (k+n)$$

      is divisible by FACT(n).

5. ⊕⊕ **The "density" of divisible pairs of numbers**
   LESSON: Employing a pigeonhole argument while reasoning about divisibility
   Prove the following assertion.

   Proposition 8.19 *For every positive integer n: If you remove any $n + 1$ integers from the set $S = \{1, 2, \ldots, 2n\}$, then the set of removed integers contains at least one pair p and $q > p$ such that p divides q.*

   This problem was a favorite of the renowned twentieth-century Hungarian mathematician Paul Erdös, who often used it to test students' mathematical talent.

   Hint: Use the Pigeonhole Principle to classify pairs $(p, q > p)$ with respect to divisibility.

6. **Divisibility and numerals**
   LESSONS: Reinforce understanding of interplay among numbers, numerals, and geometric summations

   a. **On generating numerals from numbers**
      LESSON: Understanding the importance of Euclidian division
      Prove the following assertion.

      Proposition 8.20 *The successive remainders from repeated Euclidian divisions of an integer n by a number-base b are the successive digits of the base b numeral for n, from the lowest-order digit to the highest.*

      In other words, if the base-*b* numeral for *n* is the string

      $$\delta_m \delta_{m-1} \cdots \delta_1 \delta_0$$

      then the successive remainders are, in order, δ_0, δ_1, ..., δ_{m-1}, δ_m.

   b. **When is integer *n* divisible by 9?**
      LESSON: Enhance appreciation of positional numerals' grounding in geometric summations
      Prove the following assertion.

      Proposition 8.21 *An integer n is divisible by an integer m if, and only if, m divides the sum of the digits in the base-$(m + 1)$ numeral for n.*

      The most familiar instance of this result states:
      An integer n is divisible by 9 if, and only if, the sum of the digits of n's base-10 numeral is divisible by 9.

      It is important to appreciate that this is not just a party game: It is a basic consequence of the definition of positional number representation.

7. **Divisibility under modular arithmetic**

   LESSON: Enhance ability to reason about modular arithmetic

   Prove that the set $\mathbb{N}_q$ is never closed under the operation of division when the modulus q is composite.

8. ⊕ **The set $\mathbb{Q}$ of rational numbers is countable**

   LESSON: Practice proving simple properties of highly structured infinite sets.

   Prove Proposition 4.6: $|\mathbb{Q}| = |\mathbb{N}|$

   Hint. Begin by reviewing Proposition 8.7, which asserts the technically simpler proposition: $|\mathbb{N} \times \mathbb{N}| = |\mathbb{N}|$. What is the relevance of that result to this problem?

Chapter 9
Recurrences: Rendering Complex Structure Manageable

Those who do not learn from history are doomed to repeat it.

George Santayana

One of the intellectually most powerful strategies for all manner of human endeavor is to "learn from the past"—i.e., to *re*-use knowledge that one has acquired earlier in order to acquire new knowledge. Within the domain of computing, this strategy is exemplified by computations that derive the value of a function f at an argument $n \in \mathbb{N}^+$ by invoking the values of f at arguments $1, 2, \ldots, n-1$.

The classical first example of such a *recurrent* mode of computing involves the *factorial function*, which is referred to as "FACT" Section 5.1.1.3. The "direct" mode of computing FACT at an argument $n \in \mathbb{N}^+$ is:

$$\text{FACT}(n) = 1 \times 2 \times 3 \times \cdots \times n$$

The *recurrent* mode of computing FACT(n) is more compact—and it better exposes the inherent structure of the function.

$$\text{FACT}(n) = \begin{cases} n \times \text{FACT}(n-1) & \text{if } n > 1 \\ 1 & \text{if } n = 1 \end{cases}$$

This chapter is devoted to deriving and solving a variety of types of recurrences. In common with the rest of this text, our treatment of this subject emphasizes exploiting recurrent structure in reasoning and analysis: Increased understanding will enable improved computing.

© Springer Nature Switzerland AG 2020
A. L. Rosenberg, D. Trystram, *Understand Mathematics, Understand Computing*,
https://doi.org/10.1007/978-3-030-58376-7_9

9.1 Linear Recurrences

The first family of recurrences that we study are the *linear* recurrences, as exemplified by the following function specifications, wherein we have four integers: a, b, c, and d, with $a > 1$ and $b > 1$.

The general form

$$f(n) = \begin{cases} af(n/b) + g(n) & \text{for } n \geq b \\ d & \text{for } n < b \end{cases} \tag{9.1}$$

The simple form

$$f(n) = \begin{cases} af(n/b) + c & \text{for } n \geq b \\ 1 & \text{for } n < b \end{cases} \tag{9.2}$$

Many basic algorithmic problems, including sorting, selection, matching, ... can be solved using linear-recurrent algorithms [36]—and such algorithms yield to specification and analysis via linear recurrences.

In the next two subsections, we analyze recurrences (9.1) and (9.2). The techniques that we use in our analyses can be adapted to analyze other members of the important family of linear-recurrent algorithms.

9.1.1 The Master Theorem for Simple Linear Recurrences

We focus first on the simpler of our sample recurrences, namely, (9.2). Happily, there is a single perspicuous proof that elegantly solves recurrences of this form.

By the time the reader has reached this paragraph, she has the mathematical tools necessary to prove and apply what is called *The Master Theorem for Linear Recurrences* [36]. The main tools we use to prove the Theorem are: summing geometric summations (Section 6.2.2) and employing elementary asymptotic notions and notations (Section 7.1).

Theorem 9.1 (The Master Theorem for the simple linear recurrence). *Let the function f be specified by the simple linear recurrence (9.2). The value of f on any argument n is given by*

$$f(n) = (1 + \log_b n) \cdot c \qquad \text{if } a = 1$$

$$= \frac{1 - a^{\log_b n}}{1 - a} \cdot c \approx \frac{c}{1 - a} \qquad \text{if } a < 1 \tag{9.3}$$

$$= \frac{a^{\log_b n} - 1}{a - 1} \cdot c \qquad \text{if } a > 1$$

Proof. We expose the pattern generated by recurrence (9.2), by beginning to "expand" the specified computation—replacing occurrences of $f(\circ)$ as mandated in recurrence (9.2). Once we discern the pattern, we jump to the general form.

$$
\begin{aligned}
f(n) &= \quad af(n/b)+c \\
&= \quad a\left(af(n/b^2)+c\right)+c \quad = \quad a^2 f(n/b^2)+(a+1)c \\
&= a^2\left(af(n/b^3)+c\right)+(a+1)c = a^3 f(n/b^3)+(a^2+a+1)c \\
&\quad\vdots \qquad\qquad\qquad\qquad\qquad \vdots \\
&= \left(a^{\log_b n}+\cdots+a^2+a+1\right)c
\end{aligned}
\tag{9.4}
$$

The segment of recurrence (9.4) that is "hidden behind" the vertical dots betokens an induction that is left to the reader. Eqs. (6.18) and (6.19) now enable us to demonstrate that Eq. (9.3) is the asserted case-structured solution to (9.2). $\square$

9.1.2 The Master Theorem for General Linear Recurrences

We now progress from the simple recurrence (9.2) to the more general recurrence (9.1). We simplify our problem in three ways, to avoid calculational complications (such as floors and ceilings) that can mask the principles that govern our analysis.

1. We employ a very simple nonrecurrent function g: We focus on the case $g(n)=n$.

   It requires only clerical effort to generalize to the slightly more ambitious function $g(n)=\alpha n+\beta$ (so that g is a general *linear* function), but such an extension teaches no new lessons.

2. We assume that the argument n to functions f and g is a power of b.

   This allows us to concentrate on the general unfolding of the recurrence without worrying about floors and ceilings.

3. We consider only the value $c=1$.

Removing these assumptions would significantly complicate our calculations, but it would not change our reasoning.

Theorem 9.2 (The Master Theorem for the general linear recurrence). *Let the function f be specified by the general linear recurrence (9.1). The value of f on any argument n is given by*

$$
f(n) = a^{\log_b n}f(1) + \left(\sum_{i=0}^{\log_b(n)-1}(a/b)^i\right)n
$$

When $a>b$, the behavior of $f(n)$ is dominated by the first term of this solution:

$$
a^{\log_b n}\cdot f(1) = n^{\log_b a}
$$

When $a < b$, the behavior of $f(n)$ is dominated by the second term of this solution:

$$n \cdot \sum_{i=0}^{\log_b(n)-1} (a/b)^i = \frac{\left(1 - (a/b)^{\log_b(n)}\right)}{1 - (a/b)} \cdot n \approx \frac{b}{b-a} \cdot n$$

Proof. As in Section 9.1.1, we expose the algebraic pattern created by the recurrence by "unfolding" recurrence (9.1). As in recurrence (9.4), once we discern this pattern, we jump to the general form (which can be verified via induction).

$$
\begin{aligned}
f(n) &= & af(n/b)+n \\
&= & a\left(af(n/b^2)+n/b\right)+n & = & a^2 f(n/b^2)+(an/b+n) \\
&= a^2\left(af(n/b^3)+n/b^2\right)+(a/b+1)n &= a^3 f(n/b^3)+(a^2/b^2+a/b+1)n \\
&\quad\quad\quad\quad\quad \vdots & & \quad\quad\quad\quad\quad \vdots \\
&= & a^{\log_b n} f(1) + \left(\sum_{i=0}^{\log_b(n)-1} (a/b)^i\right) n
\end{aligned}
$$

We thus see that solving the more general recurrence (9.1) requires only augmenting the solution to the simple recurrence (9.2) by "appending" to the simple solution a geometric summation whose base is the ratio a/b. The reader can now invoke the techniques from Section 6.2.2.2 to arrive at the announced solution to recurrence (9.1).

When one "does" mathematics, one is often interested in uncovering the *dominant behavior* of the function $f(n)$ specified via a recurrence such as (9.1). Therefore, the two assertions in the statement of the proposition about the "dominant behavior" of function f for given relative sizes of a and b is an integral part of the lessons that we learn from this proof.

One can learn yet other lessons about $f(n)$, specifically about how to compute $f(n)$ (exactly or approximately), by studying Fig. 9.1. In particular, one observes in the figure that when $a = b$, the computations in each row are perfectly balanced. When $a = b = 2$, for instance, the tree has n leaves, and the computations inside the tree evaluate to exactly n at each of the tree's $\ln(n)$ levels—so that $f(n) = n\ln(n)$ in this case. □

9.1.3 An Application: The Elements of Information Theory

By the middle of the twentieth century, the existence of electrical and electronic devices that enhanced our ability to compute and to intercommunicate, made it imperative that "the experts" understand the mathematical laws that govern these activities. In 1948, the American mathematician and engineer Claude E. Shannon revolutionized our understanding of these laws by inventing the field of *information theory* [103]. Shannon's intellectual innovations enabled us to quantify the quality

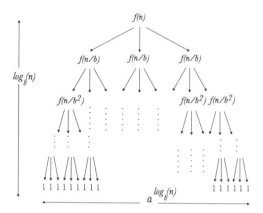

Fig. 9.1 Illustrating the calculation specified by recurrence (9.1). The total cost is obtained by the summation on each row: $a \times f(n/b)$ in the first row, $a^2 \times f(n/b^2)$ in the second row, and so on. This leads to the final value $a^{\log_b(n)}$

of our handling of data: speed of transmission, efficiency of encoding, vulnerability to errors. The evolution of data science and the ever-increasing importance of information-related topics such as *encryption* and *data security* make it important that everyone who has contact with the world of computing and communication—which means pretty much everyone—understand at least the rudiments of information theory. This section is devoted to taking one small step in that direction, a step that hints at the fundamental role that logarithms play in the theory.

One of the most basic results within information theory highlights the role of logarithms in measuring "the amount of information" that a string can hold.

Proposition 9.1 *In any assignment of distinct string-labels to n items, using strings over an alphabet of a symbols, at least one string-label must have length $\geq \lceil \log_a n \rceil$.*

Proof. Let A be an alphabet of a symbols. For each integer $k \geq 0$, let $A^{(k)}$ denote the set of all length-k strings over A; note that $A^{(1)} = A$. The bound stated in Proposition 9.1 follows by counting the numbers of strings of various lengths over A, because each such string can label at most one item. Let us, therefore, inductively evaluate the cardinality $|A^{(k)}|$ of each set $A^{(k)}$.

- $|A^{(0)}| = 1$

  This is because the null string, which is commonly denoted ε, is the unique string in $A^{(0)}$; symbolically: $A^{(0)} = \{\varepsilon\}$.

- $|A^{(k+1)}| = |A| \times |A^{(k)}|$

  This reckoning follows from the following recipe for building the set $A^{(k+1)}$ from the set $A^{(k)}$.

  $$A^{(k+1)} = \{\sigma x \mid \sigma \in A \ \text{and} \ x \in A^{(k)}\}$$

In other words, every length-$(k+1)$ string over A is obtained by taking a length-k string x over A and *prepending* to it a symbol from A.

This recipe is correct because:

- Each string in $A^{(k+1)}$, as constructed, has length $k+1$
  —because the recipe adds a single symbol to a length-k string.
- For each string $x \in A^{(k)}$, there are $|A|$ distinct strings in $A^{(k+1)}$, as constructed
  —because each string in $A^{(k+1)}$ begins with a distinct symbol from A.
- $A^{(k+1)}$, as constructed, contains all strings of length $k+1$ over A
  —because every length-$(k+1)$ over A has the form σx for some unique $\sigma \in A$ and some unique $x \in A^{(k)}$; and the set $A^{(k+1)}$ contains every such string.

We thus have the following recurrence.

$$|A^{(0)}| = 1$$
$$|A^{(k+1)}| = |A| \times |A^{(k)}| \quad \text{for } k \geq 0$$

Theorem 9.1 tells us that, for each $\ell \in \mathbb{N}$,

$$|A^{(\ell)}| = \frac{|A|^{\ell+1} - |A|}{|A| - 1} \leq c' \cdot |A|^\ell$$

for some constant $c' > 0$. In order for this quantity to reach the value n, we must have

$$n = |A^{(\ell)}| \leq c' \cdot |A|^\ell$$

We can rewrite this inequality in the form

$$|A|^\ell \geq \frac{1}{c'} n$$

Because $1/c' > 0$, we can now take logarithms of both sides of the preceding inequality. Using the logarithm to the base $|A|$ yields the most perspicacious result:

$$\ell \cdot \log_{|A|} |A| = \ell \geq \log_{|A|} n - \log_{|A|} c'$$

Because n grows without bound, while c' is a fixed constant, we can find a value n_0 for n such that, for all $n > n_0$, there exists a constant $d' > 0$ such that

$$\ell > d' \cdot \log_{|A|} n$$

In other words, the string-length ℓ must grow logarithmically with the number of strings. $\square$

The following result can be considered another way of looking at Proposition 9.1.

Proposition 9.2 *The number of distinct strings of length k over an alphabet of a symbols is a^k.*

Proof. As in Proposition 9.1, we focus on an a-letter alphabet $A = \{\sigma_1, \sigma_2, \ldots, \sigma_a\}$ and we argue by induction on string-length k.

Bases. The induction we develop can start either with the unique string ε of length $k = 0$ or with strings of length $k = 1$. In the former case, the uniqueness of the null string validates the case $k = 0$ of the proposition. In the latter case, there are $a^1 = a$ such strings over A, one for each symbol $\sigma \in A$; this validates the case $k = 1$ of the proposition.

The inductive hypothesis. Say that for all string-lengths k up through n, there are a^k distinct words of length k over A.

Extending the induction. We take each length-n string x over A, and *append* to it, in turn, each of A's a symbols. Each appendage adds an additional rightmost symbol to x. We thereby replace each string $x \in A^{(n)}$ by a distinct new strings, $x\sigma_1, x\sigma_2, \ldots, x\sigma_a$. We have thus created a^{n+1} distinct length-$(n+1)$ strings over A from A's a^n distinct length-n strings.

The induction is thus extended, which completes the proof. $\square$

9.2 Bilinear Recurrences

9.2.1 Binomial Coefficients and Pascal's Triangle

In Section 5.1.2.4, we introduced and briefly discussed the *binomial coefficients* or *triangular numbers*,

$$\Delta_{n,k} \overset{\text{def}}{=} \binom{n}{k}$$

within the context of binary operations on integers; see Eq. (5.7). Also, we established in Proposition 5.3 the summation rule

$$\binom{n}{k} + \binom{n}{k+1} = \binom{n+1}{k+1}$$

for these integers. We now *define* binomial coefficients via the *bilinear recurrence* that underlies the summation rule. This change in viewpoint is the topic of the current subsection.

9.2.1.1 The formation rule for Pascal's Triangle

Let us define the bivariate integer function[1] $\hat{\Delta}(n,k)$ via the bilinear recurrence

---

[1] We alter our notation for binomial coefficients in deference to our change in viewpoint: We promote the integer pair $\langle n, k \rangle$ from a subscript to an argument, and we embellish Δ with a "hat".

$$\hat{\Delta}(n,k) = \begin{cases} 1 & \text{if } [n=1, k=0] \\ 1 & \text{if } [n=1, k=1] \quad (9.5) \\ \hat{\Delta}(n-1,k-1) + \hat{\Delta}(n-1,k) & \text{otherwise} \end{cases}$$

We claim that the function $\hat{\Delta}(n,k)$ thus defined is, in fact, the binomial coefficient $\binom{n}{k}$. We establish this claim with the help of Pascal's Triangle, the two-dimensional array of binomial coefficients which we defined in Section 5.1.2.4 and illustrated in Fig. 5.1. Recall that the *formation rule of the array* is that the array-entry at (row $n+1$, column $k+1$) is the sum of the array-entries at (row n, column k) and at (row n, column $k+1$).

If you compare the formation rule for Pascal's Triangle with Eq. (5.9), then you will anticipate the following result.

Proposition 9.3 *The entries of Pascal's Triangle are the binomial coefficients. Specifically, for all n,k, the entry at (row n, column k) of the Triangle is $\binom{n}{k}$.*

Proof. We note by observation and direct calculation (see Fig. 5.1) that the proposition is true for $n=1$ and $k \in \{0,1\}$. A double induction which we leave to the reader verifies that every binomial coefficient appears in the Triangle and every Triangle entry is a binomial coefficient. ☐

Figures 9.2–9.4 provide the skeleton of an alternative proof of the summation formula for binomial coefficients. This proof, which we leave for the interested reader, is our first evidence of the intellectual importance of Pascal's Triangle.

The relation between binomial coefficients and Pascal's Triangle leads us to the following *a priori* non-obvious fact.

Proposition 9.4 *Every binomial coefficient is an integer.*

Proof. By the formation rule for Pascal's Triangle, every entry in that array is obtained from integers via repeated additions—an operation that maps integers to integers. Proposition 9.4 therefore follows from Proposition 9.3's proof that the elements of the Triangle are precisely the binomial coefficients. ☐

9.2.1.2 Summing complete rows of Pascal's Triangle

We conclude this section with a very consequential result about the binomial coefficients, specifically about the sum of all the coefficients that share the same bottom argument. Of course, we know now that this summation is equivalent to summing complete rows of Pascal's Triangle.

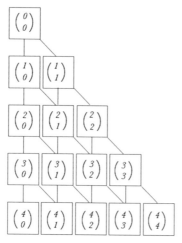

Fig. 9.2 A graph-like representation of Pascal's Triangle. Note that each entry $\hat{A}(n,k)$ equals the number of paths in the graph from entry $\hat{A}(n,0)$ to $\hat{A}(n,k)$

Fig. 9.3 Outlining an alternative proof of the summation formula for binomial coefficients: The number of paths from entry $\hat{A}(n,0)$ of the Triangle to entry $\hat{A}(n+1,k)$ equals the sum of the number of paths to entry $\hat{A}(n,k)$ and the number of paths to entry $\hat{A}(n,k-1)$

Proposition 9.5 *For every positive integer n,*

$$\sum_{i=0}^{n} \binom{n}{i} = \binom{n}{0} + \binom{n}{1} + \cdots + \binom{n}{n-1} + \binom{n}{n} = 2^n$$

Proof. This result is an immediate consequence of the Binomial Theorem (Theorem 5.3). That seminal result tells us that, for all $n \in \mathbb{N}$,

$$(x+y)^n = \sum_{i=0}^{n} \binom{n}{i} x^{n-i} y^i$$

If we instantiate this polynomial equation with the values $x = y = 1$, then we obtain the present result. □

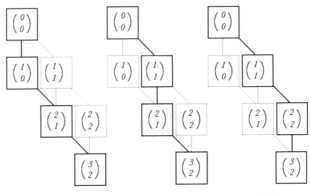

Fig. 9.4 Three distinct paths lead from entry $\hat{\Delta}(0,0)$ to entry $\hat{\Delta}(3,2) = \begin{pmatrix} 3 \\ 2 \end{pmatrix}$

We shall observe numerous applications of this result as we explore a broad variety of topics, ranging from counting discrete structures and calculating probabilities to deriving basic properties of other recursively defined families.

9.2.2 The Fibonacci Number Sequence

This section is devoted to one of the most storied topics in the world of mathematics—in terms of the topic's manifestation in the real world and in terms of the multiple names used to refer to its discoverer,[2] the thirteenth-century Italian mathematician known variously as:

Fibonacci	(abbreviated Italian for: son of Bonaccio)
Leonardo of Pisa	(his hometown)
Leonardo Pisano	(variant of "of Pisa")
Leonardo Pisano Bigolo	(his hometown plus family name)
Leonardo Fibonacci	(for: son of Bonaccio Bigolo)
Leonardo Bonacci	(for: son of Bonaccio Bigolo)

The sequence discovered by this multi-named genius is defined as follows.

The *Fibonacci sequence* or *the Fibonacci numbers* is an infinite sequence

$$F(0),\ F(1),\ F(2),\ \ldots$$

of *positive* integers. As just denoted, we see that the numbers in the sequence are traditionally indexed by the *nonnegative* integers and are often written using functional notation ($F(i)$) rather than as subscripts (F_i). The classical definition of the

---

[2] Not surprisingly, this marvelous sequence was discovered many times, in many places. Our story refers only to its discovery in the West.

sequence is via the following bilinear recurrence.

$$F(0) = 1$$
$$F(1) = 1 \qquad\qquad (9.6)$$
$$F(n) = F(n-1) + F(n-2) \quad \text{for all } n > 1$$

The sequence is often specified—particularly in the popular literature—just by listing its first few elements:

$$1, 1, 2, 3, 5, 8, 13, 21, 34, \ldots$$

9.2.2.1 The story of the Fibonacci numbers

Leonardo Fibonacci describes[3] discovering his eponymous sequence of integers in the course of contemplating the rate of population growth of successive generations beginning with an idealized immortal initial pair of *rabbits*.

Rabbits mature quickly and, after attaining maturity at age one month, they can spawn a new pair of progeny every subsequent month. So, let us begin "at time 0", with one pair of newborn rabbits. We continue with just the initial pair of rabbits at month 1, because there has not yet been time for the pair to spawn new rabbits. By month 2, though, there are two pairs of rabbits. At month 3, only the first pair will have spawned—because the second pair is too young—so there are three pairs of rabbits. At month 4, these three pairs are joined by two more. The reader who continues this story will discover that *the number of pairs of rabbits observed after successive months is specified by the process implicit in recurrence (9.6)*. Figure 9.5 illustrates the process up to month 5.

The Fibonacci sequence's ability to describe idealized rabbit population statistics is just a hint of its appearance elsewhere in the natural world—in structural features such as the patterns of seeds in flower heads, the numbers of petals of flowers, the growth patterns of pine cones and pineapples, and on and on; see [11].

The story of this fascinating sequence of numbers has a "universal" aspect also. Many cultures—the ancient Greeks among them—have ascribed mystical properties to (classes of) numbers: Our discussions of the *prime numbers* in Section 8.1 and the *perfect numbers* in Section 8.1.3 bear witness to this phenomenon. One specific number that has attracted such attention is the *golden ratio* Φ, an irrational real number which has the following (exact and approximate) values:

$$\Phi = \frac{1 + \sqrt{5}}{2} \approx 1.618\cdots$$

It has been alleged that rectangles whose *aspect ratios* (Length $\div$ Width) are (roughly) Φ are the most pleasing to the human eye. In fact, the aspect ratio of

---

[3] In his *Liber Abaci*, 1202

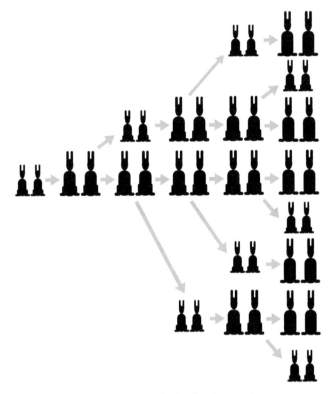

Fig. 9.5 The successive generations of rabbits for the first five months

the Parthenon in Athens is (roughly) Φ—although it is not known whether this is intentional. The relevance of Φ to the current section resides in the fact that *the sequence of ratios of successive Fibonacci numbers approaches Φ*.

You can make your own observations regarding rectangles with pleasing aspect ratios and the ratios of successive Fibonacci numbers by perusing Fig. 9.6.

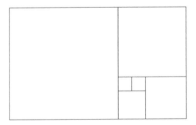

Fig. 9.6 Successive Fibonacci numbers interpreted geometrically, via a spiral of squares whose respective side-lengths form a Fibonacci sequence

The mathematical properties of this truly remarkable sequence will be our focus in the remainder of this section.

9.2.2.2 Fibonacci numbers and binomial coefficients

There is a strong, non-obvious connection between the binomial coefficients (cf. Section 9.2.1) and the Fibonacci numbers. We observe this connection by contemplating the diagonals of Pascal's Triangle. Figure 9.7 and Proposition 9.6 establish this connection.

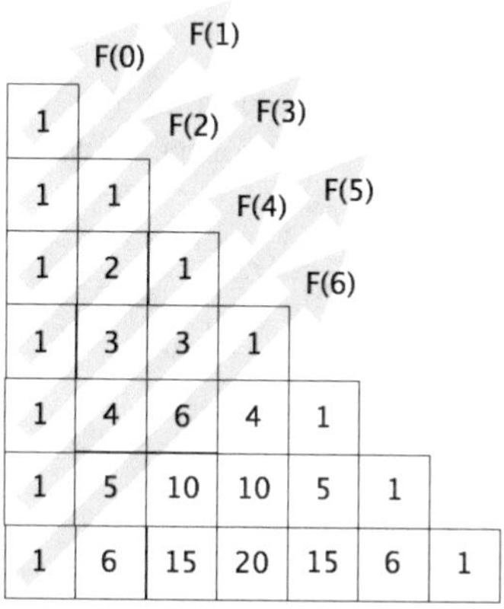

Fig. 9.7 Obtaining Fibonacci numbers as the sums of diagonal elements of the left-justified Pascal Triangle

Proposition 9.6 *For all $n \in \mathbb{N}$, the Fibonacci number $F(n)$ is the sum of the first $\lceil (n+1)/2 \rceil$ binomial coefficients $\binom{k}{i}$ such that $k + i = n$. Symbolically,*

$$F(n) = \binom{n}{0} + \binom{n-1}{1} + \cdots + \binom{\lfloor (n+1)/2 \rfloor}{\lceil (n+1)/2 \rceil - 1} \qquad (9.7)$$

Proof (Sketch). Because of the heavy calculational content of a complete proof, we provide here just a short sketch.

Figure 9.8 depicts a portion of Pascal's Triangle with shaded diagonal and horizontal annotations. The diagonal annotation depicts the three numbers in the Triangle that sum to the Fibonacci number $F(5)$.

1. Looking at the three horizontal shaded areas *individually* illustrates the relationship asserted by the proposition. Each of the three numbers on the shaded diagonal, namely, 1, 4, and 3, is a binomial coefficient; therefore, the number is

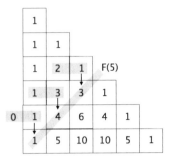

Fig. 9.8 Each term of the diagonal is obtained by summing the two preceding ones

the sum of the two numbers on the preceding row of the array: These sums are instantiations of the formation rule for Pascal's Triangle. The illustrated instance of the rule asserts that:

$$\binom{5}{0} = 0 + \binom{4}{0} = 0 + 1 = 1$$

$$\binom{4}{1} = \binom{3}{0} + \binom{3}{1} = 1 + 3 = 4$$

$$\binom{3}{2} = \binom{2}{1} + \binom{2}{2} = 2 + 1 = 3$$

2. Looking at the three horizontal shaded areas *in tandem* illustrates that the numbers along the shaded diagonal are sums of the numbers along the two diagonals that sit above the shaded one: These numbers are instantiations of the formation rule for Fibonacci numbers. The illustrated instance of the rule asserts that

$$F(5) = F(4) + F(3) = 5 + 3 = 8$$

The preceding reasoning provides the infrastructure for an induction which proves that the proposition holds for *every* Fibonacci number. As suggested by the statement of the proposition, the required calculations on indices can obscure the rather elegant basis for the result. □

9.2.2.3 Alternative origins for the Fibonacci sequence

Although the classical recurrence (9.6) is the structurally simplest generator of the Fibonacci sequence, there exist other generators that are not much more complex. We now present several alternative ways to generate the sequence: (A) two other multilinear recurrences; (B) a family of binary generating recurrences; (C) a combinatorial approach.

A. Two multilinear generating recurrences

Proposition 9.7 *For all integers $n \geq 2$,*

$$F(n) = 1 + F(0) + F(1) + F(2) + \cdots + F(n-2) \qquad (9.8)$$
$$= 1 + \sum_{k=0}^{n-2} F(k)$$

Proof. We proceed by induction.

Base. The base case, $n = 2$, holds because $F(2) = 2 = 1 + F(0)$.

Inductive hypothesis. Assume, for induction, that Eq. (9.8) holds for all arguments $2 \leq m < n$.

Inductive extension. We extend the induction as follows. Our inductive hypothesis assures us that for all $n \geq 3$,

$$F(n-1) = 1 + F(0) + F(1) + F(2) + \cdots + F(n-3)$$

Combining this with the classical recurrence (9.6), we therefore have

$$F(n) = F(n-2) + F(n-1)$$
$$= F(n-2) + 1 + F(0) + F(1) + F(2) + \cdots + F(n-3)$$

This extends the induction and completes the proof. $\square$

While recurrence (9.8) in Proposition 9.7 employs all of the Fibonacci numbers up to the desired bound, recurrence (9.9) in the next proposition employs only every other such number.

Proposition 9.8 *For all integers $n \geq 2$,*

$$F(n) = F(n-1) + F(n-3) + F(n-5) + \cdots + 1 \qquad (9.9)$$

We thereby sum every other Fibonacci number as long as we can.

Proof. We verify summation (9.9) by iteratively expanding the right-hand term — i.e., $F(n-2)$—of the classical recurrence (9.6). This expansion begins

$$F(n) = F(n-1) + F(n-2) \qquad (9.10)$$
$$= F(n-1) + F(n-3) + F(n-4)$$
$$= F(n-1) + F(n-3) + F(n-5) + F(n-6)$$

Note that all term-indices in the expanded summation have the same parity: The sequence of expanded indices begins $n-2, n-4, n-6, \ldots$, so successive expanded indices always differ by 2; therefore, the index $n-k$ that we expand always has the same parity as n.

We continue the process of expanding recurrence (9.10) as long as we can—i.e., until we encounter either $F(0)$, in the case of even n, or $F(1)$, in the case of odd n. In both cases, we will have "run out of Fibonacci numbers", so the expansion terminates—coincidentally with the final term 1.   □

B. A family of binary generating recurrences

What we have earlier called "generating recurrences" or "formation rules" for binomial coefficients and Fibonacci numbers can also be viewed as (mathematical) identities on the quantities of interest. In our usage, the line between "generating recurrence" and "identity" centers on computational issues: multilinear recurrences can feasibly be used to generate the desired numbers; nonlinear recurrences such as we expose in this subsection will likely not be used as generators. Indeed, for several of the results we cover here, we really want to stress the *methodology of proof and analysis.*

Proposition 9.9 *For all $n \in \mathbb{N}$ and $0 < k < n$*

$$F(n) = F(k) \cdot F(n-k) + F(k-1) \cdot F(n-k-1) \qquad (9.11)$$

Of course, the classical recurrence (9.6) is instance ($k = 1$) of the family of recurrent equations (9.11).

Proof. We first explain how one might guess at the existence of the family of recurrences (9.11). Then we validate the recurrences in the family.

We begin with the classical recurrence (9.6) and iteratively use this recurrence to "expand" the classical recurrence. In detail, we begin by combining the first two instances of (9.6), namely,

$$\begin{aligned} F(n) &= F(n-1) + F(n-2) \\ F(n-1) &= F(n-2) + F(n-3) \end{aligned}$$

and we combine them algebraically to produce the following.

$$F(n) = 2F(n-2) + F(n-3)$$

And then we iterate! The following table illustrates the result of the first four iterations of the process.

$$
\begin{aligned}
F(n) = F(n) + \ & F(n-1) \\
= \quad\quad & 2F(n-1) + \ F(n-2) \\
= \quad\quad\quad\quad & 3F(n-2) + 2F(n-3) \\
= \quad\quad\quad\quad\quad\quad & 5F(n-3) + 3F(n-4) \\
= \quad\quad\quad\quad\quad\quad\quad\quad & + 8F(n-4) + 5F(n-5) \\
\vdots \quad\quad \vdots \quad\quad\quad & \vdots \quad\quad\quad\quad \vdots \quad\quad\quad\quad \vdots \quad\quad\quad \vdots
\end{aligned}
$$

Note that the coefficients of the successive occurrences of the Fibonacci numbers $F(i)$ that occur in our table are themselves Fibonacci numbers. By analyzing the emerging pattern—*remember our advice in Chapter 2 to always look for patterns*—we arrive at the family (9.11) of recurrent equations.

Explanatory note.

Keep in mind that, at this point, we are still in the realm of conjecture! The observed pattern *appears to be* converging to generic recurrence (9.11). We must now verify the universal validity of the family.

We proceed by induction on the number k of iterated expansions of the classical recurrence (9.6).

Basis. The basis for our induction resides in the observation we shared right after stating the proposition: Instance ($k = 1$) of the posited family of recurrent equations is just the classical recurrence (9.6).

Inductive hypothesis. Assume that instance k of family (9.11)—i.e., the equation

$$
F(n) = F(k) \cdot F(n-k) + F(k-1) \cdot F(n-k-1)
$$

is valid. *Note that validity requires that $k < n-1$.*

Inductive extension. Let us observe, under the validity assumption, the result of producing instance $k+1$ from this instance. We algebraically combine the just-cited equation with the following instantiation of the classical recurrence:

$$
F(n-k) = F(n-k-1) + F(n-k-2)
$$

We find that

$$
\begin{aligned}
F(n) &= F(k) \cdot F(n-k) + F(k-1) \cdot F(n-k-1) \\
&= F(k) \cdot \big[F(n-k-1) + F(n-k-2)\big] + F(k-1) \cdot F(n-k-1) \\
&= \big[F(k) + F(k-1)\big] \cdot F(n-k-1) + F(k) \cdot F(n-k-2) \\
&= F(k+1) \cdot F(n-k-1) + F(k) \cdot F(n-k-2)
\end{aligned}
$$

The induction is thus extended, which establishes the proposition. $\square$

C. A combinatorial setting for the sequence

In Section 9.2.2.1, we described several settings in which one can identify the Fibonacci sequence. The settings we chose arose in nature—rabbits, pineapples, etc. We now describe a *mathematical* setting in which the sequence arises. In common with our rabbits and pineapples, this setting does not involve the solution of a system of recurrent equations. In common with our equational settings, this setting is purely mathematical.

Our setting is *combinatorial* in nature, in the sense that the Fibonacci numbers emerge as we count instances of some phenomenon. But, of course, the generating recurrence is still present; it is just camouflaged by the counting.

For each positive integer n, let S_n be the set of all length-n binary strings in which *every occurrence of bit 1 is directly preceded by an occurrence of bit 0*. Table 9.1 provides the first few instances of S_n, with their cardinalities.

Table 9.1 The first few instances of S_n, with their cardinalities

$n =$	$S_n =$	$|S_n| =$
0	$\{\varepsilon\}$	1
1	$\{0\}$	1
2	$\{00, 01\}$	2
3	$\{000, 001, 010\}$	3

Proposition 9.10 *For each $n \in \mathbb{N}^+$, the number of length-n binary strings in which each occurrence of a 1 is directly preceded by a 0 is the Fibonacci number $F(n)$.*

Proof. By definition, every binary string $w \in S_n$ ends either with 0 or with 01.

- If w ends with 0, then it has the form $w = x0$, where the prefix x is a binary string of length $n - 1$; moreover, x must belong to S_{n-1} in order for w to belong to S_n. S_n therefore contains $|S_{n-1}|$ strings of this form.

- If w ends with 01, then it has the form $w = y01$, where the prefix y is a binary string of length $n - 2$; moreover, y must belong to S_{n-2} in order for w to belong to S_n. S_n therefore contains $|S_{n-2}|$ strings of this form.

The preceding reckoning implies that the cardinalities of the sets S_n obey the following recurrence.

$$
\begin{aligned}
|S_0| &= 1 & &\text{see Table 9.1} \\
|S_1| &= 1 & &\text{see Table 9.1} \\
|S_n| &= |S_{n-1}| + |S_{n-2}| & &\text{see preceding analysis}
\end{aligned}
$$

This recurrence is just a relabeled version of the Fibonacci recurrence (9.6). $\square$

9.2.2.4 ⊕ A closed form for the nth Fibonacci number

We close our survey of the Fibonacci numbers by exposing a *closed-form expression*[4] for the numbers in this fascinating family. The detailed derivation of this expression requires advanced material that is beyond the scope of this text, so we settle for a *heuristic* explanation.

Explanatory note.

By "heuristic explanation", we mean the kind of intuitive explanation that mathematicians and scientists often use to garner intuition during the exploratory phase of studying a complex topic. As one's intuition grows, one hopes to eventually replace the heuristic explanation with a rigorous one. In the case of the sought closed-form expression for the nth Fibonacci number, the sought rigor must await rather advanced concepts and tools.

If you were to write out a sufficiently long initial sequence of Fibonacci numbers, you would observe that they grow quite fast. Indeed, by this point in the text, you have hopefully "played" with enough sequences that you might guess that the Fibonacci numbers $\{F(n)\}$ grow *exponentially* in the index n. In detail, you would be guessing that there exists a real base $\beta > 1$ and a constant of proportionality $c > 0$ such that $F(n) = c\beta^n$, at least approximately. In order to (hopefully!) garner intuition for the actual growth behavior of the Fibonacci numbers, let us observe an important corollary of the "exponential" guess. If the guessed form were accurate, then it would combine with recurrence (9.6) in the following way.

(1) $F(n) = c\beta^n$ by our guess
(2) $F(n) = F(n-1) + F(n-2)$ by recurrence (9.6)

By combining (1) and (2), we therefore find that

$$\beta^n = \beta^{n-1} + \beta^{n-2}$$

so that β^n is a root of the quadratic equation

$$x^2 - x - 1 = 0$$

By the quadratic formula—see Proposition 5.5—this polynomial has two roots:

$$\Phi = \frac{1+\sqrt{5}}{2} \quad \text{and} \quad \Phi' = \frac{1-\sqrt{5}}{2}$$

Note that Φ, which is known as the *golden ratio*, exceeds 1 while Φ' does not. Since we know that the Fibonacci numbers *grow* with n rather than shrink with n, our initial guess would assign $F(n)$ the value

---

[4] The term "closed-form expression" is defined and illustrated in Section 6.2.1.2.

$$F(n) = \Phi^n = \left(\frac{1+\sqrt{5}}{2}\right)^n$$

In fact, this guessed value of $F(n)$ is off by only a small constant factor, at least for very large values of n—as exposed by part (a) of the following result. Part (b) of the result actually provides the sought closed-form expression for $F(n)$.

Proposition 9.11 **(a)** (An approximating expression) *For all sufficiently large n,*

$$F(n) \approx \frac{1}{\sqrt{5}}\left(\frac{1+\sqrt{5}}{2}\right)^n$$

The symbol "≈" here means that the relative error incurred by approximating $F(n)$ via this expression shrinks exponentially fast as n grows.

(b) (An exact expression) *For all n,*

$$F(n) = \frac{1}{\sqrt{5}}\left(\left(\frac{1+\sqrt{5}}{2}\right)^n - \left(\frac{1-\sqrt{5}}{2}\right)^n\right)$$

9.3 ⊕ Recurrences "in Action": The Token Game

In order to truly appreciate the power of recurrences as an analysis tool, one must witness them "in action". To this end, we now describe the (*single-player*) combinatorial *Token Game*.[5] By employing recurrences to analyze plays of the game, we are able to derive an optimal strategy for playing the game.

9.3.1 Rules of the Game

The equipment. For each $n \in \mathbb{N}^+$, the order-n Token Game is played with:

- a *bank* that has n *slots*, labeled $1, \ldots, n$
- a *pile* of n tokens, each capable of completely occupying a bank slot.

Initial and terminal configurations. Each play of the game begins with the bank empty and the pile full, as depicted in Fig. 9.9.

The goal of the game. The player's goal is to transfer all n tokens from the pile into the bank.

---

[5] This game is inspired by the *Baguenaudier* introduced by Édouard Lucas [78]. It is also known as the Chinese rings.

Fig. 9.9 The initial configuration of the Token Game: Each of the n tokens appears as a grey circle, and the empty bank has n slots. In the figure, $n = 8$

The repertoire of Game moves. The player transfers tokens from the pile to the bank by executing a sequence of *moves*, each having one of the following types.

1. Change the state of bank-slot #1, which is the first (i.e., leftmost) slot in the bank:

   If slot #1 is empty, then move a token from the pile to that slot.

   If slot #1 is full (i.e., contains a token), then remove this token and return it to the pile; see Fig. 9.10.

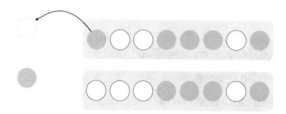

Fig. 9.10 (Top) Slot #1 contains a token. (Bottom) Therefore, remove it (i.e., move it back to the pile)

2. Change the state of the bank-slot—call it slot #s—that is immediately to the right of the first (i.e., leftmost) *empty* slot:

   If slot #s is empty, then move a token from the pile to that slot.

   If slot #s is full (i.e., contains a token), then remove this token and return it to the pile; see Fig. 9.11.

Objective of a play of the Game: To minimize the number of moves from an initially empty bank to the finally full bank.

9.3.2 An Optimal Strategy for Playing the Game

The question is how the player should choose successive moves, with the goal of filling the bank as quickly as possible. One can garner some strategic observations about how to play the Game by looking at small instances.

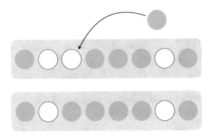

Fig. 9.11 (Top) Bank-slot #s, which is immediately to the right of the first empty slot ($s = 3$ in this example) is empty. (Bottom) Therefore, move a token from the pile into slot #3

- When $n = 1$, the player should simply fill the slot using a Type-1 move.

- When $n = 2$, the player must first play a Type-2, then a Type-1 move.

- As the Game proceeds for the next small values of n (say $n = 3, 4$, or 5), observation suggests that the player should begin with a Type-1 move when n is odd and with a Type-2 move when n is even.

- Another easy observation is: The player should not play two successive moves of the same type, because the second one just undoes the first.

A strategy is beginning to emerge:

1. Choose the initial move based on the parity of n.

2. Subsequently, alternate between the two types of moves.

This strategy is pleasingly simple, but: (a) Does it lead us to the required terminal state? (b) What is the cost of a (successful) play using this strategy? The answers to both questions can be discovered if we reformulate the strategy to a *recursive* form: We can then use recurrences to prove that the proposed strategy is optimal and to evaluate the cost of an optimal play of the order-n game.

We play the small instances of the Game—the cases $n = 1$ and $n = 2$—using the *ad hoc* strategies mandated by our earlier observations. For Game instances with $n > 2$, we devise a recursive solution, which is motivated by the following reasoning.

A token can be placed into the *last* bank-slot (i.e., the nth) via the Type-2 move

MOVE TOKEN FROM PILE INTO BANK-SLOT n

In order for this move to be eligible for execution, the bank must be in the following configuration, reading rightward from bank-slot 1:

$$\big[\text{tokens in slots } 1, 2, \ldots, n-2\big], \big[\text{no token in slot } n-1\big], \big[\text{no token in slot } n\big]$$

This configuration requires that the first $n - 2$ slots have been filled. (Note that, importantly, $n - 2$ has the same parity as n.)

Once the player has achieved this configuration and executed the mandated move, the player henceforth ignores the token in bank-slot n. Here comes the recursion!

The player is now confronting the initial configuration of the order-$(n-1)$ Game!

Thus, in this recursive formulation, the Game can be played by recursively executing the *super-steps* depicted in Figure 9.12, on successively smaller banks and piles. We summarize the super-steps:

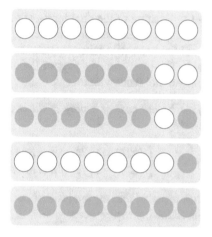

Fig. 9.12 A schematic of the recursive play of the current-sized version of the Game. The top four bank configurations indicate the four iterating super-steps in the recursions. The bottom bank configuration is the final one: The bank is entirely filled

1. *Topmost bank configuration → Second bank configuration*

   Move tokens into the leftmost $n-2$ slots of the bank, leaving the rightmost two slots empty.

2. *Second bank configuration → Third bank configuration*

   Move a token into bank-slot n, i.e., the current rightmost slot

3. *Third bank configuration → Fourth bank configuration*

   Empty bank-slots $1, 2, \ldots n-1$; i.e., leave the current rightmost slot filled, but empty all slots to its left.

4. *Final bank configuration*

   The Game is complete!

9.3.3 ⊕⊕ *Analyzing the Recursive Strategy*

Not surprisingly, our analysis of the optimal recursive strategy for playing the Token Game resides in a *recurrence* for the cost of playing the Game, as a function of the bank-size n. Also not surprisingly, the structure of the recurrence mirrors the structure of our recursive playing strategy.

For $i = 1,\ldots,n$, let $f(i)$ denote the cost of filling the bank-slots #1 through #i, measured in terms of the number of atomic moves, each of the form PLACE A TOKEN or REMOVE A TOKEN. Because of the dual forms of our atomic moves—each move fills one slot that is empty or empties one slot that is full—the cost of filling an empty length-i prefix of the bank with tokens equals the cost of emptying a full length-i prefix of the bank. The total cost of a play of the order-n Game is the cost of filling the initially empty n-slot bank with tokens.

Proposition 9.12 *Let $f(n)$ be the cost of a play of the order-n Token Game. For each $n \in \mathbb{N}^+$, $f(n)$ is given by:*

$$f(n) \;=\; \begin{cases} \left(2^{n+1} - 1\right)/3 & \text{if } n \text{ is odd} \\ \left(2^{n+1} - 2\right)/3 & \text{if } n \text{ is even} \end{cases} \tag{9.12}$$

Proof. Our discussion has revealed that the analysis of our recursive playing strategy resides in solving the following recurrence.

$$f(n) \;=\; \begin{cases} 1 & \text{if } n = 1 \\ 2 & \text{if } n = 2 \\ f(n-1) + 2f(n-2) + 1 & \text{if } n > 2 \end{cases} \tag{9.13}$$

Explanatory note.

There is a point to clarify here because the moves of the Game are not *symmetric*—we always proceed from left to right.

The recursive solution involves a super-step that empties a bank of size $n - 2$. It appears that filling and emptying a bank are mirror-image operations. Indeed, for small n, the sequence of operations that empties the bank is the reverse of the sequence that fills the bank. We now verify that the symmetry always holds.

The cost of the recursive solution is:

$$f(n) = \begin{cases} 1 & \text{if } n = 1 \\ 2 & \text{if } n = 2 \\ f(n-2) + 1 + f^{-1}(n-2) + f(n-1) & \text{if } n > 2 \end{cases} \quad (9.14)$$

Let f^{-1} denote the bank-emptying operation. If one thinks in terms of mirror-image operations, then one discovers a recursive solution for emptying the bank:

$$f^{-1}(n) = \begin{cases} 1 & \text{if } n = 1 \\ 2 & \text{if } n = 2 \\ f^{-1}(n-1) + f(n-2) + 1 + f^{-1}(n-2) & \text{if } n > 2 \end{cases} \quad (9.15)$$

Taking the difference between expressions (9.14) and (9.15), we find that the costs of $f(n)$ and $f^{-1}(n)$ are equal!

$$\begin{aligned} f(n) - f^{-1}(n) &= f(n-2) + 1 + f^{-1}(n-2) + f(n-1) \\ &\quad - f^{-1}(n-1) - f(n-2) - 1 - f^{-1}(n-2) \\ &= f(n-1) - f^{-1}(n-1) \\ &\quad \cdots \\ &= f(1) - f^{-1}(1) \\ &= 0 \end{aligned}$$

We can dramatically simplify recurrence (9.13) by focusing on the function

$$g(n) \overset{\text{def}}{=} f(n) + f(n-1) \quad \text{for } n \geq 2$$

instead of on f. Elementary calculation based on (9.13) shows that $g(n)$ satisfies the recurrence

$$g(n) = \begin{cases} 3 & \text{if } n = 2 \\ 2g(n-1) + 1 & \text{if } n > 2 \end{cases} \quad (9.16)$$

We have, thereby, replaced the *bilinear* recurrence (9.13) by the *(singly) linear* recurrence (9.16). We learned in Section 6.2.2—see Proposition 6.4—how to evaluate geometric summations that solve recurrences such as (9.16). In our case, we find that

$$g(n) = 2^{n-1} + 2^{n-2} + \cdots + 2^2 + 2 + 1 = 2^n - 1 \quad (9.17)$$

We can now return to evaluating $f(n)$ via recurrence (9.13), in the light of our analysis of $g(n)$ in recurrences (9.16) and (9.17). We find that

$$f(n) = \begin{cases} 1 & \text{if } n = 1 \\ g(n) - f(n-1) = (2^n - 1) - f(n-1) & \text{if } n > 1 \end{cases} \qquad (9.18)$$

We begin to solve the *singly* linear recurrence (9.18) for $f(n)$ using the strategy we developed in Section 6.2.2. We expand the recurrence in order to discern its pattern and then analyze the summation that the pattern leads to. In this case, we observe:

$$\begin{aligned} f(n) &= 2^n - 2^{n-1} + f(n-2) + 1 - 1 \\ &= 2^n - 2^{n-1} + 2^{n-2} - f(n-3) - 1 \\ &= 2^n - 2^{n-1} + 2^{n-2} - 2^{n-3} + f(n-4) + 1 - 1 \\ &\quad\vdots \end{aligned}$$

What we observe emerging—an inviting induction for the reader lurks in those words—is a geometric summation of powers of 2, with adjacent terms *alternating* in sign; the terminal units, ± 1, cancel after each pair of steps. We must be careful, though, because the numbers of terms in the summations differ based on the parity of n: when n is even, the last term is 0; when n is odd, the last term is -1.

We have now reached the *penultimate* step in finding the value of $f(n)$; specifically, we have derived the following parity-specified summations.

$$\text{For even values of } n: \quad f(n) = \sum_{k=1}^{n} (-1)^k 2^k \qquad (9.19)$$

$$\text{For odd values of } n: \quad f(n) = \sum_{k=0}^{n} (-1)^{k+1} 2^k \qquad (9.20)$$

Solving these summations for $f(n)$ requires a moderate bit of mathematical dexterity. For pedagogical reasons, we illustrate two quite distinct approaches for determining the value of $f(n)$ for the case of odd n: We want to expose the reader to the quite-different intuitions that each approach elicits. We shall then derive the value of $f(n)$ for the case of even n from the value for the case of odd n.

An algebraic approach for the case of odd n. We look in detail at summation (9.20), which specifies $f(n)$ when n is odd, and we invoke algebraic manipulation to determine the value of $f(n)$ in this case.

In the following chain of equalities, we: gather the positive and negative terms in summation (9.20) [line 1 in the chain], perform some elementary manipulations on the result [lines 2 and 3 in the chain], and then invoke Proposition 6.4 [line 4 in the chain, which evaluates the resulting geometric summation]. We thereby find that, for odd values of n:

$$f(n) = \left(2^n + 2^{n-2} + \cdots + 2 \right) - \left(2^{n-1} + 2^{n-3} + \cdots + 1 \right)$$

$$= 2^{n-1} + 2^{n-3} + \cdots + 1$$

$$= 2^{n-1} \cdot \left(1 + \frac{1}{4} + \frac{1}{16} + \cdots + \frac{1}{2^{n-1}} \right)$$

$$= 2^{n-1} \cdot \left(1 + \frac{1}{4} + \frac{1}{16} + \cdots + \frac{1}{4^{(n-1)/2}} \right)$$

$$= 2^{n-1} \cdot \frac{4}{3} \cdot \left(1 - \left(\frac{1}{4} \right)^{(n+1)/2} \right)$$

$$= \frac{2^{n+1}}{3} \cdot \left(1 - \frac{1}{2^{n+1}} \right)$$

$$= \frac{1}{3} \left(2^{n+1} - 1 \right)$$

A geometric approach for the case of odd n. We begin again by looking again at summation (9.20). Then, noting that 2^{n-1} is a perfect square whenever n is odd, we set out to represent $f(n)$ as the aggregated area of a shrinking sequence of squares, of successive dimensions

$$2^{(n-1)/2} \times 2^{(n-1)/2}, \quad 2^{(n-3)/2} \times 2^{(n-3)/2}, \quad 2^{(n-5)/2} \times 2^{(n-5)/2}, \quad \ldots, \quad 1 \times 1$$

Fig. 9.13 depicts such a representation of $F(7) = 64 + 16 + 4 + 1 = 85$. To facilitate

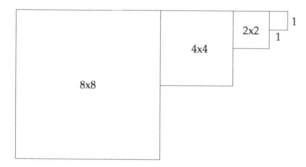

Fig. 9.13 A representation of summation (9.20) for the case $n = 7$

our upcoming manipulation of the configuration depicted in the figure, let us refer to the configuration as *the cascade of squares determined by* $f(n)$. Note that, because each cascade is associated with an odd value of n, the smallest square in the cascade (at the far right in the figure) is the unit-side square, of dimensions 1×1; hence, it contributes $+1$ to the aggregate area of the cascade.

We can now use a geometric construction to evaluate $f(n)$ on an arbitrary odd argument n. We take three copies of the cascade in Fig. 9.13 and we manipulate the copies into the form is depicted in Fig. 9.14. In detail:

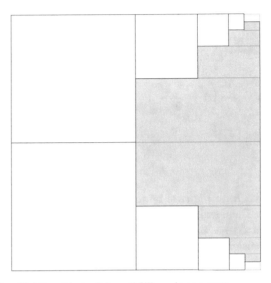

Fig. 9.14 Evaluating $f(n)$ for odd n by "almost" filling a large square

1. We choose one of the three copies as the "anchor" of the construction. We position it in space so that it serves as the upper white cascade in Fig. 9.14.

2. We then take a second copy, flip it across the horizontal axis, and abut the top edge of its largest square with the bottom edge of the largest square in the anchor cascade. It then becomes the lower white cascade in Fig. 9.14.
   Note that, importantly, the abutted white cascades fit into a $2^{(n+1)/2} \times 2^{(n+1)/2}$ square.

   > **Explanatory note.**
   > All of our observations about figures fitting within other figures are verified by direct calculations. These calculations are not too hard because all squares have side-dimensions that are powers of 2.

Indeed, the top edge of the top white cascade and the bottom edge of the bottom white cascade lie, respectively, along the top and bottom edges of the $2^{(n+1)/2} \times 2^{(n+1)/2}$ square. *However*, the cascades' edges are 1 unit shorter than the edges of the big square; i.e., they both have length

$$2^{(n-1)/2} + 2^{(n-3)/2} + 2^{(n-5)/2} + \cdots + 1 = 2^{(n+1)/2} - 1$$

3. Finally, we take the third copy, color it grey, and nest it into the abutting white cascades in the following way.

    a. Take the biggest square in the grey cascade and nest it against the abutted biggest squares in the paired white cascades, in the manner depicted in Fig. 9.14. Note that the nest places one half of its biggest grey square abutting the biggest white square of the top white cascade and one half abutting the biggest white square of the bottom white cascade. Observe (from Fig. 9.14) that the resulting configuration fits within the $2^{(n+1)/2} \times 2^{(n+1)/2}$ square, and that the fit is *exact* along the left and right edges, which are shared by the abutting white cascades and the big square.

    b. For all of the other grey squares, in decreasing order of size: We bisect—i.e., cut exactly in half—each square along its equator, and we nest the resulting two halves of that square symmetrically within the abutting white cascades, in the manner depicted in Fig. 9.14. Once again, we observe that the resulting configuration fits within the $2^{(n+1)/2} \times 2^{(n+1)/2}$ square, and that the fit is *exact* along the left and right edges, which are shared by the abutting white cascades and the big square.

The placement of the bisected squares from the grey cascade leaves two small *empty* regions within the $2^{(n+1)/2} \times 2^{(n+1)/2}$ square. The empty regions each have area $1/2$, because they are created by the "inadequate" placement of the bisected unit-side square from the grey cascade; the empty regions appear at the top right and bottom right corners of the $2^{(n+1)/2} \times 2^{(n+1)/2}$ square.

Once we have completed the described construction of the composite object depicted in Fig. 9.14, we calculate that the combined areas of the three cascades is one unit less than the area of the $2^{(n+1)/2} \times 2^{(n+1)/2}$ square (which, of course, has area 2^{n+1}). We have thus shown geometrically that $3f(n) + 1 = 2^{n+1}$, which *(of course!)* agrees with the value derived algebraically in (9.21).

We immediately derive the expression for even n using the definition of $g(n)$:

$$f(n) = 2^n - 1 - f(n-1) \quad \text{for } n \geq 2$$
$$= 2^n - 1 - \frac{1}{3}(2^n - 1)$$
$$= 2^n \left(1 - \frac{1}{3}\right) - 1 + \frac{1}{3}$$
$$= \frac{1}{3}(2^{n+1} - 2)$$

This completes the proof. □

9.4 Exercises: Chapter 9

Throughout the text, we mark each exercise with 0 or 1 or 2 occurrences of the symbol $\oplus$, as a rough gauge of its level of challenge. The 0-$\oplus$ exercises should be accessible by just reviewing the text. We provide *hints* for the 1-$\oplus$ exercises; Appendix H provides *solutions* for the 2-$\oplus$ exercises. Additionally, we begin each exercise with a brief explanation of its anticipated value to the reader.

1. **Verifying an empirically observed pattern**
   LESSON: We have advocated expanding a recurrence to gain intuition regarding its growth pattern. Of course, then one must verify one's observation.
   Verify the segment of recurrence (9.4) that is "hidden behind" the vertical dots.

2. **Completing the proof of the *general* version of the Master Theorem**
   LESSON: Practice in combining snippets of proofs.

   We noted in the text that "solving the more general recurrence (9.1) requires only augmenting the solution to the simple recurrence (9.2) by 'appending' to the simple solution a geometric summation whose base is the ratio a/b".
   Fill in the details of this observation.

3. **An unusual comparison of automobile brands.**
   LESSONS: Enhance facility with recurrences and asymptotics

   The logos of two German automakers, MERCEDES and BMW suggest a (frivolous) game whose analysis can enhance one's understanding of both recurrences and asymptotics. The game is designed around the fact that the MERCEDES logo is a *trisection* of a circle, while the BMW logo is a *quadrisection* of a circle. The

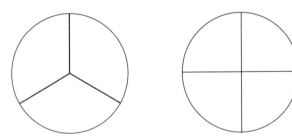

Fig. 9.15 Idealized renditions of the logos on MERCEDES™ (left) and BMW™ (right) automobiles

single-player *Logo game* proceeds as follows.

Reflecting the fact that the MERCEDES™ logo partitions the circle into *three* wedges, while the BMW™ logo partitions the circle into *four* wedges, we assign the player—who, note, has not yet acted—with the initial scores

$$M(0) = 3 \quad \text{and} \quad B(0) = 4$$

and we designate the logo-circles in Fig. 9.15 as the *stage-0* logo-circles. *Note that it is the* number *of wedges in each logo-circle that interests us, not the sizes of the wedges.*

The player now becomes active—

For each positive integer k, the player begins at the northernmost point of each stage-$(k-1)$ logo-circle and circumnavigates the circle in a clockwise sense. In the course of this circumnavigation, the player *bisects every second wedge* that she encounters. When the player regains the northernmost point of the two logo-circles, the numbers of wedges at that moment is recorded as the score at that moment: at stage k this has the form of the pair $\langle M(k), B(k) \rangle$.

Prove the following assertions about the described game.

a. *For each* $n \in \mathbb{N}^+$,

$$M(n) = B(n-1)$$

b. *For each* $n \in \mathbb{N}^+$,

$$B(n) = B(n-1) + \left\lfloor \frac{1}{2}B(n-1) \right\rfloor \tag{9.21}$$

c. $\oplus$ *The asymptotic behavior of $B(n)$ is given by*

$$B(n) = \left(\frac{3}{2}\right)^{O(n)}$$

Hint. *Solve the "continuous" version of recurrence (9.21) for $B(n)$, which is obtained by removing the floors in (9.21). Then use asymptotic reasoning to complete the proof.*

4. **Karatsuba multiplication**
   LESSON: Use the Master Theorem for Linear Recurrences (Theorem 9.2) to analyze a recursive algorithm

   Say that you are given two n-bit integers, in terms of their base-2 numerals:

   $$A = a_{n-1}a_{n-2}\cdots a_0 \quad \text{and} \quad B = b_{n-1}b_{n-2}\cdots b_0$$

   The classical elementary-school method for computing the product of A and B computes the n partial products:

   $$(a_{n-1}a_{n-2}\cdots a_0) \times b_i 2^i$$

   and then sums these partial products. This algorithm requires $\Theta(n^2)$ multiplications and a like number of additions.

   Now, multiplications are more expensive than additions on standard computing platforms. Therefore, when the common length n of the numerals for A and B

is *very* large, then one would be willing to perform somewhat more additions (while staying within the asymptotic class $\Theta(n^2)$) if, in compensation, one could save a significant number of multiplications. Happily, such a tradeoff is, indeed, available, using an algorithm that is known as *Karatsuba multiplication*, after its author, Anatoly Karatsuba; see [63].

Karatsuba's algorithm builds upon the *divide-and-conquer* algorithmic paradigm, whereby a problem **P** is decomposed into disjoint, equal-"size"[6] subproblems whose results are accumulated to solve **P**.

Explanatory note.

The Divide-and-Conquer Paradigm:

a. Decompose the size-n problem **P** into p subproblems of size n/p each.

b. Solve the subproblems.

   Typically, this recursively invokes the **P**-solver—down to some threshold size, at which point base case(s) of the recursion govern the computation.

c. Combine the solutions of the subproblems to craft a solution of **P**.

The time-cost of the recursive computation that solves **P** is given by the recurrent expression

$$T(n) = p \cdot T(n/p) + \kappa(n) + \kappa'(n)$$

$\kappa(n)$ and $\kappa'(n)$ are, respectively, the costs of phases (a.) and (c.) of the paradigm.

Karatsuba's algorithm. Let us return to the problem of multiplying the n-bit numerals A and B. To simplify both notation and analysis, say that n is a very large power of 2, i.e., $n = 2^k$ for some large $k \in \mathbb{N}^+$.

The algorithm breaks the numerals for A and B into pairs of half-size numerals:

$$A = A_1 \cdot 2^{n/2} + A_2 \overset{\text{def}}{=} (a_n \cdots a_{n/2+1})_2 \cdot 2^{n/2} + (a_{n/2} \cdots a_1)_2$$

$$B = B_1 \cdot 2^{n/2} + B_2 \overset{\text{def}}{=} (b_n \cdots b_{n/2+1})_2 \cdot 2^{n/2} + (b_{n/2} \cdots b_1)_2$$

After this decomposition, we have[7]

$$A \times B = (A_1 \times B_1) \cdot 2^n + (A_1 \times B_2 + A_2 \times B_1) \cdot 2^{n/2} + (A_2 \times B_2) \quad (9.22)$$

A recursion (on n) that is based on instantiating Eq. (9.22) employs *four* multiplications at this level.

The way that Karatsuba brought the cost of multiplying $A \times B$ below the obvious measure of $\Theta(n^2)$ multiplications was to observe the following. If one defines

---

[6] The paradigm works for a large range of "size" measures: time, space, ...

[7] To enhance legibility, we use both the centered dot ($\cdot$) and the multiplication sign ($\times$).

$$C \stackrel{\text{def}}{=} (A_1 - A_2) \times (B_2 - B_1)$$

then

$$A \times B = (A_1 \times B_1) \cdot 2^n + (A_2 \times B_2)$$
$$+ \left(C + (A_1 \times B_1) + (A_2 \times B_2)\right) \cdot 2^{n/2}$$

One thereby executes this level of the recursion that computes $A \times B$ using only three *multiplications, rather than the four multiplications in Eq. (9.22).* (The few more additions are often a beneficial cost-tradeoff.)

Prove the following assertions.

a. ⊕ *A recursion (on n) that is based on instantiating Eq. (9.22) computes the product $A \times B$ using $\Theta(n^2)$ multiplications.*

b. ⊕⊕ *A recursion (on n) that is based on Karatsuba's algorithm computes the product $A \times B$ using asymptotically fewer than $\Theta(n^2)$ multiplications.*

Your argument should find a (real) number $\alpha < 2$ such that Karatsuba's algorithm computes the product $A \times B$ using $\Theta(n^\alpha)$ multiplications.

5. **Binomial coefficients and Pascal's Triangle**
   LESSONS: Practice with induction and intuition about binomial coefficients
   Prove the following assertions, which refer to Fig. 5.1. (The first assertion is Proposition 9.3.)

   a. *For all n, k, the entry at (row n, column k) of the Triangle is $\binom{n}{k}$.*

   b. *The successive rows of the Triangle sum to the successive powers of* 2.

   c. *The first element of every row is* 1.

   d. *The successive elements of column 2 of the Triangle are the successive integers.*

   e. *The successive elements of column 3 of the Triangle are the successive triangular numbers, Δ_k.*

   f. *The successive elements of column 4 of the Triangle are the successive tetrahedral numbers, $\widehat{\Theta}_k$.*

   g. *Build on Figs. 9.2–9.4 to generate an alternative proof of the summation formula for binomial coefficients.*

6. **Some details of the Token Game of Section 9.3**
   LESSON: The mental calisthenics of analyzing frivolous games often develop the muscles needed to tackle "real" problems.
   Verify for small values of n that the sequence of operations that empties the bank is the reverse of the sequence of operations that fills the bank—the first step becomes the last step, the second step becomes the second-to-last, and so on.

Chapter 10
Numbers III: Operational Representations and Their Consequences

What's in a name?
William Shakespeare (*Romeo and Juliet*)

10.1 Historical and Conceptual Introduction

Numbers are intangible idealizations. One must endow numbers with names before one can manipulate them and compute with them. Historically, we have employed a broad range of mechanisms for naming numbers.

Nicknames for "familiar" numbers. We have endowed several numbers that are associated with concrete entities with names that do not even hint at any aspect of the nature of the named number. A few examples:

- π: the ratio of the circumference of a circle to its diameter

- e: the base of so-called natural logarithms

- ϕ: the *golden ratio* (one of several word-names for ϕ) that can be observed in nature, e.g., in the leaf patterns of plants such as pineapples and cauliflowers

- Avogadro's number: a fundamental quantity in chemistry and physics. (This number-name indicates that not all numerical nicknames are single letters and that not all are "universally" known.)

Nickname-based numerals give no information about the named number: They do not help anyone (except the *cognoscenti*: the "in-crowd") *identify* the named number, and they do not help anyone manipulate—e.g., compute with—it. These names are valuable only for *cultural* purposes, not mathematical ones.

© Springer Nature Switzerland AG 2020
A. L. Rosenberg, D. Trystram, *Understand Mathematics, Understand Computing*,
https://doi.org/10.1007/978-3-030-58376-7_10

To clarify our intended message: It is the *names* of these special numbers that convey no operational information. Each of these is attached to valuable science and/or mathematics! We have already exposed some of this mathematics for the numbers e, π, and ϕ in earlier chapters.

Alphabet-based systems. Several cultures have developed systems for naming integers by using their alphabets in some manner. One such system that is still visible in European cultures comprises *Roman numerals*. One encounters these within constrained contexts, e.g., as hour markers on "classical" clocks and as datestamps on the cornerstones of official buildings. The numerals are formed from a subset of the Latin alphabet:

Letter	Numerical value		Letter	Numerical value
I	1		C	100
V	5		D	500
X	10		M	1000
L	50			

The formation rules for Roman numerals of length exceeding 2 are a bit complicated, but *roughly*, a letter to the right of a higher-valued letter augments the value of the numeral (e.g., DCL = 650, XVI = 16), while a letter to the left of a higher-valued letter lowers the value (e.g., MCM = 1900, XLIV = 44).

A rather different way to craft numerals from letters is observable in the Hebrew system. Assimilating ideas of the ancient Egyptians, Phoenicians, and Greeks, this system assigns the following values to the 22 letters of the Hebrew alphabet.

$$1, 2, 3, 4, 5, 6, 7, 8, 9, 10, 20, 30, 40, 50, 60, 70, 80, 90, 100, 200, 300, 400$$

It then forms numerals as strings of single occurrences of letters, by accumulating letters' numerical values. Numbers that are too large to be named via strings of single letter-instances often allow repeated letter instances or incorporate auxiliary words, in a mixed-mode manner similar to our writing $5,000$ as "5 thousand".

Alphabet-based systems for creating numerals are more useful than nickname-based systems: they *do* allow anyone to *identify* any named number. Indeed, one can (algorithmically) convert any Roman numeral or any Hebrew numeral to a decimal numeral for the same number. However, any reader who is familiar with alphabet-based systems will recognize a major drawback of such systems: It is *exceedingly difficult* to do any but the most trivial arithmetic using such systems' numerals. Two simple examples using Roman numerals will make our case:

- Square CC. This is, of course, trivial using, e.g., decimal numerals: an elementary school student can compute $200 \times 200 = 40,000$. But even in an early course on programming, one would not assign the general "multiply numbers using Roman numerals" problem as an early assignment.

- Subtract MCMXCVIII from MMII. Of course, the answer is IV, but one would likely determine this by converting to decimal numerals $(2002 - 1998)$.

Positional number systems. In our daily commerce, we deal almost exclusively with numerals that are formed within a *base-b positional number system*, known also as a *b-ary positional number system.* The word *"radix"* is often used in place of the word "base"; we shall use the word "base". The most common exemplars of b-ary positional number systems are:

base-2: the 2-ary or *binary*   system
base-8: the 8-ary or *octal*   system
base-10: the 10-ary or *decimal*   system
base-12: the 12-ary or *duodecimal* system
base-16: the 16-ary or *hexadecimal* system

Most of this chapter—specifically, Section 10.2—is devoted to studying -*ary* positional number systems in detail.

Systems developed to ensure special properties. This chapter has a dual purpose. Primarily, we want to study the mathematical properties of the number systems that are widely used. However, we want also to point the interested reader to a few other systems which satisfy the following criteria:

- They provide interesting and valuable mathematical lessons.
- They provide access to interesting and valuable applications.

We briefly discuss two families of positional systems that fit these criteria.

1. The family of numerals introduced in Section 10.2.2 enjoy a mathematical property that helps one study foundational aspects of computation. These are the *bijective number systems*, which are positional number systems in which distinct numerals name distinct numbers. This property can be significant in certain genres of numeral-based encodings. (Of course, -ary number systems do not enjoy this property because of leading and trailing 0s.)

2. Finally, Appendix E introduces positional systems of numerals whose digits are *signed*, plus or minus. Such systems enable a *carry-free* addition algorithm, which significantly reduces the asymptotic time-cost of performing elementary arithmetic on a computer.

   In contrast to *carry-ripple* addition algorithms, wherein adding really long numerals takes time *linear* in the lengths of numerals, and to *carry-save* addition algorithms, wherein adding really long numerals takes time *logarithmic* in the lengths of numerals, the *carry-free* addition algorithm adds all numerals in *fixed-constant* time. The time-savings of such algorithms is of significance mainly in situations such as when doing very high-precision arithmetic.

   Because their benefits are important only in specialized circumstances, signed-digit systems are usually studied only in courses on computer architecture. Also, their underlying mathematics is usually found only in specialized studies of computer arithmetic. We mention such systems here for completeness, and we relegate them to an appendix because of their high degree of specialization.

Systems formed using special families of numbers. The final genre of number system that we discuss in this chapter are those based on special families of numbers,
such as the binomial coefficients and the Fibonacci numbers. Such systems are most
useful when some property of the underlying family of numbers can be exploited to
calculational benefit. Because of such systems' specialized interest, we discuss only
one such system, in Section C.3.

10.2 Positional Number Systems

We turn now to the immensely successful family of *operational* numerals, the *b*-
ary number systems. Each such system is built upon a *number base b*, which is
an integer $b > 1$.[1] The numerals in the system are strings of *digits* from the set
$\{0, 1, 2, \ldots, \overline{b-1}\}$, often embellished with other symbols, such as a *radix point*,[2] and
sometimes a leading "+" or "−" to indicate, respectively, the denoted number's pos-
itivity or negativity.

Explanatory note.

We employ the overline notation, as in "$\overline{b-1}$", to remind ourselves that "$b-1$"
is a digit here, not a string; e.g., when $b = 10$ (the *decimal* or 10-*ary* base), $\overline{b-1}$
is the digit 9.

We begin to discuss these systems with a few examples:

- Most of our daily activities employ the number system usually called *decimal* or
  base-10, or, unusually but also correctly, 10-*ary*. This system's digits comprise
  the set $\{0, 1, 2, 3, 4, 5, 6, 7, 8, 9\}$; its radix point is called a *decimal point.*

- Because electrical and electronic circuitry are (for the most part) built using
  bistable devices—e.g., switches that are either *on* or *off*—the system most often
  employed when dealing with such circuitry and its end products (say, computers) is the *base*-2 system, which is also called *binary* or 2-ary. The digits of this
  system comprise the set $\{0, 1\}$. Each digit is called a *bit*—a contraction of *binary
  digit.*

- Because of its small repertoire of digits, the binary system's numerals are quite
  long—roughly three times longer than decimal numerals. For instance, denoting
  the base-*b* as a subscript to the numeral (a common convention):

$$32,768_{10} \;=\; 1,000,000,000,000,000_2$$

In order to make base-2 numerals easier for humans to deal with, we often aggregate small sequences of bits to form larger number bases—but still powers of
2. Two aggregations have been particularly popular:

---

[1] In rather specialized contexts one may encounter number bases that are not positive integers.

[2] In the US, the radix point is usually denoted by a period; in much of Europe, a comma is used.

- By aggregating length-3 sequences of bits, one converts base-2 numerals to *base*-8 numerals, known also as *octal* or 8-*ary* numerals; the octal digits comprise the set $\{0, 1, 2, 3, 4, 5, 6, 7\}$.

- By aggregating length-4 sequences of bits, one converts base-2 numerals to base-16 numerals, known also as *hexadecimal* or 16-*ary* numerals;the hexadecimal digits comprise the set

$$\{0, 1, 2, 3, 4, 5, 6, 7, 8, 9, \overline{10}, \overline{11}, \overline{12}, \overline{13}, \overline{14}, \overline{15}\}.$$

Note: We have written the hexadecimal digits in decimal, to make them easy to read, but we have placed overlines above the two-decimal-digit numerals "10", "11", "12", "13", "14", and "15" as a reminder that each represents a single hexadecimal digit, not a two-digit numeral.

The success of these aggregations in achieving shortened numerals is attested to by the following chain of equations:

$$32,768_{10} \;=\; 1,000,000,000,000,000_2 \;=\; 100,000_8 \;=\; 8,000_{16}$$

10.2.1 *b-ary Number Systems*

The *b-ary* number systems are by far the most commonly used positional number systems. The names of the different instances of the system—each formed by choosing a specific number base b—derive from the *Latin* name of the base number. Regrettably, from a denotational point of view, the systems associated with certain bases end with the suffix *-ary* (as in "binary") while those associated with other bases end with the suffix *-al* (as in "decimal"). The following table codifies the multiple names of the systems associated with the most commonly used bases (for mathematicians, scientists, and engineers).

Base	-*ary* System	Names
2	binary	2-ary
4	quaternary	4-ary
8	octal	8-ary
10	decimal	10-ary
12	duodecimal	12-ary
16	hexadecimal	16-ary

The formation rules for b-ary numerals. As we turn to the *formation rules* for b-ary numerals, we want to emphasize that these rules build in an essential way on the ideas relating to summing *geometric summations*. This is, therefore, a good time to review Section 6.2.2.

A b-ary numeral is a string having three segments.

1. The numeral begins with its *integer part*, which is a *finite* string of base-b digits, i.e., digits from the set[3]

$$B_b \overset{\text{def}}{=} \{0, 1, 2, \ldots, \overline{b-1}\} \tag{10.1}$$

We denote the integer-part string as: $\alpha_n \alpha_{n-1} \cdots \alpha_1 \alpha_0$.

2. The numeral continues with a single occurrence of the *radix point*, which is denoted "." in the U.S. and "," in many other countries.

3. The numeral ends with its *fractional part*, which is a string—*finite or infinite*—of base-b digits. We denote the fractional-part string as: $\beta_0 \beta_1 \beta_2 \cdots$.

Our completed numeral now has the form

$$\alpha_n \alpha_{n-1} \cdots \alpha_1 \alpha_0 \cdot \beta_0 \beta_1 \beta_2 \cdots \tag{10.2}$$

This numeral has the *numerical value*[4]

$$\text{VAL}_b(\alpha_n \alpha_{n-1} \cdots \alpha_1 \alpha_0 \cdot \beta_0 \beta_1 \beta_2 \cdots) \overset{\text{def}}{=} \sum_{i=0}^{n} \alpha_i \cdot b^i + \sum_{j \geq 0} \beta_j \cdot b^{-j} \tag{10.3}$$

For emphasis, we review:

- The numerical value of the integer part of the numeral in expression (10.2) is:

$$\text{VAL}_b(\alpha_n \alpha_{n-1} \cdots \alpha_1 \alpha_0) = \sum_{i=0}^{n} \alpha_i \cdot b^i$$

- The numerical value of the fractional part of the numeral in expression (10.2) is:

$$\text{VAL}_b(.\beta_0 \beta_1 \beta_2 \cdots) = \sum_{j \geq 0} \beta_j \cdot b^{-j}$$

- By prepending a "minus sign" or "negative sign" $(-)$ to a numeral or a number, one renders the thus-embellished entity as negative.

Note that *two types of sequences of 0s do not affect the value of the number represented by a b-ary numeral:*

- an *initial* sequence of 0s that reside to the *left* of the radix point and of all non-0 digits;

- a *terminal* sequence of 0s that reside to the *right* of the radix point and of all non-0 digits.

---

[3] Recall that "$\overline{b-1}$" represents the integer $b-1$ as a single digit.

[4] The notation "$\text{VAL}_b(x)$" in expression (10.3) denotes an operator that produces the *numerical value* of the base-b numeral x.

One consequence of this fact is that we lose no generality by insisting that every numeral have the following *normal form:*

- it begins with a finite sequence of digits,
- it then has one occurrence of the radix point,
- it ends with an infinite sequence of digits

We finish this section with an important consequence of the definition of real numbers in Section 4.5.2, in terms of the summations in expression (4.6).

Proposition 10.1 *A number n is real, i.e., belongs to the set* $\mathbb{R}$, *if, and only if, it is the numerical value of a numeral of the form in expression (10.2).*

10.2.2 A Bijective *Number System*

This section introduces a positional number system in which distinct numerals name distinct numbers. Of course, *b*-ary systems do not enjoy this property because of the value neutrality of leading 0s for integer numerals and trailing 0s for fractional numerals. The systems we discuss now are often termed *bijective* because their unique numerals for integers arise from a bijection between the integer numerals and the set $\mathbb{N}^+$ of positive integers. *The price that these systems pay for their bijectiveness is that they cannot represent the number 0; they can, however, represent all other integers.* Each bijective base-*b* system is sometimes called the *b-adic number system.* One also finds some -adic number systems being named using Greek-inspired names for the base *b*, in imitation of the Latin-inspired names of -ary systems. The most commonly encountered -adic number system is the base-2, *dyadic,* system, which is the -adic analogue of the binary system.

For any number base $b > 1$, the base-*b* bijective system's numerals are formed in exactly the same way as are the numerals of the *b*-ary number system—see expression (10.3). But the bijective system's numerals are formed using the digit-set $B'_b = \{1, 2, \ldots, b\}$, rather than the *b*-ary set B_b of (10.1). In order to lend the reader some intuition, we display the dyadic numerals that have one or two digits together with their numerical values.

Dyadic numerals & their numerical values	
Numeral	*Value*
$x = \quad 1$	$\text{DYADIC\_VAL}(x) = 1$
2	2
11	$2 + 1 = 3$
12	$2 + 2 = 4$
21	$4 + 1 = 5$
22	$4 + 2 = 6$

The preceding table indicates how the numerical values of dyadic numerals track the lexicographic order of the numerals.

Formally verifying the bijectiveness of b-adic number systems is a valuable exercise in manipulating numerals. The first appearance of bijective number systems was in [50], where the base-10 system is introduced and shown to be bijective. A proof of bijectiveness for arbitrary b-adic systems appears in [105], where the term *b-adic* is introduced. The motivating application in [105] was to the allied fields of Mathematical Logic and Computation Theory: Encoded versions of a program's computations play a central role in these theories. While crafting the required encodings using strings of symbols accomplished many of the goals of the theories, the overarching reach of the theories was fully appreciated only when mathematical logician Kurt Gödel, whom we have met earlier, showed, in 1931, that the encodings could be achieved using integers and simple arithmetic operations—see the discussions of *Gödel numbers* in the primary sources [53, 108] or in texts such as [92]. Because of -adic systems' bijectiveness, they significantly simplify certain of the central proofs of logic-based theories such as Computation Theory.

We turn now to the issue of the bijectiveness of -adic number systems. We provide a proof only for the dyadic case, $b = 2$: this case provides all of the ideas necessary for the general result.

Proposition 10.2 *Distinct b-adic numerals name distinct positive numbers.*

Proof (For dyadic integers). Our proof that distinct dyadic numerals denote distinct integers begins by exposing the largest and smallest integers that admit d-digit dyadic numerals.

Lemma 10.1. *The* smallest *integer representable by a d-digit dyadic numeral is*

$$\text{MIN\_INTEGER}_d \; = \; 2^{d-1} + 2^{d-2} + \cdots + 1 \; = \; 2^d - 1$$

The largest *integer representable by a d-digit dyadic numeral is*

$$\text{MAX\_INTEGER}_d \; = \; 2 \times (2^{d-1} + 2^{d-2} + \cdots + 1) \; = \; 2^{d+1} - 2$$

Proof (Lemma). We derive the values of both extremal integers by invoking the summation techniques in Section 6.2.2. By definition,

- The *smallest* d-digit dyadic numeral is $11 \cdots 1$ (d digits). Its value is:

$$\text{DYADIC\_VAL}_2(11 \cdots 1) \; = \; 2^{d-1} + 2^{d-2} + \cdots + 1 \; = \; 2^d - 1$$

- The *largest* d-digit dyadic numeral is $22 \cdots 2$ (d digits). Its value is:

$$\text{DYADIC\_VAL}_2(22 \cdots 2) \; = \; 2 \times (2^{d-1} + 2^{d-2} + \cdots + 1) \; = \; 2^{d+1} - 2$$
□-Lemma

We are now ready to prove the proposition. To this end, let us be given distinct dyadic numerals,

$$x = \gamma_r \gamma_{r-1} \cdots \gamma_1 \quad \text{and} \quad y = \delta_s \delta_{s-1} \cdots \delta_1$$

(By definition, each dyadic digit (γ_i or δ_j) belongs to the set $\{1, 2, \ldots, b\}$.)

(a) Assume first that $r \neq s$. With no loss of generality, say that $r = s + c$ for some $c \geq 1$. In this case, we have, by Lemma 10.1:

$$\text{DYADIC\_VAL}(x) \geq 2^r - 1 = 2^{s+c} - 1$$

while

$$\text{DYADIC\_VAL}(y) \leq 2^{s+1} - 2$$

It follows, therefore, that

$$\text{DYADIC\_VAL}(x) \geq \text{DYADIC\_VAL}(y) + 1$$

In particular, numerals x and y thus behave in consistency with the proposition.

(b) Assume next that $r = s$. Because x and y are distinct numerals, there must be a largest index $m \leq r$ such that $\gamma_m \neq \delta_m$; say, with no loss of generality, that $\gamma_m = \delta_m + c$ for some base-b dyadic digit c. We can then rewrite numeral y as

$$y = \gamma_r \cdots \gamma_{m+1} \delta_m \delta_{m-1} \cdots \delta_1$$

Invoking Lemma 10.1, we can infer the following bounds on the difference between $\text{DYADIC\_VAL}(x)$ and $\text{DYADIC\_VAL}(y)$.

$$\text{DYADIC\_VAL}(x) - \text{DYADIC\_VAL}(y)$$

$$\begin{aligned}
&= 2^{m-1} \cdot (\gamma_m - \delta_m) + 2^{m-2} \cdot (\gamma_{m-1} - \delta_{m-1}) + \cdots + (\gamma_1 - \delta_1) \\
&\geq 2^{m-1} \cdot (\gamma_m - \delta_m) - \left(2^{m-2} + 2^{m-3} + \cdots + 1\right) \\
&= 2^{m-1} \cdot (\gamma_m - \delta_m) - \left(2^m - 1\right) \\
&= 2^{m-1} \cdot (c - 1) + 1 \\
&\geq 1
\end{aligned}$$

Once again, numerals x and y behave in consistency with the proposition. $\quad\square$

We have thus verified the property that makes the b-adic system valuable in certain computational environments.

10.3 Recognizing Integers and Rationals from Their Numerals

We have provided an adequate, albeit inelegant, characterization of the real numbers: a number r is real if, and only if, it can be represented by an infinite-length numeral in a positional number system. Because every rational number—hence, also, every integer—is also a real number, every rational number and every integer can also be written as a b-ary numeral, in the form (10.2). For rational numbers and integers, we can make much stronger statements about the possible forms of their positional numerals.

10.3.1 Positional Numerals for Integers

The following result slightly alters the usual way that we write numerals for integers, in order to render explicit the familial relationship between reals and integers.

Proposition 10.3 *A real number is an integer if, and only if, it can be represented by a finite-length numeral all of whose nonzero digits are to the left of the radix point.*

Proof. The result follows from definition (10.3). In the indicated form, if any β_i is nonzero, then the numerical value of the numeral is non-integral. That is, digit β_i witnesses that the numeral has a nonzero fractional part, hence is not an integer. □

We can go beyond the simple statement of Proposition 10.3 and develop an efficient algorithm that computes the base-b numeral for an integer n via a sequence of integer divisions.

To compute a (finite) base-b numeral for integer n. If we ignore the radix point and all of the 0s to the right of it in the base-b numerals given by expression (10.2), then we see that the base-b numeral $a_d a_{d-1} \cdots a_1 a_0$ for an integer n is a polynomial

$$P(x) = a_0 + a_1 x + a_2 x^2 + \cdots + a_{d-1} x^{d-1} + a_d x^d$$

evaluated at the point $x = b$:

$$\mathrm{VAL}_b(x) = a_d b^d + a_{d-1} b^{d-1} + \cdots + a_1 b + a_0.$$

10.3.1.1 Horner's Rule and fast numeral evaluation

Since the problem of computing an integer n from a d-digit positional numeral for n reduces to the problem of evaluating a degree-d univariate polynomial, we now digress to show how to perform such evaluations efficiently. We shall adapt the efficient polynomial-evaluation scheme which we describe now to an efficient procedure for producing a d-digit base-b numeral for n.

One can clearly evaluate the degree-d univariate polynomial $\mathrm{VAL}_b(x)$ using $\Theta(d^2)$ integer multiplications. It appears at first glance that $\Theta(d^2)$ multiplications are also *necessary* for this evaluation—which would suggest that one cannot evaluate general polynomials very efficiently. However, one can develop an $O(d)$-multiplication procedure for evaluating degree-d univariate polynomials, by adapting a method of rewriting univariate polynomials which is known as *Horner's rule* (the name we shall use) or *Horner's scheme* [60]. We now describe Horner's rule on a generic degree-d polynomial, with particular detail on the case $d = 3$.

Historical note.

The challenge of minimizing the number of multiplications in common computations has been a subject of study since the earliest days of digital computers. Even with modern hardware, multiplication of large integers is a rather costly operation.

- The "standard" way of writing the polynomial $P(x)$.
  General degree d:
  $$P(x) = a_0 + a_1 x + a_2 x^2 + \cdots + a_{d-1} x^{d-1} + a_d x^d$$
  Degree 3:
  $$P(x) = a_0 + a_1 x + a_2 x^2 + a_3 x^3$$

- Rewriting $P(x)$ using Horner's rule.
  General degree d:
  $$P(x) = a_0 + x \cdot (a_1 + x \cdot (a_2 + \cdots + x \cdot (a_{d-2} + x \cdot (a_{d-1} + a_d x)) \cdots))$$
  Degree 3:
  $$P(x) = a_0 + x \cdot (a_1 + x \cdot (a_2 + a_3 x))$$

We now describe, by example, an efficient procedure for computing a base-b numeral for a given integer n, which is based on Horner's rule. The procedure iteratively divides n by b, *using Euclidean division* (see Section 4.3.2.1). We claim that the *remainder* upon each consecutive division is the next lowest-order digit in the base-b numeral for n. (We leave this verification as an exercise.)

Illustrating the procedure. We produce the base-2 (binary) numeral for $n = 143$, (binary) digit by (binary) digit.

Step	Current Quotient	Current Remainder
1.	143	1
2.	71	1
3.	35	1
4.	17	1
5.	8	0
6.	4	0
7.	2	0
8.	1	1

We have thus derived the following base-2 numeral for $n = 143$.

$$143_{10} = 10001111_2$$

10.3.1.2 Horner's Rule and fast exponentiation

If the cost of *multiplication* of long integers can be a computational concern, the cost of *exponentiation*—which is *iterated* multiplication—can be even more troubling. Thus motivated, we now develop a procedure for exponentiating which is "fast" (the popular term), in the sense that it uses unexpectedly few multiplications. Specifically, the procedure computes the power

$$b^n$$

where both b and n are positive integers ($b, n \in \mathbb{N}^+$) using a number of multiplications that is *logarithmic* in n, rather than using the *linear* number of multiplications that a direct approach would use.

Explanatory note.

Although the fast-exponentiation procedure falls strictly within the domain of *algorithmics*, we discuss it in our mathematics text because:

(a) it provides a vivid example of the importance of data *representation*;

(b) it illustrates a direct, consequential application of Horner's rule, which is thereby seen to be a practically important technique.

The setup. Our "fast" exponentiator proceeds as follows:

1. Compute the (shortest) base-2 numeral for the power n that we are raising the base b to.

    As a running example, if $n \leq 31$, then this numeral has the form

$$a_4 a_3 a_2 a_1 a_0$$

    where each $a_i \in \{0, 1\}$ and

$$n = a_4 \cdot 2^4 + a_3 \cdot 2^3 + a_2 \cdot 2^2 + a_1 \cdot 2^1 + a_0$$

2. Use Horner's Rule to rewrite the polynomial for n in the form

$$n = a_0 + 2 \cdot (a_1 + 2 \cdot (a_2 + 2 \cdot (a_3 + 2 \cdot a_4)))$$

3. Interpret the "ground-level" computation of the numeral-polynomial within the "exponent level" of base b.

    Keep in mind that this "promotes" the operations of the polynomial, in the following sense:

    - Each addition of the polynomial spawns a multiplication at the new ground level, because $b^{c+d} = b^c \cdot b^d$.

- Each multiplication of the polynomial spawns an exponentiation at the new ground level, because $b^{(c \cdot d)} = (b^c)^d$.

Using our running example, we have

$$b^n = b^{a_0 + 2 \cdot (a_1 + 2 \cdot (a_2 + 2 \cdot (a_3 + 2 \cdot a_4)))}$$

$$= b^{a_0} \times \left(b^{(a_1 + 2 \cdot (a_2 + 2 \cdot (a_3 + 2 \cdot a_4)))} \right)^2$$

$$= b^{a_0} \times \left(b^{a_1} \times \left(b^{(a_2 + 2 \cdot (a_3 + 2 \cdot a_4))} \right)^2 \right)^2$$

$$= b^{a_0} \times \left(b^{a_1} \times \left(b^{a_2} \times \left(b^{(a_3 + 2 \cdot a_4)} \right)^2 \right)^2 \right)^2$$

$$= b^{a_0} \times \left(b^{a_1} \times \left(b^{a_2} \times \left(b^{a_3} \times (b^{a_4})^2 \right)^2 \right)^2 \right)^2$$

Note that the importance of our using base-2 numerals is that *the operation of squaring thereby requires only a single multiplication*.

Two simple examples should round out this section:

$$b^{19_{10}} = b^{10011_2}$$
$$= b^{16+2+1}$$
$$= b^{16} \times b^2 \times b$$
$$= ((((b^2)^2)^2)^2) \times b^2 \times b$$

$$b^{31_{10}} = b^{11111_2}$$
$$= b^{16+8+4+2+1}$$
$$= b^{16} \times b^8 \times b^4 \times b^2 \times b^1$$
$$= ((((b^2)^2)^2)^2) \times (((b^2)^2)^2) \times ((b^2)^2) \times b^2 \times b$$

With only a few additional details, we could flesh out the preceding discussion to a formal proof of the following result.

Proposition 10.4 (Fast Exponentiation) *Given positive integers b and n, one can compute the number b^n with $O(\log n)$ multiplications.*

10.3.2 Positional Numerals for Rationals

We can completely characterize the positional numerals that represent rational numbers in terms of the following auxiliary notion. An infinite sequence S of digits is *ultimately periodic* if there exist two *finite* sequences of digits, A and B, such that S can be written in the following form (we have added spaces to enhance legibility):

$$S = A\ B\ B\ B \cdots B\ B \cdots \tag{10.4}$$

The intention here is that the sequence B is repeated *ad infinitum*.

Proposition 10.5 *A positional numeral denotes a rational number if, and only if, it is ultimately periodic.*

Proof. 1. Part 1: the "if" clause (sufficiency).
Say first that the real number r has an ultimately periodic infinite base-b numeral, as in (10.4).

Since the exact lengths of the finite sequences A and B are not germane to the argument, we can simplify notation by arbitrarily denoting r via the following normal-form numeral (spaces added to enhance legibility):

$$a_2 a_1 a_0 \ .\ b_0 b_1 \ c_0 c_1 c_2 \ c_0 c_1 c_2 \cdots c_0 c_1 c_2 \cdots$$

so that

$$A = a_2 a_1 a_0 \ .\ b_0 b_1$$
$$B = c_0 c_1 c_2$$

Explanatory note.

We elaborate on our comment about "simplifying notation".

By choosing specific lengths for sequences A and B, we cut down on the number of "ellipsis dots" we need to denote the numeral, as in "$123123 \cdots 123 \cdots$". We thereby enhance legibility.

If we now invoke the evaluation rules of (10.3), we find that

$$
\begin{aligned}
r &= \mathrm{VAL}_b(a_2 a_1 a_0 \ .\ b_0 b_1 \ c_0 c_1 c_2 \ c_0 c_1 c_2 \cdots c_0 c_1 c_2 \cdots) \\
&= \mathrm{VAL}_b(a_2 a_1 a_0) + \mathrm{VAL}_b(b_0 b_1) \cdot b^{-2} + \mathrm{VAL}_b(c_0 c_1 c_2) \cdot b^{-5} \\
&\quad + \mathrm{VAL}_b(c_0 c_1 c_2) \cdot b^{-8} + \mathrm{VAL}_b(c_0 c_1 c_2) \cdot b^{-11} + \cdots \\
&= \mathrm{VAL}_b(a_2 a_1 a_0) + \mathrm{VAL}_b(b_0 b_1) \cdot b^{-2} + \mathrm{VAL}_b(c_0 c_1 c_2) \cdot \sum_{i=1}^{\infty} b^{-2-3i}
\end{aligned} \tag{10.5}
$$

We learned in Section 6.2.2 that infinite summations such as $\sum_{i=1}^{\infty} b^{-2-3i}$ in (10.5) *converge*—meaning that *they have finite rational sums*—and we learned how to compute these sums. For the purposes of the current proof, we just accept this fact, and we denote the summation's finite rational sum by p/q.

Collecting all of this information, we find that there exist *integers m, n, p, and q* such that

$$r = m + n/b^2 + p/q = \frac{mqb^2 + nq + pb^2}{qb^2}.$$

The number r is, thus, the ratio of two integers; hence, by definition, it is rational.

2. Part 2: the " only if" clause (necessity).

Say next that the real number r is rational—specifically, say that

$$r = s + \frac{t}{q}$$

for nonnegative integers $t < q$ and s. It is only the fraction $t/q < 1$ that can produce an infinite numeral, so it suffices for us to verify only the special case

$$r = \frac{t}{q} < 1$$

of the proposition.

We prove that $r = t/q$ has an ultimately periodic infinite numeral by using *synthetic division*—the procedure taught in elementary school—to compute the ratio t/q. As we proceed, keep in mind that we are working in base b. Each of the following successive divisions produces one digit to the right of the radix point, in addition to a possible *remainder r_i* from the set $\{0, 1, \ldots, q-1\}$.

Division step	Current numeral	Current remainder
$b \cdot t \ \ = a_0 \cdot q + r_0$	$t/q \ = \ .a_0 \cdots$	$r_0 < q$
$b \cdot r_0 \ = a_1 \cdot q + r_1$	$t/q \ = \ .a_0 a_1 \cdots$	$r_1 < q$
$\vdots$	$\vdots$	$\vdots$
$b \cdot r_i \ = a_{i+1} \cdot q + r_{i+1}$	$t/q \ = \ .a_0 a_1 \cdots a_{i+1} \cdots$	$r_{i+1} < q$
$b \cdot r_{i+1} = a_{i+2} \cdot q + r_{i+2}$	$t/q \ = \ .a_0 a_1 \cdots a_{i+1} a_{i+2} \cdots$	$r_{i+2} < q$
$\vdots$	$\vdots$	$\vdots$

$$(10.6)$$

Because of the possible values the remainders r_j can assume, no more than q of the divisions in the (infinite) system (10.6) are distinct. (*This is an application of the pigeonhole principle (Section 2.2.3).*) Because of the way the system proceeds, once we have encountered two remainders, say, r_i and r_{i+k}, that are equal—i.e., $r_i = r_{i+k}$—we must thenceforth observe periodic behavior:

$$
\begin{aligned}
r_i &= r_{i+k} = r_{i+2k} = r_{i+3k} = \cdots \\
r_{i+1} &= r_{i+k+1} = r_{i+2k+1} = r_{i+3k+1} = \cdots \\
\vdots \quad & \quad \vdots \qquad\quad \vdots \qquad\quad \vdots \\
r_{i+k-1} &= r_{i+2k-1} = r_{i+3k-1} = r_{i+4k-1} = \cdots
\end{aligned}
$$

This periodicity will engender periodicity in the digits of r's base-b numeral:

$$[\text{INITIAL SEGMENT}][a_i a_{i+1} \cdots a_{i+k-1}][a_i a_{i+1} \cdots a_{i+k-1}] \cdots [a_i a_{i+1} \cdots a_{i+k-1}] \cdots$$

We are, thus, observing the claimed ultimately periodic behavior in r's base-b numeral.
This completes the proof. □

We end this section by illustrating the process of generating numerals for rationals via synthetic division. We employ the fraction $t/q = 4/7$ and base $b = 10$.

Division step	Current numeral	Current remainder
$10 \cdot 4 = 5 \cdot 7 + 5$	$4/7 = .5 \cdots$	5
$10 \cdot 5 = 7 \cdot 7 + 1$	$4/7 = .57 \cdots$	1
$10 \cdot 1 = 1 \cdot 7 + 3$	$4/7 = .571 \cdots$	3
$10 \cdot 3 = 4 \cdot 7 + 2$	$4/7 = .5714 \cdots$	2
$10 \cdot 2 = 2 \cdot 7 + 6$	$4/7 = .57142 \cdots$	6
$10 \cdot 6 = 8 \cdot 7 + 4$	$4/7 = .571428 \cdots$	4
$\vdots$	$\vdots$	$\vdots$

The remainder 4 in the last illustrated division step cycles us back to the initial division step, where the "4" came from the numerator of the target fraction. This repetition signals that the entire process cycles from this point on. In other words, we have determined that

$$\frac{4}{7} = .[571428]\,[571428]\,[571428]\,\cdots$$

Propositions 10.3 and 10.5 show us that the three sets of numbers we have defined, the sets $\mathbb{Z}$, $\mathbb{Q}$, and $\mathbb{R}$, are a strictly nested progression:

$$\mathbb{Z} \subset \mathbb{Q} \subset \mathbb{R}$$

Verbally:

- *Every integer is a rational number; there exist rationals that are not integers.*

- *Every rational number is a real number; there exist reals that are not rational.*

Those interested in the (philosophical) foundations of mathematics might quibble about the verbs "is" and "are" in the highlighted sentences, but for the purpose of "doing mathematics", we can accept the sentences as written.

10.4 Sets That Are Uncountable, Hence, "Bigger than" $\mathbb{Z}$ and $\mathbb{Q}$

This section completes the topic of the cardinalities of infinite sets, which we began in Section 8.2.3.1. We discussed *countable* sets in that section; we turn here to the complementary topic of *uncountable* sets.

10.4.1 The Set of Binary Functions $\mathbb{F}$ Is Uncountable

Let $\mathbb{F}$ denote the set of all functions from $\mathbb{N}$ to $\{0, 1\}$. The main result of this section establishes the uncountability of the set $\mathbb{F}$. We thereby have an argument that the infinitude of this set is *of a higher order* than the infinitude of the set of integers. In fact, Georg Cantor used variants of this result as the base of his study of orders of infinity.

This section is dedicated to proving the following result.

Proposition 10.6 *The set* $\mathbb{F}$ *of binary-valued functions from* $\mathbb{N}$ *into* $\{0, 1\}$ *is not countable. In particular, there is no injection* $f : \mathbb{F} \to \mathbb{N}$.

In the next subsection, we prove the qualitatively similar but technically more complicated companion of Proposition 10.6 which establishes that the set $\mathbb{R}$ of real numbers is also uncountable.

Proof. Our multi-step proof of Proposition 10.6 shows that assuming the countability of $\mathbb{F}$ leads to a contradiction. Being built around Georg Cantor's revolutionary *diagonalization construction*, the upcoming argument—and its companion argument about the set $\mathbb{R}$—provides the most sophisticated proof by contradiction in our text. The reader might want to review the reasoning underlying such argumentation, in Section 2.2.2, as a "warm-up".

10.4.1.1 Plotting a strategy to prove uncountability

Invoking the definition of countability, our proof begins with the assumption that $|\mathbb{F}| \le |\mathbb{N}|$ and demonstrates that this assumption leads to a contradiction. In fact, we can simplify our goal by recasting the problem in terms of *bijections*. We begin with two simplifying lemmas.

Lemma 10.2. *There exists an injection* $f : \mathbb{N} \to \mathbb{F}$; *i.e.,* $|\mathbb{N}| \le |\mathbb{F}|$.

Verification. For each nonnegative integer $n \in \mathbb{N}$, define the function $g_n : \mathbb{N} \to \{0, 1\}$ as follows: for each integer $k \in \mathbb{N}$,

$$g_n(k) \;=\; \textbf{if } [k = n] \textbf{ then } 1 \textbf{ else } 0$$

Clearly, specifying the integer n uniquely identifies the function f_n. This means that the defined correspondence specifies an injection from $\mathbb{N}$ into $\mathbb{F}$.   □

Lemma 10.3. *If there exists an injection* $g : \mathbb{F} \to \mathbb{N}$, *then there exists a bijection* $h : \mathbb{N} \leftrightarrow \mathbb{F}$. *In other words, if* $|\mathbb{F}| \le |\mathbb{N}|$, *then* $|\mathbb{N}| = |\mathbb{F}|$.

Verification. This is an immediate consequence of the Schröder-Bernstein Theorem (Theorem 3.2).   □

It may surprise you that our proof is simplified if we replace our initial assumption

$$|\mathbb{F}| \leq |\mathbb{N}|$$

by the stronger assumption

$$|\mathbb{F}| = |\mathbb{N}|,$$

but Cantor's diagonalization argument deals quite gracefully with the latter assumption. (There is a methodological lesson here which is as true for mathematicians as for carpenters: *Know your tools!*)

A. Seeking a bijection $g : \mathbb{N} \leftrightarrow \mathbb{F}$

We assume, for contradiction, that there exists a bijection $g : \mathbb{N} \leftrightarrow \mathbb{F}$. As part of this two-way mapping, there exists an *injection*

$$h : \mathbb{N} \rightarrow \mathbb{F}$$

We view h as an *enumeration*—i.e., an ordered listing—of the elements of $\mathbb{F}$. Specifically, for each integer $k \in \mathbb{N}$, we can think of $h(k)$ as the "kth binary-valued function in the set $\mathbb{F}$". We thereby view h as producing an "infinite-by-infinite" matrix Δ of bits, whose kth row is the infinite string of bits that is the characteristic vector ξ of the function $h(k)$. Recall that ξ is an infinite binary vector whose ith element specifies the value of h at argument i; i.e., $h(i) = \xi_i$. Let us visualize Δ:

$$
\Delta \;=\;
\begin{array}{l}
\delta_0 = \delta_{0,0}\ \delta_{0,1}\ \delta_{0,2}\ \delta_{0,3}\ \delta_{0,4}\ \cdots \\
\delta_1 = \delta_{1,0}\ \delta_{1,1}\ \delta_{1,2}\ \delta_{1,3}\ \delta_{1,4}\ \cdots \\
\delta_2 = \delta_{2,0}\ \delta_{2,1}\ \delta_{2,2}\ \delta_{2,3}\ \delta_{2,4}\ \cdots \\
\delta_3 = \delta_{3,0}\ \delta_{3,1}\ \delta_{3,2}\ \delta_{3,3}\ \delta_{3,4}\ \cdots \\
\delta_4 = \delta_{4,0}\ \delta_{4,1}\ \delta_{4,2}\ \delta_{4,3}\ \delta_{4,4}\ \cdots \\
\vdots\quad \vdots\quad \vdots\quad \vdots\quad \vdots\quad \vdots\quad \ddots
\end{array}
$$

We summarize, to reinforce our strategy:

- Each row of Δ consists of the characteristic vector of a function in the set $\mathbb{F}$— which can be thought of as a *name* for the function.

- Each function in the set $\mathbb{F}$ contributes its characteristic vector as a row of Δ.

We can, thus, view the successive rows of Δ, $h(0)$, $h(1)$, $\ldots$, as an enumeration of all of the functions in the set $\mathbb{F}$. In other words, we can view Δ as "containing" each function in $\mathbb{F}$ precisely once.

B. Every bijection "misses" some function from $\mathbb{F}$

We are finally poised to find the contradiction to our assumption that $|\mathbb{F}| \leq |\mathbb{N}|$. Specifically, we define from Δ an infinite bit-string

$$\Psi = \psi_0\, \psi_1\, \psi_2\, \psi_3\, \psi_4 \cdots,$$

that *does not* appear in Δ. For each index $i \in \mathbb{N}$, we define the ith (binary) digit ψ_i of Ψ from the ith *diagonal (binary) digit*[5] $\delta_{i,i}$ of Δ in the following manner.

$$\psi_i \overset{\text{def}}{=} \overline{\delta}_{i,i} = \left[\textbf{if}\ [\delta_{i,i} = 0]\ \textbf{then}\ 1\ \textbf{else}\ 0 \right]$$

The important feature of the definition is the following.

Lemma 10.4. *The bit-string Ψ does not occur as a row of Δ.*

Verification. This is true because the bit-string Ψ differs from each row k of Δ in the kth position; i.e., $\psi_k \neq \delta_{k,k}$.   □

10.4.1.2 The denouement: There is no bijection $h : \mathbb{N} \leftrightarrow \mathbb{F}$

The infinite binary string Ψ differs from every row of Δ, even though Ψ is the characteristic vector of a binary valued function, i.e., a member of $\mathbb{F}$. Therefore, Δ *does not* contain as a row *every* element of $\mathbb{F}$, i.e., the characteristic vector of every binary-valued function. But this contradicts Δ's assumed defining characteristic!

Where could we have gone wrong? Every step of our argument, save one, is backed up by a proof—so the one step that is not so bolstered must be the link that has broken the argument. This one unsubstantiated step is our assumption that the set $\mathbb{F}$ is countable. Since this assumption has led us to a contradiction, we must conclude that the set $\mathbb{F}$ is *not* countable!   □

10.4.2 $\oplus \mathbb{R}$ *Is Uncountable*

This section, in which we establish the uncountability of the set $\mathbb{R}$ of real numbers, is a companion to Section 10.4. We thereby have an argument that the infinitude of the set of real numbers is *of a higher order* than the infinitude of the set of integers. We accomplish this by developing a proof of the following result.

Proposition 10.7 *The set $\mathbb{R}$ of real numbers is not countable. In particular, there is no injection $g : \mathbb{R} \to \mathbb{N}$.*

---

[5] Our use of Δ's diagonal digits in this definition is the origin of the term "*diagonalization argument*" to describe this proof and its intellectual kin.

The flow of the proof of Proposition 10.7 follows the flow of the proof of Proposition 10.6 but differs in one crucial technical step. When we specified the "diagonal" bit-string Ψ in the proof of Proposition 10.6, it was totally clear that the function $f \in \mathbb{F}$ specified by Ψ did not appear in matrix Δ's allegedly complete enumeration of $\mathbb{F}$. As we now craft an analogue of Ψ that works for real numbers rather than binary-valued functions, we must exercise much more care—because we have to deal with redundancy in real numerals. The delicate part of the present argument occurs in Lemma 10.7.

Proof. We simplify the exposition by making extensive use of the proof of Proposition 10.6. Our proof begins with the following lemmas.

Lemma 10.5. *There exists an injection* $f : \mathbb{N} \to \mathbb{R}$*; i.e.,* $|\mathbb{N}| \leq |\mathbb{R}|$*.*

Verification. For each nonnegative integer $n \in \mathbb{N}$, let $\text{NAME}_b(n)$ denote the shortest base-b numeral for n, i.e., a numeral with no leading 0s. The mapping that associates each $n \in \mathbb{N}$ with the infinite string

$$\text{NAME}_b(n) \, . \, 00 \cdots$$

is an injection from $\mathbb{N}$ into $\mathbb{R}$. To see this, let us be given a real numeral that has only 0s to the right of its radix point. We produce the integer n by: (1) stripping the numeral of its radix point and all 0s to the right of the point; (2) evaluating the remaining string of digits, which is $\text{NAME}_b(n)$, according to Section 4.5.2's rules for evaluating integer numerals. □

Lemma 10.6. *If there exists an injection* $g : \mathbb{R} \to \mathbb{N}$*, then there exists a bijection* $h : \mathbb{N} \leftrightarrow \mathbb{R}$*. In other words, if* $|\mathbb{R}| \leq |\mathbb{N}|$*, then* $|\mathbb{N}| = |\mathbb{R}|$*.*

Verification. This is an immediate consequence of the Schröder-Bernstein Theorem (Theorem 3.2). □

We now make some technical assumptions that simplify our proof without weakening the result.

We henceforth focus on the proper subset of $\mathbb{R}$ *that consists of the set* $\mathbb{R}_{(0,1)}$ *of real numbers between* 0 *and* 1*.*

This simplifies our argument because every real number in the set $\mathbb{R}_{(0,1)}$ has an infinite decimal numeral of the form

$$0 \, . \, \delta_0 \delta_1 \delta_2 \delta_3 \cdots$$

where each δ_i is a decimal digit: $\delta_i \in \{0,1,2,3,4,5,6,7,8,9\}$. We employ *decimal* numerals for our eventual convenience in Lemma 10.7. We could equally conveniently use any other number $b > 2$. What we really want to avoid is the lengthy clerical detail that we would need if we were to employ the base $b = 2$. The reader should attempt the argument for base $b = 2$ as we proceed with base $b = 10$.

Of course, if we prove that the proper subset $\mathbb{R}_{(0,1)} \subset \mathbb{R}$ is uncountable, then it will follow that $\mathbb{R}$ is uncountable. (In informal terms which can be made formal, any putative injection $f : \mathbb{R} \to \mathbb{N}$ "contains" an injection $f_{(0,1)} : \mathbb{R}_{(0,1)} \to \mathbb{N}$.)

Now we begin to seek a bijection $g : \mathbb{N} \leftrightarrow \mathbb{R}_{(0,1)}$.

Assume, for contradiction, that the targeted bijection g exists. By definition, g consists of two *injections*:

- one—call it h—which maps $\mathbb{N}$ into $\mathbb{R}_{(0,1)}$ in a one-to-one fashion;
- the inverse, h^{-1}, of h, which maps $\mathbb{R}_{(0,1)}$ into $\mathbb{N}$ in a one-to-one fashion.

We view h as an *enumeration* of the elements of $\mathbb{R}_{(0,1)}$. Specifically, for each integer $k \in \mathbb{N}$, we can think of $h(k)$ as a base-10 numeral for the "kth number in the set $\mathbb{R}_{(0,1)}$". We thereby view h as producing an "infinite-by-infinite" matrix $\Delta^\star$ of decimal digits, whose kth row is the infinite string of decimal digits $\text{NAME}_{10}(h(k))$.

Explanatory note.

Because the present proof, which focuses on the set $\mathbb{R}$, so closely tracks the proof of Proposition 10.6, which focuses on the set $\mathbb{F}$, we make a notational change which we hope will keep the reader oriented.

	The proof of Proposition 10.6	The current proof
The $\infty \times \infty$ matrix	Δ	$\Delta^\star$
Rows of the matrix	δ_i	$\delta_i^\star$
Entries of the matrix	$\delta_{i,j}$	$\delta_{i,j}^\star$
The diagonal string	Ψ	$\Psi^\star$
Entries of the string	ψ_i	$\psi_i^\star$

Let us visualize $\Delta^\star$:

$$\Delta^\star = \begin{array}{l} \delta_0^\star = \delta_{0,0}^\star \ \delta_{0,1}^\star \ \delta_{0,2}^\star \ \delta_{0,3}^\star \ \delta_{0,4}^\star \ \cdots \\ \delta_1^\star = \delta_{1,0}^\star \ \delta_{1,1}^\star \ \delta_{1,2}^\star \ \delta_{1,3}^\star \ \delta_{1,4}^\star \ \cdots \\ \delta_2^\star = \delta_{2,0}^\star \ \delta_{2,1}^\star \ \delta_{2,2}^\star \ \delta_{2,3}^\star \ \delta_{2,4}^\star \ \cdots \\ \delta_3^\star = \delta_{3,0}^\star \ \delta_{3,1}^\star \ \delta_{3,2}^\star \ \delta_{3,3}^\star \ \delta_{3,4}^\star \ \cdots \\ \delta_4^\star = \delta_{4,0}^\star \ \delta_{4,1}^\star \ \delta_{4,2}^\star \ \delta_{4,3}^\star \ \delta_{4,4}^\star \ \cdots \end{array}$$

$$\vdots \quad \vdots \quad \vdots \quad \vdots \quad \vdots \quad \vdots \quad \ddots$$

We summarize, for emphasis:

- Each row of $\Delta^\star$ consists of a decimal numeral for a number in the set $\mathbb{R}_{(0,1)}$.

- Each number in the set $\mathbb{R}_{(0,1)}$ contributes at least one numeral to the rows of $\Delta^\star$.

- A number may contribute more than one numeral because of an artifact of positional number systems, which is exemplified by equalities such as

$$0.25 = 0.24999\cdots = \cdots 0.25 = 0.2500\cdots$$

and their kin, which exploit salient characteristics of base-10 numerals.

We can, thus, view the successive rows of $\Delta^\star$:

$$h(0) = \Delta_0^\star, \quad h(1) = \Delta_1^\star, \quad h(2) = \Delta_2^\star, \ldots$$

as an ordered listing (with repetitions) of all of the real *numbers* in the set $\mathbb{R}_{(0,1)}$.

We show now that every bijection $h : \mathbb{N} \leftrightarrow \mathbb{R}_{(0,1)}$ "misses" some real $x \in \mathbb{R}_{(0,1)}$.

We are finally poised to find the contradiction to our assumption that $|\mathbb{R}_{(0,1)}| \leq |\mathbb{N}|$. Specifically, we define from $\Delta^\star$ an infinite decimal numeral

$$\Psi^\star = \psi_0^\star \; \psi_1^\star \; \psi_2^\star \; \psi_3^\star \; \psi_4^\star \cdots,$$

that *does not* appear as a row of $\Delta^\star$, even though $\text{VAL}_{10}(\Psi^\star)$ belongs to $\mathbb{R}_{(0,1)}$. For each index $i \in \mathbb{N}$, we define the ith digit $\psi_i^\star$ of $\Psi^\star$ from the ith *diagonal digit* $\Delta_{i,i}^\star$ of $\Delta^\star$, in the following manner.

$$\psi_i^\star \overset{\text{def}}{=} \begin{cases} 0 & \text{if } \delta_{i,i}^\star > 5 \\ 9 & \text{if } \delta_{i,i}^\star \leq 5 \end{cases}$$

The important feature of the definition is the following.

Lemma 10.7. *The numeral $\Psi^\star$ does not occur as a row of $\Delta^\star$.*

Verification. Focus on an arbitrary row of $\Delta^\star$, say row k, and on the numeral, $\delta_k^\star$, in that row.

If $\Delta_{k,k}^\star > 5$ then $\text{NAME}_{10}(\delta_k^\star) - \text{NAME}_{10}(\psi^\star) > 4 \cdot 10^{-k}$

If $\Delta_{k,k}^\star \leq 5$ then $\text{NAME}_{10}(\psi^\star) - \text{NAME}_{10}(\delta_k^\star) > 4 \cdot 10^{-k}$

In either case, we have $\text{NAME}_{10}(\Psi^\star) \neq \text{NAME}_{10}(\delta_k^\star)$ so that $\Psi^\star$ does not appear as row k of $\Delta^\star$. Since $k \in \mathbb{N}$ is an arbitrary row-index of $\Delta^\star$, we conclude that $\Psi^\star$ does not occur as any row of $\Delta^\star$. □

We are now ready for the denouement of our argument: There is no bijection $h : \mathbb{N} \leftrightarrow \mathbb{R}$.

Because the infinite decimal string $\Psi^\star$ differs from every row of $\Delta^\star$, even though $\text{NAME}_{10}(\Psi^\star)$ is a numeral for some number in $\mathbb{R}_{(0,1)}$, we have shown that $\Delta^\star$ *does not* contain as a row a numeral for *every* number in $\mathbb{R}_{(0,1)}$. But this fact contradicts $\Delta^\star$'s assumed defining characteristic!

Where could we have gone wrong? Once again, we find that every step of our argument, save one, is backed up by a proof—so the one step that is not so bolstered must be the link that has broken. This one unsubstantiated step is our assumption that the set $\mathbb{R}_{(0,1)}$ is countable.

Since this assumption has led us to a contradiction, we must conclude that the set $\mathbb{R}_{(0,1)}$, and hence the set $\mathbb{R}$, is *not* countable! □

10.5 Scientific Notation

There is a familiar game in which one is challenged to guess how many beans there are in a jar. The wild ranges of guesses that players make indicate eloquently what is one of the main starting points in the popular-mathematics book *Innumeracy* [86]: While we "know" a lot about even *very* large numbers and *very* small numbers, we often lack *operational* command of the numbers. This fact can be illustrated in at least two ways.

1. Our (lack of) ability to compare the magnitudes of numbers, especially ones that are either very large or very small.

- Can you compare the probabilities of a person's being hit by lightning (say, in Mexico City) or by a car (say, crossing Seventh Avenue in Times Square at 3pm)?

- Do you know whether you have more hairs on your body than there are grains of sand on the beach at Ipanema, Brazil?

- Do you know whether there were more humans alive on December 31, 1999, than had lived from the moment of the Big Bang until December 18, 1945?

- Can you compare the number of rhinoviruses that can populate a square of side-length 1mm with the number of stars visible on a clear night at the summit of Mount Everest?

2. The ability to delineate "how much information" a number tells us: Many of us know—or can calculate—that (in some sense) the distance between Earth and its closest star, the Sun, is, very roughly,

$$93,000,000 \text{ miles}$$
$$148,800,000 \text{ km}$$
$$491,040,000,000 \text{ feet}$$
$$5,892,480,000,000 \text{ inches}$$

An aside.

One might read an argument in favor of the metric system into the preceding listing of "miles" and "feet" and "inches", whose interrelationships require a lexicon, in contrast to the singular "km" whose relationships to "cm" and "meter" are transparent.

All of these numbers are coarse approximations. In some sense, they all convey exactly the same information, since all are obtained from the first number (the number of miles to the Sun) by simple scaling. Yet, while the first number projects a modest two (decimal) digits of accuracy, the others project, respectively, four digits, five digits, and six digits. Do all of these numbers convey the same (level of) truth?

Scientists and pedagogues and philosophers have grappled throughout time with the problems engendered by numbers that are *very large* or *very small*.[6] One ingenious approach within the domain of astronomy has been to establish a new standard unit of distance to express the *very* large distances from Earth to stars beyond our solar system: A *light-year* is the distance that light travels in an Earth-year, roughly 9.4607×10^{12} km. By using this measure, one can describe enormous (well, astronomical) numbers without unwarranted appearances of inflated accuracy. The notion of light-year plays an important role for astronomy, but it does not port gracefully to other domains, for two reasons: (1) The use of the speed of light as a frame of reference has no meaning when one is, for instance counting grains of sand or numbers of viruses. (2) The scaling factor inherent in a light-year is not appropriate for other domains. The widely accepted general alternative to a new scaling unit is *scientific notation.*

Within scientific notation, one specifies an arbitrary number of arbitrary magnitude via a *rational approximation* of the form

$$.\beta_0\beta_1\beta_2\cdots\beta_{a-1} \times b^s$$

The interpretation is that

- $\beta_0\beta_1\beta_2\cdots\beta_{a-1}$ represents the a base-b *digits of accuracy* that are warranted by the accuracy of one's level of knowledge about the number being specified.

- b^s is the base-b *scaling factor* that adjust the digits of accuracy relative to the radix point.

Within this system of specification, we thus have

.93	$\times 10^8$	miles from Earth to the Sun
.94607	$\times 10^{13}$	kilometers traveled by light in an Earth-year
.31415	$\times 10$	value of π to five digits of accuracy
.166	$\times 10^{-23}$	grams of weight of a proton, to three digits of accuracy

10.6 Exercises: Chapter 10

Throughout the text, we mark each exercise with 0 or 1 or 2 occurrences of the symbol $\oplus$, as a rough gauge of its level of challenge. The 0-$\oplus$ exercises should be accessible by just reviewing the text. We provide *hints* for the 1-$\oplus$ exercises; Appendix H provides *solutions* for the 2-$\oplus$ exercises. Additionally, we begin each exercise with a brief explanation of its anticipated value to the reader.

1. **An application of the Fundamental Theorem of Arithmetic**
   LESSON: Enhance understanding of the implications of the Fundamental Theorem of Arithmetic (Theorem 8.1)

---

[6] See, e.g., Blaise Pascal's essay "The Two Infinities", in his *Pensées*.

Prove that the following function f is a bijection between $\mathbb{N}^+ \times \mathbb{N}^+$ *and* $\mathbb{N}^+$.

$$(\forall \langle x,y \rangle \in \mathbb{N}^+ \times \mathbb{N}^+) \quad f(x,y) \stackrel{\text{def}}{=} 2^{x-1} \cdot (2y-1)$$

2. **The average length of a carry in a binary (or ternary, or ...) counter**
   LESSON: Enhance understanding of the role of summations in positional number representations

   Say that you are adding from 1 to n, in increments of 1, using a binary counter which employs a carry-ripple adder. Each time you increment the counter, there is a *carry*. These carries have varying lengths; for instance, when $n = 32 = 100000_2$, the carry-lengths range from 0—whenever you increment an even integer—to 5—when you increment $31 = 11111_2$ to achieve $32 = 100000_2$.
   Prove that the average carry as you proceed from 1 to n has length 2.

   Hint. Use techniques from Chapter 6 to quantify the lengths of carries engendered by increments to numbers whose base-2 representations end with a 0, with 01, with 011, and so on. You will thereby generate an infinite series which converges, with the sum 2.

3. ⊕⊕ **The Josephus Problem**
   LESSON: Experience with formulating a problem mathematically, and then solving the problem

   The *Josephus Problem* appears in the 1894 book *Récréations mathématiques* by French mathematician François Édouard Anatole Lucas (usually known as Édouard Lucas) [78]. Lucas attributed the problem to a story by the first-century Jewish historian Flavius Josephus during the Jewish-Roman war of his era.

   In the story, Flavius was among a band of 41 rebels trapped in a cave by the Roman army. Preferring suicide to capture, the rebels decided to form a circle and to proceed around the circle, killing every second living person. The last remaining person would kill himself. According to the legend, Flavius decided to use his mathematical skills to avoid dying. Specifically, he calculated where he should stand in the circle in order to become the last survivor.

   The mathematical Josephus Problem. We are given a circle with the numbers $1, \ldots, n$ inscribed along its perimeter. Proceeding clockwise around the circle, beginning from 1, we erase every second not-yet-erased number. We end when there is only one *survivor*, i.e., one not-erased number. We denote this survivor by $J(n)$, in honor of Josephus.

   The aim of this exercise is to determine the survivor $J(n)$ as a function of n.[7]

   Because determining $J(n)$ is not straightforward, we provide a guided walk toward a solution.

   The process's early erasures, depicted in Fig 10.1, should help you start your solution for $J(n)$.

---

[7] Not surprisingly, there are generalizations of this problem that erase every third number or every fourth number, ...

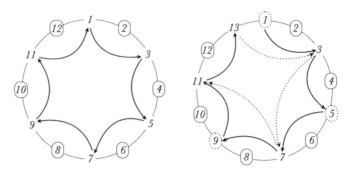

Fig. 10.1 (Left) First round of the Josephus erasure procedure for $n = 12$. (Right) First two rounds of the procedure when $n = 13$; the first round is depicted by solid lines, the second by dotted lines. In both figures, the encircled integers are the ones that have been removed

The following sequence of results provides a roadmap to the solution.

We analyze *rounds* of the process—sequences of elementary steps that return to the initial position, 1. If position 1 has been erased, then its role is assumed by the then-smallest survivor.

Easily, the first round of the process takes $\lceil n/2 \rceil$ steps. Each subsequent round then takes a number of steps that is "half" the number of its preceding round. We place "half" in quotes to emphasize the required rounding-up of each halving.

Fig. 10.2, which describes the final rounds of the process, lends intuition for the number of rounds before $J(n)$ stands alone.

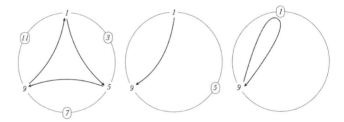

Fig. 10.2 The final rounds of the Josephus erasure process when $n = 12$

The following four-part proposition now leads to the solution.

Proposition 10.8 *The value of $J(n)$, as determined by n:*

Condition on n	Value of $J(n)$
(a) *For all n*	$J(n)$ *is odd.*
(b) *n is even; i.e., $n = 2m$*	$J(2m) = 2J(m) - 1$
(c) *n is odd; i.e., $n = 2m + 1$*	$J(2m+1) = 2J(m) + 1$
(d) *$n = 2^m + k$, with $k < 2^m$*	$J(2^m + k) = 2k + 1$

Your ultimate solution for $J(n)$ will emerge from analyzing what the preceding propositions yield as the value of $J(n)$.

Provide a closed-form expression for $J(n)$ in terms of n's base-2 numeral.

Hint. Consider the binary *numerals* for $n = 2^m + k$ and k. We first note that, *numerically*,

$$n = 2^m b_m + 2^{m-1} b_{m-1} + \cdots + 2b_1 + b_0$$
$$k = 2^{m-1} b_{m-1} + \cdots + 2b_1 + b_0$$

where each bit $b_i \in \{0, 1\}$.

4. **A "trick" for squaring integers whose decimal numeral ends in** 5

   LESSON: Solving a fun problem by relating numbers to their numerals.

   The uninitiated might misperceive the topic of this problem as an amusing "trick". Of course, we would not present the "trick" if there were not a serious mathematical message behind it. Hopefully you will be inspired to design your own "tricks".

   Say that someone presents you with an integer n, by means of two-digit decimal numeral that ends in 5. You can compute n^2 virtually instantaneously.

   For instance: if someone says "$n = 75$",
       then you can instantly respond "$n^2 = 5625$";
   if they say "$m = 95$",
       then you can instantly respond "$m^2 = 9025$".

   The "trick" that leads to your responses underlies the following proposition.

   Proposition 10.9 *If the integer n has the decimal numeral $\delta 5$, where $\delta \in \{1, 2, \ldots, 9\}$, then*

   $$n^2 = 100 \times \delta \times (\delta + 1) + 25$$

   (Note how our two examples satisfy this rule.)

   Prove Proposition 10.9.

   Hint. Review how numerals represent numbers.

5. ⊕⊕ **An alternative to Horner's Rule**

   LESSON: Experience with operation-counting and asymptotics

   In Section 10.3.1.1, we described Horner's Rule, a method for calculating polynomials faster than the common way of writing polynomials would suggest. This exercise is devoted to an alternative streamlined polynomial evaluation procedure, called *Estrin's method*, after its inventor, Gerald Estrin [48].

   Estrin's method begins with a polynomial of degree d with real coefficients:

   $$P(x) = a_0 + a_1 x + a_2 x^2 + \cdots + a_{d-1} x^{d-1} + a_d x^d$$

   To avoid complicated expressions, we apply the method to the degree-$(d = 7)$ polynomial:

$$P_7(x) \;=\; a_0 + a_1x + a_2x^2 + a_3x^3 + a_4x^4 + a_5x^5 + a_6x^6 + a_7x^7$$

We begin by rewriting P as follows, via repeated factorizations.

$$P(x) \;=\; a_0 + x\cdot(a_1 + x\cdot(a_2 + \cdots + x\cdot(a_{d-2} + x\cdot(a_{d-1} + a_dx))\cdots))$$

Applied to $P_7(x)$, this rewriting yields:

$$P_7(x) \;=\; a_0 + a_1x + x^2\left(a_2 + a_3x\right) + x^4\left(a_4 + a_5x + x^2\left(a_6 + a_7x\right)\right)$$

We then introduce the auxiliary recursive expressions.

$$C_i^{(0)} = a_i + xa_{i+1}$$

$$\vdots$$

$$C_i^{(n)} = C_i^{(n-1)} + x^{2^n}C_{i+2^n}^{(n-1)}$$

The method is completed by expressing $P(x)$ in terms of the auxiliary $C_i^{(k)}$
Example:

$$P_7(x) = C_0^{(0)} + x^2\,C_2^{(0)} + x^4\left(C_4^{(0)} + x^2\,C_6^{(0)}\right)$$
$$= C_0^{(1)} + x^4\,C_4^{(1)}$$

To do:

a. *Write $P(x)$ using the auxiliary expressions C_i. To simplify this task, say that the degree d is such that $d+1$ is a power of 2.*

b. *Determine how many additions and multiplications are required to evaluate $P(x)$.*

6. **A simple, good, rational approximation to** e
   LESSON: Enhance understanding of the two main ways of representing rational numbers

   It is well known that the decimal expansion of Euler's constant e is infinite and non-periodic and that it begins

   $$e = 2.718281828\cdots$$

   Therefore, Proposition 10.5 assures us that the infinite ultimately periodic decimal expansion (with added spaces to highlight the periodicity)

   $$2.7\ 1828\ 1828\ 1828\ \cdots\ 1828\ 1828\ \cdots$$

   represents a rational number r which agrees with e for ten decimal places.

Find a representation of r as a fraction. In more detail, find positive integers p and q such that

$$\frac{p}{q} = r = 2.7\ 1828\ 1828\ 1828 \cdots 1828\ 1828 \cdots$$

7. ⊕⊕ **The uniqueness of Zeckendorf numerals**
LESSON: Experience with Fibonacci numbers and with number representations

This problem refers to Zeckendorf numerals, which are representations of positive integers by sums of Fibonacci numbers. The framework appears in detail in Section C.3.

Prove that every positive integer has a unique Zeckendorf numeral.

Chapter 11
The Art of Counting: Combinatorics, with Applications to Probability and Statistics

> *Let me count the ways*
> Elizabeth Barrett Browning (*Sonnets from the Portuguese* 43)
>
> *Once you eliminate the impossible, whatever remains,*
> *no matter how improbable, must be the truth.*
> Arthur Conan Doyle (*Sherlock Holmes stories*)

We have named this chapter in honor of the great German mathematician Gottfried Leibniz, whose 1666 doctoral thesis[1] gave us the now-common phrase, "the art of counting". The word "counting" in this context must be understood much more broadly than in the vernacular: In days past, the phrase encompassed much of the field known nowadays as *combinatorics*.

The chapter is devoted to the basics of three closely related subfields of mathematics: combinatorics, probability theory, and the mathematical underpinnings of statistics. These subfields are treated within the text in an order that allows each topic to flow gracefully from its predecessors. The four major sections in the chapter unfold as follows.

Combinatorics (Section 11.1). Our introduction to combinatorics can be viewed in many ways as a return to Chapter 3's study of sets, especially finite sets. Indeed, the first topics we cover involve looking at a set S "from the inside", to determine how many elements set S contains, especially elements of specific designated types.

Combinatorial probability (Section 11.2). The tools we develop for determining the cardinalities of finite sets give us access to the important (and, often, fun!) field

---

[1] G.W. Leibniz (Leibnitz) (1666): *Dissertatio de Arte Combinatoria. Sämtliche Schriften und Briefe*. Akademie Verlag, Berlin.

While Leibniz is best known (at least among students of mathematics) for his never-to-be-settled dispute with Isaac Newton over prior discovery of calculus, Leibniz's life was in fact dedicated to a broad range of topics in mathematics and philosophy.

© Springer Nature Switzerland AG 2020
A. L. Rosenberg, D. Trystram, *Understand Mathematics, Understand Computing*,
https://doi.org/10.1007/978-3-030-58376-7_11

known as *combinatorial probability*. In reality, probability theory is an *applied* spin-off of combinatorics. In its most elementary form, combinatorial probability uses counting to answer a broad range of significant questions about the relative frequencies of occurrences that range from the important—Is one more likely to die while crossing the street or while crossing the ocean?—to those some might call frivolous—Why does the game of (five-card) poker value three-of-a-kind more highly than two-pair? By the end of the section, you will have the wherewithal to answer many questions of this ilk.

The elements of probability theory and statistics infuse many aspects of life—and every area of computing.

• The practicality of many algorithms that are experientially efficient often results from the *distributions* of the inputs they encounter in "real" situations.

• Design methodologies for complex electronic circuits must be aware of statistics such as the *mean times to failure* of the critical components of the circuits.

• Sophisticated searching algorithms—and heuristic search strategies—must take into account the relative *likelihoods* of finding one's goal by following the various search directions that one has access to.

• Analyzing and understanding large corpora of data require methodologies that build on the (related) concepts of *clustering* and *decomposition*.

Every reader whose life will be touched by computing—which nowadays means just about every reader—needs at least an introduction to the foundations of probability theory to even understand, all the more so to master, the terms highlighted in the preceding bulleted items.

The structures of probability theory (Section 11.3). Just as engineering can be viewed as the "applied" sibling of science, *statistics* can be viewed as the "applied" sibling of probability. Whereas combinatorial probability gives the reader the ability to calculate the likelihoods of various events happening, statistics looks as the events "in the large". Statisticians use parameters—aptly called "statistics"—that can be viewed as *summarizing* large corpora of data. Such summarizing measures include:

• *means, medians*, and *modes*: different ways to embody concepts such as *averages* and *likelihood*. These are attempts to find single values that capture the essence of many observed instances; they are often called the *first statistical moments*.

• *variances* and *standard deviations*: two closely related ways of measuring the error incurred if one employs only *first* statistical moments; they are often called the *second statistical moments*.

• *higher moments*: successive ways of measuring the errors incurred by employing lower statistical moments to describe observed measurements.

A second mechanism used in statistics is the *(probability) distribution*. Probability distributions can be viewed as *families* of functions that measure probabilities of kindred events. Most often, a distribution is viewed as a curve that describes the probabilities of the items that are produced by an event of interest, where (*a*) each item is associated with a numerical value; (*b*) all items are laid out in order of value. The items could be, for illustration, the observed diameters of roller bearings produced by a particular manufacturing process.

Perhaps the most familiar probability distribution is the *normal distribution*, which is readily recognized from its *bell-shaped curve*. The shape of the bell curve tells us: the probabilities of the items having extreme values—very small or very large—are very small; and, the probabilities of the items are distributed (roughly) symmetrically around the "average" item. Returning to our roller bearings: (1) Very few bearings have diameters that are very far from the target; (2) deviations from the target are as likely to produce bearings that are too large as bearings that are too small.

The elements of empirical reasoning (Section 11.4). In earlier chapters, we have discussed at length the *deductive paradigm* for uncovering truth. We all know, though, that only a small fraction of situations we encounter in "real life" are amenable to the deductive paradigm. For all other situations, we use some form of empirical reasoning—*we learn from experience*. Fortunately, even "experience" has mathematical underpinnings, at least over "large" numbers of trials. One simple, but illustrative, example of the principle of *experientia docet*–Latin for "experience teaches"—is that if a repeatedly tossed coin does not come out HEADs roughly half the time over a very long sequence of tosses, then we can infer that the coin is not unbiased. More advanced techniques enable one to quantify the adjective "large" in the preceding sentences. Since much of the mathematics relevant to empirical reasoning is beyond the scope of a beginning text, most of our effort in this section is devoted to imbuing the reader with a level of literacy adequate to understand relevant concepts and their significance: Rather than supplying proofs of truly advanced theorems, we supply only introductory discussions and pointers to more advanced treatment.

11.1 The Elements of Combinatorics

This section contains three quite distinct, but closely related, lessons about counting. The first lesson is in the form of a single important example. In Section 11.1.1, we exploit the special structure of *strings* to count the number of length-*n* strings over fixed-size alphabets; and we extend this ability to other objects whose structures can be *encoded* as strings—notably the power sets of finite sets. In Section 11.1.2, we illustrate how to count within sets that are "complex" in the sense of being formed from other sets by using the algebraic operations on sets that we discussed in Section 3.2.2. Section 11.1.3 lays the foundations for *combinatorial probability*: the subarea of probability theory that is based on *counting*. The section introduces two operations on sets—the *selection* and *arrangement* of the objects in sets—that are at the center of probability-via-counting.

11.1.1 Counting Binary Strings and Power Sets

Proposition 11.1 *For every integer $b > 1$, the number of b-ary strings of length n is b^n.*

Proof. The asserted numeration follows most simply by noting that there are always b times as many b-ary strings of length n as there are strings of length $n-1$. This is because we can form the set of b-ary strings of length n as follows. Take the set A_{n-1} of b-ary strings of length $n-1$, and make b copies of it, call them $A_{n-1}^{(0)}, A_{n-1}^{(1)}, \ldots, A_{n-1}^{(b-1)}$. Now, append 0 to every string in $A_{n-1}^{(0)}$, append 1 to every string in $A_{n-1}^{(1)}$, ..., append $\bar{b} = b-1$ to every string in $A_{n-1}^{(b-1)}$. The thus-amended sets $A_{n-1}^{(i)}$ are mutually disjoint (because of the terminal letters of their respective strings), and they collectively contain all b-ary strings of length n.   □

Proposition 11.1 has two corollaries which are much more important than the proposition's statement.

The first corollary follows the proposition so closely that we leave its proof to the reader.

Proposition 11.2 *If a particular reproducible event (think of tosses of a fair coin) has b possible outcomes, then a sequence of n independent repetitions of the event has b^n possible outcomes.*

The second corollary of Proposition 11.1 requires the auxiliary device of the *characteristic sequence* of a set, which we introduced in Section 3.3.5. Using this device, we invoke the case $b = 2$ of the proposition to determine how many subsets a finite set S has—i.e., to count the elements of S's *power set*. (All technical terms come from Chapter 3.)

Proposition 11.3 *The power set, $\mathscr{P}(S)$, of a finite set S contain $2^{|S|}$ elements.*

Proof. We begin by taking an arbitrary finite set S—say of n elements—and laying its elements out in a line. We thereby establish a one-to-one correspondence between S's elements and the first n positive integers: There is the first element, which we associate with the integer 1, the second element, which we associate with the integer 2, and so on, until the last element along the line gets associated with the integer n.

Next, we note that we can specify any subset S' of S by specifying a length-n *binary (i.e., base-2) string*, i.e., a string of 0s and 1s. The translation is as follows. If an element s of S appears in the subset S', then we look at the integer we have associated with s (via our linear ordering of S), and we set the corresponding bit-position of our binary string to 1; otherwise, we set this bit-position to 0. In this way, we get a distinct subset of S for each distinct binary string, and a distinct binary string for each distinct subset of S. *We thus have a* bijection *between the set of length-n bit-strings and the power set of S.*

Let us pause to illustrate our correspondence between sets and strings by focussing on the set $S = \{a,b,c\}$. Just to make life (a little) more interesting, let us lay S's elements out in the order b,a,c, so that b has associated integer 1, a has associated integer 2, and c has associated integer 3. We depict the elements of $\mathscr{P}(S)$ and the corresponding binary strings in the following table.

Binary string	Set of integers	Subset of S
000	$\emptyset$	$\emptyset$
001	$\{3\}$	$\{c\}$
010	$\{2\}$	$\{a\}$
011	$\{2,3\}$	$\{a,c\}$
100	$\{1\}$	$\{b\}$
101	$\{1,3\}$	$\{b,c\}$
110	$\{1,2\}$	$\{a,b\}$
111	$\{1,2,3\}$	$\{a,b,c\} = S$

Back to the Proposition: We have verified the following: *The number of length-n binary strings is the same as the number of elements in the power set of S.* The desired numeration thus follows by the ($b = 2$) instance of Proposition 11.1. $\square$

Explanatory note.

The binary string that we have constructed to represent each set of integers $N \subseteq \{0,1,\ldots,n-1\}$ is the *(length-n) characteristic vector of the set N*. Of course, the finite set N has characteristic vectors of all finite lengths. Generalizing this idea, *every* set of integers $N \subseteq \mathbb{N}$, whether finite or infinite, has an *infinite* characteristic vector, which is formed in precisely the same way as are finite characteristic vectors, but now using the set $\mathbb{N}$ as the base set.

11.1.2 Counting Based on Set Algebra

In this section, we assume that we know the cardinalities of certain *finite* sets—call them A, B, and C—and we want to know the cardinality of a new set which is formed from these sets by the basic operations of the algebra of sets, as discussed in Section 3.2.2. There are a few commonly invoked counting laws which should be in your toolkit.

- *The bijection rule*

  If the elements of set A can be put into bijective correspondence with the elements of set B, then sets A and B have the same cardinality. Symbolically, $|A| = |B|$.

- *The addition rule*

  The addition rule is also known as the *Law of inclusion and exclusion*.

  One can compute the cardinalities of unions of sets by adding and/or subtracting the cardinalities of the individual sets. For any sets A and B,

$$|A \cup B| = |A| + |B| - |A \cap B|$$

Specifically:

- If A and B are *disjoint*—i.e., have no common elements (symbolically, $A \cap B = \emptyset$)—then $|A \cup B| = |A| + |B|$.

- if A and B *intersect*—i.e., share some elements (symbolically, $A \cap B \neq \emptyset$)—then $|A \cup B| = |A| + |B| - |A \cap B|$.

  This formula *includes* the elements of both sets and then *excludes* the sets' shared elements, which are double-counted by the inclusion. (You can see the origin of the name "the law of inclusion and exclusion".)

Figures 11.1 and 11.2 illustrate how to generalize the preceding equations to collections of three sets; going beyond three adds complexity that is clerical but not conceptual. In order to draw explicit expressions that express the content of

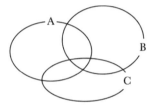

Fig. 11.1 Three sets, A, B and C, in "general" position, i.e., with all possible overlaps

Fig. 11.2 The union of sets A, B, and C (left), and their intersections (right)

Fig. 11.2, one must apply the law of *inclusion and exclusion* in multiple ways, as we compensate for the pairwise and triple intersections among sets A, B, and C. A careful reckoning using the figure indicates that

$$|A \cup B \cup C| = \big(|A| + |B| + |C|\big) - \big(|A \cap B| + |A \cap C| + |B \cap C|\big) + |A \cap B \cap C|.$$

In particular, we begin by *including* the union; then we *exclude* the pairwise intersections, which were double-counted; and we finally *include* the triple intersection, which was first counted three times and then excluded three times while removing the pairwise intersections.

- *The multiplication rule*

This rule tells us how to count the cardinalities of Cartesian products of sets, as illustrated in Fig. 3.6.

$$|A \times B| = |A| \cdot |B|$$

The multiplication rule extends immediately to multiplicities of sets; e.g., for three sets:

$$|A \times B \times C| = |A| \cdot |B| \cdot |C|$$

As an important special case, we note that

$$|A \times A \times \cdots \times A| \ (p \text{ occurrences of } A) = |A|^p$$

The exciting feature in this last equality is that the set

$$A \times A \times \cdots \times A \ (p \text{ occurrences of } A)$$

is a (nonstandard) way of representing the set of length-p strings of elements of A. Thereby, the multiplication rule gives us an alternative way to view—and to prove—Proposition 11.1.

As we move forward, we shall encounter many other ways of counting the elements of sets, most importantly by using operations involving *selection* and *rearrangement*. As we proceed with this new material, we will encounter old friends, often wearing new clothing. We will renew our acquaintance with the factorial operation and with binomial coefficients; we will have cause to recall the pigeonhole principle in new settings. Also, we will discover new domains of application, as we move beyond "pure" mathematics to the applications of mathematical laws in domains related to probability and statistics.

Let us begin our journey.

11.1.3 Counting Based on Set Arrangement and Selection

This section introduces the primary operations that are used for arranging finite sets as we establish a base for studying combinatorial probability. The importance of set arrangement in this context results from the common practice of defining discrete probability/likelihood as a counting problem—most specifically, of defining the ratio

$$\frac{\text{number of ways of achieving a targeted event } E}{\text{number of possible events}} \tag{11.1}$$

as the *probability of event E*.

We focus on three notions of arranging, and selecting from, a set of n objects:

$$\{x_1, x_2, \ldots, x_n\}$$

When useful for explaining some idea, we sometimes identify the x_i as some specific type of objects, such as numbers, but generally we do not exploit any specific characteristics of the x_i.

Permutations. A permutation of an n-element set A is a *fixed ordering* of the elements of A: each element of A appears precisely once in the ordering. We denote a permutation of an n-element set via the notation

$$(x_1, x_2, \ldots, x_n)$$

where each x_i appears precisely once. Note that we use parentheses for grouping, as in "$(3, 2, 1, 4)$", instead of braces, as in "$\{3, 2, 1, 4\}$", to emphasize the ordering. In particular,

$$\{3, 2, 1, 4\} = \{1, 2, 3, 4\} \quad \text{but} \quad (3, 2, 1, 4) \neq (1, 2, 3, 4)$$

In terms of problems relating to *selecting m* elements from a set of $n \geq m$ elements, permutations give rise to the most demanding genre of selection: *They demand accounting for both the* identities *of the selected elements and the* orders *in which the elements were selected.*

It is a straightforward exercise to count the number of permutations of n items.

Proposition 11.4 *The number of permutations, $P(n)$, of an n-element set equals n!*

Proof. Just for the practice, we describe two inductive proofs of this simple result, based on two quite distinct ways of constructing permutations. We merely sketch the inductions underlying the proofs, leaving the details for the reader.

1. To construct a permutation of the n-element set $\{x_1, x_2, \ldots, x_n\}$:

- We can select the first element of the permutation in n ways: any x_i will work.

- Having selected the first item, we are left with the $(n-1)$-element version of the problem.

We thereby find that $P(n)$ is specified recursively as follows.

$$P(n) = \begin{cases} 1 & \text{if } n = 1 \\ n \times P(n-1) & \text{if } n > 1 \end{cases}$$

Elementary reasoning now shows that $n!$ is the value of $P(n)$.

2. To construct a permutation of the n-element set $A = \{x_1, x_2, \ldots, x_n\}$:

- Assume, inductively, that you are given a fixed permutation $(x_1,\ldots,x_{n-1})$ of an $(n-1)$-element subset $A' \subset A$, chosen somehow from among the $(n-1)!$ possible orderings of set A'. (Note the thinly veiled use of induction.)

- Take element x_n, which does not belong to A', and place it into the ordering of A' that you are given. You can place x_n in any of n positions in the new permutation:

  - before (i.e., to the left of) all of the already placed elements of A';

  - after (i.e., to the right of) all of the already placed elements of A';

  - in between any adjacent already placed elements of A'.

  Each of the preceding n choices creates a unique permutation of the n-element set A.

Using arithmetic virtually identical to case **1**, we verify that $P(n) = n!$.

Other variants on the preceding proof themes will undoubtedly occur to you. □

Combinations. Within the context of selection problems, combinations are the "next step down" in strictness from permutations. When one selects m items from a set of $n \geq m$ items, combinations are concerned with the *identities* of the elements selected, but not with the order in which their elements are chosen.

(This willingness to ignore order gives a big hint about how to count the number of combinations of m items chosen from n. Can you figure out how to use the hint before we get to Proposition 11.5?)

Factorials play as essential a role in combinational selection as in permutational selection, but now it plays a *dual* role: choosing the identities of the selected elements *and* factoring out the order in which the elements were selected. The binomial coefficients that we introduced in Section 5.1.2.4 and that have arisen several times since then throughout our journey are (literally!) tailor-made for this genre of selection problem. Recall that

$$\binom{n}{m} = \frac{n \times (n-1) \times \cdots \times (n-m+1)}{m!} = \frac{n!}{m!(n-m)!}$$

Proposition 11.5 *The number of ways of selecting m elements from a set of $n \geq m$ elements, while ignoring the order in which the $m > 0$ elements were selected, is* $\binom{n}{m}$.

Proof. The number of *ordered* ways of selecting $m > 0$ elements from a set of $n \geq m$ elements is

$$n \times (n-1) \times \cdots \times (n-m+1) = \frac{n!}{(n-m)!} \tag{11.2}$$

One can select the first element in any of n ways. Having selected this element, there are: $n-1$ ways to choose the second element; $n-2$ ways to choose the third element; $n-3$ ways to choose the fourth element; and so on.

There is a bit of complexity here, so let us say the same thing in a few ways: This method of numeration accounts for both the identities of the m elements and the order in which they are selected. For each position i, the method accounts for both the identity of the ith chosen element *and* the fact that it was the ith element chosen. In more detail, if element x_1 is the first element chosen, and element x_2 is the second one chosen, then the two orders of selection

$$x_1, x_2, \text{ (some fixed order for the remaining } m-2 \text{ selections)}$$

and

$$x_2, x_1, \text{ (some fixed order for the remaining } m-2 \text{ selections)}$$

are counted as separate events. It is easy to see that this overcounting uniformly expands each of the events that we *do* want to account for by the factor $m!$, for this is the number of orders in which we could have selected the finally chosen m elements.

This reasoning indicates that we can compensate for our overcounting by dividing the *ordered* tally (11.2) by $m!$. This compensation replaces tally (11.2) by the binomial coefficient $\binom{n}{m}$, whence the result.   □

Explanatory note.

Binomial coefficients have thus made yet another appearance in our journey! The many situations that are counted by this construct are only suggested by their occurrence here, in respect to probabilities, as well as in Chapter 6 within the context of evaluating summations and in Chapter 9 within the context of solving recurrences. The multiple occurrences of binomial coefficients within these apparently unrelated contexts provide a powerful argument for the value of abstracting beyond the superficial details of phenomena in the direction of their deep essentials. This insight will only be reinforced as we proceed with our study of combinatorial probabilities.

Derangements. Derangements represent one of the simplest forms of *avoidance problems*. Here is a well-known version of such a problem, in preparation for the general definition.

Professor X views it as a win-win strategy for the students in her class to grade each others' essays on *The Essential Truth in the Universe*. The essays thereby get graded faster, because the grading process becomes a parallel rather than sequential process. Moreover, each student gets a chance to see how another student has interpreted some basic component of the human experience. The only complication is: How should Professor X allocate essays among the students. *The process must ensure that no student is assigned her own essay to critique.*

The challenge that Professor X is facing is known as a *derangement problem*.

A *derangement* of a (finite) set A is a *bijection* $f : A \leftrightarrow A$ that has no *fixed point*. In other words, for every $a \in A$, we must have $f(a) \neq a$.

Clearly, derangements always exist (for $n > 1$). One can just label the elements of set A by the numbers $0, 1, \ldots, |A| - 1$ and specify $f(a) = a + 1 \bmod |A|$.

However, playing around with some sets should convince you that derangements are not so common! In fact, the set $A = \{0, 1, 2\}$ admits six self-bijections, but only two are derangements, namely:

$$f(a) = a + 1 \bmod 3 : \text{which maps } (0 \to 1), (1 \to 2), (2 \to 0)$$
and
$$g(a) = a - 1 \bmod 3 : \text{which maps } (0 \to 2), (1 \to 0), (2 \to 1)$$

How many derangements does an arbitrary n-element set A have? We denote this quantity by $d(n)$.

While the general problem of counting derangements is beyond the scope of an introductory text, the reader will benefit from observing the process of deriving a recurrence for $d(n)$.

We compute $d(n)$ for arbitrary $n \in \mathbb{N}^+$ via the following recursion:

- For $n = 1$: $d(1) = 0$.

  The unique bijection in this case consists only of a fixed point.

- For $n = 2$: $d(2) = 1$.

  There are two bijections in this case, the identity, which has two fixed points, and the swap, which is a derangement.

- For $n > 2$: $d(n) = (n-1)(d(n-1) + d(n-2))$:

  To see this, note first that in any derangement, the first element of A, call it a, must map to some $b \neq a$.

  Note next that there are $n - 1$ ways to choose b.

  - There are $d(n-2)$ derangements under which b maps to a. In those cases, we know everything about a and b, so we need worry only about the remaining elements of A. These $n - 2$ elements can "derange" in all possible ways.

  - There are $d(n-1)$ derangements under which element b does not map to a. In detail, as the process of "choosing an image" passes through A, every element has two forbidden choices: it cannot choose itself (for that would be a fixed point) and it cannot choose one other element (for element b, this is element a). In some sense, the notion of a forbidden element gets passed around, changing its identity at every step.

The preceding reasoning verifies the following recurrence

$$d(n) = \begin{cases} 0 & \text{if } n = 1 \\ 1 & \text{if } n = 2 \\ (n-1)(d(n-1) + d(n-2)) & \text{if } n > 2 \end{cases}$$

Interestingly, as the number of objects in the set to be deranged grows without bound, the proportion of bijections that are derangements tends to the limit $1/e$, where e is Euler's constant (the base of natural logarithms).

11.2 Introducing Combinatorial Probability via Examples

Perhaps the easiest and most engaging way to introduce "combinatorial probability", i.e., probability via counting, is by calculating game-related likelihoods—deals of cards, rolls of dice, and guessing games of various sorts. Why is one specific genre of deal in the game of poker (say, a "straight") worth more than another (say, "three of a kind")? The arithmetic required for such a discussion is elementary, and the references to *gedanken gambling* (gambling via thought experiment) are easily understood—and, they are usually of some interest even to non-gamblers. One can also introduce in such a setting concepts such as randomness and bias, which are so important in the design of experiments and the analysis of their outcomes.

This, then, will be our approach to the subject of combinatorial probability. The reader has already seen more than enough "dry" facts about sets and how to count their elements. It is time to reap some rewards from the work of acquiring that knowledge.

Before we begin to play, though, we need some groundwork, beginning with the combinatorics-based definition of "probability" that sets the tone for this section.

As noted earlier, the *(discrete) probability* or *likelihood* of an event is given by the ratio (11.1), which measures the number of targeted events as a fraction of the number of possible events.

This approach to (*discrete*) probabilities has important technical consequences.[2]

- The probability of any targeted event E is ≤ 1. This means:
  - If the probability of E is 1, then event E is a *certainty*.
  - If the probability of E is < 1, then event E is possible but not certain.

- The probability of any targeted event is ≥ 0. This means:
  - If the probability of E is 0, then event E is an *impossibility*.
  - If the probability of E is > 0, then event E is a possibility.

- The more likely event E is, the closer its probability is to 1.

- By the addition rule for counting sets, the joint probability of two *disjoint* events equals the sum of the events' probabilities.

---

[2] We emphasize the word "discrete" because the asserted dichotomy does not generally hold in the world of events that assume continuous values.

- As a special case of the preceding rule: the joint probability of two *complementary* events equals 1. This is because the complement of an event whose probability is p has probability $(1 - p)$.

Explanatory note.

In our formulation, a probability is a number p in the range $0 \le p \le 1$.

It is also very common, though, to view a probability as a *percentage*, as in

"The likelihood of rain this morning is 50%."

In order to understand these alternative locutions, the reader should recall that the word "percent" derives from the Latin "*per centum*", which means "per hundred". Therefore, one can always interchange the following locutions

"Event E occurs with probability p."

"Event E occurs with probability $(100p)\%$."

We are going to focus on *games of pure chance*—with no complicating notion of strategy—because our underlying interest is in probability, not game theory.

Enrichment note.

There exists an exciting mathematical field of study that is dedicated to all games: games of pure chance (which is our focus), games that combine skill and chance (e.g., *backgammon* or *bridge* or *poker*), and games of pure skill (e.g., *chess* or *go*). But, alas, the issues that one must deal with when studying the broad spectrum of games go beyond the scope of our introductory text.

We recommend to the reader just two classical works that may be of interest for historical and cultural reasons, as well to supply a foundation for further study. The first reference is to the classic [110] by American polymath John von Neumann and German economist Oskar Morgenstern, which established the connection between the mathematical subject of game theory and the field of economics. (Now-familiar concepts such as "rational game" and "zero-sum game" originated in [110].) The second classic is [83], by the American mathematician John Nash, which introduced the now-familiar notion of "Nash equilibrium". (The story behind [83] is movingly related in the movie "*A Beautiful Mind*".)

11.2.1 The Game of Poker: Counting Joint Events

We focus on a typical deck of 52 playing cards: The cards are partitioned into

$*$ four *suits*:   CLUBS, DIAMONDS, HEARTS, SPADES (in increasing value)

$*$ 13 *face values*: $\begin{cases} \text{number cards}: 2, 3, 4, 5, 6, 7, 8, 9, 10 \\ \text{picture cards}: \text{JACK, QUEEN, KING, ACE (in increasing value)} \end{cases}$

In a popular version of the game of poker, each player is dealt a *hand* consisting of five cards. The total number of possible five-card hands, i.e., of (unordered sets) of five cards, is just the number of ways of choosing a five-element set from a 52-element set. Invoking the counting techniques from Section 11.1.3, we note that this "astronomical" number is

$$\binom{52}{5} = \frac{52!}{5!47!} = 2,598,960$$

The size of this number explains much of what of makes the game of poker so interesting.

Of course, it is the *patterns* of the cards in various hands that determines which hands beat other ones in the competition that is at the heart of poker. It is a reasonable conjecture that

Conjecture. *The value of a pattern in poker accurately tracks the likelihood of a player's receiving the pattern in a random deal.*

(Of course, *random* deals are the embodiment of a *fair* game.)

We now calculate the likelihoods of a variety of patterns' arising in a fair game of poker, in order to study the validity of the preceding conjecture. As we derive the likelihoods of the various patterns, observe the many invocations of the multiplication rule for the probabilities of independent events.

- A *royal flush* is a poker hand of the form

  10, JACK, QUEEN, KING, ACE   of the same suit

  There are only four ways to form a royal flush—one for each suit. Therefore, the probability that a fair deal will produce such a hand is

$$Pr(\text{royal flush}) = \frac{4}{2,598,960} = \frac{1}{649,740} \approx 0.00000154 = 1.54 \times 10^{-6}$$

- A *straight flush* is a poker hand in which the cards share the same suit and follow one another in value. Thus, a royal flush is a straight flush whose highest-rank card is an ACE.

  Let us calculate the number of straight flushes.

  The counting is similar to that for a royal flush: For each suit, there are nine sequences of five cards that are consecutive in rank. From lowest rank to highest:

$$
\begin{aligned}
&\text{Hand \#1}: \quad 2 \ \ 3 \ \ 4 \ \ 5 \ \ 6 \\
&\text{Hand \#2}: \quad 3 \ \ 4 \ \ 5 \ \ 6 \ \ 7 \\
&\text{Hand \#3}: \quad 4 \ \ 5 \ \ 6 \ \ 7 \ \ 8 \\
&\text{Hand \#4}: \quad 5 \ \ 6 \ \ 7 \ \ 8 \ \ 9 \\
&\text{Hand \#5}: \quad 6 \ \ 7 \ \ 8 \ \ 9 \ \ 10 \\
&\text{Hand \#6}: \quad 7 \ \ 8 \ \ 9 \ \ 10 \ \ J \\
&\text{Hand \#7}: \quad 8 \ \ 9 \ \ 10 \ \ J \ \ Q \\
&\text{Hand \#8}: \quad 9 \ \ 10 \ \ J \ \ Q \ \ K \\
&\text{Hand \#9}: \quad 10 \ \ J \ \ Q \ \ K \ \ A
\end{aligned}
$$

The probability of being dealt a royal flush is, thus, precisely $1/9$ the probability of being dealt a straight flush. The probability of getting a straight flush is, therefore,

$$
Pr(\text{straight flush}) \;=\; \frac{9}{649,740} \;\approx\; 0.0000139 \;=\; 1.39 \times 10^{-5}
$$

- *Four-of-a-kind* is a poker hand of the form

    $$X,\ X,\ X,\ X,\ Y$$

    where $X, Y \in \{2, 3, 4, 5, 6, 7, 8, 9, 10, J, Q, K, A\}$, and $X \neq Y$.

    There are 13 possible face values; each is a candidate for being card X. Having chosen card X, there are 48 possible choices for card Y. Each selection of cards X and Y specifies a unique *four-of-a-kind* poker hand. The multiplication principle thus tells us that there are precisely $13 \times 48 = 624$ *four-of-a-kind* poker hands, so the probability of being dealt such a hand is

    $$
    Pr(\text{four of a kind}) \;=\; \frac{624}{2,598,960} \;\approx\; 2.4 \times 10^{-4}.
    $$

- We look at the next two poker hands together because the analyses of their respective patterns are so similar.

    – A *Full house* is a poker hand of the form

    $$X,\ X,\ X,\ Y,\ Y$$

    where $X, Y \in \{2, 3, 4, 5, 6, 7, 8, 9, 10, J, Q, K, A\}$, and $X \neq Y$.

    – *Three of a kind* is a poker hand of the form

    $$X,\ X,\ X,\ Y,\ Z$$

    where $X, Y, Z \in \{2, 3, 4, 5, 6, 7, 8, 9, 10, J, Q, K, A\}$, and $|\{X, Y, Z\}| = 3$. The final equation, about set-cardinality, is a "cute", succinct, way of saying that Y and Z differ both from X and from each other.

For both of these patterns, the analysis begins by noting that the face value X can be chosen in 13 ways. Having chosen face value X, we select three of the four cards with that face value: this choice can be made in $\binom{4}{3} = 4$ ways. The X component of the hand has now been selected.

With the *full-house* pattern, we must now choose the face value Y. This can be done in 12 ways. The selection of the hand is completed once we select the $\binom{4}{2} = 6$ needed cards that have face-value Y.

Summing up, the *full-house* pattern can be assembled in

$$13 \cdot \binom{4}{3} \cdot 12 \cdot \binom{4}{2} = 3,744$$

ways, so the probability of being dealt such a hand is

$$Pr(\text{full house}) = \frac{3,744}{2,598,960} \approx 1.44 \times 10^{-3}$$

To finish up the *three-of-a-kind* poker hand, we remark that once we have selected the X component of the hand, we must select the remaining cards by choosing the (distinct) face values of Y and Z; this can be done in $\binom{12}{2} = 66$ ways. Then, for each of these choices we must select a specific card, which can be done in four ways. We thereby find that the *three-of-a-kind* pattern can be assembled in

$$13 \cdot \binom{4}{3} \cdot \binom{12}{2} \cdot 4^2 = 54,912$$

ways, so the probability of being dealt such a hand is

$$Pr(\text{three of a kind}) = \frac{54,912}{2,598,960} \approx 2.11 \times 10^{-2}$$

- The final poker hand that we consider is *two-pair*. We remarked earlier in this section that three-of-a-kind beats two-pair. Comparing the likelihood of two-pair versus the just-computed likelihood of three-of-a-kind will be a good test of our poker-valuation conjecture.

  The *two-pair* hand in poker is a deal that is *a permutation of* the hand

$$X, X, Y, Y, Z$$

where X, Y, Z are distinct face values.

You can show (*in an exercise*) that the probability of receiving a two-pair hand in a fair poker deal is

$$Pr(\text{two-pair}) = \frac{123,552}{2,598,960} \approx 4.75 \times 10^{-2}$$

This is a bit more than twice the probability of receiving a three-of-a-kind— which explains why the latter hand beats the current one in a poker game.

One can find in many sources (including the Internet) tables that: enumerate all possible poker hands; describe all ways of assembling each hand; compute the probabilities of being dealt each hand.

Of course, our interest has been in exposing how to perform these calculations, not in their results—but it is fun to determine why the various poker hands have their relative values in the game.

11.2.2 The Monty Hall Puzzle: Conditional Probabilities

The following real-life situation illustrates that reasoning probabilistically is not easy and that it can be quite unintuitive.

We recall a popular US TV show from the 1970s and 1980s, *Let's Make a Deal*. In one segment of each episode of the show, a contestant was confronted by three doors: door (1), door (2), and door (3) in Fig. 11.3. The contestant was told that a

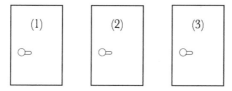

Fig. 11.3 The three doors of *Let's Make a Deal*

jackpot was hidden behind one of the doors and that the other two doors hid less desirable outcomes. The show's host, Monty Hall, invited the contestant to choose one of the doors: the contestant would receive whatever prize was behind the selected door.

Since the doors' names are irrelevant to the story, let us assume that the contestant chose door (3). Before that selection was considered final, Monty Hall would open *one of* doors (1) or (2), i.e., one of the two *unchosen* doors. Monty made sure—unbeknownst to the contestant—to choose a door that did *not* hide the jackpot. Again for illustration, let us assume that Monty opened door (1). When the jackpot did not appear behind door (1), Monty now gave the contestant the option of changing her initial choice, from door (3) to door (2).

What should the contestant do?

Let us begin with intuition. As the story began, the contestant had a one-in-three chance of choosing the door that hid the jackpot. As the story progresses, is it conceivable that these odds could have changed after door (1) is shown *not* to hide the jackpot? How could this be?

Well, the popular (well-named) newspaper columnist Marilyn vos Savant published an analysis that shows that *it is better to change one's choice!* Here is her reasoning, enhanced by the illustration in Fig. 11.4. There are two possible cases.

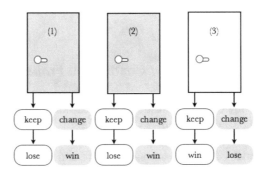

Fig. 11.4 Suppose the jackpot is behind door (3). If the contestant selected door (3) initially, then *not switching is a win*, while *switching is a loss*. If the contestant selected either door (1) or door (2) initially, then *switching is a win*, while *not switching is a loss*

- If the contestant had selected the jackpot door initially, then holding on to this choice leads to the same probability of success as she had from the beginning, namely, $1/3$.

- If the contestant had selected a wrong door initially—which occurs with probability $2/3$—then she will win the jackpot if she now changes her initial (erroneous) choice.

Thus, quite unintuitively, the probabilities *have changed*, and the contestant should change her original choice based on the information she has now.

You can have fun by confronting your friends with this unintuitive analysis.

Of course, the Monty Hall game and its analysis are not just a trick! Our story is a demonstration that you can improve your chances of winning by taking advantage of all available information! The "science" behind this analysis builds upon the topic of *conditional probability*. Rather than looking only at the probability of an event E in isolation, one focuses on the probability of event E, *given that some event F has occurred*. The *notation* that traditionally indicates this shift in focus is the following: If one uses the notation $Pr[E]$ to denote the probability of event E in isolation, then one uses the embellishment $Pr[E \mid F]$ to denote the probability of the compound event: "E, given F".

In order to supply a complete development of the theory that explains the Monty Hall puzzle, we would have to employ conditional probabilities, which would require invoking Bayes's Theorem, a seminal result within the theory of conditional probabilities—which would take us beyond the limits of an introductory text.

We refer the interested reader to [75] for a general presentation of this area, and to [12] for the original work on the Theorem.

Also, the reader should see the sites on the Internet that interactively perform the computation of (conditional) likelihoods!

11.2.3 The Birthday Puzzle: Complementary Counting

This section presents another application of the rules of enumeration to the calculation of probabilities. The main lesson associated with this application is the following.

It is sometimes easier to calculate the probability that event E does not occur than the probability that it does occur.

(Of course, this is not a formal proposition. But, we do invite you to replicate the conclusions of this section via a direct argument.)

A common technique that teachers employ to establish rapport with a new class is to determine whether any students in the class share a birthday. We now demonstrate that when the class has (at least) around 30 students, the answer is usually *yes*. The fact that we can make such a specific claim suggests that the presence of birthday-sharers is not a mere accident—there are probabilities at work here!

We study the following question.

How many people must gather before the probability that some two share a birthday reaches 50%?

We develop an answer to this question, based on the assumption that all dates are equally probable potential birthdays.

Let us focus on a common year, with 365 days; a simple modification of the coming calculation accommodates leap years. A first, obvious, remark is that when there are 366 people in the class—and only then—the pigeonhole principle *guarantees* (with probability 100%) that some two people share a birthday. When we lower our birthday-sharing target to a *50% probability* of a shared birthday, how much does this lower the required population of 366? Somewhat unintuitively, the required number is lowered dramatically!

We simplify the calculational component of our solution considerably by focusing on the *dual* form of our problem. That is:

Instead of determining a *lower bound* on the population that would guarantee a 50% probability of a shared birthday,

we determine an *upper bound* on the population that would guarantee a 50% probability of *no* shared birthday.

Since we assume that all birthdays are equally likely, the probability that any particular person has any particular birthday is $1/365$. In other words, the space of all possible assignments of birthdays to all n people has 365^n assignments—all of which are equally likely.

Let us perform the following *gedanken* assignment of birthdays.

Say that we have an assemblage of n "proto-people"—i.e., people who do not yet have birthdays. Our task is to assign birthdays to these folks in such a way that no two of them share a birthday.

- We can select a birthday for person #1 from any of the 365 (indistinguishable) days of the year.

- We can select a birthday for person #2 only from the 364 (indistinguishable) days of the year that person #1 has not "occupied".

  Otherwise persons #1 and #2 would share a birthday!

- Proceeding by similar reasoning: We can select person #3's birthday from only the 363 days that the first two people have not "occupied"; we have 362 days available for person #4, and so on, until person #*n*'s birthday must be selected from among $365 - (n - 1)$ days.

Our procedure indicates that among the 365^n possible assignments of birthdays to n people, only

$$365 \times 364 \times 363 \times \cdots \times (365 - n + 1) = \frac{365!}{(365 - n)!}$$

assignments have no two people sharing a birthday.

Worded as a probability: Letting $Pr(n)$ denote the probability that there are no shared birthdays in a population of n people:

$$Pr(n) = \frac{\text{number of birthday assignments with no shared birthday}}{\text{total number of birthday assignments}}$$

$$= \frac{365!}{(365 - n)! \times 365^n}$$

$$= \frac{n!}{365^n} \binom{365}{n}$$

Let us return to our original problem, which focuses on assignments in which there exists at least one shared birthday. Since the event some two people share a birthday

is the complement of the event

no two people share a birthday

the probability of a shared birthday is precisely $1 - Pr(n)$. For $n = 23$, the desired probability is 0.5073.

Proposition 11.6 *In any group of* 23 *people, the likelihood that two people share the same birthday is slightly greater than* 50%, *assuming a year with* 365 *equally likely days.*

When our teacher's class grows to 30 students, the probability of a shared birthday grows to 70.632%.

11.2.4 Sum-of-Spots Dice Games: Counting with Complex Events

Archaeology suggests that games have fascinated our species since our origins. The games that have endured span a spectrum from those that depend almost entirely on thinking and strategy to those that depend on pure chance. A number of games that have endured involve the rolling of six-sided dice, each die[3] having distinct numbers (usually denoted by *spots*) on its six faces. For convenience, we refer to each die's face-numbers via the numerals $1, 2, 3, 4, 5, 6$. (Ancient Egyptians played such games, as did Roman legionaries; and such games persist in present-day casinos.)

We discuss two ancient dice games: The first game, known as "*craps*" in North America, involves rolling two dice at a time; the second game, which we call "*sum-of-three*", involves rolling three dice at a time. With both games, the outcome of interest is the *sum* of the spots on the upward-oriented faces of the rolled dice. (For example, the sums "7" and "11" are winning outcomes in craps; "2", "3", and "12" are losing outcomes; all other sums are "roll-again" outcomes.) We highlight the word "sum" to stress that the order in which the dice reveal their spots is not relevant. This means that the outcomes we distinguish consist of pairs of the form ab for the game craps and triples of the form abc for the game sum-of-three, where:

- each of a, b, c is an integer from the set $\{1, 2, 3, 4, 5, 6\}$

- $a \le b \le c$. This is just a convenient way of saying that the order in which the numbers appear is irrelevant.

We call each such pair (in craps) or triple (in sum-of-three) a *configuration*.

Under the assumption of a *fair* game—i.e., one in which all faces of all dice are equally likely to appear—we can analyze both craps and sum-of-three by enumerating the 21 possible outcomes of a roll in the game of craps and the 56 possible outcomes of a roll in the game of sum-of-three; see Fig. 11.5.

---

[3] The word "dice" is plural; the singular form is "die".

						sum
111						3
112						4
113	122					5
114	123	222				6
115	124	133	223			7
116	125	134	224	233		8
126	135	144	225	234	333	9
136	226	145	244	235	334	10
146	236	155	245	335	344	11
156	246	336	255	345	444	12
166	256	346	445	355		13
266	356	446	455			14
366	456	555				15
466	556					16
566						17
666						18

			sum
11			2
12			3
13	22		4
14	23		5
15	24	33	6
16	25	34	7
26	35	44	8
36	45		9
46	55		10
56			11
66			12

Fig. 11.5 The possible outcomes of craps (left) and of sum-of-three (right)

As one might expect, the analysis of craps is very similar to, and much simpler than, the analysis of sum-of-three. Therefore, we provide only the latter analysis here, leaving the analysis of craps as an exercise.

We develop the enumeration of sum-of-three in two distinct ways, in the hope of encouraging the reader to seek multiple ways to think about such enumerations.

1. Our first technique of enumeration lists the 16 possible sums of a roll of the three dice and, for each, lists all possible configurations that yield that sum.

   The 16 possible sums arise from the fact that dice are labeled with the integers $1, 2, 3, 4, 5, 6$, so the smallest possible sum is 3 (which occurs when all three dice show the number 1), and the largest possible sum is 18 (which occurs when all three dice show the number 6). All possible sums are listed in Fig. 11.5(right). The depicted table allows us to easily compute, e.g., that the probability of obtaining the sum 6 in a roll of three dice is $Pr(6) = 4/56 = 0.071\cdots$; in other words, we expect to observe this sum in just a bit more than 7% of rolls.

2. A rather different way of enumerating sums is to partition configurations abc based on the number of distinct numbers in the set $\{a, b, c\}$.

   - At one extreme, there are six configurations in which $a = b = c$, based on the six possible values on the face of a die.

   - In the intermediate case, either $a = b$ or $b = c$ but the third value is distinct.
     - The situation $a = b$ can occur in five ways (we must leave room for c). For each way, c can assume $6 - a$ values (because c is the biggest value). There are, therefore, 15 such configurations.

     - The situation $b = c$ is the mirror image of the preceding situation, hence engenders the same number of configurations.

- Finally, when a, b, and c are all distinct (but appear in increasing order), a can assume the values $1, 2, 3, 4$. For each choice of a, b can assume the values $a+1, a+2, \ldots, 5$. And, for each choice of b, c can assume the values $b+1, b+2, \ldots, 6$. A simple calculation identifies 20 configurations in which a, b, and c are all distinct.

Of course, we again identify 56 possible configurations.

This completes our (exhaustive) analysis of the sum-of-three game.

To close this section, we briefly consider the variant of the sum-of-three game in which the order of observing outcomes a, b, and c *does matter*. To expose the difference that attention to order makes, let us look again at the table in Fig. 11.5(right). We note, as we focus for illustration just on the sum 6, that three different configurations produce this (unordered) sum:

1. one die of each value 1, 2, 3

2. two dice of value 1 and one of value 4

3. three dice of value 2

However, if the order in which values occur matters, then the sum 6 arises in *ten* different ways. Specifically, we now roll the dice one at a time instead of as a group of three, and we observe the sum 6 arising in the ten ways enumerated in Fig. 11.6; in this figure, each column represents a roll of die #1, then of die #2, then of die #3.

First die	1	1	4	1	1	2	2	3	3	2
Second die	1	4	1	2	3	3	1	1	2	2
Third die	4	1	1	3	2	1	3	2	1	2

Fig. 11.6 Achieving sum 6 via *ordered* rolls of three dice

Sum	3, 18	4, 17	5, 16	6, 15	7, 14	8, 13	9, 12	10, 11
number of configurations	1	3	6	10	15	21	25	27

Fig. 11.7 Achieving all possible sums via *ordered* rolls of three dice

We leave it to the reader to verify the table in Fig. 11.7, which gives the number of configurations that engender each possible sum (from 3 to 18) in the ordered-roll version of sum-of-three. Note that each possible sum k is engendered by the same number of configurations as is sum $21 - k$. (*Can you figure out why?*) The total number of configurations is $6^3 = 216$; it is easy to see that this number is equal to

$$2 \times (1 + 3 + 6 + 10 + 15 + 21 + 25 + 27)$$

Summing up. We have provided several examples which illustrate how to compute the probability of an event (e.g., observing a "heads" when tossing a coin), by taking the ratio of the number of favorable instances of the event divided by the number of all possible instances. This approach can be adapted to situations defined by repeated occurrences of an event, such as repeated rolls of a die or repeated tosses of a coin. More ambitious application of these concepts add conditioning to the scenario—as we did with the Monty Hall Puzzle. Combining such concepts can lead to back-of-the-envelope strategies for playing games of chance successfully.

11.3 Probability Functions: Frequencies and Moments

Applications of probability theory often operate with sets whose elements represent the outcomes of individual trials in an experiment—think, for instance, of the deal of a single card or the roll of a single die. The range of possible outcomes is termed the *sample space*; a variable that ranges over the frequencies of outcomes from the sample space is called a *random variable*. The goal of a probabilistic analysis is to capture important elements of *"the truth"*, as exposed via experiments that produce sequences of *outcomes*, each being an instantiation of a random variable with a value from its sample space.

11.3.1 Probability Frequency Functions

Let us abstract the preceding environment, by focussing on the following concepts.

Sample spaces and outcomes. We focus on sequences $S = \langle s_1, s_2, \ldots, s_n \rangle$ of n *outcomes* from an underlying *sample space* Σ. Each s_i represents one outcome of the n produced by an n-trial experiment.

Exemplary sample spaces Σ:

1. Rolls of a six-sided die:
   $\Sigma = \{1, 2, 3, 4, 5, 6\}$.

2. The noontime temperatures (in Celsius) observed in Grenoble, France:
   $\Sigma = \{t \mid -10 \leq t \leq 45\}$

3. The observed diameters of n roller bearings after a given run at Factory X:
   Σ = the range of diameters of roller bearings.

   Another way to look at this setup is that the sample space Σ comprises the range of values of a random variable X and that the outcome sequence S announces the values that X achieves during the course of the subject experiment.

Frequency functions and probability frequency functions. A *frequency function* f_S exposes the multiplicity of each outcome $s_i \in \Sigma$ within the sequence S.

For illustration, if

$$\Sigma = \{1,2,3,4\} \quad \text{and} \quad S = \langle 1,2,2,3,3,3,4,4,4,4 \rangle \tag{11.3}$$

then for each $i \in \Sigma$, $f_S(i) = i$.

Note that when $|\Sigma| = m$, we have $f_S(s_1) + f_S(s_2) + \cdots + f_S(s_m) = |S|$.

Now, let us focus on a random variable X that ranges over a sample space Σ. Each value $k \in \Sigma$ has an associated probability, which we denote $Pr[X = k]$. (Note that the identity of the relevant sample space is traditionally assumed to be known from context, hence is not mentioned explicitly.)

The function f on Σ which is defined as follows

$$f(x) = Pr[X = x]$$

is called a *probability frequency function (for sequence S)*. (We have already seen probability frequency functions, but we have not yet used the term.)

Summarization tools. Probability frequency functions give very local information about sequences and their associated frequency functions. If a sequence of interesting events results from a very long experiment or from a massive collection of observations, then one needs some sort of mechanism for *summarizing and partially analyzing* the resulting data—i.e., for making the data suitable for human consumption. We provide a short survey of summarizations that have proven useful over time.

The simplest possible framework for summarizing a sequence S is via a single number. The reader should be forewarned that any attempt to summarize a long sequence of numbers S via a small set of numbers whose sizes are commensurate with those in S must lead to loss of information—and/or to possible frustration. The inevitability of losing information is clear just from the shrinkage in the number of bits used to represent the outcomes of the experiment.

Explanatory note.

This might be an opportune time for the reader to review the discussions of *encodings* (Section 8.1.1.4) and *pairing functions* (Section 8.2). Both of these topics involve creating small sets of numbers to encode large sets of numbers. The fact that the small sets must retain all of the information resident in the large sets guarantees that the encoding numbers must become large. You cannot shrink bits!

The possible frustration results from the fact that acceptable summarizations may not always exist. That being said, several summarizing numbers have been introduced, and each is valuable within certain domains of inquiry.

11.3.2 The (Arithmetic) Mean: Expectations

For any outcome-sequence S over a sample space Σ with associated frequency function f_S, the *arithmetic mean* of S is the weighted average of S's elements:

$$\frac{1}{n} \sum_{k \in \Sigma} k \cdot f_S(k)$$

This quantity is often denoted $\text{MEAN}(S)$.

For illustration: Given the 10-element sequence S of Eq. (11.3), we have:

$$\begin{aligned}
\text{MEAN}(S) &= \frac{1}{10} \left(1 \cdot f_S(1) + 2 \cdot f_S(2) + 3 \cdot f_S(3) + 4 \cdot f_S(4) \right) \\
&= \frac{1}{10} \left(1 \cdot 1 + 2 \cdot 2 + 3 \cdot 3 + 4 \cdot 4 \right) \\
&= \frac{1 + 4 + 9 + 16}{10} \\
&= 3
\end{aligned}$$

In this example, $\text{MEAN}(S)$ is an element of S. In general, it needs not be.

We close our introduction to means by exposing one of the most important attributes of the measure $\text{MEAN}(X)$, where X is a random variable—namely, the *linearity* of the measure. We leave to the reader the task of proving the following formal manifestations of this attribute.

Proposition 11.7 *Let X be a random variable over a given sample space S.*

(a) *For any constants a and b,*

$$\text{MEAN}(aX + b) = a \cdot \text{MEAN}(X) + b$$

Thus, the mean is a linear *measure.*

(b) *For any function g of variable X,*

$$\text{MEAN}(g(X)) = \sum_{k=0}^{n} g(k) \cdot Pr[X = k]$$

(c) *Let Y be another random variable over sample space S.*

$$\text{MEAN}(X + Y) = \text{MEAN}(X) + \text{MEAN}(Y)$$

In addition to their *a posteriori* analytical value, arithmetic means can be used as aids to prediction. (We often do this, for example, when we consult mean temperatures before we pack for a trip to an unfamiliar location.) When used predictively, the mean of a sequence S is often called S's *expected value* or, more succinctly, its *expectation*.

11.3.3 Additional Single-Number "Summarizations"

The major source of discontent with the arithmetic mean as a summarization tool is that, while the mean correctly identifies the "center of gravity", or, "balancing point" of a sequence S, it gives no information about the "denseness" (er, complementarily, the "sparseness") of the sequence or about its "skewness" . We illustrate these deficiencies using two simple sequences.

- Regarding "denseness"/"sparseness": Consider the following $2n$-element sequence S_1 chosen from the sample space $\Sigma_1 = \{1, 2, \ldots k\}$.

$$S_1 = \langle 1, 1, \ldots, 1, k, \ldots, k, k \rangle$$

where there are n occurrences of 1 and n occurrences of k. One would imagine that any analysis of an experimental setting that yielded the sequence S_1 of outcomes would *prominently* mention the sequence's sparseness—the fact that all outcomes are at the extremes of their range. The arithmetic mean

$$\text{MEAN}(S_1) = \frac{(k+1)n}{2n} = \frac{k+1}{2}$$

gives no information other than the "center of gravity" of S_1.

- Regarding "skewness": Consider n-element sequences that share the following construction formula, which we illustrate for the case $n = 31$.

$$S_2 = \langle 1, 3, 3, 5, 5, 5, 7, 7, 7, 7, 7, 7, 9, 9, 9, 9, 9, 9, 9, 11, 11, 11, 11, 11, 11, 11, 11, 11, 11, 11 \rangle$$

S_2 is constructed via the following formula. For $i \in \{1, 3, 5, 7, 9, 11\}$

$$f_{S_2}(i) = \begin{cases} 1 & \text{if } i = 1 \\ i - 1 & \text{if } i > 1 \end{cases}$$

The following calculation verifies that $\text{MEAN}(S_2) \approx 8.1$.

$$\begin{aligned} \text{MEAN}(S_2) &= \frac{1}{31}(1 \cdot 1 + 3 \cdot 2 + 5 \cdot 4 + 7 \cdot 6 + 9 \cdot 8 + 11 \cdot 10) \\ &= \frac{1}{31}(1 + 6 + 20 + 42 + 72 + 110) \\ &= \frac{251}{31} \end{aligned}$$

One would expect the analysis of an experiment that yielded the sequence S_2 of outcomes to focus *prominently* on the "skew" of the outcomes toward the larger outcomes. While $\text{MEAN}(S_2)$ correctly identifies the sequence's "center of gravity", it provides no information about how the outcome values are spread out within the sequence.

The median of a sequence. To cope with the second of the preceding complaints, namely, the issue of "skewness", many empiricists advocate reporting on a sequence's *median*, in addition to its mean.

The *median* of an n-element sequence S, denoted MEDIAN(S), determines the "midpoint" of sequence $S = \langle s_1, s_2, \ldots, s_n \rangle$:

- When n is odd, sequence S has a single midpoint:

  $$\text{MEDIAN}(S) = s_{\lceil n/2 \rceil}$$

- When n is even, sequence S has two "midpoints", namely, $s_{n/2}$ and $s_{n/2+1}$. In this case,

  - some empiricists say that S has two medians:

    $$\text{MEDIAN}(S) = \langle s_{n/2}, s_{n/2+1} \rangle$$

  - others use interpolation to maintain the existence of a single median:

    $$\text{MEDIAN}(S) = \tfrac{1}{2} \left(s_{n/2} + s_{n/2+1} \right)$$

Our 31-element sequence S_2, having an odd population, has a single median, namely,

$$\text{MEDIAN}(S_2) = s_{\lceil n/2 \rceil} = s_{16} = 9$$

Explanatory note.

The median appears rather often in the news media, especially in regard to skewed economic distributions such as individual wealth or annual income. In many countries, the mass of wealth is so concentrated in the possession of a small fraction of the population that arithmetic means such as *per capita* income or wealth give no meaningful information about the economic state of the vast majority of the population. Economists find the median income or wealth to be a much more meaningful measure than the arithmetic mean in such environments.

Modes and geometric means. Two other single-number "summarizations" have garnered support in specialist communities. We mention them just for completeness, so that the reader can consider more options when seeking useful summarizations of sequences of empirical outcomes.

If there is a single element $s_i \in S$ whose frequency $f_S(s_i)$ is the maximum among all frequencies $f_S(x)$, then that element is called the *mode* of sequence S and is denoted MODE(S). The mode is typically said not to exist for sequences that have multiple maxima.

The *geometric mean* of an n-element sequence S is a weighted *product* of S's elements. As with the arithmetic mean, the geometric mean is defined in terms of S's frequency function f_S:

$$\text{GEOMETRIC\_MEAN}(S) \overset{\text{def}}{=} \left(\prod_{s_i \in S} s_i^{f_S(s_i)} \right)^{1/n}$$

A bit of arithmetic verifies that the geometric mean is defined by the following property:

The logarithm of GEOMETRIC_MEAN(S) *is the arithmetic mean of the logarithms of the elements of sequence S.*

(Of course, we do not have to specify the bases of the logarithms—as long as all have the same base.)

11.3.4 Higher Statistical Moments

We have thus far discussed measures of "midpoint" in terms of the sequence of outcomes of a fixed experiment. The more common use of means—and of other statistical measures—in daily life involves the means of sequences which are unbounded (hence, often informally thought of as infinite): the average July rainfall in Buenos Aires, the mean time to failure of transistors produced in a given silicon foundry, the average wealth of the residents of Monaco. Within this expanded framework, we distinguish between the *actual mean* of a random variable X, which is often denoted μ, and the *experimental mean* of the random variable, which is usually denoted $\bar{x}$. (Here again the identity of X is usually assumed to be known from context.)

Once we distinguish between actual and experimental means, we must address the question of the quality of predictions which arise from a given experiment: How well does the experimental mean, $\bar{x}$, match the actual mean, μ? The standard approach to this question takes the form of the *higher moments* of random variable X and of the distribution that it ranges over. Within this framework:

- The *first moment* of a distribution is its mean μ. The *first moment* of a sequence S of experimental outcomes is $\bar{x} = \text{MEAN}(S)$.

- The *second moment* of a sequence S of experimental outcomes is termed the *variance* of the sequence. It measures the *distance* between the actual and experimental means of the distribution that S represents:[4]

$$\text{VARIANCE}(S) \overset{\text{def}}{=} \frac{1}{n} \sum_{k \in S} (k - \bar{x})^2 \cdot f_S(|k|)$$

The preceding expression for variance can be replaced with the following streamlined expression.

Proposition 11.8 *For any random variable X,*

$$\text{VARIANCE}(X) = \text{MEAN}(X^2) - \big(\text{MEAN}(X)\big)^2$$

---

[4] Note that the variance and standard deviation are adaptations of the L^2-*norm* of Section 5.5.1 to the setting of probability and statistics.

Proof. (Sketch) The argument centers on the following chain of equations.

$$\text{VARIANCE}(X) = \text{MEAN}(X - \bar{x})^2$$
$$= \text{MEAN}(X^2) - 2\text{MEAN}(X) \cdot \bar{x} + \bar{x}^2$$
$$= \text{MEAN}(X^2) - \left(\text{MEAN}(X)\right)^2$$

The fact that the variance is defined in terms of the *squares* of the components of the distance measure can make it awkward to reason with: The mean is *linear* in the experimental outcomes, but the "corrector" of the mean is *quadratic* in the outcomes. To compensate for this, it is common to use the *standard deviation*

$$\text{STANDARD\_DEVIATION}(S) \overset{\text{def}}{=} \sqrt{\text{VARIANCE}(S)}$$

rather than the variance when reasoning about how well $\bar{x}$ approximates μ.

- Any finite approximation to a measure that is based on infinitely many data points is bound to have imperfections. The variance exposes some of the imperfections of the mean as a summarization tool, but it is just one step in exposing such imperfections. In response, an increasingly fine infinite sequence of moments has been developed, the kth being defined as follows.

$$\text{MOMENT}^{(k)}(S) \overset{\text{def}}{=} \frac{1}{n} \sum_{j \in \Sigma} (j - \bar{x})^k \cdot f_S(j)$$

The third and fourth moments (the cases $k = 3, 4$) report on further aspects of the "spread" of the distribution, including "skew". In general, each higher moment exposes some of the weaknesses of lower moments.

11.4 The Elements of Empirical/Statistical Reasoning

In the "real world", we pragmatically employ a variety of modalities of reasoning. When the setting is appropriate, we can employ (usually simplified and relaxed) versions of the tools for strict *deductive* reasoning that dominate this text. Suitable settings for such tools must be predictably small and stable. When such a setting is either unavailable or inconvenient, we employ reasoning tools that are *inferential* rather than deductive. This is the world of *empirical reasoning*, which we discuss in this section.

Empirical reasoning does not convey the certitude that deductive reasoning does: Empirical truths are established only up to *levels of confidence*—which are established based on mathematical analyses. The aspiring empirical researcher must learn how to:

- design experiments whose outcomes will supply data appropriate for studying a desired phenomenon

- analyze experimental outcomes in order to draw inferences and establish levels of confidence for these inferences.

The problem of how to *design* experiments is completely outside the scope of this text, but there are aspects of the *analysis* of experiments which do build on mathematical foundations, and hence can legitimately be discussed here. While there exist computational statistical packages that relieve one from the burden of mastering the mathematics that underlies much of statistics, one must acquire a *basic* understanding of how such packages work in order to derive justifiable conclusions from their use. Absent such understanding, one can all too easily fall prey to the pitfalls of fallacious statistical reasoning. Here are just two such pitfalls.

1. *Does your experiment deal with the same population as your analysis tools presume?*

   An important instance: Medical researchers may pronounce that a particular substance is safe for human consumption. An elderly patient reports symptoms that seem to follow consuming the substance. It turns out that the safety result was biased by the predominance of *young* experimental subjects—and the results without these young subjects is inconclusive.

2. *Do your analysis tools adequately distinguish between* positive correlation *and* causation?

   Instances of positive correlations that are mistaken for causation abound. Entire websites are devoted to weird coincidences that the unaware confuse with causations.

This section is devoted to introducing a small number of major mathematical concepts and results that impact the related fields of statistics and computational experimentation. We select material which underlies the powerful mathematical and computational tools that statisticians have developed for analyzing experiments and extracting the truths they reveal. We refer readers who wish to delve more deeply into this rich field to sources such as [59], which focuses on the mathematical aspects of the field, and [24], which provides a comprehensive introduction.

Empirical reasoning strives to draw reliable conclusions from the analysis of large corpora of (often numerical) data which have been generated by some type of experiment. The experiments in question usually involve measuring either the outcomes of some process or instances of some natural or artificial phenomenon.

1. An example of the former situation: An instructor in an introductory Physics course derives an estimate of the value of Earth's gravitational force by having students repeatedly drop a ball of known mass and time the duration of the ball's fall before it hits the floor.

2. An example of the latter situation: An observational experiment records, over fixed periods of time, the percentage of human births in which the newborn has a specific gender.

Our ability to derive reliable truths from mountains of observed data relies on two monumental theorems on which the edifice of empirical reasoning stands. We approach these results by introducing conceptual tools that aggregate (outcome sequence)-(frequency function) pairs into well-behaved *probability distributions*. Such distributions:

- help us (approximately) expose analyzable structure within the data
- enable us go beyond the observed data, via extrapolation and interpolation.

We begin our study by introducing a small number of distributions that are among the most useful to experimental researchers.

11.4.1 Probability Distributions

We now take an important step in the direction of abstraction, by noting that many probability frequency functions can fruitfully be viewed as belonging to the same mathematical family. There are often sets of parameters whose values determine a particular family member's probability frequency function or its summarizing measures (means or variances or . . .). These families are called *probability distributions*. Rather than discuss this topic abstractly, we present and analyze a small set of distributions that are very useful in empirical studies.

11.4.1.1 The binomial distribution

Focus on an experiment that involves a sequence of repeated identical events. (A sample event might be: (a) a single toss of a coin or (b) a single roll of a die or (c) a single roll of a pair of dice.) Choose one or more outcomes—say: (a) HEADS for the coin toss; (b) a 3 for the roll of a single die; (c) a 7 or 11 for the roll of a pair of dice—to be *success*, while any other outcome is *failure*. Our sample space Σ is the count of the number of *successes*, so we can discuss and analyze a random variable X that ranges over this space.

By convention, we use the parameter p to represent the probability of a *success* from any event, and we use the parameter q to represent the probability of a *failure*; of course, we insist that $q = 1 - p$.

Consider an experiment that consists of n independent trials. If the probability of k *successes* in this experiment is given by the probability frequency function

$$f_{n,p}(k) \;=\; Pr[X = k] \;=\; \binom{n}{k} p^k q^{n-k} \;=\; \binom{n}{k} p^k (1-p)^{n-k} \qquad (11.4)$$

then we say that random variable X obeys a *binomial distribution* or, equivalently, that X is a *binomially distributed random variable*. In more detail, the subscript on the probability frequency function $f_{n,p}$ identifies this as *the binomial distribution on n items, with probability p of success.*

We remark, as a "sanity check", that the Binomial Theorem (Theorem 5.3) ensures that the joint event

$$[X = 0] \vee [X = 1] \vee \cdots \vee [X = n]$$

has probability

$$\sum_{k=0}^{n} \binom{n}{k} p^k (1-p)^{n-k} = (p + (1-p))^n \equiv 1$$

hence is certainly TRUE.

One can derive simple explicit expressions for the mean and variance of binomially distributed random variables. Let us continue with the random variable X.

Proposition 11.9 *Consider the binomially distributed random variable X that varies over n trials with per-trial success probability p.*

(a) $\text{MEAN}(X) = np$

(b) $\text{VARIANCE}(X) = np(1-p)$

Proof. The proof is by direct calculation.

(a) For the mean of X:

$$
\begin{aligned}
\text{MEAN}(X) &= \sum_{k=1}^{n} k \binom{n}{k} p^k (1-p)^{n-k} \qquad \text{(the } (k=0) \text{ term vanishes)} \\
&= \sum_{k=1}^{n} k \cdot \frac{n!}{k!(n-k)!} p^k (1-p)^{n-k} \\
&= np \cdot \sum_{k=1}^{n} \frac{(n-1)!}{(k-1)!(n-k)!} p^{k-1} (1-p)^{n-k} \\
&= np \cdot \sum_{j=0}^{n-1} \frac{(n-1)!}{j!(n-j-1)!} p^j (1-p)^{(n-j-1)} \\
&= np \cdot (p + (1-p))^{n-1} \\
&= np
\end{aligned}
$$

(b) For the variance of X:

We invoke the properties of means and variances that we have derived thus far, including Proposition 11.7 and the mean-formula of part **(a)**. We find:

$$\text{VARIANCE}(X) = \text{MEAN}(X^2) - (\text{MEAN}(X))^2$$

$$= \left(\sum_{k=0}^{n} k^2 \binom{n}{k} p^k (1-p)^{n-k} \right) - (np)^2$$

$$= \left(\sum_{k=0}^{n} k(k-1) \binom{n}{k} p^k (1-p)^{n-k} \right) + np - (np)^2$$

$$= \left(\sum_{k=2}^{n} k(k-1) \frac{n!}{k!(n-k)!} p^k (1-p)^{n-k} \right) + np - (np)^2$$

$$= \left(\sum_{k=2}^{n} \frac{n!}{(k-2)!(n-k)!} p^k (1-p)^{n-k} \right) + np - (np)^2$$

$$= \left(\sum_{j=0}^{n-2} \frac{n!}{j!(n-(j+2))!} p^{(j+2)} (1-p)^{n-(j+2)} \right) + np - (np)^2$$

$$= n(n-1)p^2 \left(\sum_{j=0}^{n-2} \frac{(n-2)!}{j!((n-2)-j)!} p^j (1-p)^{(n-2)-j} \right) + np - (np)^2$$

$$= n(n-1)p^2 + np - (np)^2$$
$$= n^2 p^2 - np^2 + np - (np)^2$$
$$= np(1-p)$$

This completes the proof.   □

The binomial distribution is the discrete avatar of a *bell-shaped* distribution. We observe this most easily by focusing on the probability frequency function associated with the *coin-tossing* game. We focus on determining the probability of achieving exactly k HEADs via n tosses of a fair coin. This problem is clearly an instance of a binomial distribution with success probability $p = 1/2$; hence, it obeys the probability frequency function

$$f_{n,1/2}(k) = \binom{n}{k} 2^{-n}$$

cf. Eq. (11.4). This simple example is a good starting point for observing the "bell-shape" in such distributions.

Table 11.1 illustrates the probabilities associated with the possible numbers of HEADs achieved in two instances of the coin-tossing game, namely, $n = 10$ tosses and $n = 15$ tosses. Figure 11.8 is a companion to Table 11.1 which presents histograms depicting the values of $\binom{n}{k}$, i.e., the unnormalized probabilities in the table; these values are, of course, rows $n = 10$ and $n = 15$ of Pascal's triangle.

It is worth noting in Fig. 11.8 that the set of binomial coefficients

$$\left\{ \binom{n}{k} \mid k = 0, 1, \ldots, n \right\}$$

Table 11.1 Probabilities (to three significant figures) associated with achieving k HEADS in the coin-tossing game, for $n = 10$ coin tosses and $n = 15$ coin tosses

Value of k	$\binom{10}{k}$	$p_{10}(k)$	$\binom{15}{k}$	$p_{15}(k)$
0	1	0.000977	1	0.0000305
1	10	0.00977	15	0.000458
2	45	0.0439	105	0.00320
3	120	0.117	455	0.0139
4	210	0.205	1365	0.0417
5	252	0.246	3003	0.0916
6	210	0.205	5005	0.153
7	120	0.117	6435	0.196
8	45	0.0439	6435	0.196
9	10	0.00977	5005	0.153
10	1	0.000977	3003	0.0916
11	–	–	1365	0.0417
12	–	–	455	0.0139
13	–	–	105	0.00320
14	–	–	15	0.000458
15	–	–	1	0.0000305

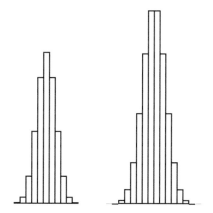

Fig. 11.8 Histograms of the binomial coefficients in the 10th row (left) and the 15th row (right) of Pascal's triangle. The right-hand histogram has been scaled—in the proportion 1:20—in order to fit on the page

hence, also the set of probabilities within a binomial distribution, has a single maximum when n is even but *two* maxima when n is odd. This is important to keep in mind during analyses, so that one does not inadvertently miscount cases.

11.4.1.2 Normal distributions; the *Central Limit Theorem*

Although our focus is squarely on *discrete* mathematical phenomena, it is important to point out concepts and insights that emerge from related *continuous* phenomena. This subsection and the next focus on two of the most important continuous probability distributions.

The most familiar bell curve probability distributions is the continuous *normal* or *Gaussian* distribution. The term "normal" here does not refer to a specific probability distribution but rather to a large family of distributions all of whose "higher moments"—meaning all moments other than the mean and variance (or, equivalently, the standard deviation) are identically 0. Despite this rigorous specification, most non-statisticians identify a normal probability distribution as any one whose probability frequency function has the form

$$f(x) = \frac{1}{\sqrt{2\pi}\sigma} e^{\frac{(x-\mu)^2}{2\sigma^2}}$$

In this formula μ is the *mean* of the distribution, and σ^2 (resp., σ) is its *variance* (resp., its *standard deviation*). In this setting, the parameters μ and σ *specify* each individual normal distribution, which one can unambiguously denote $ND(\mu, \sigma)$. Many people focus on a *standardized* normal distribution as the default, namely $ND(0, 1)$; this distribution has the simpler functional form

$$f(x) = \frac{1}{\sqrt{2\pi}} e^{x^2/2}$$

To those who focus on the standardized normal distribution, $ND(0, 1)$, as the default, the more general family of distributions $ND(\mu, \sigma)$, specified by the values of μ and σ^2, is called a *generalized normal distribution*. Of course, the values μ and σ^2 shift the specified bell curve and change the slope of its ascent and descent.

Cultural aside.

Note that the normal probability frequency function involves two of the fundamental constants of mathematics: e and π. The ubiquity of these constants throughout disparate areas in mathematics testifies eloquently to how fully they deserve to be labeled *fundamental*.

The normal distribution is of immense importance in empirical studies of phenomena whose true distribution is unknown. For such studies, the *Central Limit Theorem* of probability theory and statistics guarantees the following (stated informally). Say that we design an experiment whose outcomes comprise the sample space of interest. Say further that the trials from the experiment are mutually independent—so that their outcomes can be viewed as independently drawings of random variables from the sample space. Then:

No matter how the original random variables are distributed—in particular, they *need not* be distributed normally!—*the averages of the outcomes become normally distributed when the number of outcomes is sufficiently large.*

⊕ **Enrichment note.**

For those readers who would like to see how one can formally state a sweeping result such as the Central Limit Theorem, we provide the following annotated semi-formal statement, of course without proof.

The Central Limit Theorem

Premise: *Let X_1, X_2, ..., X_n be random variables (over $\mathbb{R}$) which are all drawn from a distribution that has mean μ and finite variance σ^2.*

Commentary: Note that we do not demand any uniformity among the X_i other than that they all come from *the same* distribution. When these variables represent outcomes of distinct trials from the same experiment, this demand will be satisfied automatically.

Conclusion: *As the number of X_i, call it n, grows without bound—i.e., for sufficiently large values of n—the random variable*

$$S_n = \frac{X_1 + X_2 + \cdots + X_n}{n}$$

(which is the sample mean) converges to the actual mean μ of $ND(\mu, \sigma^2)$.

There are many versions of the Central Limit Theorem which instantiate certain details of its premise in various ways. The earliest version seems to date to a 1738 edition of *The Doctrine of Chances* [47], the first text on probability theory. In this major work, the French mathematician Abraham de Moivre conceived a brilliant idea while observing the pattern of the various summands in Newton's binomial formula (Theorem 5.3). He observed that as the Theorem expands the polynomial $(x + y)^n$, one can view what is happening as an analogue of quite a different manner of "expansion". (Recall our call in Chapter 2 to look for nonobvious, sophisticated patterns throughout the world.) De Moivre noted that one can think of the polynomial being expanded as a two-event system that one is observing. Viewed this way, the variables x and y become disjunctive sub-events of the compound event $x + y$, and the power n becomes a measure of the number of observations that one has made. This bold intuition suggests that *maybe*, over sufficiently many observations, one might observe the regularity in the two-event system that the Binomial Theorem guarantees in the bivariate polynomial.

This intuition turns out to be correct—as verified by the Central Limit Theorem.

Back to the world of probability distributions . . .

Cumulative distribution functions. In addition to the point-wise information that a probability frequency function yields, we are often interested in the *cumulative* information yielded by the *(cumulative) distribution function*[5] F on S, which is defined in one of the following dual forms for $x \in S$:

$$F^{(\leq)}(x) = Pr[X \leq x] \quad \text{or} \quad F^{(\geq)}(x) = Pr[X \geq x]$$

In *discrete* settings, cumulative distributions are usually defined via summation:

$$F^{(\leq)}(x) = \sum_{y \leq x} f(y) \quad \text{or} \quad F^{(\geq)}(x) = \sum_{y \geq x} f(y)$$

The computational relation of a probability frequency function f to its associated cumulative distribution functions can be quite complicated. For one probability distribution that is very important in empirical studies of many phenomena, this relation is very simple. We refer to the so-called *exponential distributions*.

11.4.1.3 Exponential distributions

Distributions whose probabilities of success fall off at an exponential rate are roughly as familiar to experimenters as are normal distributions, because, as we shall discuss momentarily, such distributions are *memoryless*—as, it turns out, are many natural processes.

A probability frequency function describes an exponential probability distribution just when there exists a *rate parameter*—traditionally denoted by λ—for which

$$f(x) \overset{\text{def}}{=} Pr[X = x] = \lambda \cdot e^{-\lambda x}$$

The parameter λ is often termed the *decay rate* of the distribution because of the behavior of the distribution's *cumulative frequency function F*—specifically its exponential drop-off, or, decay, with increasing λ.

$$F(x) \overset{\text{def}}{=} Pr[X \leq x] = 1 - e^{-\lambda x}$$

The detailed derivations of the properties of the exponential distribution employ continuous mathematical concepts that are outside the scope of this discrete text. But the properties themselves are worth mentioning, both for their (scientific and mathematical) cultural value and as an enticement to acquire the mathematical tools that will give access to this material, which is indispensable to empirical studies in computational domains.

---

[5] Mathematicians usually refer to function F simply as a distribution function; statisticians usually add the qualifier "cumulative".

Proposition 11.10 *For any exponentially distributed random variable X with rate parameter* λ*:*

1. MEAN$(X) = 1/\lambda$.

2. VARIANCE$(X) = 1/\lambda^2$.

3. STANDARD_DEVIATION$(X) = $ MEAN(X).

4. *The memoryless property: For all* $x, y \geq 0$,

$$Pr[X > x+y \mid X > y] = Pr[X > x]$$

Note our invocation in Part 4 of *conditional* probabilities; cf. Section 11.2.2: We are concerned with the probability that the random variable X exceeds the quantity $x+y$ *given that X* exceeds the quantity y.

11.4.2 A Historically Important Observational Experiment

An observational experiment that dates to the early eighteenth century played a historically significant role in the development of the operational methodology of the field of statistics. The experiment arose from the curiosity of the Scottish polymath John Arbuthnot.

Arbuthnot's curiosity, so the story goes, was piqued by a popular belief of the period that more boys are born than girls. In an effort to gauge the accuracy of this belief, Arbuthnot studied an archive that he had access to, which reported on births in London during the 82-year period 1629–1710. As reported in his 1710 paper [8], Arbuthnot observed results such as the following (citing figures just from the first and last years that he reported on):[6]

- In 1629, male births were 52.7% of the total: 5,218 males versus 4,683 females;

- in 1710, male births were 51.8% of the total: 7,640 males versus 7,288 females.

In fact, to Arbuthnot's surprise, the archives exposed that male births *consistently* outnumbered female births throughout the 82 years covered by the archives!

Roughly a century after Arbuthnot, the French mathematician Pierre-Simon Laplace, who invented much of the mathematics underlying probability theory and statistics, reported in [73] that the total number of male births outnumbered the total number of female births in the proportion of roughly 22 (51.2%) to 21 (48.8%).

So ... Are male births actually more common than female births?

Arbuthnot would have responded YES to this query, based on the following reasoning, which is inspired by how one would analyze a coin-tossing experiment. If

---

[6] London's population was growing during this period, which accounts for the roughly 50% growth in the total number of births.

the "coin" is unbiased—so that the genders of newborns occur as a random process with two equiprobable outcomes—then after many trials (births), we would expect to see roughly half of new births be male and roughly half female. However, over the entire span of 82 years, with (very roughly) $10,000$ births per year, Arbuthnot consistently observed more male births than female births, year by year. This convinced him that he was observing a true deviation from randomness—i.e., that there *is* an actual bias toward male births!

Would we draw the same conclusion today as Arbuthnot did roughly three centuries ago? Is the ratio $22/43 \approx 0.512$ of male births that Laplace observed sufficiently larger than the random expectation $1/2 = 0.5$ for us to reject the hypothesis that this deviation could be pure chance?

This is precisely the kind of question that the field of statistics specializes in!

11.4.3 The Law of Large Numbers

An operationally sound answer to the Arbuthnot-Laplace question can be derived from a second major probability-related theorem. This result, which is known (bilingually) as the *Law of Large Numbers* or *la Loi des grands nombres*, focuses on the sequence of random variables $\langle S_1, S_2, \ldots \rangle$, where each S_n is the average of the first n observed outcomes from an experiment of interest. The Law asserts that the variables S_n tend progressively closer to (i.e., *converge to*) the actual idealized mean of all possible outcomes of the experiment—no matter how the outcomes are distributed probabilistically. Said differently:

> *The more times one repeats a random experiment, the more closely the observed outcomes trend toward the actual mean outcome: Repetition improves accuracy!*

The *Law* and the *Central Limit Theorem* are really kindred results. Both yield information about the sums of random variables, which is as significant in practice as it is interesting mathematically, and both yield to essentially the same formal reasoning. Of course, the Law is really the weaker of the two results, in that it yields guarantees only about the *means* of sums of random variables, whereas the Theorem yields guarantees about the sums' *variances* also.

Historical note.

Much of our knowledge about the early history of the Law is due to the 1767 book on statistics by the prolific German writer Jakob von Bielfeld. The Law first appears as a theorem in the 1713 book [16] by the Swiss mathematician Jacques (Latin: Jacobi) Bernoulli. Because of Bernoulli's death, this work was completed by his nephew Nicholas, and the book was published posthumously (the notation "*opus posthumum*" in the citation).

⊕ **Enrichment note.**

As we did with the Theorem, we provide the following annotated semi-formal statement of the sweeping Law of Large Numbers.

The Law of Large Numbers

Premise: *Let $X_1, X_2, \ldots, X_n$ be random variables which are all drawn from a distribution that has mean μ.*

Commentary: The Law and the Central Limit Theorem make similar demands on their inputs.

Conclusion: *As the number n of X_i grows without bound—i.e., for sufficiently large values of n—the random variable*

$$S_n = \frac{X_1 + X_2 + \cdots + X_n}{n}$$

(which is the sample mean) converges to the actual mean μ.

The technical details concerning the Law—including, notably, its proof—are beyond the scope of an introductory text. However, in the anticipation of the readers' ongoing educations and careers, we close our discussion of probability theory and statistics with simple examples of the Law "in action", plus some final words on the Arbuthnot-Laplace question.

We begin with two illustrative examples.

1. We know from earlier sections the probabilistic behavior of experiments involving rolls of a six-sided die. Since each of the six outcomes is equiprobable, we can argue that as we roll the die many times, we expect to see the *average* outcome

$$\frac{1+2+3+4+5+6}{6} = 3.5$$

Of course, we cannot expect to see this average roll by roll—*that is why we need higher moments to truly understand the behavior of experiments*! But the Law guarantees that as we roll the die over and over, we should observe the sequence of average roll-values tending to 3.5.

2. We can play an illustrative (but uninteresting) "game" using a barrel that contains both white and black balls. In the game, we repeatedly draw a ball from the barrel, record its color, and replace the ball into the barrel. What will our records show about the relative numbers of black and white balls we shall observe?

Say for illustration that the barrel contains 300 white balls and 200 black balls, for a 60%–40% split.

Clearly, there is always a greater chance of pulling a white ball from the barrel than a black ball. The Law of Large Numbers tells us quite a bit more: It tells us that as we play the "game" for longer and longer, the proportion of white balls drawn will get closer and closer to 60%.

Let us put this result in perspective.

Of course, if we draw only a single ball—a single trial in the "game"—then the result is simply either one white ball and no black balls or vice versa. It is only after many trials that we will encounter interesting tallies. How many trials do we have to do before we begin to see the probability of a white ball edge toward 60%? Will we observe close to 60% white balls after 10 trials? 100 trials? 1,000 trials? In fact the proof of the Law will answer this question, to within whatever *level of confidence* we want! If, for instance, we want to be sure that the proportion of white balls drawn is between 59.9% and 60.1%, i.e., is 60% *with a confidence level of* 99%, then the analysis underlying the Law will tell us how many trials we have to do to achieve this.

As an aside, in commercial applications, a 95% confidence level is often considered an acceptable compromise between the (probabilistic) guarantee of the Law and the work required to achieve that guarantee.

We return in closing to the Arbuthnot–Laplace question in the light of the Law of Large Numbers. Imagine that a researcher had set out in 1629 to test the hypothesis that male births and female births were equally likely in London (i.e., each had a probability of 1/2). After collecting data on the literally hundreds of thousands of actual births during the period 1629–1710, the researcher would calculate that male births exceeded female births. The Law tells us that the hypothesis of a balanced probability is false.

The samples are large enough to refine this analysis. Using results from the 82-year span, the researcher observed a ratio of male births to female births that was 22/43 (i.e., 22 male births for every 21 female births). The Law of Large Numbers therefore told the researcher that the actual mean ratio of male births to female births was very close to the observed ratio of 22/43.

Interestingly, the gender imbalance in births exposed by Arbuthnot can still be observed today—however with ratios that may not be exactly 22 to 21, depending on the country and the periods of observations. The reason for the imbalance has not yet been clearly identified.

11.5 Exercises: Chapter 11

Throughout the text, we mark each exercise with 0 or 1 or 2 occurrences of the symbol $\oplus$, as a rough gauge of its level of challenge. The 0-$\oplus$ exercises should be accessible by just reviewing the text. We provide *hints* for the 1-$\oplus$ exercises; Appendix H provides *solutions* for the 2-$\oplus$ exercises. Additionally, we begin each exercise with a brief explanation of its anticipated value to the reader.

1. Counting arrangements

LESSON: Experience with combinatorial calculations

a. Counting repetitions

Prove Proposition 11.2:
The number of possible outcomes when a fair six-sided die is rolled n times is 6^n.

b. Counting permutations

Prove Proposition 11.4: The number of permutations, $P(n)$, of a n-element set equals n!

c. Counting arrangements

Prove that the number of ways of selecting k unordered items out of a set of n items is $\binom{n}{k}$. Your proof must be based on Proposition 11.4.

2. ⊕ Counting replicated triangles

LESSONS: Experience with detecting sophisticated patterns and using recurrences.

Let us begin with a single isosceles triangle, call it T_1, as in Fig. 11.9(left).

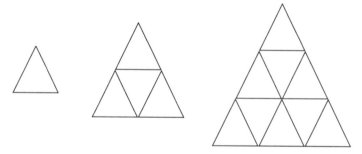

Fig. 11.9 The first three stages of triangle-replication.

Next, let us create the isosceles triangle T_2 of Fig. 11.9(center) by taking four copies of T_1 and placing them together in the manner depicted in the figure. Now, T_2 is similar to T_1, in the geometric sense of the word, and its height and base are twice those of T_1. What really interests us, though, is that the *drawing* of T_2 in the figure contains *five* triangles—namely, the four copies of T_1 *plus* the isosceles triangle that contains these four copies.

As our next step, we iterate the replication process by appending to T_2 an additional layer consisting of five copies of T_1 arranged as indicated in Fig. 11.9(right). The resulting isosceles triangle, T_3, has a base and a height that are three times those of T_1. Once again, what really interests us is that the *drawing* of T_3 in the figure contains *thirteen* triangles—namely, the three copies of T_2 *plus* the isosceles triangle that contains these three copies. Specifically, we can discern *nine* copies of T_1, *three* copies of T_2, *plus* the all-encompassing big isosceles triangle. Of course, we can now append to T_3 an additional layer consisting of seven copies of T_1, and we can then append an additional layer consisting of nine copies of T_1, and on and on.

As we make these successive augmentations, how does the progression of numbers of triangles grow? It is clear that there is a recurrence lurking here. Your challenge is to discover it. To this end, let N_k denote the number of triangles within triangle T_k. We have just seen that $N_1 = 1$, $N_2 = 5$, and $N_3 = 13$.

   a. ⊕ *Compute N_4.*

   b. ⊕⊕ *Develop a recurrence for N_k* (i.e., the general case).

3. **More about birthdays**
   LESSON: Experience with combinatorial probability

   a. *Determine the probability that a person you meet on the street was born on February 28.*

   b. *Determine the probability that a person you meet on the street was born on February 29.*

   c. *Find a number n that satisfies the following. With probability at least $1/2$, any collection of n people contains ≥ 3 people who share the same birthday.*

4. ⊕ **Further analysis of poker hands**
   LESSONS: Experience analyzing simple combinatorial situations
   Determine the probability of being dealt a two-pair hand in a fair game of poker.
   Hint. As noted in the text, a two-pair hand is any permutation of

$$X, X, Y, Y, Z$$

where X, Y, and Z are distinct face values. If you try to pick these three values separately, you will be subjecting yourself to a complex case analysis. You are much better off to select the three distinct values from the 13 possible face values in a single conceptual step. Then you must select which of the three face values will play the role of Z (the unpaired card). At that point, you should be well on your way to the answer!

5. **Further analysis of simple dice-rolling games**
   LESSONS: Experience analyzing simple combinatorial situations

   a. *Analyze the game of craps in the same way as we did for the (unordered) sum-of-three game.*

b. One notes in Fig. 11.7 that in the ordered version of the sum-of-three game, the sum k is engendered by the same number of configurations as is the sum $21 - k$. *Explain why this is true.*

6. Some statistical calculations

LESSON: Experience with combinatorial calculations

Prove Proposition 11.7(a), which asserts the linearity of the mean.

7. Monge shuffle: mathematical party tricks

LESSON: Experience with analyzing combinatorial patterns

We consider two ways of shuffling cards which are associated with the eighteenth-nineteenth-century French mathematician Gaspard Monge. We employ the shuffle-techniques as vehicles for honing your skills in detecting combinatorial patterns. Others have used them as party tricks.

For both shuffle-techniques. we begin with a deck of $2n$ distinct cards. We employ the running example of the following small deck, where $n = 5$:

$$(1, 2, 3, 4, 5, 6, 7, 8, 9, 10)$$

For both shuffles of the cards, we cut the deck in the middle, to create two n-card decks:

$$(1, 2, 3, 4, 5), (6, 7, 8, 9, 10)$$

We then merge the two n-card decks to again obtain a single $2n$-card deck. The two shuffles differ in their merging techniques.

a. ⊕ The riffle shuffle

The first shuffle, the *riffle shuffle*, alternates the "top" cards from the right-hand and left-hand n-card decks; see Fig 11.10. The merged deck has the form

$$(6, 1, 7, 2, 8, 3, 9, 4, 10, 5)$$

In the party-game incarnation of the riffle shuffle, you demonstrate the cut-then-merge process of Fig 11.10, and you suggest to your audience that this process is a good first step in really mixing up the cards. If this were so, then repeating the step several times should be a good way to obtain a "random" mixture of the cards.

Let's see what happens after several steps of this cut-then-merge process:

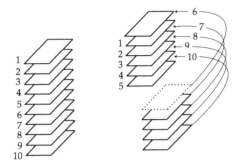

Fig. 11.10 The riffle shuffle for ten cards ($n = 5$)

$$(1\ 2\ 3\ 4\ 5\ 6\ 7\ 8\ 9\ 10) \to (1\ 2\ 3\ 4\ 5)\ (6\ 7\ 8\ 9\ 10)$$
$$\swarrow$$
$$(6\ 1\ 7\ 2\ 8\ 3\ 9\ 4\ 10\ 5) \to (6\ 1\ 7\ 2\ 8)\ (3\ 9\ 4\ 10\ 5)$$
$$\swarrow$$
$$(3\ 6\ 9\ 1\ 4\ 7\ 10\ 2\ 5\ 8) \to (3\ 6\ 9\ 1\ 4)\ (7\ 10\ 2\ 5\ 8)$$
$$\swarrow$$
$$(7\ 3\ 10\ 6\ 2\ 9\ 5\ 1\ 8\ 4) \to (7\ 3\ 10\ 6\ 2)\ (9\ 5\ 1\ 8\ 4)$$
$$\swarrow$$
$$(9\ 7\ 5\ 3\ 1\ 10\ 8\ 6\ 4\ 2) \to (9\ 7\ 5\ 3\ 1)\ (10\ 8\ 6\ 4\ 2)$$
$$\swarrow$$
$$(10\ 9\ 8\ 7\ 6\ 5\ 4\ 3\ 2\ 1) \to (10\ 9\ 8\ 7\ 6)\ (5\ 4\ 3\ 2\ 1)$$
$$\swarrow$$
$$(5\ 10\ 4\ 9\ 3\ 8\ 2\ 7\ 1\ 6) \to (5\ 10\ 4\ 9\ 3)\ (8\ 2\ 7\ 1\ 6)$$
$$\swarrow$$
$$(8\ 5\ 2\ 10\ 7\ 4\ 1\ 9\ 6\ 3) \to (8\ 5\ 2\ 10\ 7)\ (4\ 1\ 9\ 6\ 3)$$
$$\swarrow$$
$$(4\ 8\ 1\ 5\ 9\ 2\ 6\ 10\ 3\ 7) \to (4\ 8\ 1\ 5\ 9)\ (2\ 6\ 10\ 3\ 7)$$
$$\swarrow$$
$$(2\ 4\ 6\ 8\ 10\ 1\ 3\ 5\ 7\ 9) \to (2\ 4\ 6\ 8\ 10)\ (1\ 3\ 5\ 7\ 9)$$
$$\swarrow$$
$$(1\ 2\ 3\ 4\ 5\ 6\ 7\ 8\ 9\ 10)$$

So, this does not look too random! After *five* (which, "coincidentally", equals n) cut-then-merge steps, the deck has been reversed, and after $2n$ cut-and-merge steps, it has been replicated in its original order!

As mathematicians, we do not like "coincidences"!

Prove the following assertion.

Proposition 11.11 *Let p be a prime ($p > 2$), and let $n = \frac{1}{2}(p - 1)$. Let us begin with a deck of $2n$ distinct cards and perform the riffle shuffle on the* *deck. After some number m of cut-then-merge steps, where m divides p − 1, the riffle shuffle replicates the initial deck.*

Studying our sample sequence of cut-then-merge steps should provide a valuable hint.

b. ⊕⊕ **The Monge shuffle**

We now provide another merge procedure, which yields a sophisticated variant of the riffle shuffle. In each stage of this variant—each corresponding to a cut-then-merge step of the simple shuffle—we rearrange the $2n$-card deck directly, from the middle outward, via the following regimen:

- Card #1 of the original deck is placed in position $n + 1$ of the shuffled deck
- Card #2 of the original deck is placed in position n of the shuffled deck
- Card #3 of the original deck is placed in position $n + 2$ of the shuffled deck
- Card #4 of the original deck is placed in position $n − 1$ of the shuffled deck
  ... and so on, until the original deck is empty.

The first few steps of a stage of the Monge shuffle are illustrated for the case $n = 4$ in Fig. 11.11; one complete stage is illustrated in the table following the figure.

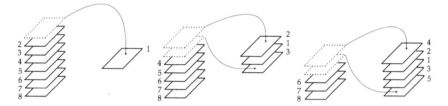

Fig. 11.11 The Monge shuffle for eight cards ($n = 4$): Step 1 (left), Step 2 (center), Step 3 (right)

Step	Original deck		Shuffled deck
1	$(1\ 2\ 3\ 4\ 5\ 6\ 7\ 8)$	$\rightarrow$	$(-\ -\ -\ -\ 1\ -\ -\ -)$
2	$(-\ 2\ 3\ 4\ 5\ 6\ 7\ 8)$	$\rightarrow$	$(-\ -\ -\ 2\ 1\ -\ -\ -)$
3	$(-\ -\ 3\ 4\ 5\ 6\ 7\ 8)$	$\rightarrow$	$(-\ -\ -\ 2\ 1\ 3\ -\ -)$
4	$(-\ -\ -\ 4\ 5\ 6\ 7\ 8)$	$\rightarrow$	$(-\ -\ 4\ 2\ 1\ 3\ -\ -)$
5	$(-\ -\ -\ -\ 5\ 6\ 7\ 8)$	$\rightarrow$	$(-\ -\ 4\ 2\ 1\ 3\ 5\ -)$
6	$(-\ -\ -\ -\ -\ 6\ 7\ 8)$	$\rightarrow$	$(-\ 6\ 4\ 2\ 1\ 3\ 5\ -)$
7	$(-\ -\ -\ -\ -\ -\ 7\ 8)$	$\rightarrow$	$(-\ 6\ 4\ 2\ 1\ 3\ 5\ 7)$
8	$(-\ -\ -\ -\ -\ -\ -\ 8)$	$\rightarrow$	$(8\ 6\ 4\ 2\ 1\ 3\ 5\ 7)$

Prove the following assertion.

Proposition 11.12 *For any integer $n \in \mathbb{N}^+$: If $4n + 1$ is prime, then performing $2n$ steps of the Monge shuffle on a deck of $2n$ distinct cards restores the deck to its original order.*

Chapter 12
Graphs I: Representing Relationships Mathematically

Oh what a tangled web we weave ...
Sir Walter Scott (*Marmion*)

Graphs provide one of the richest technical and conceptual frameworks in the world of computing. They provide tangible structural representations of the manifold data interrelationships that are needed to craft sophisticated algorithms. Indeed, they embody valuable abstractions of relationships of all sorts, hence must be well understood in order to discuss entities as varied as web-search engines and social networks with precision and rigor. As with most of the areas we cover in this text, graph-oriented concepts must be studied "in layers", with each reader delving to a depth that is appropriate for their goals. The fact that so much of modern society depends on understanding interconnections of all sorts suggests that every reader should master at least the basic concepts of graph theory, as exposed by the first two sections of this chapter. The remainder of this chapter, and the entirety of the next, provide a buffet of more specialized and/or advanced topics, with delicacies for the aspiring mathematician, computer scientist, computer engineer, computational scientist, data scientist, ... Please peruse, sample, and enjoy.

Getting more technical: Many developments in computing technology over recent decades have made it imperative that graphs no longer be viewed as the static objects introduced in the early decades of computational studies. For instance, while it was innovative in the 1960s to employ graphs and trees computationally as abstractions of data structures, such a view is standard today. Similar remarks, perhaps with differing dates, can be made about graphs as vehicles for representing the flow of control and information and as vehicles for representing interconnectivity among both concepts and populations. Applications ranging from databases to web-search engines to social networks demand an appreciation of graphs as dynamic objects. The ways in which we manipulate graphs have broadened dramatically, as the technologies underlying computation and communication have evolved. We must now study graphs that grow, in order to understand communication within social media.

© Springer Nature Switzerland AG 2020
A. L. Rosenberg, D. Trystram, *Understand Mathematics, Understand Computing*,
https://doi.org/10.1007/978-3-030-58376-7_12

We must understand how to decompose graphs, in order to impose structure on the way (both hardware and software) networks grow to hitherto unencountered sizes. We must understand how to color the component vertices and edges of graphs in order to orchestrate complex parallel and distributed processes. We must understand how to proceed beyond the (binary) point-to-point connections that the simplest genres of graph prescribe, in order to understand complex nonbinary group dynamics.

This chapter and the next provide the basic concepts and tools necessary to incorporate all manner of graph-theoretic reasoning into our intellectual toolkits.

12.1 Generic Graphs: Directed and Undirected

12.1.1 The Basics of the World of Graphs

The basic components of a graph $\mathscr{G}$ are *vertices*—objects which represent things one wants to interrelate—and *edges*—each edge connecting two of $\mathscr{G}$'s vertices. Formally, we view an edge as a two-element set of $\mathscr{G}$'s vertices. For each graph $\mathscr{G}$, we denote the set of $\mathscr{G}$'s vertices by $\mathscr{N}_\mathscr{G}$ and the set of $\mathscr{G}$'s edges by $\mathscr{E}_\mathscr{G}$.[1]

Explanatory notes.
The singular form of "vertices" is "*vertex*".
The word "*node*" is often used as a synonym of "vertex". We usually employ the somewhat more common "vertex".

A *subgraph* $\mathscr{G}'$ of a graph $\mathscr{G}$ is a graph whose vertices are a subset of $\mathscr{G}$'s and whose edges are a subset of $\mathscr{G}$'s that interconnect only vertices of $\mathscr{G}'$.

Bipartite graphs. A graph $\mathscr{G}$ is *bipartite* if its vertices can be partitioned into (by definition of "partition", disjoint) sets X and Y in such a way that every edge of $\mathscr{G}$ has one endpoint in X and the other in Y.

Among the many scenarios that benefit from modeling via bipartite graphs are those in which set X represents a base set—say, the members of an organization—while set Y represents a collection of aggregates of elements of X—say, the various committees that organization members can serve on. In the depicted scenario, an edge that connects $x \in X$ with $y \in Y$ could expose x's membership on committee Y. An illustrative such situation can be found in the proof of Lemma 13.2.

Directed graphs. Each of a graph's edges connotes some sort of sibling-like relationship among vertices of "symmetric" status. Of course, there are relationships that do not enjoy such symmetry: parenthood between people and dependence between computational procedures exemplify such asymmetries. We model such asymmetry

---

[1] We often omit the subscript "$\mathscr{G}$" in notation such as $\mathscr{N}_\mathscr{G}$ or $\mathscr{E}_\mathscr{G}$ unless there is danger of ambiguity.

by invoking a *directed*, asymmetric analogue of the graphs we have discussed to this point.

A *directed graph* (*digraph*, for short) $\mathcal{G}$ is given by a set of *vertices* $\mathcal{N}_\mathcal{G}$ and a set of *arcs* (or *directed edges*) $\mathcal{A}_\mathcal{G}$. Each arc of $\mathcal{G}$ has the form $(u \to v)$, where $u, v \in \mathcal{N}_\mathcal{G}$; we say that this arc goes *from u to v*.

In many situations involving directed graphs, it is important to deal not only with a digraph $\mathcal{G}$ but also with the *reverse digraph* of $\mathcal{G}$, which is the digraph obtained by *reversing* all of $\mathcal{G}$'s arcs (so that they point in the opposite direction). There is no universally accepted notation for the reverse of digraph $\mathcal{G}$, but one often encounters notational embellishments of "$\mathcal{G}$", such as $\widehat{\mathcal{G}}$ or $\widetilde{\mathcal{G}}$, used for this purpose. One sometimes encounters situations wherein only clerical details need be changed in order to convert an argument about, or an operation on, digraph $\mathcal{G}$ to the analogous argument about, or operation on, $\mathcal{G}$'s reverse digraph.

Undirected graphs are usually the default concept, in the following sense: When $\mathcal{G}$ is described as a "graph," with no adjectival discriminator "directed" or "undirected", it is understood that $\mathcal{G}$ is an *undirected* graph.

Paths and cycles. A *path* in an undirected graph is a sequence of vertices within which every adjacent pair is connected by an edge. A path is *simple* if all of its constituent vertices are distinct. A path is a *cycle* if all vertices in the sequence are distinct except for the first and last, which are identical.

Paths and cycles in directed graphs are defined similarly, except that now every adjacent pair of vertices must be connected by an *arc*, and all arcs must "point in the same direction".

In detail: A *path* in a digraph $\mathcal{G}$ is a sequence of arcs that share adjacent endpoints, as in the following $(n-1)$-arc path in $\mathcal{G}$ from vertex u_1 to vertex u_n:

$$(u_1 \to u_2), (u_2 \to u_3), \ldots, (u_{n-2} \to u_{n-1}), (u_{n-1} \to u_n) \qquad (12.1)$$

The path (12.1) is often written in the more succinct form

$$u_1 \to u_2 \to u_3 \to \cdots \to u_{n-2} \to u_{n-1} \to u_n$$

The just-described path makes sense only when every vertex u_i belongs to $\mathcal{N}_\mathcal{G}$ and every one of its arcs $(u_i \to u_{i+1})$ belongs to $\mathcal{A}_\mathcal{G}$.

When $u_1 = u_n$, then the path (12.1) is a *cycle* that contains vertices $u_1, \ldots, u_{n-1}$.

12.1.1.1 Locality-related concepts

- Let $\mathcal{G}$ be an undirected graph having vertex-set $\mathcal{N}$ and edge-set $\mathcal{E}$: For each edge $\{u, v\} \in \mathcal{E}$, we say that vertices u and v are *neighbors* (in $\mathcal{G}$) or, equivalently, that they are *adjacent* (in $\mathcal{G}$).

  The *degree* of vertex $u \in \mathcal{N}$ is the number of neighbors that u has.

$\mathcal{G}$ is *regular* (or is a *regular graph*) if all vertices have the same degree.

- Let $\mathcal{H}$ be a directed graph having vertex-set $\mathcal{N}_{\mathcal{H}}$ and arc-set $\mathcal{A}_{\mathcal{H}}$: For each arc $(u \to v) \in \mathcal{A}$, we say that vertex u is a *predecessor* of v (in $\mathcal{H}$) or, equivalently, that vertex v is a *successor* of u (in $\mathcal{H}$).

The words "*source*" and "*target*" sometimes replace the words "*predecessor*" and "*successor*".

The *out-degree* of vertex $u \in \mathcal{N}$ is the number of successors that u has. The *in-degree* of vertex $v \in \mathcal{N}$ is the number of predecessors that v has.

$\mathcal{G}$ is *in-regular* if all vertices have the same in-degree; it is *out-regular* if all vertices have the same out-degree.

We can observe a few important facts that can be useful when analyzing a broad range of computation-related issues involving graphs (either as auxiliary notions or as subjects of discourse).

Proposition 12.1 (a) *An n-vertex digraph $\mathcal{G}$ has no more than n^2 arcs.*
(b) *An n-vertex undirected graph $\mathcal{G}$ has no more than $\binom{n}{2}$ edges.*

Proof. (a) The set $\mathcal{A}$ of arcs of $\mathcal{G}$ is a subset of the set of ordered pairs of vertices of $\mathcal{G}$. This latter number is clearly n^2, because one can choose the first vertex of a pair in n ways and then *independently* choose the second vertex in n ways.

(b) The stated quantity is the number of two-vertex subsets of $\mathcal{N}$. To see this, start by listing the n^2 ordered pairs of vertices of $\mathcal{G}$. First, eliminate from the list all n pairs whose first and second elements are equal: a set of the form $\{u, u\}$ has only one element, hence is not an edge of $\mathcal{G}$. Then, for each distinct pair of vertices $u, v \in \mathcal{N}$, eliminate one of the two ordered pairs, $\langle u, v \rangle$ and $\langle v, u \rangle$: both of these ordered pairs lead to the same unordered doubleton set $\{u, v\}$, hence to the same edge of $\mathcal{G}$. After these eliminations, we are left with

$$\frac{n^2 - n}{2} = \binom{n}{2}$$

two-element subsets of $\mathcal{N}$, from which we choose the edges of $\mathcal{G}$. □

Proposition 12.2 *In any undirected graph, the number of vertices of odd degree is even.*

We leave the proof of Proposition 12.2 as an exercise.

We sometimes use the term *neighbor* also when speaking of a *directed* graph, $\mathcal{G}$, by using an obvious analogy to the undirected version of $\mathcal{G}$ (in which arcs lose their directionality). In detail, when we say that vertices u and v are "neighbors" in the digraph $\mathcal{G}$, we mean that $\mathcal{A}$ contains at least one of the arcs $(u \to v)$ or $(v \to u)$.

More typically, we use terminology that is more faithful to digraph $\mathscr{G}$'s direction-ality. If $\mathscr{A}$ contains the arc $(u \rightarrow v)$, then we call v a *(direct) successor* of u, and we call u a *(direct) predecessor* of v. The term *parent* often replaces "predecessor vertex", and the term *child* often replaces "successor vertex", especially when $\mathscr{G}$ is a directed *tree*.

12.1.1.2 Distance-related concepts

Path-length and distance. The *length* of a path or cycle in an undirected graph is the number of edges in the path or cycle; the analogous length in a digraph is the number of arcs.

The *distance* in graph $\mathscr{G}$ between vertex u_1 and vertex u_n is the length of a short-est path that connects u_1 with u_n. When $\mathscr{G}$ is a digraph, then the *distance* from vertex u_1 to vertex u_n is the length of a shortest directed path that leads from u_1 to u_n. Note that the existence of the path (12.1) in digraph $\mathscr{G}$ means that the distance from u_1 to u_n is *no greater than* $n-1$: there might exist shorter paths in $\mathscr{G}$ from u_1 to u_n.

Distance and diameter in a digraph. Extrapolating from our discussion of path (12.1): The *distance* from vertex u_1 to vertex u_n in the digraph $\mathscr{G}$ is the smallest number of arcs in any path from u_1 to u_n. In detail:

$$\text{DISTANCE}(u_1, u_n) \begin{cases} = 0 & \text{if } u_1 = u_n \\ \leq n-1 & \text{if there is a path } (12.1) \text{ from } u_1 \text{ to } u_n \\ = \infty & \text{if there is no path } (12.1) \text{ from } u_1 \text{ to } u_n \end{cases} \quad (12.2)$$

The *diameter* of a directed graph $\mathscr{G}$ is the largest distance between two vertices of $\mathscr{G}$, i.e., the largest number d for which there exist vertices $u_1, u_n \in \mathscr{N}$ such that $\text{DISTANCE}(u_1, u_n) = d$. Note that when discussing digraphs, we always use *directed* paths when defining distance.

Distance and diameter in an undirected graph. It is easy to extrapolate from our discussion of *directed* paths in *directed* graphs, as illustrated in (12.1), to an anal-ogous notion of *undirected* path in an *undirected* graph. Informally, one derives each undirected notion by removing the arrowheads in from the arcs in a directed graph. We leave details to the reader. Having thus converted a directed graph into an undirected one, call it $\mathscr{G}$: The *distance between* vertex u_1 and vertex u_n in $\mathscr{G}$ is the smallest number of edges in any path from u_1 to u_n. In detail:

$$\text{DISTANCE}(u, v) \begin{cases} = 0 & \text{if } u = v \\ \leq n-1 & \text{if there is a path } (12.1) \text{ between } u \text{ and } v \\ = \infty & \text{if there is no path } (12.1) \text{ between } u \text{ and } v \end{cases} \quad (12.3)$$

The *diameter* of an undirected graph is the largest distance between two vertices, i.e., the largest number d for which there exists a pair of vertices u_1, u_n such that $\text{DISTANCE}(u_1, u_n) = d$. Note that when discussing undirected graphs, we always use *undirected* paths when defining distance.

Explanatory note.

Our discussion of inter-vertex distances within graphs has focused on shortest (or on longest) path problems in *unweighted* graphs. A variety of important applications can be modeled via path-distance problems in graphs $\mathscr{G}$ each of whose edges $\{u,v\}$ is *weighted* with a number that measures the cost of going between vertices u and v in $\mathscr{G}$. Of course, when graph $\mathscr{G}$ is directed, then the arcs $(u \to v)$ and $(v \to u)$ can have different weights, to model situations wherein going from u to v is easier/cheaper than going from v to u. Happily, determining shortest (or longest) paths in a directed or undirected graph $\mathscr{G}$ can be accomplished "efficiently"—which in the world of algorithmics means "in a number of steps that is polynomial in the size of $\mathscr{G}$".

We have thus far considered only "valid", i.e., non-redundant, paths, i.e., sequences of vertices with no repetitions. At times, say, when devising algorithms, one often wishes to employ path-related concepts without checking sequences of vertices for repetition-freeness. One can always avoid problems via the following ploy. When dealing with a sequence of vertices of unknown validity, simply enforce validity by proceeding along a given sequence of vertices and skipping every subsequence that appears between consecutive occurrences of the same vertex. Of course, each such subsequence is a cycle.

Even at this early stage of our discussion of graphs, we can state and prove an important nontrivial fact. Every graph $\mathscr{G}$ that has at least one edge contains at least one path; if $\mathscr{G}$ has sufficiently many edges, then it also contains at least one cycle. We word this result in the language of undirected graphs only for simplicity; the analogous result for digraphs is much more complicated.

Proposition 12.3 *If every vertex of a graph $\mathscr{G}$ has degree ≥ 2, then $\mathscr{G}$ contains a cycle.*

Proof. Let us assume that we have a cycle-free graph $\mathscr{G}$ all of whose vertices have degree ≥ 2. We invoke the Pigeonhole Principle to find a cycle in $\mathscr{G}$.

Let us view graph $\mathscr{G}$ as a park: Every vertex of $\mathscr{G}$ is a statue, and every edge is a lane between two statues. (We use the word "lane" rather than the more common "path" because of the technical meaning of "path" within the world of graphs.) The fact that every vertex of $\mathscr{G}$ has degree ≥ 2 means that if we take a stroll through Park $\mathscr{G}$, then every time we leave a vertex $v \in \mathcal{N}\mathscr{G}$, we can use a *different* edge/lane than we used when we came to v.

Consider now the following *gedanken* experiment ("thought" experiment). Say that we initially paint every statue in Park $\mathscr{G}$ *green* and that as we stroll through the park, we repaint every statue we encounter *red*. So, we begin our stroll at some statue v_0, and we paint that statue *red*. Then we leave v_0 via some lane, and we encounter some other statue v_1. Because we have not encountered v_1 before on our stroll, it is *green*. Now that we encounter v_1 on our stroll, we paint it *red*. Next, we leave v_1 via a different lane from the one we used to get to it. (The degree property of $\mathscr{G}$ guarantees that we can do this.) This new lane leads us to a statue v. Now, if statue v is *red*, then we have discovered a cycle in $\mathscr{G}$—we are encountering v for the second time! If statue v is *green*, then we rename v as v_2, and we paint v_2 *red*. We continue out stroll in the described manner. In detail: At stage m of this process, we leave the newly-painted statue v_m via a lane that is distinct from the lane we used to get to v_m, and we reach a statue v. If statue v is *red*, then we have discovered a cycle in $\mathscr{G}$! If statue v is *green*, then we rename v as v_{m+1}, and we paint v_{m+1} *red*, and we continue our stroll.

Because Park $\mathscr{G}$ is *finite*, having precisely n statues/vertices, we can hope to encounter *green* statues along our stroll no more than n times. After this point, every lane will lead us to a *red* statue—i.e., to a cycle in $\mathscr{G}$. $\square$

Labeled graphs and digraphs. It is sometimes useful to endow the arcs of a digraph with labels from an alphabet Σ. When so endowed, the path (12.1) would be written in a form such as

$$\left(u_1 \overset{\lambda_1}{\to} u_2\right), \left(u_2 \overset{\lambda_2}{\to} u_3\right), \ldots, \left(u_{n-2} \overset{\lambda_{n-2}}{\to} u_{n-1}\right), \left(u_{n-1} \overset{\lambda_{n-1}}{\to} u_n\right)$$

where the λ_i denote symbols from Σ. Labeled paths also are often written in a succinct manner, as:

$$u_1 \overset{\lambda_1}{\to} u_2 \overset{\lambda_2}{\to} u_3 \overset{\lambda_3}{\to} \cdots \overset{\lambda_{n-3}}{\to} u_{n-2} \overset{\lambda_{n-2}}{\to} u_{n-1} \overset{\lambda_{n-1}}{\to} u_n$$

12.1.1.3 Connectivity-related concepts

The reader will note that we have nowhere guaranteed that there is always a path that connects each vertex u with each other vertex v. Indeed, we have important adjectives that describe when such *inter-vertex access* is possible. We say that a graph $\mathscr{G}$ is *connected* if every pair of vertices $u, v \in \mathscr{N}_\mathscr{G}$ is connected by a path in $\mathscr{G}$. If graph $\mathscr{G}$ is *not* connected, then it is the disjoint union of some number c of connected subgraphs, usually called $\mathscr{G}$'s *(connected) components*. Of course, $\mathscr{G}$ is connected just when $c = 1$; i.e., there is a single connected component.

There are, of course, analogous connectivity-related adjectives for directed graphs. Most notions involving connectivity and inter-vertex accessibility are discussed using the same adjectives for undirected graphs and directed graphs—but with the word "directed" prominently asserted in the latter case. *One must be meticulous in distinguishing the directed and undirected versions of accessibility and connectedness when discussing digraphs.* There is one exception to the terminological looseness we have just mentioned: We say that a digraph $\mathscr{G}$ is *strongly connected* if there is a directed path from every vertex of $\mathscr{G}$ to every other vertex.

There is an important connection between graph-connectivity and equivalence relations. The proof of the following important result is left as an exercise.

Proposition 12.4 *The property of inter-vertex accessibility in graphs is an equivalence relation.*

12.1.2 Graphs as a Modeling Tool: The 2SAT Problem

Now that we have the basic notions relating to connectivity in graphs, we can develop the proof of Proposition 3.6. For the reader's convenience, we restate the Proposition here.

Proposition 12.5 (Restatement of Prop. 3.6) *The* 2SAT *problem can be solved in polynomial time.*

That is, given any instance Φ of 2SAT, *one can determine in time polynomial in the number of literals in Φ whether there exists a satisfying assignment of truth-values to the variables of Φ.*

We develop the proof of Proposition 12.5 by focusing on an instance of the 2SAT problem: the following POS expression for a propositional formula Φ:

$$\Phi \quad = \quad C_1 \wedge C_2 \wedge \cdots \wedge C_m \qquad (12.4)$$
$$\text{where: } \bullet \text{ each clause } C_i = \ell_{i,1} \vee \ell_{i,2}$$
$$\bullet \; \Phi \; \text{ has } n \text{ logical variables}$$

We transform Φ into a directed graph $\mathscr{G}(\Phi)$ that has $2n$ vertices and $2m$ arcs.

- For each logical variable x there is one vertex that represents the TRUE literal form, x, of variable x, and a second vertex that represents the FALSE literal form, $\bar{x}$, of the variable.

- Each clause $C_i = (\ell_{i,1} \vee \ell_{i,2})$ is represented by a pair of arcs. Say that literal $\ell_{i,1}$ comes from variable x_1, and literal $\ell_{i,2}$ comes from variable x_2. These arcs represent "instructions" for assigning truth-values in a way that maximizes the number of clauses that receive the value TRUE.

  - There is an arc $(\bar{x}_1 \to x_2)$.

    This arc indicates that, if variable x_1 is assigned truth-value FALSE, then variable x_2 should be assigned truth-value TRUE.

  - Symmetrically, there is an arc $(\bar{x}_2 \to x_1)$.

    This arc indicates that, if variable x_2 is assigned truth-value FALSE, then variable x_1 should be assigned truth-value TRUE.

All paths in $\mathscr{G}(\Phi)$ represent logical implications.

The core of our proof is the following result.

Proposition 12.6 *The POS formula $\Phi = C_1 \wedge C_2 \wedge \cdots \wedge C_m$ is satisfiable if, and only if, no strongly connected component of $\mathscr{G}(\Phi)$ contains both the positive form (x) and the negated form $(\bar{x})$ of any variable x of Φ.*

The proof is a consequence of the two following elementary results.

Lemma 12.1. *If $\mathscr{G}(\Phi)$ contains a path from vertex x to vertex y, then it contains a path from vertex $\bar{y}$ to vertex $\bar{x}$.*

The proof, which is an induction on the length of the shortest path from vertex x to vertex y, is left to the reader.

Lemma 12.2. *If $\mathscr{G}(\Phi)$ contains a path from vertex x to vertex y, then for every truth assignment t that satisfies formula Φ (i.e., evaluates Φ to TRUE), if t assigns variable x the truth-value TRUE, then t also assigns variable y the truth-value TRUE.*

Proof. Assume that Φ is satisfied by a truth assignment t which assigns variable x the truth-value TRUE. Say, for contradiction, that along the path from x to $\bar{x}$ in $\mathscr{G}(\Phi)$, there exists an arc—call it $(u \to v)$—such that assignment t assigns the value TRUE to u and the value FALSE to v. Because of the way we constructed $\mathscr{G}(\Phi)$, the existence of this arc means that Φ contains the clause $\bar{u} \vee v$. Moreover, under truth assignment t, this clause of Φ evaluates to FALSE because both of its literals are assigned the value FALSE. This contradicts the assumption that t is a satisfying assignment for Φ. $\square$

We finally prove the main result of this section, Proposition 12.6.

Proof (Proposition 12.6). The overall form of the assertion we wish to prove is that of a *biconditional*, i.e., an assertion of the form

$$P \Leftrightarrow Q$$

This is, of course, a shorthand for the conjunction

$$\big[P \Rightarrow Q\big] \ \text{and} \ \big[Q \Rightarrow P\big]$$

We are able to derive a simplified proof here by the stratagem of replacing one of the two implications by its *contrapositive*; i.e., instead of proving the implication

$$P \Rightarrow Q$$

for the *necessity* component of the biconditional, we prove the *logically equivalent* implication

$$\neg Q \Rightarrow \neg P$$

The reader should read the *necessity* component of the proof with an eye toward identifying the contraposed implication.

Focus on a POS formula Φ and its associated digraph $\mathscr{G}(\Phi)$.

Necessity. Say first that $\mathscr{G}(\Phi)$ has a strongly connected component which contains vertices arising from a variable x in both positive (x) and negated $(\bar{x})$ forms. We claim that Φ is not satisfiable.

By definition of "strongly connected", $\mathscr{G}(\Phi)$ must contain paths between the vertices corresponding to x and to $\bar{x}$. By Lemma 12.2, therefore, any truth assignment that could satisfy formula Φ would have to assign literals x and $\bar{x}$ the same truth-value. Any such truth assignment to formula Φ's *literals* would not be a valid truth assignment to Φ's *variables* (specifically to variable x). We conclude that no valid truth assignment could satisfy formula Φ.

Sufficiency. Say next that $\mathscr{G}(\Phi)$ has no strongly connected component which contains vertices arising from a variable x in both positive (x) and negated ($\bar{x}$) forms. We construct a truth assignment t to Φ's variables under which Φ evaluates to TRUE. Assignment t witnesses Φ's satisfiability.

We construct a satisfying truth assignment t for Φ as follows.
1. Say that graph $\mathscr{G}(\Phi)$ has k mutually disjoint strongly connected components. We label these components in *topological order*, as $S_1, S_2, \ldots, S_k$. The phrase *"topological order"* means the following.

- $\mathscr{G}(\Phi)$ contains no arc of the form $(u \to v)$ where vertex u belongs to some component S_i, and vertex v belongs to some component S_j with $j < i$.

We know that this labeling is possible because any such arc would make all vertices of S_j accessible from all vertices of S_i, and conversely. This would mean that S_i and S_j would belong to the same strongly connected component—which would contradict the components' assumed disjointness.

2. We assign truth-values to variables of Φ by scanning the vertices/literals of $\mathscr{G}(\Phi)$ in decreasing order of the topological indices of $\mathscr{G}(\Phi)$'s strongly connected components.

- The *first time* that we encounter a vertex/literal ℓ, in true or negated form, we assign the truth-value to ℓ's associated variable that makes literal ℓ TRUE. This strategy also makes the clause that this instance of literal ℓ occurs in evaluate to TRUE.

- If we encounter an instance of a vertex/literal ℓ whose associated variable has already been assigned a truth-value, then we assign to this instance a truth-value that is consistent with the variable's assignment: i.e., a positive literal gets the same assignment, while a negative instance gets the negated version of the assignment.

Proceeding in this fashion, we develop a truth assignment that satisfies all of Φ's clauses. Assume for contradiction that some clause of Φ, say $(\xi \vee \eta)$, is not satisfied under our procedure. This means, in particular, that our assignment t assigns the truth-value FALSE to vertex/literal ξ. But, this can happen only if t has assigned the truth-value TRUE to vertex/literal $\bar{\xi}$, within a strongly connected component of $\mathscr{G}(\Phi)$ whose index is higher than that of the strongly connected component that ξ occurs in. The same observation applies to vertex/literal η. But this is impossible, because within our construction of $\mathscr{G}(\Phi)$, the clause $(\xi \vee \eta)$ in formula Φ would add the arc $(\bar{\xi} \to \eta)$ to graph $\mathscr{G}(\Phi)$. It follows that we have found a truth assignment that satisfies formula Φ, as was claimed.   □

We close with sample illustrations of Proposition 12.6.

1. Figure 12.1 illustrates $\mathscr{G}(\Phi_1)$ for the formula

$$\Phi_1 \;=\; (a \vee \bar{b}) \wedge (b \vee \bar{c}) \wedge (c \vee \bar{a})$$

The graph has the two strongly connected components illustrated in the figure.

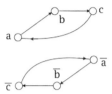

Fig. 12.1 The digraph $\mathscr{G}(\Phi_1)$ and its two strongly connected components

One thereby observes two satisfying truth-value assignments t for formula Φ_1:

$t(a) = t(b) = t(c) = \text{TRUE}$

or symmetrically

$t(a) = t(b) = t(c) = \text{FALSE}.$

2. Figure 12.2 illustrates $\mathscr{G}(\Phi_2)$ for the formula

$$\Phi_2 \;=\; (a \vee \bar{b}) \wedge (b \vee \bar{c}) \wedge (c \vee \bar{a}) \wedge (a \vee c) \wedge (\bar{a} \vee \bar{c})$$

Because the graph is strongly connected and contains directed paths in both

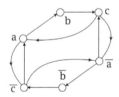

Fig. 12.2 The digraph $\mathscr{G}(\Phi_2)$

directions between vertex a and vertex $\bar{a}$, Proposition 12.6 assures us that formula Φ_2 does not admit any satisfying truth assignment.

Even without invoking formal algorithmic concepts, it is clear that the processes of
- constructing digraph $\mathscr{G}(\Phi)$ from formula Φ;
and
- isolating and investigating the strongly connected components of $\mathscr{G}(\Phi)$

can be accomplished in a number of computational steps that is polynomial in the size of Φ.

12.1.3 Matchings in Graphs

Matching is fundamental to many situations that can be modeled using graphs. A *matching* in an undirected graph $\mathscr{G}$ is a set of edges of $\mathscr{G}$ that have no vertices in common. In many computational settings, matchings are, thus, a convenient formal mechanism for pairing a graph's vertices. The broad range of activities which can be modeled using graph matching include: pairing competitors for a tennis tournament; helping a person select a potential spouse (which even in the vernacular is often termed "matchmaking"); determining (near-)optimal layouts for a keyboard in language X (based on the relative "affinities" of various pairs of letters for one another in X); selecting persons to command the police stations in city Y (based on the perceived "match" between a candidate's qualifications and the needs of specific stations). Even this small sampler makes it clear that there are many significant variations on this formal theme. This section is devoted to describing, and briefly discussing, a few of the most commonly encountered versions of matching in graphs.

Explanatory note.

Although the definitions of the various versions of matching are readily accessible to even the mathematical novice, much of the more sophisticated mathematical knowledge about matchings is beyond any beginning text. The interested reader might consult a more advanced source, such as [14], to get a feeling for what is known about this conceptually simple, yet rich, topic.

Matchings in unweighted graphs. The most straightforward notion of matching involves an undirected graph $\mathscr{G}$ with unlabeled edges. The optimization criterion most often invoked with this genre of matching is to find a matching that involves as many edges of $\mathscr{G}$ as possible.

The target in this "vanilla-flavored" matching problem is often a matching that is *maximal*, in the sense that adding any further edge of $\mathscr{G}$ to the matching leaves one with a set of edges that is no longer a matching.

Among maximal matchings in a graph $\mathscr{G}$, the "ultimate treasure" is a matching that is *perfect*, in the sense that every vertex of $\mathscr{G}$ belongs to some edge of the matching.

Proposition 12.7 *Maximal matchings exist for any graph $\mathscr{G}$. One can find such a matching in a number of steps proportional to $|\mathscr{N}|$.*

Proof. We leave to the reader the challenge of verifying that the following *greedy*[2] process satisfies the conditions of the Proposition.

The Process:
Begin by laying the vertices of $\mathscr{G}$ out, left to right, in any way.

---

[2] In the world of algorithmics, the term "greedy" describes any process that seeks local optimizations, with no consideration of how such a myopic strategy might limit future options.

Repeat the following process until no vertices remain in the layout.

Select the leftmost vertex, u, in the remaining layout of $\mathcal{N}$.

- If we succeed in finding such a neighbor of u (taken from left to right)), call it vertex v, then add edge $\{u,v\}$ to the matching we are building. Then, remove both u and v from the layout.

- If there is no neighbor v of u, then remove vertex u from the layout.

Of course, the real challenge here is to find a data structure that allows an efficient search for a "remaining" neighbor-vertex v at each step of the selection process. □

In contrast to the situation with maximal matchings, many simple graphs do not admit any perfect matching. Contemplating, for instance, matchings within any cycle with an odd number of vertices may prepare the reader for the challenge of verifying the following necessary condition for a graph to admit a perfect matching.

Proposition 12.8 *Let $\mathcal{G}$ be a graph that admits a perfect matching. Then:*

- *$\mathcal{G}$ has an even number of vertices.*

- *The cardinality of the (perfect) matching—i.e., the number of edges in the matching—is exactly $\frac{1}{2}|\mathcal{N}|$.*

Matchings in weighted graphs. The other very popular genre of matching problem focuses on graphs each of whose edges, say $\{u,v\}$, is weighted with a number that measures the "affinity" of vertices u and v for each other. The challenge is to find a matching that is *maximal* in the sense of having a cumulative sum of edge-weights that is not exceeded by any other matching's.

We note in closing that, while edge-weightings often complicate computational processing of graphs, they need not render such computations practically infeasible. For instance, the problem of discovering a perfect matching of minimal weight in an edge-weighted graph can be solved moderately efficiently—i.e., in a number of steps that is polynomial in the size of the graph. (One algorithm that achieves this efficiency can be based on the colorfully named *Hungarian assignment method*; see the original source [70] or the encyclopedic algorithms text [36].)

12.2 Trees and Spanning Trees

The special class of graphs called *trees* occupy a place of honor within both the mathematical field called *graph theory* and within the vernacular. Trees are identified mathematically as graphs that contain no cycles (as subgraphs) or, equivalently, as graphs in which each pair of vertices is connected by a unique path: a tree is thus the embodiment of "pure" connectivity: it provides the minimal interconnection structure (in number of edges) that provides paths that connect every pair of

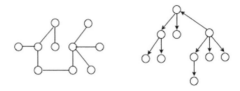

Fig. 12.3 (Left) An undirected tree with 10 vertices. (Right) A rooted directed out-tree. (The root is not where you might expect it to be.)

vertices; see Fig. 12.3. As one would expect from the vernacular, a set of trees is called a *forest*.

We have asserted implicitly that our two definitions of "tree" define the same class of graphs. Of course, we do not accept this on faith—we must prove it.

Proposition 12.9 *The two definitions of "tree" are equivalent: they define the same class of graphs. Stated more directly:*

The following assertions about a connected graph $\mathcal{T}$ are logically equivalent.

1. The graph $\mathcal{T}$ is cycle-free.

2. Each pair of distinct vertices of $\mathcal{T}$ is connected by precisely one path.

We leave the proof of Proposition 12.9 to the reader.

The following property of connected trees provides a valuable insight into many tree-related phenomena. The property is a subtle corollary of Proposition 12.9.

Proposition 12.10 *An n-vertex connected tree $\mathcal{T}$ has precisely $n - 1$ edges.*

We precede the proof by an essential lemma.

Lemma 12.3. (a) *Let $\mathcal{G}$ be a connected graph having $n \geq 2$ vertices. Every vertex of $\mathcal{G}$ has degree ≥ 1.*

(b) *Let $\mathcal{T}$ be a connected tree having $n \geq 1$ vertices. At least one vertex of $\mathcal{T}$ has degree 1.*

Proof (Lemma 12.3). **(a)** A vertex of $\mathcal{G}$ that had degree 0 would have no neighbors; hence, $\mathcal{G}$, having $n \geq 2$ vertices, would not be connected if it had such a vertex.

(b) By Proposition 12.3 every graph whose vertices all have degree ≥ 2 contains a cycle. Since $\mathcal{T}$, being a tree, is cycle-free, not every vertex of $\mathcal{T}$ can have degree ≥ 2. By part **(a)**, then, at least one vertex of tree $\mathcal{T}$ has degree 1.   □

Proof (Proposition 12.10). We proceed by induction on n.

Base case. The case $n = 2$ is obvious, because a single edge is both necessary and sufficient to connect two vertices.

Inductive hypothesis. Assume that the indicated tally is correct for all trees having no more than k vertices.

Inductive extension. Consider, to extend the induction, any tree $\mathcal{T}$ on $k + 1$ vertices.

By Lemma 12.3, $\mathcal{T}$ must contain at least one vertex v of degree 1. If we remove v and its (single) incident edge, we now have a tree $\mathcal{T}'$ on k vertices. By induction, $\mathcal{T}'$ has $k - 1$ edges. When we reattach vertex v to $\mathcal{T}'$, we restore $\mathcal{T}$ to its original state. Because this restoration adds one vertex and one edge to $\mathcal{T}'$, we see that $\mathcal{T}$ has $k + 1$ vertices and k edges.

This extends the induction, hence completes the proof. $\square$

Just as with graphs, there is a *directed* version of trees, which is formed by replacing the (nonoriented) edges of an undirected tree by (oriented) arcs; see Fig. 12.3(right). Within a directed tree $\mathcal{T}$, one often says that an arc goes *from* a *parent vertex to* a *child vertex*. Extending this anthropomorphic metaphor, one often talks about the *ancestor(s)* and *descendant(s)* of a tree-vertex. We single out two special classes of vertices: A *root (vertex)* of $\mathcal{T}$ is defined by having no entering arcs, i.e., by having in-degree 0; a *leaf (vertex)* of $\mathcal{T}$ is defined by having no exiting arcs, i.e., by having out-degree 0.

Explanatory note.

Note that the directed tree in Fig. 12.3(right) is rooted, but we have not drawn the tree in the usual way, with the root either above or below all other vertices. This was a purposeful decision, so that the reader would note that root-hood is a property of graph structure, not of drawing.

The reader is certainly familiar with the use of rooted directed trees to represent family trees and corporate hierarchies, as described in the following examples.

Explanatory note.

Sociologically, the historical *atomic family tree* has two roots, representing the matriarch and patriarch of the family. The entirety of the tree represents a single family generationally, before any children form their own families. When a child pairs with a partner and the pair spawns children, the family tree is expanded to accommodate the additions: each (new-partner)-plus-(old child) pair become twin roots of its own family tree, which grows from the former tree. The leaves of any such tree are the childless descendants of the roots.

Note that, while we are using anthropomorphic language here, we could be discussing other genres of "family", such as, e.g., many types of biological taxonomies.

Among rooted directed trees, an important subclass comprises those that have a *single root*, which has a directed path to every other vertex. The length of each such directed path is often used to label the *generation* of the vertex at the end of the path: root, child, grandchild, great-grandchild, etc. Every vertex of the tree is the root of a singly rooted directed subtree of the entire tree. All subtrees that are rooted at vertices of the same generation are mutually disjoint.

Explanatory note.

A singly rooted tree represents a *hierarchy*. Given two distinct directed subtrees within a hierarchy, either the root of one of the subtrees is a descendant of the root of the other, or the two subtrees are mutually disjoint.

Spanning trees. One of the major uses of trees and forests is as a way of succinctly "summarizing" the connectivity structure inherent in an undirected graph. This role is inherent in the notion of a *spanning tree* of a connected graph $\mathcal{G}$. A spanning tree of $\mathcal{G}$ is a tree $\mathcal{T}(\mathcal{G})$ whose vertex-set is identical to $\mathcal{G}$'s:

$$\mathcal{N}_{\mathcal{T}(\mathcal{G})} = \mathcal{N}_{\mathcal{G}}$$

and all of whose edges are edges of $\mathcal{G}$:

$$\mathcal{E}_{\mathcal{T}(\mathcal{G})} \subseteq \mathcal{E}_{\mathcal{G}}$$

Not surprisingly, a connected graph $\mathcal{G}$ typically has *many* spanning trees. All such trees share $\mathcal{G}$'s vertex-set, but they may choose quite different sets of edges.

For a graph $\mathcal{G}$ that is not connected, we replace the notion of a spanning tree of $\mathcal{G}$ with the kindred notion of a *spanning forest*. We discuss only spanning trees in this section, but we urge the reader to extrapolate the discussion to spanning forests of unconnected graphs.

Within applications, as a spanning tree "summarizes" the connectivity structure of a graph, it does the same for any entities that the graph models, such as a map, the layout of a museum, etc. This modeling role makes spanning trees invaluable as the basis for algorithms that involve connectivity-related notions. To encompass an even greater range of such notions, we often *weight* the edges of spanning trees, in order to model the "cost" of incorporating that edge into the tree. The types of computational problem modeled via edge-weighted spanning trees include: the optimal placement of firehouses or hospitals in a town and the optimal deployment of security mechanisms in an art museum. Reflecting problems wherein edge-weights measure transit costs, it is a classical computational problem to seek a *minimum-weight spanning tree* (usually abbreviated in the vernacular to *minimum spanning tree*). Happily, this classical optimization problem can be solved within a number of steps that is linear in the number of edges of $\mathcal{G}$ [36].

12.3 Computationally Significant "Named" Graphs

The area of mathematics called graph theory is an important source of formal aids for designing, analyzing, utilizing, modeling, and verifying systems that are used for computation and/or communication. Of course, such systems are designed by humans. Among the manifold consequences of this fact is the observation that the (families of) graphs that are among the most commonly used to model systems tend to be rather uniform in structure, in a variety of ways: Such graphs, when drawn, often exhibit a lot of structural regularity. We single out one particularly popular form of structural regularity to watch for as we prepare to describe five families of graphs that have proven so significant as models of a variety of computational and communicational processes that they are usually referred to just by their given names. (We noted a similar phenomenon early in Chapter 10, as we discussed numbers such as e and π and Avogadro's number.) The notion of regularity that we single out is so popular that it is familiarly known as just *"regularity"*, even though its complete name is *"degree regularity"*.

- *Undirected version.* An undirected graph $\mathscr{G}$ is *(degree-)regular* if all of its vertices have the same degree.

- *Directed version.* A directed graph $\mathscr{G}$ is *(degree-)in-regular* if all of its vertices have the same in-degree. $\mathscr{G}$ is *(degree-)out-regular* if all of its vertices have the same out-degree.

We now describe five families of regular graphs that have proven useful for understanding and implementing both computation and communication over the entire digital era, and we expose some basic properties of each. Each of these graphs is available in both a directed and an undirected version, although, as we note, in each case, one of these versions is more commonly encountered. We have selected these specific graphs for rather different reasons.

- The first two graphs, the *cycle-graph* of Section 12.3.1 and the *complete graph* (or *clique*) of Section 12.3.2 were selected to represent, respectively, the lowest-degree and highest-degree graphs that share two properties:

1. Every vertex of each graph is accessible from every other vertex.

2. All vertices of each graph "look alike" to someone traversing the graph.

   To elaborate: If someone places you on a vertex of either graph, there is no way that you can determine the identity of that vertex.

   This is an important feature to ponder, because it is a simple instance of the *anonymity* problem that plagues many modern distributed computing environments: *How does one orchestrate cooperative activities when all agents are "indistinguishable"?*

- The remaining graphs, the *mesh and torus networks*[3] of 12.3.3, the *hyper-cube network* of Section 12.3.4, and the *de Bruijn network* of Section 12.3.5, were selected for their importance within the world of parallel and distributed computing—as abstract platforms for developing efficient computational and communicational processes, and as abstract versions of the networks that underlie parallel architectures by interconnecting their processors.

The hypercube network and the de Bruijn network are known also as indispensable aids in constructing codes that are used in activities as varied as cryptology and the testing of electronic circuits.

Throughout this section, each parameter n that identifies an instance from our graph families ranges over either $\mathbb{N}$ or $\mathbb{N}^+$; we shall always indicate which.

12.3.1 The Cycle-graph $\mathscr{C}_n$

For each positive integer $n \in \mathbb{N}^+$, both the *undirected order-n cycle-graph* $\mathscr{C}_n$ and the *directed order-n cycle-graph* $\widehat{\mathscr{C}}_n$ have *vertex-set*

$$\mathscr{N}_{\mathscr{C}_n} = \mathscr{N}_{\widehat{\mathscr{C}}_n} = \{0, 1, \ldots, n-1\}$$

- $\mathscr{C}_n$ has n edges; its *edge-set* is

$$\mathscr{E}_{\mathscr{C}_n} = \left\{ \{i, i+1 \bmod n\} \mid i \in \{0, 1, \ldots, n-1\} \right\}$$

Figure 12.4 depicts the eight-vertex cycle $\mathscr{C}_8$.

  - $\mathscr{C}_n$ is degree-regular: each vertex has degree 2.
    Specifically, each vertex i of $\mathscr{C}_n$ has its *predecessor $i-1$* mod n and its *successor $i+1$* mod n.

  - $\mathscr{C}_n$ has diameter $\lfloor n/2 \rfloor$.
    Direct calculation shows that $\mathscr{C}_n$'s diameter is no larger than $\lfloor n/2 \rfloor$. The fact that this is, in fact, the graph's diameter is witnessed by the distance between each vertex $k \in \mathscr{N}_{\mathscr{C}_n}$ and its antipodal vertex $k + \lfloor n/2 \rfloor$ mod n.

- $\widehat{\mathscr{C}}_n$ has *arc-set*

$$\mathscr{A}_{\widehat{\mathscr{C}}_n} = \left\{ (i \rightarrow i+1 \bmod n) \mid i \in \{0, 1, \ldots, n-1\} \right\}$$

  - $\widehat{\mathscr{C}}_n$ is a regular directed network: each vertex has in-degree and out-degree 2.

---

[3] These two structures, though distinct, are usually discussed together because they share so many important properties.

– $\widehat{\mathscr{C}}_n$ has (directed) diameter $n-1$.

Of course, $n-1$ is an upper bound on the diameter of any n-vertex digraph. The fact that this is exactly $\widehat{\mathscr{C}}_n$'s diameter is witnessed by the directed distance from each vertex k of $\widehat{\mathscr{C}}_n$ to its predecessor vertex $k-1$ mod n.

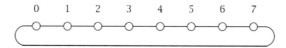

Fig. 12.4 The eight-vertex cycle $\mathscr{C}_8$

12.3.2 The Complete graph (Clique) $\mathscr{K}_n$

For each positive integer $n \in \mathbb{N}^+$, we denote by $\mathscr{K}_n$ the *undirected* order-n *complete-graph* (or, *clique*), and by $\widehat{\mathscr{K}}_n$ the *directed* order-n *complete-graph* (or, *clique*). Both $\mathscr{K}_n$ and $\widehat{\mathscr{K}}_n$ have *vertex-set*

$$\mathscr{N}_{\mathscr{K}_n} = \mathscr{N}_{\widehat{\mathscr{K}}_n} = \{0, 1, \ldots, n-1\}$$

- $\mathscr{K}_n$ has $\binom{n}{2}$ edges; its *edge-set* is

$$\mathscr{E}_{\mathscr{K}_n} = \{\{i, j\} \mid i, j \in \{0, 1, \ldots, n-1\}, i \neq j\}$$

  – $\mathscr{K}_n$ is a regular network: each vertex has degree $n-1$; every vertex $i \in \mathscr{N}_{\mathscr{K}_n}$ is a neighbor of every other vertex.
  – $\mathscr{K}_n$ has diameter 1.
    $\mathscr{K}_n$'s diameter is a direct consequence of its vertex-degrees, and vice versa.

- $\widehat{\mathscr{K}}_n$ has $(n-1)n$ arcs; its *arc-set* is

$$\mathscr{A}_{\widehat{\mathscr{K}}_n} = \{(i \to j) \mid i, j \in \{0, 1, \ldots, n-1\}, i \neq j\}$$

  – $\widehat{\mathscr{K}}_n$ is a regular directed network: each vertex has in-degree and out-degree $n-1$.
  – $\widehat{\mathscr{K}}_n$ has (directed) diameter 1.
    $\widehat{\mathscr{K}}_n$'s diameter and its (in- and out-) vertex-degrees determine one another.

Recalling our discussion of matchings in (unweighted) graphs: The structure of the set of perfect matchings in general graphs is decidedly nontrivial. For clique-graphs, though, the structure is much easier to discuss.

Proposition 12.11 *The number of perfect matchings admitted by the clique-graph $\mathcal{K}_n$ is either 0—if n is odd—or exponential in n—if n is even.*

Proof. The assertion about cliques with odd numbers of vertices is immediate (see Proposition 12.8): one vertex can never been paired.

We verify the assertion about cliques of the form $\mathcal{K}_{2k}$ by induction on k. To this end, let M_n denote the number of perfect matchings that $\mathcal{K}_n$ admits.

We are going to fashion this proof to reflect the way that a mathematician would reason, rather than presenting the highly structured form of induction that we have been using until now. After all, we are already in Chapter 12!

The base case $k = 1$ is immediate: Because $\mathcal{K}_2$ consists of a single edge, it admits precisely one perfect matching; i.e., $M_1 = 1$.

To garner intuition, we also explicitly solve the case $k = 2$, which is illustrated in Fig. 12.5. As the figure illustrates, $\mathcal{K}_4 = \mathcal{K}_{2 \cdot 2}$ can be viewed as a 4-cycle (drawn

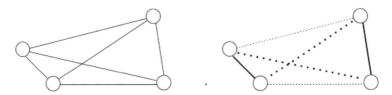

Fig. 12.5 $\mathcal{K}_4$ (left) and its three different perfect matchings (right): The matchings' edges are drawn, respectively, with bold lines, dashed lines, and dotted lines

with bold and dashed lines), augmented by two "cross-edges" (drawn with dotted lines). It is easy to see, then, that $\mathcal{K}_4$ admits three different perfect matchings, which can be identified (and specified) by the edge that contains the northwesterly vertex—call it v—in the figure. Vertex v has the choice of three vertices to "boldly" match with. (In the figure, v has chosen the southwesterly vertex as its "bold" match.) Once v has chosen its match, there is only one viable choice for the second edge in the matching. Thus, $M_2 = 3$.

We jump now to the case of any arbitrary $k > 2$. We remark that there are precisely $2k - 1$ vertices of $\mathcal{K}_{2k}$ that vertex 1 can "choose" as its mate in a perfect matching. Once we match vertex 1 to its mate, we confront an independent instance of the perfect-matching-counting problem with parameter $k - 1$, i.e., the problem of counting the number of perfect matchings in $\mathcal{K}_{n-2} = \mathcal{K}_{2k-2}$. We thereby note that as k grows, the quantity M_k obeys the following recurrence:

$$M_k = (2k - 1) \times M_{k-1}$$

In other words:

> M_k is the product of the first k odd numbers.

To gauge the growth rate of M_k, we concentrate on cases $k > 2$ and ignore the $\lfloor k/2 \rfloor$ smallest odd numbers. We then replace each of the remaining odd numbers by its smallest possible value. We thereby find that

$$M_k \;=\; \prod_{i=1}^{k} (2i-1) \;\geq\; \prod_{i=\lceil k/2 \rceil}^{k} (2i-1) \;\geq\; (2\lceil k/2 \rceil - 1)^{k/2} \;>\; k^{k/2}$$

In summary, M_k grows exponentially with the parameter k, as claimed. □

The two families of graphs we have discussed thus far—cycles and cliques—are recommended by their structural simplicity: They epitomize, respectively, the most sparse way (the cycle) and the most dense way (the clique) to completely interconnect n vertices. The remainder of this section is devoted to three families of graphs which have been found to be very useful in computational and communicational scenarios. These graphs' structures are simple enough to be amenable to rigorous reasoning, while being rich enough to support a broad range of procedures that have been shown to aid one in the crafting of algorithms that exploit the potential efficiencies—in computation and communication—that one can achieve using modern technologies.

12.3.3 Sibling Networks: the Mesh ($\mathcal{M}_{m,n}$); the Torus ($\widetilde{\mathcal{M}}_{m,n}$)

For positive integers $m, n \in \mathbb{N}^+$, both the $m \times n$ mesh (network) $\mathcal{M}_{m,n}$ and the $m \times n$ toroidal network (torus, for short) $\widetilde{\mathcal{M}}_{m,n}$ have vertex-set

$$\mathcal{N}_{\mathcal{M}_{m,n}} = \mathcal{N}_{\widetilde{\mathcal{M}}_{m,n}} = \{1, 2, \ldots, m\} \times \{1, 2, \ldots, n\}$$
$$= \{\langle i, j \rangle \mid i \in \{1, 2, \ldots, m\}, \; j \in \{1, 2, \ldots, n\}\}$$

- $\mathcal{M}_{m,n}$ has $(m-1)n + (n-1)m$ edges; its edge-set is

$$\mathcal{E}_{\mathcal{M}_{m,n}} = \{\{\langle i,j \rangle, \langle i+1,j \rangle \mid 1 \leq i < m, \; 1 \leq j \leq n\}$$
$$\cup \; \{\langle i,j \rangle, \langle i,j+1 \rangle \mid 1 \leq i \leq m, \; 1 \leq j < n\}\}$$

  – For $i \in \{1, 2, \ldots, m\}$, the subgraph of $\mathcal{M}_{m,n}$ defined by the vertex-set

$$V_i^{(\mathrm{row})} \;\overset{\text{def}}{=}\; \{\langle i, j \rangle \mid j \in \{1, 2, \ldots, n\}\}$$

  and all edges both of whose endpoints belong to $V_i^{(\mathrm{row})}$ is the ith row of $\mathcal{M}_{m,n}$. Dually, for $j \in \{1, 2, \ldots, n\}$, the subgraph of $\mathcal{M}_{m,n}$ defined by the vertex-set

$$V_j^{(\mathrm{col})} \stackrel{\mathrm{def}}{=} \{\langle i, j \rangle \mid i \in \{1, 2, \ldots, m\}\}$$

and all edges both of whose endpoints belong to $V_j^{(\mathrm{col})}$ is the jth *column* of $\mathcal{M}_{m,n}$.

– $\mathcal{M}_{m,n}$ is *not* a regular graph. Its four corner vertices each have degree 2; its non-corner extreme edge vertices each have degree 3; its *internal vertices* each have degree 4.

– The diameter of $\mathcal{M}_{m,n}$ is $m + n - 2$, as witnessed by the distance between vertices $\langle 1, 1 \rangle$ and $\langle m, n \rangle$.

- $\widetilde{\mathcal{M}}_{m,n}$ has $2mn$ arcs; its *arc-set* is

$$\mathcal{A}_{\widetilde{\mathcal{M}}_{m,n}} = \{(\langle i, j \rangle \to \langle i+1 \bmod m, j \rangle) \mid 1 \le i \le m, \ 1 \le j \le n\}$$
$$\cup \ \{(\langle i, j \rangle \to \langle i, j+1 \bmod n \rangle) \mid 1 \le i \le m, \ 1 \le j \le n\}$$

– $\widetilde{\mathcal{M}}_{m,n}$ is a regular network; each vertex has degree 4. Despite the fact that $\widetilde{\mathcal{M}}_{m,n}$ is an *undirected* graph, its edges are commonly referred to via an anthropomorphic directional labeling, as "up, down, left, and right" or as "north, south, west, and east".

– $\widetilde{\mathcal{M}}_{m,n}$'s diameter is $\lfloor m/2 \rfloor + \lfloor n/2 \rfloor$. This can be verified by analogy to the diameter of the cycle-graph $\mathcal{C}_n$.

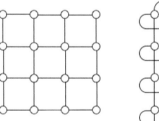 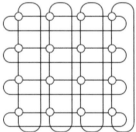

Fig. 12.6 $\mathcal{M}_{4,4}$ (left) and $\widetilde{\mathcal{M}}_{4,4}$ (right)

Enrichment Note.

The mesh and torus networks are among the simplest instances of graphs that are built from simpler graphs. In this case, the underlying operation is the *direct product*—or *Cartesian product*—of graphs. We now define the general case of this notion.

As you read this definition, keep our notational conventions in mind:

$\langle u, v \rangle$ is the ordered pair with first element u and second element v
(u, v) is the undirected edge connecting vertices u and v
$(u \rightarrow v)$ is the directed edge from vertex u to vertex v

The *direct product* of graphs $\mathscr{G}_1$ and $\mathscr{G}_2$ is denoted $\mathscr{G}_1 \times \mathscr{G}_2$ and is defined as follows.

- The vertex-set of $\mathscr{G}_1 \times \mathscr{G}_2$ is the direct product

$$\mathscr{N}_{\mathscr{G}_1} \times \mathscr{N}_{\mathscr{G}_2} = \{ \langle u, v \rangle \mid u \in \mathscr{N}_{\mathscr{G}_1}, v \in \mathscr{N}_{\mathscr{G}_2} \}$$

- Each edge of $\mathscr{G}_1 \times \mathscr{G}_2$ comes either from $\mathscr{G}_1$ or from $\mathscr{G}_2$, but *not from both*.

  - If (u_1, v_1) is an edge of $\mathscr{G}_1$, then for each $w \in \mathscr{N}_{\mathscr{G}_2}$, $(\langle u_1, w \rangle, \langle v_1, w \rangle)$ is an edge of $\mathscr{G}_1 \times \mathscr{G}_2$.

  - If (u_2, v_2) is an edge of $\mathscr{G}_2$, then for each $x \in \mathscr{N}_{\mathscr{G}_1}$, $(\langle x, u_2 \rangle, \langle x, v_2 \rangle)$ is an edge of $\mathscr{G}_1 \times \mathscr{G}_2$.

  - Those are all of the edges of $\mathscr{G}_1 \times \mathscr{G}_2$.

As specific examples: *Each mesh-graph $\mathscr{M}_{m,n}$ is the direct product of path-graphs*

$$\mathscr{M}_{m,n} = \mathscr{P}_m \times \mathscr{P}_n$$

and each torus-graph $\widetilde{\mathscr{M}}_{m,n}$ is the direct product of cycle-graphs

$$\widetilde{\mathscr{M}}_{m,n} = \mathscr{C}_m \times \mathscr{C}_n$$

12.3.4 The (Boolean) Hypercube $\mathscr{Q}_n$

The graphs we focus on in this section have had a major impact on the world of coding, especially in regard to codes that are *error correcting* [87], and on the world of computing, especially in regard to parallel and distributed computing [37, 62, 96, 101]. The cited sources give a range of perspectives on the importance of *hypercube networks*.

The *order-n Boolean hypercube*, traditionally denoted $\mathscr{Q}_n$, is the 2^n-vertex graph defined via one of the following (equivalent) definitions.

- *The recursive definition.*

– The order-0 Boolean hypercube, $\mathscr{Q}_0$, has a single vertex, and no edges.

– The order-$(k+1)$ Boolean hypercube, $\mathscr{Q}_{k+1}$, is obtained by taking two copies of $\mathscr{Q}_k$, call them $\mathscr{Q}_k^{(1)}$ and $\mathscr{Q}_k^{(2)}$, and creating an edge that connects each vertex of $\mathscr{Q}_k^{(1)}$ with the corresponding vertex of $\mathscr{Q}_k^{(2)}$.

For illustration:

– $\mathscr{Q}_1$ consists of two vertices connected by a single edge.

– $\mathscr{Q}_2$ can be viewed as a "square", or equivalently, a copy of the 4-cycle $\mathscr{C}_4$.

– $\mathscr{Q}_3$ can be viewed as a "cube", i.e., as two copies of $\mathscr{C}_4$ with edges connecting corresponding vertices: Each of the following pairs of vertices are connected by an edge:

> the upper right corner-vertices
> the upper left corner-vertices
> the lower right corner-vertices
> the lower left corner-vertices

- *The direct definition.* For each $n \in \mathbb{N}$, the vertices of the order-n Boolean hypercube, $\mathscr{Q}_n$, are all length-n binary strings. For illustration:

$$\mathscr{N}_{\mathscr{Q}_0} = \{\varepsilon\}, \quad \text{the length-0 } null \; string$$
$$\mathscr{N}_{\mathscr{Q}_1} = \{0, 1\}$$
$$\mathscr{N}_{\mathscr{Q}_2} = \{00, 01, 10, 11\}$$
$$\mathscr{N}_{\mathscr{Q}_3} = \{000, 001, 010, 011, 100, 101, 110, 111\}$$

The iteration-based construction of big hypercubes from the next smaller ones is illustrated in Fig. 12.7.

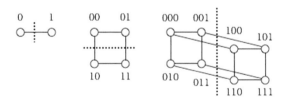

Fig. 12.7 The iteration-based construction of order-n hypercubes: (1) Take two copies of the order-$(n-1)$ hypercube. (2) Prepend a 0 to the vertex-labels of the first copy and a 1 to the vertex-labels of the second copy

It is easy to see that each $\mathscr{Q}_n$ has 2^n vertices, for this is the number of length-n binary strings.

For each value of n, each edge of $\mathscr{Q}_n$ connects two vertex-strings that differ in precisely one bit-position. This means that $\mathscr{Q}_n$ has $n2^{n-1}$ edges: Each of its 2^n

vertices has n neighbors, so the quantity $n2^n$ counts each of $\mathcal{Q}_n$'s edges twice—one for each endpoint. For illustration:

$$\mathcal{E}_{\mathcal{Q}_1} = \{\{0,\ 1\}\}$$
$$\mathcal{E}_{\mathcal{Q}_2} = \{\{00,\ 01\},\ \{00,\ 10\},\ \{01,\ 11\},\ \{10,\ 11\}\}$$
$$\mathcal{E}_{\mathcal{Q}_3} = \{\{000,\ 001\},\ \{000,\ 010\},\ \{000,\ 100\},\ \{001,\ 011\},$$
$$\{001,\ 101\},\ \{010,\ 011\},\ \{010,\ 110\},\ \{100,\ 101\},$$
$$\{100,\ 110\},\ \{101,\ 111\},\ \{011,\ 111\},\ \{110,\ 111\}\}$$

It is easy to observe $\mathcal{Q}_n$'s basic structural properties.

- $\mathcal{Q}_n$ is a regular network: each of its 2^n vertices has degree n.

  This follows from the fact that each edge of $\mathcal{Q}_n$ rewrites a single bit-position in the length-n binary string that is the edge's source vertex.

- $\mathcal{Q}_n$ has diameter $n = \ln(|\mathcal{N}_{\mathcal{Q}_n}|)$.[4]

  We verify this fact formally.

Proposition 12.12 *For all $n \in \mathbb{N}^+$, $\mathcal{Q}_n$ has diameter $n = \ln(|\mathcal{N}_{\mathcal{Q}_n}|)$.*

Proof. We prove this diameter bound by construction. Focus on two arbitrary vertices of $\mathcal{Q}_n$:

$$x = \alpha_1 \alpha_2 \cdots \alpha_n \quad \text{and} \quad y = \beta_1 \beta_2 \cdots \beta_n$$

One of the several paths in $\mathcal{Q}_n$ from x to y is described schematically as the following left-to-right bit-by-bit "rewriting" of x as y using edges of $\mathcal{Q}_n$.

$$x = \alpha_1 \alpha_2 \cdots \alpha_{n-1} \alpha_n \to \beta_1 \alpha_2 \cdots \alpha_{n-1} \alpha_n$$
$$\to \beta_1 \beta_2 \cdots \alpha_{n-1} \alpha_n$$
$$\vdots \qquad \vdots$$
$$\to \beta_1 \beta_2 \cdots \beta_{n-1} \alpha_n$$
$$\to \beta_1 \beta_2 \cdots \beta_{n-1} \beta_n = y$$

Since each bit of each string is rewritten at most once—bit-position i is rewritten precisely when $\alpha_i \neq \beta_i$—the bound follows. $\square$

The fact that $\mathcal{Q}_n$'s diameter is *logarithmic* in its number of vertices makes $\mathcal{Q}_n$ an efficient network for many tasks related to parallel computing and communication.

A powerful approach to understanding the structure of a given family of graphs is to understand how the perceived "shapes" of graphs in the family can apparently change just by relabeling/renaming the vertices, or the edges, of the graphs. The formal mechanism for studying such relabelings/renamings is the concept of *graph isomorphism*. Let $\mathcal{G}$ and $\mathcal{H}$ be undirected graphs that have the same numbers of

---

[4] Recall that $\ln n = \log_2 n$; see Section 5.4.

vertices and edges. (The following definition can easily be adapted to deal with *directed* graphs.) An *isomorphism* between $\mathscr{G}$ and $\mathscr{H}$ is a *bijection*[5]

$$f : \mathscr{N}_{\mathscr{G}} \leftrightarrow \mathscr{N}_{\mathscr{H}}$$

such that

- For each edge $\{u,v\}$ of $\mathscr{G}$ (i.e., $\{u,v\} \in \mathscr{E}_{\mathscr{G}}$), the doubleton set $\{f(u),f(v)\}$ is an edge of $\mathscr{H}$ (i.e., $\{f(u),f(v)\} \in \mathscr{E}_{\mathscr{H}}$).
- For each edge $\{x,y\}$ of $\mathscr{H}$ (i.e., $\{x,y\} \in \mathscr{E}_{\mathscr{H}}$), the doubleton set $\{f^{-1}(x),f^{-1}(y)\}$ is an edge of $\mathscr{G}$ (i.e., $\{f^{-1}(x),f^{-1}(y)\} \in \mathscr{E}_{\mathscr{G}}$).

We can immediately exemplify this notion via the following example.

Proposition 12.13 *The hypercube $\mathscr{Q}_4$ is isomorphic to the torus $\widetilde{\mathscr{M}}_{4,4}$.*

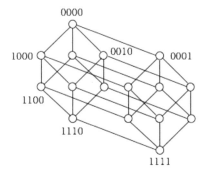

Fig. 12.8 The hypercube $\mathscr{Q}_4$ with a partial vertex-labeling by bit-strings

The isomorphism of Proposition 12.13 is depicted in Figs. 12.8 and 12.9. We challenge the reader in Exercise 12.8. to craft a formal proof of this result, using as a hint the coding scheme (or bijection) depicted in Fig. 12.10.

12.3.5 The *de Bruijn Network* $\mathscr{D}_n$

While the family of hypercube networks has few competitors in the world of parallel and distributed computing, in terms of performance and ease of designing algorithms, it does have one major shortcoming that relates to its realizability in

---

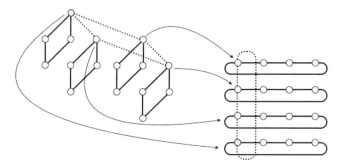

Fig. 12.9 Transforming $\mathcal{Q}_4$ to $\widetilde{\mathcal{M}}_{4,4}$. The bold edges correspond to horizontal cycles, while the dashed edges correspond to vertical cycles

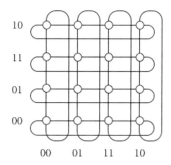

Fig. 12.10 Hinting at the coding scheme that yields an isomorphism between $\mathcal{Q}_4$ and $\widetilde{\mathcal{M}}_{4,4}$

hardware. The basic problem is that each vertex of $\mathcal{Q}_n$ has vertex-degree n; i.e., hypercubes' vertex-degrees are logarithmic in their numbers of vertices. This feature makes the hypercube's actual performance much slower than its theoretical performance: each step of a vertex v of $\mathcal{Q}_n$ is slowed down by v's having to get inputs from and supply outputs to its n (directed) neighbors. The phenomenon we are discussing can actually be described *geographically*! The physical area that $\mathcal{Q}_n$ occupies grows *exponentially* with the common degrees of the network's vertices—because $\mathcal{Q}_n$ has 2^n vertices which are no more than n edge-traversals from one another. (This is the "inverse" way of talking about logarithmic vertex-degrees.) In contrast, the space in which we (and our computers) live grows only *cubically* with linear distance. (That is, we live in three-dimensional space!) The resulting disparity in growth rates means that the wires in large hypercube-connected computing platforms must inevitably be *very* long—in contrast to the unit size of idealized network-edges. Consequently, electrical signals within a large hypercube must travel long distances in physical space—which means that the physical computer is much slower than its idealized version. (One finds a more technical discussion of this phenomenon in, e.g., [109].)

The just-described shortcoming of hypercubes led researchers for decades (beginning in the 1970s) to seek a family of networks whose vertex-degrees stay constant even as one deploys successively larger instances of the network. We now describe such a family of networks—which combines constant vertex-degrees with logarithmic diameters. The network family we focus on in this subsection was discovered within the domain of *coding theory*—as, coincidentally, was the hypercube.

In the mid-twentieth century, Dutch mathematician Nicolaas Govert de Bruijn discovered a way to generate compact sequences that contain all possible strings of a prespecified length. Focusing on *binary* strings—although de Bruijn's strategy works for any finite alphabet—de Bruijn could generate a string of length $2^n + n - 1$ which contains every length-n binary string as a substring. Quite appropriately, such a string is called an order-n *de Bruijn sequence*.

It is not obvious that order-n de Bruijn sequences exist for every n, but we now plant the seeds of a proof that they do. We begin by illustrating two sample sequences in (12.5).

n	LENGTH-n BINARY STRINGS	ORDER-n DE BRUIJN SEQUENCE
1	00, 01, 10, 11	00110
2	000, 001, 010, 011, 100, 101, 110, 111	0001110100

(12.5)

The table in (12.5) spawns several interesting questions:

- Do de Bruijn sequences exist for every n?

- If so,

  - How does one compute them?

  - Can one always find a de Bruijn sequence of length $2^n + n - 1$?

  - Can one find de Bruijn sequences of length $< 2^n + n - 1$?

The answers to all of these questions—and the connection of de Bruijn sequences to the current chapter—reside in the family of directed graphs called *de Bruijn graphs* (or *networks*). (The term used varies by intended application—mainly, coding theory and [the interconnection networks of] parallel computer architectures. We use the names interchangeably.)

For every integer $n \in \mathbb{N}^+$, the *order-n de Bruijn network* is the *directed* graph $\mathscr{D}_n$ whose vertices comprise the set of length-n binary strings.[6] The sets $\mathscr{N}_{\mathscr{D}_2}$ and $\mathscr{N}_{\mathscr{D}_3}$ appear in Table (12.5).

$\mathscr{D}_n$ is a regular directed graph; its vertices all have in-degree 2 and out-degree 2. Each vertex of $\mathscr{D}_n$ is a binary string of length $n \geq 1$; hence it can be written in the form βx, where $\beta \in \{0, 1\}$ is a *bit* and x is a length-$(n-1)$ binary string.

---

[6] While *binary* de Bruijn networks are the most frequently encountered ones, one can also find de Bruijn networks whose vertices comprise all length-n strings over larger finite alphabets. Such extended families also find applications in coding theory.

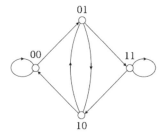

Fig. 12.11 The four-vertex, order-2 de Bruijn network

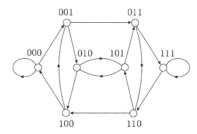

Fig. 12.12 The eight-vertex, order-3 de Bruijn network

The 2^{n+1} arcs of $\mathscr{D}_n$ come in pairs specified as follows. For each $\beta \in \{0,1\}$ and for each length-$(n-1)$ binary string x, $\mathscr{D}_n$ has the two arcs

$$(\beta x \to x0) \quad \text{and} \quad (\beta x \to x1)$$

We enumerate $\mathscr{A}_{\mathscr{D}_3}$ in Table (12.6).

SOURCE VERTEX		TARGET VERTEX		TARGET VERTEX
000		000		001
001		010		011
010		100		101
011	GOES TO	110	AND TO	111
100		001		000
101		011		010
110		101		100
111		111		110

(12.6)

For each $n \in \mathbb{N}^+$, $\mathscr{D}_n$ has diameter n. To see why this is true, note that following any one of $\mathscr{D}_n$'s arcs, say from vertex x to vertex y, consists of "rewriting" the length-n string x as the length-n string y. The diameter bound therefore follows by showing that, for any two string-vertices of $\mathscr{D}_n$, say vertex u and vertex v, one can rewrite u as v by traversing a sequence of arcs—i.e., a directed path—of length at most n. Observe, for instance, that the path in $\mathscr{D}_3$ described schematically as follows

$$000 \rightarrow 001 \rightarrow 011 \rightarrow 111$$

leads from vertex 000 to vertex 111, by rewriting string 000 as string 111. The diameter bound is now an immediate consequence of the existence in de Bruijn graphs of cycles that contain every vertex precisely once! We are not yet ready to study these cycles, although we soon will be.

A *Hamiltonian cycle* in an n-vertex graph $\mathcal{G}$ is a length-n cycle that contains every vertex of $\mathcal{G}$ precisely once. A *directed Hamiltonian cycle* in an n-vertex digraph $\mathcal{H}$ is a length-n directed cycle that contains every vertex of $\mathcal{H}$ precisely once. We study Hamiltonian cycles in some detail in Section 13.2.2, and, we prove in Section 13.2.2.2 (see Proposition 13.12(e)) that every de Bruijn network is directed-Hamiltonian!

Stepping back from the structural specifics of $\mathcal{D}_n$, we now see that de Bruijn networks provide us with a *bounded-degree*—specifically, a degree-2—family of networks each of whose constituent digraphs has diameter that is *logarithmic* in its size! In this regard, at least, de Bruijn networks have exactly the same cost-performance as hypercubes—both graphs have 2^n vertices and diameter n—*but the de Bruijn networks achieve this with bounded degrees.* Even more dramatic: It has been shown that sophisticated algorithmic techniques can achieve roughly equivalent computational efficiency, on a broad range of significant computational problems, using de Bruijn networks as using like-sized hypercubes [3, 15, 109].

Enrichment note.

At the end of Chapter 11, there is an exercise that discusses *riffle* and *Monge* suffles, a mathematical device for rearranging cards in a deck. The main message of that exercise is that both shuffles are *not* a good approach for shuffling playing cards: they don't even approach the level of randomness that a good game of chance demands.

In case that exercise got you curious about mathematical devices that *do* provide a good approach for shuffling playing cards, you may want to read the article [106], which describes a "cousin" of de Bruijn networks that is called the *perfect shuffle network*. As the name suggests, that family of networks *can* be used to obtain good shuffles. The fact that you have successfully made it through Section 12.3.5 means that you have the mathematical underpinnings necessary to understand perfect shuffle networks.

In fact, maybe you will feel ambitious enough to take a stab at formally relating the graph-theoretic structures of de Bruijn networks and perfect shuffle networks!

12.4 Exercises: Chapter 12

Throughout the text, we mark each exercise with 0 or 1 or 2 occurrences of the symbol $\oplus$, as a rough gauge of its level of challenge. The 0-$\oplus$ exercises should be accessible by just reviewing the text. We provide *hints* for the 1-$\oplus$ exercises; Appendix H provides *solutions* for the 2-$\oplus$ exercises. Additionally, we begin each exercise with a brief explanation of its anticipated value to the reader.

1. $\oplus$ **Vertex-degrees and the existence of paths**
   LESSON: Experience with graph-theoretic reasoning
   Prove the following slightly more advanced cousin of Proposition 12.3.

   Proposition 12.14 *If every vertex of graph $\mathscr{G}$ has degree $\geq d$, then $\mathscr{G}$ contains a simple path of length d.*

2. **Vertex-degrees and their distributions in graphs**
   LESSON: Experience with graph-theoretic reasoning
   Prove the following results.

   Proposition 12.15 *Let $\mathscr{G}$ be an arbitrary graph.*

   (a) *The sum of $\mathscr{G}$'s vertex-degrees is even.*

   $\oplus$ **(b)** *The number of $\mathscr{G}$'s vertices having odd vertex-degree is even. (This is thm:even-num-odd-degrees 12.2 in the text.)*

3. **The graph-theoretic formulation of the 2SAT problem**
   LESSON: Experience with graph-theoretic reasoning
   We recall the graph $\mathscr{G}(\Phi)$ that we constructed in Section 12.1.2 from a 2SAT-formula Φ.
   Prove Lemma 12.1: If $\mathscr{G}(\Phi)$ contains a path from vertex x to vertex y, then it contains a path from vertex $\bar{y}$ to vertex $\bar{x}$.

4. **Finding maximal matchings greedily**
   LESSON: Experience with graph-theoretic argumentation
   Verify the following properties of the greedy process in the proof of Proposition 12.7. When applied to any graph $\mathscr{G}$:

   a. *The process will find a maximal matching for $\mathscr{G}$.*

   b. *The process will do its job in a number of steps proportional to $|\mathscr{N}_{\mathscr{G}}|$.*

5. **A necessary condition for perfect matchings**
   LESSON: Experience with graph-theoretic reasoning
   Prove Proposition 12.8: Let $\mathscr{G}$ be a graph that admits a perfect matching. Then:

   a. *$\mathscr{G}$ has an even number of vertices.*

b. *The cardinality of $\mathcal{G}$'s (perfect) matching—i.e., the number of edges in the matching—is exactly $\frac{1}{2}|\mathcal{N}_\mathcal{G}|$.*

6. **Equivalent definitions of "tree"**
   LESSON: Experience with graph-theoretic reasoning

   Prove Proposition 12.9: The following assertions about a connected graph $\mathcal{T}$ are logically equivalent.

   - *The graph $\mathcal{T}$ is cycle-free.*

   - *Each pair of distinct vertices of $\mathcal{T}$ is connected by precisely one path.*

7. **Defining major components of mesh- and torus-graphs in detail**
   LESSON: Practice with formal description of graph structures

   One notes in Section 12.3.3 that the basics of mesh- and torus-graphs are easy to capture via drawings, once the most basic notions—vertices and edges—have been defined. But sometimes—e.g., to craft proofs or design algorithms—one needs some higher-level constructs defined in formal detail.

   Formally define the following notions for the mesh $\mathcal{M}_{m,n}$ and the torus $\widetilde{\mathcal{M}}_{m,n}$:

   a. *a row of $\mathcal{M}_{m,n}$; of $\widetilde{\mathcal{M}}_{m,n}$*

   b. *a column of $\mathcal{M}_{m,n}$; of $\widetilde{\mathcal{M}}_{m,n}$*

   c. *a rectangular submesh of $\mathcal{M}_{m,n}$*

   d. *a rectangular sub-torus of $\widetilde{\mathcal{M}}_{m,n}$*

   e. *the four quadrants of $\mathcal{M}_{m,n}$*
      Be careful—we have not specified the parity of m and n!

8. $\oplus$ **A small graph isomorphism**
   LESSON: Reasoning about graph structure

   Prove Proposition 12.13:

   The order-4 hypercube $\mathcal{Q}_4$ is isomorphic to the 4×4 torus $\widetilde{\mathcal{M}}_{4,4}$.

   Hints. Start informally. Garner intuition by asking how to view each of $\mathcal{Q}_4$ and $\widetilde{\mathcal{M}}_{4,4}$ as a redrawing of the other.

   Try to find a relation (an "encoding") between the bit-strings that name the vertices of $\mathcal{Q}_4$ and the ordered pairs of integers that name the vertices of $\widetilde{\mathcal{M}}_{4,4}$.

   Once you get an idea for how such an "encoding" might work, try to incorporate the inter-vertex names of edges for both graphs.

Chapter 13
Graphs II: Graphs for Computing and Communicating

Many hands make light work

John Heywood (sixteenth-century English writer)

Chapter 12 has laid the groundwork for an intensive study of the many ways in which graphs and graph-related concepts find application within the fields of computing and communicating. That chapter focused mainly on properties of graphs—and families of graphs—that are local and descriptive: What is the essence of a graph's being a tree? a hypercube? The current chapter takes a further step and studies how the described local structure influences some of the dynamically applicable characteristics of graphs. Two major foci in this chapter are *vertex-coloring* (Section 13.1) and *path discovery* (Section 13.2). Both of these topics are used algorithmically in a broad range of applications of graphs, from circuit layout to computation scheduling to inter-agent communication scheduling. We provide a layered presentation to both topics: We completely cover the basic aspects of both topics; we offer extended material in appendices, providing further study opportunities for the interested reader; and we provide pointers to the literature for a number of fascinating more advanced topics. We close the chapter with capsule discussions of several specialized topics which go beyond the scope of any introductory text. We hope that the reader will be excited by these preliminary excursions into some of the ways that graph theory plays a role in "real life".

13.1 Graph Coloring and Chromatic Number

A *vertex-coloring* of a graph $\mathscr{G}$ is an assignment of labels (the "*colors*") to $\mathscr{G}$'s vertices, in such a way that all of a vertex v's neighbors get different labels than v's. The *chromatic number* of a graph $\mathscr{G}$ is the smallest number of colors that one can use in crafting a legal vertex-coloring of $\mathscr{G}$. In traditional parlance, the assertions

© Springer Nature Switzerland AG 2020
A. L. Rosenberg, D. Trystram, *Understand Mathematics, Understand Computing*,
https://doi.org/10.1007/978-3-030-58376-7_13

"$\mathcal{G}$ has chromatic number c" and "$\mathcal{G}$ is *c-colorable*"
are viewed as synonymous.

Enrichment note.

The notion of graph coloring can be used to computational advantage in a broad variety of situations.

One extremely important, and illustrative, use of graph coloring is to model independence among agents in *distributed* computing: The vertices of a computation-graph $\mathcal{G}$ represent *agents*, such as, e.g., processing elements in a multicomputer; and $\mathcal{G}$'s edges represent *communication links* that enable each vertex u to check its neighbors' states before taking any action and to inform its neighbors of state-changes occasioned by an action by u.

The prohibition against "monochrome" edges—i.e., edges both of whose incident vertices have the same color—guarantees that vertex u and all of its like-colored vertices can act at the same instant with no fear of missing an important input to those actions. Indeed, one often encounters programs for distributed computing that look something like

1. All *red* vertices perform simultaneously an action
2. All *green* vertices perform simultaneously an action
3. All *blue* vertices perform simultaneously an action
...

In a quite different setting, graph coloring can be used to represent either compatibility or incompatibility between pairs of graph vertices. As but one example: The graph-theoretic formulation in Section 12.1.2 of the 2SAT problem employs adjacencies in a graph to signal incompatible literals in a POS expression.

Most of the "named" graphs in Section 12.3 have quite small chromatic numbers. This is no accident: These graphs were invented (or, at least, placed in the spotlight) because of their importance to the field of parallel and distributed computing. As such, vertex-colorings of these graphs are an important tool for identifying the dependencies, or the lack thereof, as one schedules the concurrent execution of the graphs' task-vertices.

Motivated by such applications of graph models, we devote the current section to studying graphs that have small chromatic numbers.

13.1.1 Graphs with Chromatic Number 2

We begin our study of graphs having small chromatic numbers by focusing on 2-*colorable* graphs, which enjoy the smallest nontrivial chromatic number. It is not difficult to characterize these graphs structurally.

A graph $\mathscr{G}$ is *leveled* if there exists an assignment of *level-numbers* $\{1, 2, \ldots, \lambda\}$ to the vertices of $\mathscr{G}$ in such a way that every neighbor of a vertex having level-number ℓ has either level-number $\ell + 1$ or level-number $\ell - 1$. A vertex that has level-number ℓ is said to *reside on level ℓ of $\mathscr{G}$*.

Proposition 13.1 *A graph $\mathscr{G}$ has chromatic number 2 if, and only if, it is leveled.*

Proof. Say first that $\mathscr{G}$ is a leveled graph. Then labeling each vertex of $\mathscr{G}$ with the (odd-even) parity of its level provides a valid 2-coloring of $\mathscr{G}$.

Say next that $\mathscr{G}$ is 2-colorable. Pick any vertex v_0 of $\mathscr{G}$ and assign it to be the unique vertex on level 1. Next, assign all neighbors of v_0 to level 2. Continuing iteratively, say that the largest level-number that we have employed—i.e., assigned vertices to—is ℓ. Then we now assign to level $\ell + 1$ all neighbors of level-ℓ vertices that have not yet been assigned to a level. Because $\mathscr{G}$ is 2-colorable, this process colors every vertex of $\mathscr{G}$. Moreover, for each vertex v of $\mathscr{G}$:

- Vertex v is assigned a single color.
- If the shortest path from v to vertex v_0 has length ℓ, then v is assigned to level $\ell + 1$.
- Each neighbor of level-$(\ell + 1)$ vertex v is assigned to level ℓ or level $\ell + 2$, as required.

Thus, if we color each vertex v of $\mathscr{G}$ by the parity of its assigned level each edge of $\mathscr{G}$ connects a vertex of one color with a vertex of the other color. □

We can now show that the following named graphs are 2-colorable.

Corollary 13.1 *The following graphs are leveled, hence have chromatic number 2:*

(a) *every tree (which includes any path-graph $\mathscr{P}_n$)*

(b) *every cycle-graph $\mathscr{C}_n$ that has an even number n of vertices*

(c) *every mesh-graph $\mathscr{M}_{m,n}$*

(d) *every torus-graph $\widetilde{\mathscr{M}}_{m,n}$ with m and n both even*

(e) *every hypercube $\mathscr{Q}_n$*

Proof. We provide a detailed sketch for each of the five graph families in turn.

(a) The procedure from the second half of the proof of Proposition 13.1 exposes a level structure in any tree $\mathscr{T}$, as follows. Pick any vertex v of $\mathscr{T}$ and make it the unique vertex on level 1. Let all neighbors of v be assigned to level 2. Continuing iteratively, say that the largest level-number that we have employed is ℓ. Then we now assign level-number $\ell + 1$ to all neighbors of level-ℓ vertices that have not yet been assigned to a level of $\mathscr{T}$.

Of course this process can be simplified when $\mathscr{T}$ is a path-graph $\mathscr{P}$, by choosing one of $\mathscr{P}$'s end vertices as vertex v. We thereby have precisely one vertex on each level. (Starting with an internal vertex would create some levels with two vertices.) A lesson here is that *a graph may admit distinct level structures*.

(b) When we apply the procedure of part (a) to an even-length cycle, $\mathcal{C}_{2q}$, we produce a level structure in which levels 1 and $q+1$ have one vertex apiece, while all other levels have two vertices apiece.

(c) The edge-structure of mesh-graphs ensures that the labeling of each vertex $\langle i, j \rangle$ of $\mathcal{M}_{m,n}$ with the odd-even parity of the number $i + j$ is a 2-coloring of $\mathcal{M}_{m,n}$.

(d) The labeling in part (c) provides a 2-coloring of any torus-graph $\widetilde{\mathcal{M}}_{m,n}$ with m and n both even.

(e) Each edge of a hypercube $\mathcal{Q}_n$ connects a vertex $v = \beta_1 \beta_2 \cdots \beta_n$, where each $\beta_i \in \{0, 1\}$, to a vertex $v' = \beta_1' \beta_2' \cdots \beta_n'$ where $\beta_j \neq \beta_j'$ for precisely one j. Therefore, the following aggregation of vertices of $\mathcal{Q}_n$ into sets $S_0, S_1, \ldots, S_n$ provides a valid leveling of $\mathcal{Q}_n$.

Assign vertex $v = \beta_1 \beta_2 \cdots \beta_n$ to set S_k precisely if k of the bits β_i equal 1.

The preceding explanations complete the proof. $\square$

A simple argument verifies that no odd-length cycle $\mathcal{C}_{2q+1}$ with $q \geq 1$ is 2-colorable. One can craft such an argument by trying to 2-color such a cycle. One can then extend this argument to prove that no graph $\mathcal{G}$ that *contains* an odd-length cycle (as a subgraph) can be 2-colored.

This is an example of an excluded subgraph *condition:* Containing odd-length cycles is the feature that prevents 2-colorings of many graphs, including all de Bruijn networks and all torus-networks $\widetilde{\mathcal{M}}_{m,n}$ having odd m or n.

Let us focus momentarily on de Bruijn networks because they have quite interesting cyclic sub-digraphs. One observes the directed 3-cycle

$$00 \to 01 \to 10 \to 00$$

in $\mathcal{D}_2$ in Fig. 12.11 and the 3-cycle

$$001 \to 010 \to 100 \to 001$$

in $\mathcal{D}_3$ in Fig. 12.12. In fact, these small odd-length cycles are only the proverbial tip of the iceberg for de Bruijn networks. In fact, de Bruijn networks are *directed-pancyclic*, in the sense of the following result from [117].

Proposition 13.2 *For all n, the order-n de Bruijn network $\mathcal{D}_n$ is directed-pancyclic, i.e., it contains directed cycles of all possible lengths* $1, 2, \ldots, 2^n$.

The proof of this result is beyond the scope of an introductory text, but its treatment in [117] should be accessible to the motivated reader.

13.1.2 Planar and Outerplanar Graphs

In this section, we focus on two graph families that are defined in terms of the way they can be drawn (on a two-dimensional medium, such as a piece of paper).

Enrichment note.

Paying attention to how a graph can be drawn is not just an abstract game. The process of designing and implementing circuits within the constraints of *VLSI, Very Large Scale Integrated Circuit* technology, is very similar to drawing a circuit on a two-dimensional medium.

We refer the reader to the revolutionary 1979 text by Mead and Conway [82] for an introduction to this fascinating technology; the text requires technical literacy but little specialized knowledge.

- A graph is *planar* precisely if it can be drawn *without any crossing edges*.

- A graph $\mathcal{G}$ is *outerplanar* precisely if it can be drawn by *placing its vertices along a circle in such a way that its edges can be drawn as noncrossing chords of the circle*.

  The latter condition is equivalent to demanding that $\mathcal{G}$'s edges can be drawn within the circle without any crossings.

  We encourage the reader to garner intuition about the graphs in these families by experimenting with drawing some specific, rather complex graphs.

- The first set of graphs to draw are cliques, as defined in Section 12.3.2. The cliques $\mathcal{K}_3$, $\mathcal{K}_4$, and $\mathcal{K}_5$ will help expose the nature of the planar and outerplanar graphs, because:

  - $\mathcal{K}_3$ is outerplanar;

  - $\mathcal{K}_4$ is planar but not outerplanar;

  - $\mathcal{K}_5$ is not planar.

- The *bipartite* cousins of the cliques also provide valuable insights. For positive integers m and n, the $m \times n$ *biclique* (or *bipartite clique*) $\mathcal{K}_{m,n}$ is the graph whose vertex-set comprises the ordered pairs of integers:

$$\mathcal{N}_{\mathcal{K}_{m,n}} = \big\{ \langle i,0 \rangle \mid i \in \{1,2,\ldots,m\} \big\} \cup \big\{ \langle j,1 \rangle \mid j \in \{1,2,\ldots,n\} \big\}$$

and whose edges connect each vertex $\langle i,0 \rangle$ with each vertex $\langle j,1 \rangle$ where $1 \leq i \leq m$ and $1 \leq j \leq n$.

The second set of graphs to draw are the bicliques $\mathcal{K}_{1,3}$, $\mathcal{K}_{2,3}$, and $\mathcal{K}_{3,3}$. These graphs will also help expose the nature of the planar and outerplanar graphs, because:

– $\mathcal{K}_{1,3}$ is outerplanar;

– $\mathcal{K}_{2,3}$ is planar but not outerplanar;

– $\mathcal{K}_{3,3}$ is not planar.

We selected the preceding cliques and bicliques to "play with" very carefully. Using arguments that go beyond the scope of an introductory text, one can prove the following result, which characterizes each of our graph families by identifying *forbidden subgraphs*. The notion of *graph homeomorphism* plays a fundamental role in the characterization.

Homeomorphism is a daunting technical term which is easily understood informally. A *homeomorph* of a graph $\mathcal{G}$ is obtained by adding (degree-2) vertices along one or more edges of $\mathcal{G}$.

The characterization of planar graphs via forbidden subgraphs constitutes a celebrated theorem by the Polish mathematician and logician Kazimierz Kuratowski; the analogous result for outerplanar graphs was derived by the American mathematicians Gary Chartrand and Frank Harary (who invented the name "outerplanar", for reasons described in Section 13.1.2.2).

Theorem 13.1. (a) [29] *A graph is outerplanar if, and only if, it does not have a subgraph that is homeomorphic to either $\mathcal{K}_4$ or $\mathcal{K}_{2,3}$.*

(b) [71] *A graph is planar if, and only if, it does not have a subgraph that is homeomorphic to either $\mathcal{K}_5$ or $\mathcal{K}_{3,3}$.*

13.1.2.1 On 3-coloring *outerplanar* graphs

We look first at the smaller of this section's graph families, the *outerplanar graphs*. We begin our journey toward 3-colorings of these graphs with three results that highlight basic facts about the family. We begin by exposing an important inclusion relationship between outerplanar graphs and another family of simple graphs.

Proposition 13.3 *Every tree is outerplanar.*

We leave the challenge of providing the formal proof of this property to the reader. To aid in finding such drawings, we graphically describe in Figs. 13.1–13.3 the process of developing an outerplanar drawing of a directed tree. The first two figures begin to distribute the vertices of $\mathcal{T}$ around a circle in a way that allows one to draw $\mathcal{T}$'s edges in a noncrossing manner. The third figure depicts an entire outerplanar drawing of $\mathcal{T}$.

We next broaden our focus to the family relations among outerplanar graphs.

Proposition 13.4 *Let $\mathcal{G}$ be an outerplanar graph. Then:*

(a) $\mathcal{G}$ *is planar.*

(b) *Every subgraph of $\mathcal{G}$ is outerplanar.*

Fig. 13.1 Beginning an outerplanar drawing of a rooted tree $\mathcal{T}$: Placing $\mathcal{T}$'s root on a circle

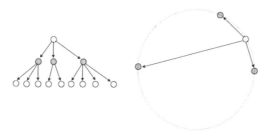

Fig. 13.2 Continuing an outerplanar drawing of $\mathcal{T}$: Placing $\mathcal{T}$'s root's children around the circle

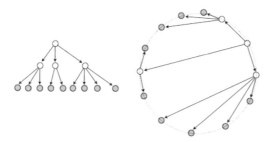

Fig. 13.3 A complete outerplanar drawing of $\mathcal{T}$

Proof. (a) $\mathcal{G}$'s planarity can be inferred from our ability to draw $\mathcal{G}$'s edges as noncrossing chords of the circle.

(b) We can produce a drawing of any subgraph $\mathcal{G}'$ of $\mathcal{G}$ that witnesses $\mathcal{G}'$'s outerplanarity by erasing some vertices and/or some edges from our outerplanarity-witnessing drawing of $\mathcal{G}$. These erasures cannot introduce any edge-crossings. □

We finally establish a technical result which is important in our quest to 3-color outerplanar graphs. The proof of this result is simplified if we restrict attention to those outerplanar graphs that are densest in edges. An outerplanar graph $\mathcal{G}$ is *maximally outerplanar* (or *maximal*, for short) if adding any new edge to $\mathcal{G}$ would destroy our ability to place $\mathcal{G}$'s vertices around a circle and draw its edges as noncrossing chords of the circle.

Lemma 13.1. *Let $\mathscr{G}$ be an n-vertex outerplanar graph.*

(a) *For maximal $\mathscr{G}$: If $n \geq 3$, then at least one of $\mathscr{G}$'s vertices has degree 2.*

(b) *For all outerplanar $\mathscr{G}$: If $n \geq 1$, then at least one of $\mathscr{G}$'s vertices has degree ≤ 2.*

Proof. We remark first that part (b) follows immediately from part (a) because adding edges to a non-maximal graph—with the goal of making it maximal—can never reduce the degree of any vertex. So we turn to crafting a proof of part (a).

We visualize each n-vertex maximal outerplanar graph $\mathscr{G}$ in terms of the drawing that witnesses its outerplanarity. We name $\mathscr{G}$'s vertices $0, 1, \ldots, n-1$, in their clockwise order around the circle in the drawing. $\mathscr{G}$'s edges come in two groups.

- There are the *ring edges*, i.e., the ones that go around the circle.

  These are edges $(0,1)$, $(1,2)$, $(2,3)$, $\ldots$, $(n-2,n-1)$, $(n-1,0)$.

- There are the *chordal edges*.

  In the drawing, these are noncrossing chords of the circle.

Small outerplanar graphs. We begin by analyzing outerplanar drawings of graphs having three, four, and five vertices, in order to help the reader develop intuition.

$n = 3$ *vertices.*   The unique 3-vertex maximal outerplanar graph has three ring edges

$$(0,1), \ (1,2), \ (2,0)$$

and no chordal edges. All three of its vertices have degree 2.

$n = 4$ *vertices.*   There are two 4-vertex maximal outerplanar graphs. Both graphs have four ring edges

$$(0,1), \ (1,2), \ (2,3), \ (3,0)$$

and one chordal edge.

The graph with chordal edge $(0,2)$ has two vertices of degree 2, namely, 1 and 3.

The graph with chordal edge $(1,3)$ has two vertices of degree 2, namely, 0 and 2.

$n = 5$ *vertices.*   There are five 5-vertex maximal outerplanar graphs. All graphs have five ring edges

$$(0,1), \ (1,2), \ (2,3), \ (3,4), \ (4,0)$$

and two chordal edges (which share an endpoint).

- The graph with chordal edges $(0,2)$ and $(0,3)$ has two vertices of degree 2, namely, 1 and 4
- The graph with chordal edges $(1,3)$ and $(1,4)$ has two vertices of degree 2, namely, 2 and 0
- The graph with chordal edges $(2,4)$ and $(2,0)$ has two vertices of degree 2, namely, 3 and 1

- The graph with chordal edges $(3,0)$ and $(3,1)$ has two vertices of degree 2, namely, 4 and 2
- The graph with chordal edges $(4,1)$ and $(4,2)$ has two vertices of degree 2, namely, 0 and 3

Large outerplanar graphs. A valuable way to think about maximal outerplanar graphs is in terms of the drawings that witness their maximal outerplanarity. Given any such drawing, one cannot add a chordal edge to the drawing without crossing an existing chordal edge. We now turn this intuition into a proof of the lemma.

Let us be given a maximal outerplanar graph $\mathscr{G}$ having $n > 3$ vertices—as usual, arrayed in the clockwise order $0, 1, \ldots, n-1$ around the circle that witnesses $\mathscr{G}$'s outerplanararity.

Because $\mathscr{G}$ has more than three vertices and because it is *maximally* outerplanar, the given drawing has chordal edges. Let us measure the *length* of each chordal edge (u, v) by the smallest number of ring edges that connect vertex u with vertex v. (In other words, we are traversing the circle between u and v in the shorter of the two possible directions.) We want to look at the shortest chordal edges according to our length measure. We distinguish two cases.

1. If the shortest chordal edge (u, v) has distance 2, then there must be a single vertex $w \in \mathscr{N}_{\mathscr{G}}$ that lies between u and v along the circle. (The shortest path of ring edges between u and v would have the form $(u - w - v)$.) It is easy to see that vertex w has degree 2.

2. If the shortest chordal edge (u, v) has distance $d > 2$, then the shortest path of ring edges between u and v would have the form

$$u - w_1 - w_2 - \cdots - w_{d-1} - v$$

for some distinct vertices $w_1, w_2, \ldots, w_{d-1} \in \mathscr{N}_{\mathscr{G}}$. Moreover, the fact that (u, v) is a *shortest* chordal edge in the drawing means that there does not exist a chordal edge that connects any pair of vertices in the set $S = \{u, v, w_1, w_2, \ldots, w_{d-1}\}$, except, of course, for vertices u and v.

Now we are "in trouble". Because there are at least two vertices w_1, w_2 between u and v around the circle, the absence of a chordal edge (other than (u, v)) with endpoints in the set S *contradicts the alleged maximal outerplanarity* of graph $\mathscr{G}$.

In more detail, the property of maximal outerplanarity guarantees that at least one (chordal) edge of one the following forms must be an edge of $\mathscr{G}$.

(u, w_i) for some $i \in \{1, 2, \ldots d-1\}$
(w_i, w_{i+j}) for some $j \in \{2, 3, \ldots, d-i-1\}$
(w_i, v) for some $i \in \{1, 2, \ldots d-1\}$

This contradiction shows that this case, $d > 2$, cannot occur.

We thus see that every maximally outerplanar graph has a vertex of degree 2. $\square$

We return now to our primary concern—vertex-colorings in outerplanar graphs.

The three-vertex cycle $\mathscr{C}_3$ witnesses the fact that not every outerplanar graph is 2-colorable. Therefore, the chromatic number for the family of outerplanar graphs can be no smaller than 3. We now show that 3 is, in fact, the chromatic number for this family. Our inductive proof of this fact can easily be turned into an efficient 3-coloring algorithm.

Proposition 13.5 (The 3-Color Theorem: Outerplanar Graphs) *Every outerplanar graph is 3-colorable.*

Proof. We proceed by induction on the number of vertices in the outerplanar graph to be colored.

Base case. It is very easy to find 3-colorings of outerplanar graphs having ≤ 3 vertices. These will form the base of our induction.

Inductive assumption. Assume that every outerplanar graph having $< n$ vertices is 3-colorable.

Inductive extension. Focus on an arbitrary n-vertex outerplanar graph $\mathscr{G}$.

By Proposition 13.4(c), $\mathscr{G}$ has a vertex v of degree ≤ 2. Let us remove vertex v from $\mathscr{G}$, along with its *incident* edges, i.e., those that connect v to the rest of $\mathscr{G}$; call the resulting graph $\mathscr{G}'$. Now:

- $\mathscr{G}'$ is clearly outerplanar.
  This will be clear once we "stitch" together the circle that we "damaged" by removing v.

- $\mathscr{G}'$ has *fewer than n* vertices.

By our inductive hypothesis, $\mathscr{G}$ is 3-colorable.

But now we can reattach vertex v to $\mathscr{G}'$ by replacing the edges that attach v to $\mathscr{G}$. Moreover, we can now color v using whichever of the three colors on $\mathscr{G}$ is *not* used for v's neighbors in $\mathscr{G}$. Once we so color v, we will have a 3-coloring of $\mathscr{G}$.

Our induction is thus extended, which completes the proof.   □

13.1.2.2 On vertex-coloring *planar* graphs

The larger of this section's two graph families comprises *planar graphs*: graphs that can be drawn (on a two-dimensional medium) with no crossing edges.

> **Historical note.**
>
> The historically original focus on planar graphs and their vertex-colorings stemmed from viewing the graphs as abstractions of geographical maps. The political units on a map (countries and/or cities and/or . . .) became the vertices of a graph $\mathscr{G}$, and political units that shared a border had an edge connecting the corresponding vertices of $\mathscr{G}$.
>
> Of course, every abstraction idealizes reality in some way. In this geographical setting, the major idealization is the assumption that "adjacent" units shared a boundary that had non-zero length. Adjacencies such as one observes with the US states of Arizona, Utah, Colorado, and New Mexico—the famous "four corners states"—which meet at a point were not allowed in the abstraction.
>
> The chromatic number of $\mathscr{G}$ then became, quite literally, the numbers of shades of ink that one would need in order to print the map in a way that assigned different colors to units that shared a border.

In a clique, every pair of vertices are mutually adjacent. Therefore:

For all n, the n-vertex clique $\mathscr{K}_n$ can be colored with n colors but no fewer.

Thus, the four-vertex clique $\mathscr{K}_4$ witnesses the fact that *not every planar graph is* 3-*colorable*. But it also raised the question, *Is every planar graph* 4-*colorable*?

A. On the Four-Color Theorem for planar graphs

A century-plus attempt to prove that four colors suffice for planar graphs culminated in one of the most fascinating dramas in modern mathematics: American mathematician Kenneth Appel and German mathematician Wolfgang Haken enlisted the help of their families—*and of their computer!*—as they crafted a proof of their renowned Four-Color Theorem for Planar Graphs. Their 1974 proof was long enough to consume *two* journal articles—both appearing in volume 21 of the *Illinois Journal of Mathematics* [4, 5].

Theorem 13.2 (The 4-Color Theorem: Planar Graphs). *Every planar graph is* 4-*colorable.*

The proof of Theorem 13.2, which first appeared in [4, 5], is beyond the scope of any textbook, even an advanced one, but the backstory of the proof is truly fascinating! And, the backstory supplies ample motivation for the proofs we present of the six-color and five-color analogues of the Theorem.

Beginning with a failed attempt in 1875 to prove that every planar map can be 4-colored, the so-called *Four-Color Problem* held the world of discrete mathematics in thrall for roughly a century before Appel and Haken announced their proof of Theorem 13.2 in 1974. But, this proof notwithstanding, the drama surrounding the 4-Color Problem persisted, because of the Appel-Haken proof's reliance—in a fundamental way—on a computer program that checked more than a thousand essential, but clerical, assertions (about forbidden subgraphs). It took the mathematics community years before the Appel-Haken proof, with its massive complexity and unprecedented employment of "collaboration" by computer, was generally accepted.

Even readers who might be (justifiably!) daunted by the primary references [4, 5] that accompany our statement of the Theorem may well enjoy the much more accessible articles [6, 7] in which the authors summarize—and, at a rather sophisticated level, popularize—this marvelous mathematical tale.

We turn now to the eminently accessible proofs of the six- and five-color analogues of Theorem 13.2. There are significant lessons within the proofs of these analogues.

- The weaker, six-color, version of the Theorem can be proved in much the same way as its outerplanar-graph cousin, Proposition 13.5. This proof appears in detail in Paragraph B.

- The proof of the stronger, five-color, version of the Theorem already requires us to break the world into multiple cases—but only a single-digit number of cases, in contrast to the four-digit list of cases required by the proof of the four-color version of the Theorem; see [4, 5]. That said, the complexity of the Five-Color Theorem has led us to include its proof only as an "Enrichment Topic", in Paragraph C.

B. The Six-Color Theorem for planar graphs

The first step in showing that every planar graph can be vertex-colored using six colors resides in the following analogue for planar graphs of Proposition 13.4(c), which asserts that every outerplanar graph has a vertex of degree 2.

Lemma 13.2. *Every planar graph has a vertex of degree ≤ 5.*

Proof. Let us focus on a planar drawing of a (perforce) planar graph $\mathscr{G}$ which has n vertices, e edges, and f *faces*. A *face* in a drawing of $\mathscr{G}$ is a polygon whose sides are edges of $\mathscr{G}$, whose points are vertices of $\mathscr{G}$, and whose interiors are "empty"—no edge of $\mathscr{G}$ crosses through a face.

Historical note.

Now that we know about faces, we can finally describe the origin of the term *outerplanar*. A graph $\mathscr{G}$ is outerplanar if it can be drawn (on a two-dimensional medium) in the following manner. All of $\mathscr{G}$'s vertices are placed around a circle, and all of $\mathscr{G}$'s edges are drawn as noncrossing chords of the circle. The region outside the vertex-bearing circle is, thus, the *outer* face of this special planar drawing of $\mathscr{G}$.

The following auxiliary result derives a celebrated "formula" attributed to Euler.

Proposition 13.6 (Euler's Formula for Planar Graphs) *Let $\mathscr{G}$ be a planar graph having n vertices and e edges. For every f-face planar drawing of $\mathscr{G}$, we have*

$$n - e + f = 2 \tag{13.1}$$

We defer proving Proposition 13.6 so that we can proceed with our ongoing proof of Lemma 13.2. We devote Section 13.1.2.3 to two quite different proofs of Euler's formula.

As we approach the next step in the proof of Lemma 13.2, we simplify the setting by assuming henceforth that $\mathscr{G}$ is *connected* and that it is a *maximal* planar graph—meaning that one cannot add any new edge to the drawing without crossing an existing edge (and, thereby, destroying planarity). Figure 13.4 illustrates the notion of maximality by providing planar drawings of two connected *maximal* planar graphs. The assumption of maximality only strengthen's the Lemma's conclusion

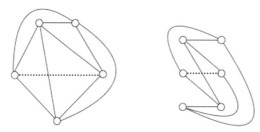

Fig. 13.4 Planar drawings of a maximal planar graph (left) and a maximal bipartite planar graph (right). One cannot add an edge to either drawing without destroying planarity—by introducing crossing edges. For reference, we have added dotted edges in the figure, which would augment the depicted planar graphs to the (inherently nonplanar) graphs $\mathscr{K}_5$ (on the left) and $\mathscr{K}_{3,3}$ (on the right)

by (apparently) making it more difficult to find a small-degree vertex.

With the maximality assumption in place, we now adapt a pedagogical tool from [14], in order to make the following counting argument easier to follow. We construct a *directed bipartite* graph **G** which exposes certain features of $\mathscr{G}$'s structure. On one side of **G** are the f faces of $\mathscr{G}$; on the other side are $\mathscr{G}$'s e edges. **G** contains an arc from each face of $\mathscr{G}$ to each edge of $\mathscr{G}$ that forms a "side" of the polygonal drawing of the face. Because $\mathscr{G}$ is a *maximal* planar graph, we have:

- Each face of $\mathscr{G}$ is a 3-cycle, hence involves three vertices.

- Each edge of $\mathscr{G}$ touches two faces.

- Each edge of $\mathscr{G}$ touches two vertices.

Let us now put these facts together, and assume, for contradiction, that every vertex of $\mathscr{G}$ had degree ≥ 6. We would then find that

$$f \leq \frac{2}{3}e \quad \text{and} \quad e \geq 3n$$

Incorporating these two bounds into Euler's Formula (13.1), we arrive at the following contradiction.

$$2 = n - e + f \leq \frac{1}{3}e - e + \frac{2}{3}e = 0$$

This contradiction proves that every planar graph must have a vertex of degree ≤ 5. □-Lemma 13.2

We finally have the tools to color any planar graph using six colors.

Proposition 13.7 (The 6-Color Theorem: Planar Graphs) *Every planar graph is 6-colorable.*

Proof. The Two-Color Theorem for Outerplanar Graphs (Proposition 13.5) and this result follow via almost-identical inductions on the number of vertices in the graph $\mathcal{G}$ that is being colored. Both arguments:

1. remark that the target number of colors is adequate for small graphs

   For outerplanar graphs, "small" means "≤ 3 vertices". For planar graphs, it means "≤ 4 vertices".

2. remove from $\mathcal{G}$ a vertex v of smallest degree d_v, together with all incident edges

   For outerplanar graphs, we guarantee that $d_v \leq 2$ (Proposition 13.4(c)). For planar graphs, we guarantee that $d_v \leq 5$ (Lemma 13.2).

3. inductively color the vertices of the graph left after the removal of v

   Let us denote by $\mathcal{G}'$ the graph obtained by removing v from $\mathcal{G}$. Then:
   For outerplanar graphs, we color $\mathcal{G}'$ with ≤ 3 colors (Proposition 13.5). For planar graphs, we use our inductive assumption that $\mathcal{G}'$ can be colored with ≤ 6 colors.

4. reattach v via its d_v edges and then color v.

   Note that the coloring guarantee in both results—Proposition 13.5 for outerplanar graphs and the current result for planar graphs—allows us to use $d_v + 1$ colors to color $\mathcal{G}$. Because v has degree d_v, it can have no more than d_v neighboring vertices in $\mathcal{G}'$, so our access to $d_v + 1$ colors guarantees that we can successfully color v.

The proofs of the 3-colorability of outerplanar graphs and the 6-colorability of planar graphs thus differ only in the value of d_v. □

C. ⊕ The Five-Color Theorem for planar graphs

Proposition 13.8 (The 5-Color Theorem: Planar Graphs) *Every planar graph is 5-colorable.*

Enrichment note.

The case analysis in the following proof is a bit more complex than in most of the results in the text, but a methodical reading should make the proof accessible to the reader. Moreover, the *roadmap* of the case analysis is a valuable lesson in how mathematics is really done!

The motivated reader will be able to recast the totally positive proof we present into the form of a proof by contradiction. The positive version should be more to the taste of a computation-oriented reader—but both proofs are equally correct.

Proof. This result appeared first in [57].

Throughout this proof, we discuss, without explicit mention, only *valid* vertex-colorings—in which neighboring vertices get different colors. We proceed by induction.

Base case. Because the 5-clique $\mathcal{K}_5$ is obviously 5-colorable, so also must be all graphs having ≤ 5 vertices. Therefore, we know that any non-5-colorable graph would have ≥ 6 vertices.

Inductive hypothesis. Assume, for induction, that every planar graph having $\leq n$ vertices is 5-colorable.

Inductive extension. If the proposition were false, then there would exist a planar graph $\mathcal{G}$ having $n + 1$ vertices which is not 5-colorable. By Lemma 13.2, $\mathcal{G}$ would have a vertex v of degree ≤ 5. The remainder of the proof focuses on the graph $\mathcal{G}$, its minimal-degree vertex v, and on v's $(d_v \leq 5)$ neighbors in $\mathcal{G}$.

Assume that there were a coloring of $\mathcal{G}$'s vertices which used ≤ 4 colors to color v's neighbors. We could, then, produce a 5-coloring of $\mathcal{G}$ by using the following analogue of the coloring strategy we used to prove Proposition 13.7.

1. Remove vertex v and its incident edges from $\mathcal{G}$, thereby producing the n-vertex planar graph $\mathcal{G}'$.

2. Produce a 5-coloring of $\mathcal{G}'$ that uses only four colors for the vertices that are neighbors of v in $\mathcal{G}$.

3. (*a*) Reattach vertex v and its edges to $\mathcal{G}'$, thereby reconstituting $\mathcal{G}$.

   (*b*) Color v with a color that is not used for v's neighbors; there must exist one.

In order to proceed toward a contradiction, we must understand what structural features of $\mathcal{G}$ make it impossible to use only four colors on v's neighbors while 5-coloring $\mathcal{G}$. There are three important situations to recognize.

Case 1.   Vertex v has degree ≤ 4.

By definition, we need at most four colors to color v's neighbors in this case.

In all remaining cases, vertex v has precisely five neighbors—or else, we would have invoked Case 1 to color $\mathcal{G}$ with five colors.

Case 2.   For some 5-coloring of $\mathcal{G}$, at least two neighbors of v get the same color.

In this case, only four colors are used to color v's (exactly five) neighbors. Then, we can use the remaining available color for v, which gives us a valid 5-coloring.

In all remaining cases, the five neighbors of v receive distinct colors. The situation therefore appears as in Fig. 13.5, which depicts vertex v and its neighbors. In

Fig. 13.5 A partially colored vertex of degree 5

this figure and its companions that illustrate the current proof, we use integer-label i to denote, ambiguously, vertex v_i and its assigned color c_i. The question mark "?" that labels vertex v indicates that we do not yet know what color to assign to v.

Case 3.   For some 5-coloring of $\mathcal{G}$, some two neighbors of v—call them v_1 and v_2—reside in distinct components of $\mathcal{G}$ once it is decomposed by removing v and its incident edges.

Let $\mathcal{G}'$ be the (in this case, disconnected) graph which results when v and its incident edges are removed from $\mathcal{G}$. For $i = 1, 2$, say that vertex v_i resides in component $\mathcal{G}_i$ of $\mathcal{G}'$.

Say that, under the 5-coloring of $\mathcal{G}$ that we are constructing, v_1 is colored *grey* and v_2 is colored *black*, as depicted in Fig. 13.6.

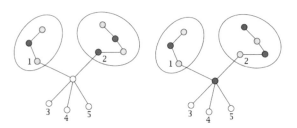

Fig. 13.6 A vertex-coloring for Case 3

We recolor the vertices of $\mathscr{G}_2$ so that vertex v_2 is now *grey*. (One needs only interchange the colors *black* and *grey* in the existing coloring to achieve this.) Because $\mathscr{G}_1$ and $\mathscr{G}_2$ are mutually disjoint, we can always do this without affecting the valid coloring of $\mathscr{G}_1$.

Once we have thus recolored $\mathscr{G}_2$, we obtain a 5-coloring of $\mathscr{G}$ for which Case 2 holds. (We can color v with the color that is not used when we reattach v to $\mathscr{G}'$.)

We now see that Cases 1–3 cannot prevent us from 5-coloring $\mathscr{G}$, so we are left with the following minimally constrained situation.

Case 4. • Every minimum-degree vertex of $\mathscr{G}$ has five neighbors.

For the minimum-degree vertex v, let us call these neighbors v_1, v_2, v_3, v_4, v_5, *in clockwise order within the planar drawing of* $\mathscr{G}$.

• In every 5-coloring of $\mathscr{G}$, the neighbors of every minimum-degree vertex receive distinct colors.

For vertex v, let us say that each neighbor v_i receives color c_i.

To analyze Case 4, we focus on vertices v_1 and v_3 in Fig. 13.7. Importantly, these

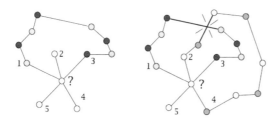

Fig. 13.7 A vertex-coloring for Case 4.

vertices have received distinct colors (c_1 and $c_3 \neq c_1$, respectively), and these vertices are not adjacent to one another along a clockwise sweep around vertex v.

Now take $\mathscr{G}$ and focus only on the vertices that are colored c_1 or c_3 (as are v_1 and v_3, respectively) and on the vertices that are colored c_2 or c_4 (as are v_2 and v_4, respectively). One sees from Fig. 13.7 that:

• $\mathscr{G}$ can—*but need not*—contain a path whose vertices alternate colors c_1 and c_3. Call this a "c_1-c_3 path" between vertices v_1 and v_3.

• $\mathscr{G}$ can—*but need not*—contain a path whose vertices alternate colors c_2 and c_4. Call this a "c_2-c_4 path" between vertices v_2 and v_4.

• $\mathscr{G}$ *cannot* contain both of the paths just described, i.e., a c_1-c_3 path between v_1 and v_3 *and* a c_2-c_4 path between v_2 and v_4.

These two paths, if they existed, would cross—see Fig. 13.7—which is forbidden because $\mathscr{G}$ is *planar*.

It follows that *either* $\mathscr{G}$ does not contain a c_1-c_3 path between v_1 and v_3 *or* $\mathscr{G}$ does not contain a c_2-c_4 path between v_2 and v_4. Say, with no loss of generality, that the former path does not exist. Then we can switch colors c_1 and c_3 beginning with vertex v_1 and obtain a coloring of $\mathscr{G}$ in which v_1 and v_3 both receive the color c_3. We can then proceed as in Case 3 to get a 5-coloring of $\mathscr{G}$.

This four-case analysis shows that we can always produce a 5-coloring of $\mathscr{G}$. □

We close our study of vertex-colorings of planar graphs with an overview of the intellectual cost-benefit tradeoff that we have exposed:

- A straightforward recursive coloring strategy suffices if one is willing to settle for a six-color palette when coloring planar graphs (Proposition 13.7).

- A four-case analysis, in which one case comprises several subcases, is needed in order to eliminate one of the colors from our palette, i.e., to achieve a 5-coloring strategy (Proposition 13.8).

- An analysis involving close to 2,000 cases is needed in order to achieve the provable optimal, a four-color palette that always works (Theorem 13.2).

13.1.2.3 Two validations of Euler's Formula

We now develop two quite different proofs of Proposition 13.6, which expose different ways to think about the identity.

Validation via structural induction. Our first approach validates formula (13.1) by growing a planar graph $\mathscr{G}$ inductively, edge by edge. Note that we formulate our induction a bit differently than our earlier, simpler ones, particularly in the inductive hypothesis.

Base case. The Formula clearly holds for the smallest planar graphs, including the smallest interesting one, $\mathscr{C}_3$, which has $n = e = 3$ and $f = 2$ (the inner and outer faces of the "triangle").

Inductive hypothesis. Assume that the Formula holds for a given graph $\mathscr{G}$.

Inductive extension. We extend our induction by growing the current version of $\mathscr{G}$, by adding a new edge. Two cases arise.

- *The new edge connects existing vertices.* In this case, this augmentation of $\mathscr{G}$ increases the number of edges (e) and the number of faces (f) by 1 each, while keeping the number of vertices (n) unchanged.

- *The new edge adds a new vertex, which is appended to a preexisting vertex.* In this case, this augmentation of $\mathscr{G}$ keeps the number of faces (f) unchanged while it increases by 1 both the number of edges (e) and the number of vertices (n).

In both cases Euler's Formula (13.1) continues to hold as we augment $\mathscr{G}$.

The augmentation thus extends the induction, hence validates the Formula. □

Validation via deconstruction. Let us be given a planar graph $\mathscr{G}$ that has n vertices, e edges, and f faces. We validate Formula (13.1) by deconstructing $\mathscr{G}$ and showing that each step in the process preserves as *invariant* the expression

$$\phi(n,e,f) \overset{\text{def}}{=} n - e + f \tag{13.2}$$

Explanatory note.

Invariants are extremely important conceptual tools when crafting proofs. The idea underlying their use is to find an expression $\phi(\circ)$ whose value is preserved—i.e., it "holds invariant"—as some relevant process proceeds.

With many mathematical concepts—including the notion of invariant—it is easier to grasp the concept by experiencing examples than by internalizing abstract explanations. Therefore, we recommend that the reader follows this second validation of Euler's Formula while keeping track of the invariance of expression $\phi(n,e,f)$ of (13.2).

Focus on the following two-phase process.

Phase 1. Iterate the process of removing edges from $\mathscr{G}$ until some edge-removal reduces $\mathscr{G}$ to a graph with a single face.

The reader should verify that this termination condition is equivalent to stopping when the remaining graph is a (connected) tree.

The action. If the graph remaining at some step contains an edge that is shared by two distinct faces, then remove any such edge.

Figure 13.8 illustrates Phase 1 for a (residual) graph having $n = 8$ vertices, $e = 11$ edges, and $f = 5$ faces.

The analysis.

- The graph remaining after an edge-removal is still planar, so we can continue the process.

- The process preserves the value of function ϕ; i.e.,

$$\phi(n,e,f) = \phi(n,e-1,f-1)$$

This is because n is unchanged, while e and f are each reduced by 1.

Phase 2. Iterate the following process of removing vertices from the tree produced by Phase 1, until only one vertex remains.

The action. Remove any leaf of the current tree, together with its incident edge.

Figure 13.9 illustrates the action of Phase 2 for a tree having $n = 8$ vertices— hence, $e = 7$ edges and $f = 1$ face.

The analysis. The value of ϕ remains invariant under each leaf-removal:

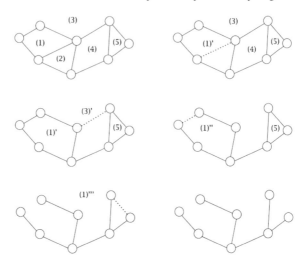

Fig. 13.8 Illustrating Phase 1: From top left to bottom right, each transformation preserves the invariant $\phi(n, e, f)$. The first transformation removes the edge shared by faces (1) and (2), creating a new, merged, face (1)'

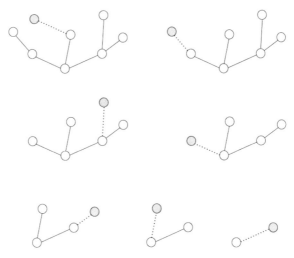

Fig. 13.9 Illustrating Phase 2: We remove a sequence of leaves, each with its incident edge, until we reach a single vertex. Here again, ϕ remains invariant after each leaf-removal

$$\phi(n, e, f) \;=\; \phi(n-1, e-1, f)$$

That is, f is unchanged ($f = 1$ for a tree), while n and e are each reduced by 1.

The cumulative analysis.

- The described process executes a number of steps exactly equal to the number of edges of the graph $\mathscr{G}$ that we begin with. Specifically, the edge-removing Phase 1 removes $e - n$ edges, while the leaf-removing Phase 2 removes $n - 1$ vertices.

- At the end of the process, $e = 0$ and $n = f = 1$, so that the residual value of ϕ is 2. Because each action prescribed by the process preserves the value of ϕ, we know that ϕ has the value 2 also *before* each edge-removal and leaf-removal.

We conclude, in summation, that at the start of the process, $\phi(n, e, f)$ had the value 2. In other words, $n - e + f = 2$, as asserted by Euler's Formula. $\square$

Enrichment note.

Our study of vertex-coloring in graphs has focused almost entirely on planar and outerplanar graphs. This is because—aside from the importance of such graphs in applications—their vertex-coloring problems can be studied entirely in terms of the mathematical structure of the graphs being colored. Vertex-coloring for broader families of graphs usually cannot be studied solely by focusing on the target graph's structure: Algorithmic issues arise quite early in the coloring process (although of course, mathematics is employed to analyze the algorithms).

In closing our treatment of this topic, though, we remark that there are many interesting algorithmic problems relating to vertex-coloring arbitrary graphs. The following such result dates to the earliest days of the study of NP-completeness [52, 64].

For any fixed $k \geq 3$, the problem "Is graph $\mathscr{G}$ k-colorable" is NP-*complete.*

This complexity-theoretic result suggests that *optimal* vertex-coloring is computationally inherently difficult. The "sting" of the result is moderated by the existence of "greedy", hence efficient, vertex-coloring algorithms that give quite good results in practice [36]. The term "greedy" here refers to algorithms that allocate a yet-unused color to a vertex only when no uncolored vertex can be validly colored with an already-used color. (See Footnote 2 in Chapter 12.)

13.2 Path- and Cycle-Discovery Problems in Graphs

Just as various genres of *spanning trees* are used to "summarize" aspects of the *connectivity* structure of a graph $\mathscr{G}$, various genres of *paths* and *cycles* in $\mathscr{G}$ are often useful to "summarize" aspects of $\mathscr{G}$'s *traversal* structure. This section is devoted to a range of problems related to determining the existence in a graph $\mathscr{G}$ of a path or a cycle that *completely* "summarizes" $\mathscr{G}$'s traversal structure—either by containing every edge of $\mathscr{G}$ precisely once (the *Eulerian* version of "traversal summarization") or by containing every vertex of $\mathscr{G}$ precisely once (the *Hamiltonian* version of "traversal summarization").

We generally focus in this section only on *undirected* paths and cycles in *undirected, unweighted* graphs. Extending the notions we discuss to their directed analogues in directed and/or weighted graphs will be accomplished via carefully crafted exercises. In a similar, but simpler, vein, we generally discuss only problems concerning *cycles*, leaving to the reader the analogous path-related notion. We begin by delimiting the two main classes of cycle-discovery problems that we study.

Eulerian cycles (or tours). A cycle in a graph $\mathscr{G}$ that traverses each of $\mathscr{G}$'s *edges* precisely once is called an *Eulerian cycle* (or, often, an *Eulerian circuit*). The edge-exhausting cycles/circuits/tours referred to by these several names were introduced as a topic of study in 1736 by the Swiss mathematician Leonhard Euler, whose name we have already encountered multiple times. Euler allegedly identified the topic while contemplating how to devise a tour of the town of Königsberg that would cross each of the town's bridges precisely once. The quest for Eulerian cycles in graphs is certainly one of the oldest problems—perhaps literally the oldest problem—in the fields now called *operations research* and *graph theory*. (An edge-exhausting *path* in $\mathscr{G}$ is referred to in the obvious analogous way.) When one views an Eulerian cycle as a "map" for traversing a graph—as did Euler when contemplating this problem—one often calls the cycle an *Eulerian tour*. Traditionally, a graph that admits an Eulerian cycle is said to be *Eulerian*.

Hamiltonian cycles (or circuits, or tours). A cycle in a graph $\mathscr{G}$ that encounters each of $\mathscr{G}$'s *vertices* precisely once is called a *Hamiltonian cycle* (or, often, a *Hamiltonian circuit*). This cycle-discovery problem is named in honor of the Irish mathematician Sir William Rowan Hamilton, who is credited with inventing the concept in the mid-nineteenth century. (In fact, Hamilton's work on his eponymous cycles was anticipated by many decades in the work of English mathematician, Thomas Pennington Kirkman.) When one views a Hamiltonian cycle as a "map" for traversing a graph, one often calls the cycle a *Hamiltonian tour*. Traditionally, a graph that admits a Hamiltonian cycle is said to be *Hamiltonian*. (A vertex-exhausting *path* in a graph is referred to in the obvious analogous way.)

Despite the conceptual duality between the edge-exhausting goal that underlies Eulerian paths/cycles and the vertex-exhausting goal that underlies Hamiltonian paths/cycles, these two graph-traversing goals differ from one another in virtually every significant mathematical and algorithmic respect. It is rather easy to characterize the family of graphs that admit Eulerian tours and to find such a tour if it exists (Section 13.2.1); in sharp contrast, there is no known characterization of the family of graphs that admit Hamiltonian tours, and the computational problem of efficiently determining whether a graph admits such a tour is one of the major classical problems in the field of computational complexity (Section 13.2.2).

13.2.1 Eulerian Cycles and Paths

The main results in this section characterize the families of directed and undirected graphs that admit *Eulerian cycles* or *paths*. The proofs of these characterizations are constructive: they are built upon algorithms that efficiently find such a cycle or such a path if one exists. We focus on graphs that are *connected*—but our algorithms can actually be adapted to find an Eulerian cycle or path in each connected component of a general graph.

We open our adventure with the following amusing fact.

Enrichment note.
The problem of discovering an Eulerian cycle in a graph $\mathscr{G}$ is (algorithmically) equivalent to the problem of drawing $\mathscr{G}$ "cyclically"—i.e., starting and ending with the same vertex—without ever lifting one's pencil.

The next two propositions develop—for both cycles and paths, in both undirected and directed graphs—simple, elegant characterizations of graphs that admit Eulerian cycles and paths. These characterization are validated via simple and efficient algorithms that determine whether a given graph admits such a cycle or such a path and that discover such a cycle or path when one exists. To avoid "special cases" that disrupt the flow of a proof, we restrict attention to graphs and digraphs that are *nontrivial* in the sense that

- each (di)graph has at least three vertices;
- each (di)graph is connected.

Within this domain, we have the following characterizations.

Proposition 13.9 (Eulerian Cycles) (a) *A connected undirected graph $\mathscr{G}$ admits an Eulerian cycle if, and only if, every vertex of $\mathscr{G}$ has even degree.*
(b) *A connected directed graph $\mathscr{G}$ admits a directed Eulerian cycle if, and only if,* IN-DEGREE(v) = OUT-DEGREE(v) *for every vertex v of $\mathscr{G}$.*

Proof. **Necessity.** The conditions asserted in parts (a) and (b) are *necessary* because of the following fact. Every time any cycle traverses a vertex v of a graph $\mathscr{G}$, that encounter accounts for *two* edges that are incident to v: the cycle must "enter" v using one edge and "exit" v using a different edge. For part (b), $\mathscr{G}$ is a digraph, and the encounter with vertex v traverses arcs rather than edges.

Sufficiency. We now verify that the Proposition's conditions are also *sufficient*. We provide a complete proof of sufficiency only for part (a), for undirected graphs; we provide only hints regarding the sufficiency of the conditions for part (b).

We have a choice of proving sufficiency either by induction on the number of vertices in a graph or by induction on the number of edges in the graph. While we invite the reader to develop the vertex-based proof, we focus only on the technically more streamlined edge-based proof.

Focus on a connected undirected graph $\mathscr{G}$ that has m edges and n vertices, each vertex having nonzero even degree. We prove by induction on m that $\mathscr{G}$ is Eulerian.

Base case. We begin with the base case $m = 3$. (You should consider why graphs with $m = 1$ or $m = 2$ edges cannot be Eulerian.) For $m = 3$, there is only one Eulerian graph, namely, $\mathscr{C}_3 = \mathscr{K}_3$.

Inductive hypothesis. Assume that the Proposition holds for any number of edges $k \leq m$.

Inductive extension. Consider a connected $(m + 1)$-edge graph $\mathscr{G}$ all of whose vertices have nonzero even degree. By Proposition 12.3, $\mathscr{G}$ contains a nonempty cycle $\mathscr{C}$ (see Fig. 13.10). Focus on the subgraph $\mathscr{G}'$ of $\mathscr{G}$ that is obtained by removing

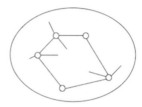

Fig. 13.10 A cycle in graph $\mathscr{G}$ (ignoring its other vertices and their incident edges)

cycle $\mathscr{C}$. Clearly $\mathscr{G}'$ contains no more than $m - 2$ edges, because $\mathscr{G}$ has $m + 1$ edges and $\mathscr{C}$ contains at least three edges.

Now, graph $\mathscr{G}'$ may have multiple connected components because the process of removing cycle $\mathscr{C}$ could have disconnected the residual graph $\mathscr{G}'$; see Fig. 13.11. However, no matter how many components $\mathscr{G}'$ has, each such component has no

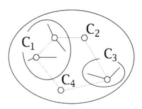

Fig. 13.11 Subgraph $\mathscr{G}'$ and its connected components after we removed the edges of cycle $\mathscr{C}$ (whose edges are dotted lines). In this example, there are four connected components, C_1, C_2, C_3, C_4; two of these, namely C_2 and C_4, are isolated vertices

more than $m - 2$ edges. This means that each connected component of $\mathscr{G}'$ is either an isolated vertex—i.e., a graph with one vertex and no edges—or a *nontrivial component*. In the latter case, our inductive hypothesis guarantees that $\mathscr{G}'$ has an Eulerian cycle.

The final step in extending the induction is to "knit" the structure we have derived—namely, cycle $\mathscr{C}$ plus graph $\mathscr{G}'$—into a single Eulerian cycle for $\mathscr{G}$. We accomplish this via the following *gendanken* process. We begin to traverse cycle $\mathscr{C}$, continuing to walk until we encounter a nontrivial component, call it $\mathscr{G}''$, which we have never before encountered. We now pause in our walk along $\mathscr{C}$ to traverse the Eulerian cycle in $\mathscr{G}''$ that the inductive hypothesis guarantees. When we conclude the traversal of $\mathscr{G}''$—perforce (since it is a cycle) at the same vertex where the traversal began—we resume our walk along $\mathscr{C}$. Eventually, the walk along $\mathscr{C}$— punctuated by traversals of Eulerian cycles in connected components—terminates. At this point, we are guaranteed to have crossed every edge of $\mathscr{G}$ exactly once.

The described walk exposes an Eulerian cycle in $\mathscr{G}$, as claimed. $\square$

What we have not *done.* (1) Adapting our proof of Proposition 13.9 to directed graphs requires only ensuring that we always enter a vertex v along an in-arc and leave via an out-arc. Because the in-degree of each vertex v of $\mathscr{G}$ equals v's out-degree, the required changes to the proof are clerical in nature.
(2) The challenge of algorithmically producing an Eulerian cycle in graph $\mathscr{G}$ requires nontrivial data structuring that is outside the scope of a mathematics text. However, our proof of Proposition 13.9 does give major hints about how to produce a *recursive* algorithm that constructs such a cycle.

The simplicity of the preceding characterization degrades a trifle when one seeks Eulerian *paths* rather than *cycles*.

Proposition 13.10 (Eulerian Paths) (a) *A connected undirected graph $\mathscr{G}$ admits an Eulerian path if, and only if, at most two vertices of $\mathscr{G}$ have odd degree.*
(b) *A connected directed graph $\mathscr{G}$ admits an Eulerian path if, and only if: either $\mathscr{G}$ admits an Eulerian cycle, or $\mathscr{G}$ contains one vertex u such that*
$$\text{IN-DEGREE}(u) \;=\; \text{OUT-DEGREE}(u) + 1$$
and one vertex v such that
$$\text{IN-DEGREE}(v) \;=\; \text{OUT-DEGREE}(v) - 1.$$

The proof of Proposition 13.10 shares its overall structure with the proof of Proposition 13.9, with one major difference. Whereas a cycle has neither beginning nor end, a path has both. Proposition 13.10(a) asserts that an undirected graph which admits an Eulerian path but not an Eulerian cycle has precisely two vertices of odd degree, and these odd-degree vertices play the roles of the endpoints of the Eulerian path. In similar fashion, Proposition 13.10(b) asserts that a directed graph which admits an Eulerian path but not an Eulerian cycle contains one vertex, u, whose out-degree exceeds its in-degree and one vertex, v, whose in-degree exceeds its out-degree: vertex u plays the role of the beginning vertex of the Eulerian path, and vertex v plays the role of the end vertex of the path. With these hints, we invite the reader to adapt the proof of Proposition 13.9 to obtain a proof of Proposition 13.10.

We close this subsection by applying Proposition 13.9 to the "named graphs" of Section 12.3.

Corollary 13.2 *The following facts can be derived from Proposition 13.9.*

Graph	*Eulerian?*	*Explanation*
Cycle	*Yes*	*Each vertex has degree* 2
Clique	*Odd indices*	$\mathcal{K}_{2n}$ *has odd vertex-degrees*
2-Dim. Mesh	*No*	*Each non-corner top, side vertex has degree* 3
2-Dim. Torus	*Yes*	*Each vertex has degree* 4
Hypercube	*Even indices*	*Each vertex of* $\mathcal{Q}_n$ *has degree n*
de Bruijn		
directed	*Yes*	*Vertices have equal* IN-DEGREE *and* OUT-DEGREE
undirected	*Yes*	*For* $\mathcal{D}_n$*: Vertices* $\bar{0}$, $\bar{1}$ *have degree* $2n - 2$;
		Each other vertex has degree $2n$

13.2.2 Hamiltonian Paths and Cycles/Tours

We turn now to the problem of determining when a connected graph $\mathcal{G}$ has a *Hamiltonian cycle*—and the allied problem of finding such a cycle when one exists. Supplementary material related to cycles in general graphs, as well as in our "named" graphs can be found in the survey [89].

One can envision a number of benefits rendered accessible by the presence of a Hamiltonian cycle in a graph $\mathcal{G}$. Most obviously, the cycle specifies a tour of $\mathcal{G}$ (or of a map whose structure $\mathcal{G}$ abstracts) which visits each of $\mathcal{G}$'s vertices precisely once. This is the sense in which the cycle "summarizes" $\mathcal{G}$'s traversal structure.

13.2.2.1 More inclusive notions of Hamiltonicity

Many graphs—even ones with "nice" structures—do not admit Hamiltonian cycles. As an exercise, the reader can generate *mesh-graphs* (Section 12.3.3) that admit no Hamiltonian cycle. The existence of such non-Hamiltonian graphs has spawned several independent paths of inquiry. One path seeks "modest" ways to weaken the property of *Hamiltonicity* in a way that retains many of the benefits of Hamiltonicity's while encompassing a broader range of graph structures. We describe two avenues toward weakened, more inclusive, notions of Hamiltonicity.

Be satisfied with paths, rather than cycles. A *Hamiltonian path* in a graph $\mathcal{G}$ is a path that passes through each of $\mathcal{G}$'s vertices precisely once. Hamiltonian paths can easily be shown to be a strictly weaker notion than Hamiltonian cycles: Clearly, every graph which admits a Hamiltonian cycle also admits a Hamiltonian path: one just drops any single edge of such a cycle to obtain such a path. However, there are many graphs which admit a Hamiltonian path but *not* any Hamiltonian cycle. Mesh-graphs can supply easy examples. You have produced (in the preceding paragraph) mesh-graphs which admit no Hamiltonian cycle, even though every mesh-graph easily admits a Hamiltonian path—just traverse rows one after another, in alternating directions.

Be satisfied with paths constructed from short paths, rather than edges. A Hamiltonian cycle in graph $\mathscr{G}$ is a circular enumeration of $\mathscr{G}$'s vertices in which adjacent vertices are at unit distance from one another—i.e., are connected by an edge. We can weaken (or, generalize) this notion by introducing the notion of a *Hamiltonian k-cycle* in $\mathscr{G}$, for any positive integer k: This is a circular enumeration of $\mathscr{G}$'s vertices in which adjacent vertices are at distance $\leq k$ from one another—so a Hamiltonian 1-cycle is what we have been calling a Hamiltonian cycle. The following result shows that one need not let k be very big before one finds that all connected graphs admit Hamiltonian k-cycles. The proof of this result is based on traversals of a graph's spanning trees, an algorithmic stratagem which is beyond the current text.

Proposition 13.11 **(a)** [30] *Every connected graph admits a Hamiltonian 3-cycle.*
(b) [49] *Let $\mathscr{G}$ be any graph that is* 2-connected *(or, biconnected) in the sense that, for every two vertices, $u, v \in \mathscr{N}_{\mathscr{G}}$, there exist at least two vertex-disjoint paths in $\mathscr{G}$ that connect u and v. Every biconnected graph $\mathscr{G}$ admits a Hamiltonian 2-cycle.*

13.2.2.2 Hamiltonicity in "named" graphs

Yet another direction of inquiry is to determine whether specific graphs of interest are Hamiltonian. We illustrate this avenue by reviewing the five "named" families of graphs in Section 12.3.

Proposition 13.12 **(a)** *Every cycle-graph $\mathscr{C}_n$ is Hamiltonian.*
(b) *Every clique-graph $\mathscr{K}_n$ is Hamiltonian.*
(c).1. *For all m, n: the mesh-graph $\mathscr{M}_{m,n}$:*
  (i) is path-Hamiltonian.
  (ii) contains no odd-length cycle; hence, is not Hamiltonian if mn is odd.
  (iii) is Hamiltonian whenever mn is even.
(c).2. *For all m, n: the torus-graph $\widetilde{\mathscr{M}}_{m,n}$ is Hamiltonian.*
(d) *Every hypercube $\mathscr{Q}_n$ is Hamiltonian.*
(e) *Every de Bruijn network $\mathscr{D}_n$ is both Hamiltonian and directed-Hamiltonian.*

Before we embark on our proof Proposition 13.12, we remark that some of the assertions in the Proposition can be strengthened to assert the presence in the target graph of cycles of many lengths; cf. [89]. In particular, we devote Section F.1 to proving that the 2^n-vertex de Bruijn network $\mathscr{D}_n$ is *directed-pancyclic*—meaning that it contains directed cycles of *every* length, from length 1 to length 2^n.

Proof (Proposition 13.12). **(a)** This is a tautology, by definition of $\mathscr{C}_n$.

(b) This is immediate because, by definition, $\mathscr{K}_n$ contains every n-vertex graph—including $\mathscr{C}_n$—as a subgraph.

(c).1.i. As we noted earlier (when discussing Hamiltonian paths vs. cycles), one can "snake" a path through $\mathcal{M}_{m,n}$, row by row, traversing adjacent rows in alternating directions.

(c).1.ii. This is a consequence of the fact that $\mathcal{M}_{m,n}$ is *bipartite*, so that every cycle has even length.

(c).1.iii. We sketch the construction of a Hamiltonian cycle in $\mathcal{M}_{m,n}$ when mn is even. Say, with no loss of generality, that m is even, so that $\mathcal{M}_{m,n}$ has an even number of rows. Temporarily remove column 1 of $\mathcal{M}_{m,n}$, and construct the "snaking" Hamiltonian path described in part **(c).1.i** of this proof. Because m is even, this path begins and ends in column 2 of $\mathcal{M}_{m,n}$. One can, therefore, reinstate column 1 and use it to connect the ends of the "snaking" Hamiltonian path. This describes a Hamiltonian cycle in $\mathcal{M}_{m,n}$.

(c).2. When mn is even, the Hamiltonicity of $\widetilde{\mathcal{M}}_{m,n}$ follows from the fact that $\mathcal{M}_{m,n}$ is a spanning subgraph of $\widetilde{\mathcal{M}}_{m,n}$. (Think about it!) When mn is odd, one need just traverse $\widetilde{\mathcal{M}}_{m,n}$'s vertices row by row, going to the cyclically next vertex after completing each row (see Fig. 13.12). Details are left to the reader.

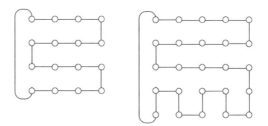

Fig. 13.12 Illustrating the principles for building Hamiltonian cycles in $m \times n$ tori when both m and n are even (left) and when both are odd (right)

(d) One can craft a Hamiltonian cycle in $\mathcal{Q}_n$ by generating an *order-n binary reflected Gray code*—so named for its inventor, Bell Laboratories researcher Frank Gray; see [87]. Such a "code" is a cyclic enumeration of all 2^n binary strings of length n, having the property that cyclically adjacent strings differ in only one bit-position. Length-n strings x_i and x_j are *cyclically adjacent* in the Gray code $\langle x_0, x_1, \ldots, x_{2^n-1} \rangle$ if $j = i + 1 \bmod 2^n$.

It is computationally easy to generate an order-n Gray code from an order-$(n-1)$ Gray code, as follows.

We note first that the order-1 code is just the sequence $\langle 0, 1 \rangle$.

Inductively, to generate the order-$(k+1)$ Gray code from the order-k code:

- Concatenate the order-k code with a *reversed* copy of itself. (It is the code-sequence that is reversed, not the individual strings. For instance, as we go from the order-2 code $\langle x_0, x_1, x_2, x_3 \rangle$ to the order-3 code, we concatenate that sequence with $\langle x_3, x_2, x_1, x_0 \rangle$.)

- Augment each length-k string in one copy of the order-k Gray code to length $(k+1)$ by prepending a 0 to each string; augment each length-k string in the other (reversed) copy of the order-k Gray code to length $(k+1)$ by prepending a 1 to each string.

Table 13.1 illustrates Gray codes of the first few orders.

Table 13.1 Gray codes of orders 1–4

Order 1	Order 2	Order 3	Order 4
0	00	000	0000
1	01	001	0001
	11	011	0011
	10	010	0010
		110	0110
		111	0111
		101	0101
		100	0100
			1100
			1101
			1111
			1110
			1010
			1011
			1001
			1000

$$(13.3)$$

We now sketch a proof that for each index $n \in \mathbb{N}^+$, the order-n Gray code sequence specifies a Hamiltonian cycle in $\mathcal{Q}_n$; exercises will give the reader the opportunity to fill in details. We verify the following two assertions in turn:

1. *The order-n Gray code contains all 2^n length-n binary strings.*

2. *Every pair of cyclically adjacent strings in the order-n Gray code differ in a single bit-position.*

Verification.
Assertion 1. We sketch an inductive verification.

The order-$(n=1)$ Gray code consists of the two distinct strings 0 and 1.

Assume that the assertion holds for all orders $n \le k$.

The order-$(k+1)$ code is obtained by taking two copies of the order-k code and prepending 0 to the strings in one copy and 1 to the strings in the other copy. The 2^k distinct binary strings from the order-k code thereby produce 2^{k+1} distinct binary strings in the order-$(k+1)$ code.

Assertion 2. We distinguish three situations. Let the adjacent strings be string x, which appears in position i of the code, and string y, which appears in position $i+1 \bmod 2^n$ of the code. ·

- Say that $i = 2^n - 1$. In this case x is the last string in the code, and y is the first string. By the "reflective" nature of the construction of the code, we know that $x = 1z$ and $y = 0z$ for some length-$(n-1)$ binary string z. Strings x and y therefore differ in precisely one bit-position, namely, bit-position 0.

- Precisely the same argument shows that when $i = 2^{n-1} - 1$, strings x and y again differ precisely in bit-position 0.

- In all other cases—i.e., when $i \in \{0, 1, \ldots, 2^n - 1\} \setminus \{2^{n-1} - 1, 2^n - 1\}$—strings x and y share the same first bit-position. In fact, for some bit $\beta \in \{0, 1\}$, $x = \beta u$ and $y = \beta v$ for length-$(n-1)$ binary strings u and v which are cyclically adjacent in the order-$(n-1)$ Gray code. By an inductive argument which we leave to the reader, u and v differ in precisely one bit-position—which means that x and y also differ in precisely one bit-position.

The previous analysis is summarized in Fig. 13.13 for $n = 4$.

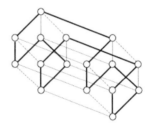

Fig. 13.13 A Hamiltonian cycle (bold) in $\mathcal{Q}_4$ built using a binary reflected Gray code

(e) de Bruijn networks require the most complex analysis of Hamiltonicity among our "named" graphs . We begin to deal with them by restating the result in more detail in order to establish some notation.

For all $n \in \mathbb{N}^+$, $\mathcal{D}_n$ contains a directed Hamiltonian cycle, i.e., a length-2^n directed cycle of the form

$$x \to y_1 \to y_2 \to \cdots \to y_{2^n-1} \to x \qquad (13.4)$$

which contains every vertex of $\mathcal{D}_n$ precisely once; i.e.:

- $\{x, y_1, y_2, \ldots, y_{2^n-1}\} = \mathcal{N}_{\mathcal{D}_n}$.

- *All of the vertices y_j that appear in cycle (13.4) differ from x and from each other.*

The simplest proof of this result has two steps. The first step introduces a significant, rather sophisticated, concept, the *line (di)graph* of a (di)graph.

(1) For any directed graph $\mathscr{G}$, the *line digraph* of $\mathscr{G}$, denoted $\Lambda(\mathscr{G})$, is the following directed graph.

- The vertices of $\Lambda(\mathscr{G})$ are the arcs of $\mathscr{G}$:

$$\mathcal{N}_{\Lambda(\mathscr{G})} = \mathcal{A}_{\mathscr{G}}$$

- For each pair of arcs of $\mathscr{G}$ of the form

$$\left[a_{x,y} = (x \to y)\right] \quad \text{and} \quad \left[a_{y,z} = (y \to z)\right]$$

  i.e., arcs such that the target of the first arc is the source of the second arc, $\Lambda(\mathscr{G})$ contains an arc $(a_{x,y} \to a_{y,z})$.

The relevance of this concept to this section is that the line digraph of every de Bruijn network $\mathscr{D}_n$ is the "next bigger" de Bruijn network, $\mathscr{D}_{n+1}$. Let us verify this claim.

Lemma 13.3. *For all $n \in \mathbb{N}^+$, $\mathscr{D}_{n+1}$ is the line digraph of $\mathscr{D}_n$; symbolically,*

$$\mathscr{D}_{n+1} = \Lambda(\mathscr{D}_n)$$

Proof (Lemma 13.3). Each vertex of $\Lambda(\mathscr{D}_n)$ is an arc of $\mathscr{D}_n$, hence has the form

$$(\beta x \to x\gamma)$$

where x is a length-$(n-1)$ binary string and $\beta, \gamma \in \{0,1\}$. Let us associate vertex $\beta x\gamma$ of $\mathscr{D}_{n+1}$ with this vertex of $\Lambda(\mathscr{D}_n)$. See Fig. 13.14.

Fig. 13.14 Illustrating how to label each arc of a de Bruijn network by concatenating the labels of the vertices incident to the arc and compacting the common intermediate bits. In the depicted example, the vertex-labels 01 and 11 combine to yield the arc-label 011

Note first that each arc of $\mathscr{D}_{n+1}$ has the form

$$(\delta y\varepsilon \to y\varepsilon\varphi),$$

where y is a length-$(n-2)$ binary string and $\delta, \varepsilon, \varphi \in \{0,1\}$. By our association of vertices of $\mathscr{D}_{n+1}$ with arcs of $\mathscr{D}_n$, this arc of $\mathscr{D}_{n+1}$ does, indeed, correspond to two successive arcs of $\mathscr{D}_n$. The first of these successive arcs *enters* vertex $y\varepsilon$ of $\mathscr{D}_n$; the second *leaves* that vertex.

Note next that, given any two successive arcs of $\mathscr{D}_n$, say

$$(\rho\sigma z \rightarrow \sigma z\tau) \quad \text{and} \quad (\sigma z\tau \rightarrow z\tau\xi)$$

where z is a length-$(n-2)$ binary string and $\rho, \sigma, \tau, \xi \in \{0, 1\}$, there is, indeed, an arc of $\mathscr{D}_{n+1}$ of the form

$$(\rho\sigma z\tau \rightarrow \sigma z\tau\xi)$$

This means that the digraph $\mathscr{D}_{n+1}$ is identical to the digraph $\Lambda(\mathscr{D}_n)$, except for a renaming of vertices and arcs.[1]

The just-described correspondence between the vertices and arcs of digraphs $\mathscr{D}_{n+1}$ and $\Lambda(\mathscr{D}_n)$ completes the proof.   □-Lemma 13.3

(2) The table following Proposition 13.9 contains the seeds of a proof of the following corollary to the proposition.

Corollary 13.3 *Every de Bruijn network $\mathscr{D}_n$ admits a (directed) Eulerian cycle.*

This corollary combines with Lemma 13.3 to complete the proof of Proposition 13.12(e). Each $\mathscr{D}_{n+1}$ is the line digraph of $\mathscr{D}_n$. Therefore, by definition of "line (di)graph", the fact that $\mathscr{D}_{n+1}$ is (directed)-Eulerian (Corollary 13.3) means that $\mathscr{D}_n$ is (directed)-Hamiltonian.   □

13.2.2.3 Testing general graphs for Hamiltonicity

The techniques we use in Section 13.2.2.2 to investigate the Hamiltonicity of our "named" graphs exploit the detailed structures of the individual graphs. Thus, we cannot expect the proof of Proposition 13.12 to suggest avenues for determining whether an arbitrary given graph is Hamiltonian. In fact, quite sophisticated results proved in the early 1970s make a strong mathematical argument that no set of case studies is likely to have a major impact on the problem of testing general graphs for Hamiltonicity. This is because, in common with the Satisfiability problem 3SAT of Section 3.4.3, the Hamiltonicity-detection problem is NP-complete. We repeat from our discussion in Section 3.4.3 that the details of the theory of NP-completeness are beyond the scope of this text, but we do want the reader to recognize the following.

The problem of deciding, given a graph $\mathscr{G}$ that is presented via a list of vertices and a list of edges, whether $\mathscr{G}$ admits a Hamiltonian path or a Hamiltonian cycle is NP-*complete.*

---

[1] Technically, we are asserting that the digraphs $\mathscr{D}_{n+1}$ and $\Lambda(\mathscr{D}_n)$ are *isomorphic* to one another. The topic of graph isomorphism is beyond the scope of this text, but our informal description provides all the details one would need to formalize the described isomorphism.

13.3 Exercises: Chapter 13

Throughout the text, we mark each exercise with 0 or 1 or 2 occurrences of the symbol $\oplus$, as a rough gauge of its level of challenge. The 0-$\oplus$ exercises should be accessible by just reviewing the text. We provide *hints* for the 1-$\oplus$ exercises; Appendix H provides *solutions* for the 2-$\oplus$ exercises. Additionally, we begin each exercise with a brief explanation of its anticipated value to the reader.

1. **Leveling—hence, 2-coloring, of torus-graphs**
   LESSON: Practice reasoning about graphs

   Provide a detailed proof of Corollary 13.1(d): The torus-graph $\widetilde{\mathcal{M}}_{m,n}$ is leveled—hence, 2-colorable—if, and only if, both m and n are even.

2. **Appreciating de Bruijn networks**
   LESSON: Practice reasoning about graphs

   Reasoning about de Bruijn networks quickly goes beyond the elementary, but many notions are quite accessible in small, specific examples. Here are two such, which refer to the order-3 de Bruijn network $\mathcal{D}_3$ and the order-4 de Bruijn network $\mathcal{D}_4$.

   a. $\oplus$ *Identify, by coloring the edges of $\mathcal{D}_3$ and $\mathcal{D}_4$, how each network can be viewed as two trees which are "embracing" one another.*

      These drawings are very reminiscent of the lithograph *Drawing Hands* by the Dutch artist M. C. Escher. Studying the drawing may be useful in solving this problem.

   b. *Prove Proposition 13.2 for $\mathcal{D}_3$ and $\mathcal{D}_4$: Describe the eight directed paths in $\mathcal{D}_3$ and the sixteen directed paths in $\mathcal{D}_4$.*

3. **Colored-triangle subgraphs**
   LESSON: Practice reasoning about graphs
   Prove the following assertion.

   Proposition 13.13 *Let us be given a copy $\mathcal{G}$ of $\mathcal{K}_6$ each of whose edges has been colored either* red *or* blue. *Graph $\mathcal{G}$ contains either a* red *"triangle" or a* blue *"triangle" as a subgraph. (A "triangle" is a copy of $\mathcal{C}_3$.)*

4. **Fundamental insights into outerplanarity in graphs**
   LESSON: Practice reasoning about graph-theoretic properties.
   Prove the following assertions.

   a. $\oplus$ **Toward understanding the forbidden-subgraph characterization**

      Proposition 13.14 *The complete bipartite graph $K_{3,2}$ is not outerplanar.*

Fig. 13.15 $K_{3,2}$: Its two types of vertices are identified by color—white and shaded

Hint: Use Fig. 13.15 to garner intuition. Perform a case-by-case analysis of all possible ways to distribute $K_{3,2}$'s vertices around a circle.

b. **The outerplanarity of trees**

   Prove Proposition 13.3: Every tree is outerplanar.

   Use the hints in the text.

c. **Outerplanarity and Hamiltonicity**

   Prove the following assertion.

   Proposition 13.15 *Every outerplanar graph is a subgraph of a Hamiltonian graph.*

5. **Elements of Euler's Formula**
   LESSON: Experience with following graph-theoretic arguments

   Phase 1 in our verification of Euler's Formula "via deconstruction" iterates a process of edge-removals until some edge-removal reduces the initial graph $\mathscr{G}$ to a graph with a single face.

   Verify that this termination condition is equivalent to stopping when the residual graph is a (connected) tree.

6. **Experiences with Eulerian graphs and Hamiltonian graphs**
   LESSONS: Experience with graph-theoretic argumentation

   a. **Characterizing Eulerian graphs**

   Adapt the proof of Proposition 13.9 to obtain a proof of Proposition 13.10.

   b. **Non-Hamiltonian meshes**

   Exhibit a mesh-graph that does not enjoy a Hamiltonian cycle.

   c. **Pre-Hamiltonian meshes**

   Prove that every mesh-graph can be converted to a Hamiltonian graph by adding a single edge.

Appendix A
A Cascade of Summations

In the course of introducing the many mathematical topics that have occupied us throughout the core of our text, we have encountered many instances of "order" in the world, which some might call "beauty". This chapter builds upon instances of beauty that we have uncovered in studying relationships among classes of numbers.

In Chapter 2, we observed that there is a very compact formula for the sum of the first n positive integers, in the form of a degree-2 polynomial which we later named Δ_n. In Chapter 6, we discovered that the elegance of the expression for Δ_n extends to the problem of summing the first n perfect squares, and the first n perfect cubes, and so on to higher powers: For each fixed power d, the sum of the first n perfect dth powers of positive integers is a polynomial of degree $d + 1$. So, our first search for beauty in numbers uncovered an extension that is based on the degrees of simple polynomials.

We now focus on summing the first n numbers Δ_i and looking at the sums of those sums—another way to extend beyond the sum of the first n integers. And we find yet more order and beauty!

The reader has probably guessed by this time that our quest for order and beauty is just the "spoonful of sugar" which accompanies our primary goal—to develop the methodology which enables our quest. You have surely noted by this point in our joint journey that the activity of solving mathematical problems encompasses several sub-activities which are challenging in their own right. One must develop the insights that help one settle on an *approach* to the problem—should one strive to represent the problem textually or pictorially …? Then, having chosen an approach, one must select appropriate *tools*: Should one seek to argue from small instances to larger ones (induction)? Should one strive to derive a contradiction based on assuming the negation of one's targeted assertion? And so on.

We hope that our focus in this section on highly structured problems will help the reader gain valuable experience in proceeding through the entire spectrum of activities needed to tackle sophisticated mathematical problems.

© Springer Nature Switzerland AG 2020
A. L. Rosenberg, D. Trystram, *Understand Mathematics, Understand Computing*,
https://doi.org/10.1007/978-3-030-58376-7

A.1 Revisiting the Triangular Numbers Δ_n

Recall that for each positive integer n, the *order-n triangular number* is the sum of the first n positive integers: $\Delta_n = 1 + 2 + \cdots + n$. In Chapter 6, we derived the value Δ_n, namely $\frac{1}{2}n(n+1)$, in several ways, including, notably,

- Gauss's "trick" that builds upon the following invariant:
  For all k, the kth and $n-k+1$th terms of Δ_n's defining summation sum to $n+1$
- Fubini's double-counting principle, which represents the kth integer by k tokens:
  Merging two triangle-shaped collections of tokens yields a rectangle of area $n(n+1)$. Indeed, the representation of Δ_n by a triangle of tokens is the origin of its "triangular" name.

Figure A.1 depicts a stratagem that melds Gauss's and Fubini's. The figure's use of invariants will serve as the basic reasoning principle in our expedition beyond Δ_n.

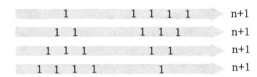

Fig. A.1 Computing Δ_n for $n = 4$: Two copies of a triangle are arranged so that all rows have the same sum; i.e., the row-sum is an *invariant*

A.2 Tetrahedral Numbers

This section extends our work on Δ_n. The extension emerges from the problem of summing the first n triangular numbers, i.e., to evaluate the summation

$$\widehat{\Theta}_n \overset{\text{def}}{=} \Delta_1 + \Delta_2 + \cdots + \Delta_n$$

Each such sum is called a *tetrahedral number*: the nth sum is denoted $\widehat{\Theta}_n$; see Footnote 8 in Section 6.4. In common with Δ_n's geometric epithet, "triangular", the "tetrahedral" name of $\widehat{\Theta}_n$ calls to mind the number's three-dimensional representation as a tetrahedron of tokens.

A.2.1 An Algebraic Evaluation of $\widehat{\Theta}_n$

The most obvious way to evaluate the summation underlying $\widehat{\Theta}_n$ is to replace each Δ_k by its value $\frac{1}{2}k(k+1)$ and to proceed by algebraic manipulations to evaluate the sum of the n triangular numbers. We leave it to the reader to follow this strategy and verify that

$$\widehat{\Theta}_n = \sum_{k=1}^{n} \Delta_k$$

$$= \sum_{k=1}^{n} \frac{1}{2}(k^2 + k)$$

$$= \frac{1}{2}\left(\sum_{k=1}^{n} k^2 + \sum_{k=1}^{n} k\right)$$

$$= \frac{1}{6}n(n+1)(n+2)$$

The final value comes from invoking Eq. (6.44) and the value of Δ_n in the penultimate expression.

All of this is fine, but not very intuitive. However, there is also a perspicuous pictorial evaluation of $\widehat{\Theta}_n$. We develop this evaluation technique now.

A.2.2 A Pictorial Evaluation of $\widehat{\Theta}_n$

We employ the double-counting stratagem of Fubini's principle. We begin by representing $\widehat{\Theta}_n$ by means of triangular piles of tokens, in the manner of Fig. A.1. However, we now replace the 1s in the figure by successive integers.

We employ geometric intuition. A tetrahedral number is the sum of triangular numbers; therefore, it can be arranged as a triangle, as in Fig. A.2. Because triangles

```
        1
      1   2
    1   2   3
  1   2   3   4
```

Fig. A.2 The basic pattern for computing $\widehat{\Theta}_n$ (for $n = 4$)

have three sides, we have three ways to represent them—by rotating the sides in the manner of Fig. A.3. Once we exploit these multiple views, we invoke Fubini's principle to sum up the successive rows of the figure. Our strategy is guided by a search for an *invariant*. In this case, we find that the sum of the elements along the rows of the three triangles is proportional to $n + 2$.

Finally, we sum the numbers in each row in Fig. A.4.

- The sum along the first row is $1 + 1 + n = n + 2$.

Fig. A.3 A schematic view of the three ways of organizing the triangles. The numbers are the same on each solid line: the bold one contains the 1s

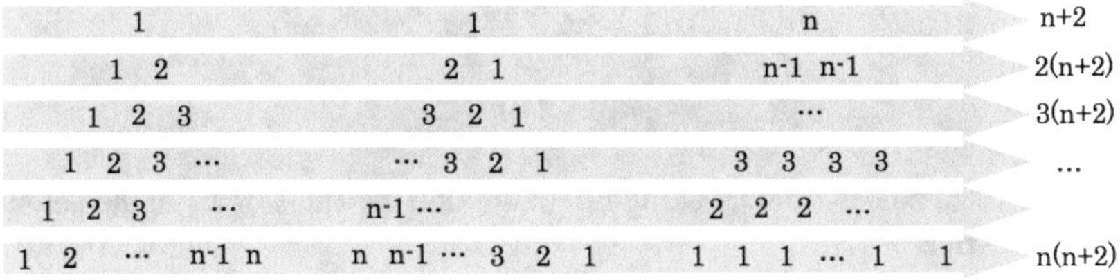

1	1	n	n+2
1 2	2 1	n-1 n-1	2(n+2)
1 2 3	3 2 1	...	3(n+2)
1 2 3 ...	... 3 2 1	3 3 3 3	...
1 2 3 ...	n-1 ...	2 2 2 ...	
1 2 ... n-1 n	n n-1 ... 3 2 1	1 1 1 ... 1 1	n(n+2)

Fig. A.4 Detail of the computation of the invariant

- The sum along the second row is $3 + 3 + 2(n-1) = 2(n+2)$.
- Let use sum up the elements in row k:
  $$\Delta_k + \Delta_k + k(n-k+1) = k(k+1) + kn - k^2 + k = k(n+2).$$

Thus, the sum of all rows equals $\Delta_n \cdot (n+2)$.

But this summation has computed $\widehat{\Theta}_n$ three times! We conclude that

$$\widehat{\Theta}_n = \frac{1}{3}\Delta_n \cdot (n+2)$$

We thereby get a more intuitive path to the value of $\widehat{\Theta}_n$.

> **A summarizing note**.
>
> It is important to recognize the natural progression in this section, because this will guide further extensions of these ideas.
>
> We have thus far proved the following:
>
> $$\begin{aligned} Id_n &= 1+1+\cdots+1 &&= n \\ \Delta_n &= 1+2+\cdots+n &&= \tfrac{1}{2}Id_n\cdot(n+1) = \tfrac{1}{2}n(n+1) \\ \widehat{\Theta}_n &= \Delta_1+\Delta_2+\cdots+\Delta_n = \tfrac{1}{3}\Delta_n\cdot(n+2) = \tfrac{1}{3!}n(n+1)(n+2) \end{aligned}$$
>
> With each increase in rank, we take one additional copy of the structure that characterizes the preceding rank:
>
> - With rank 1 (Id_n), we take *one* copy of n tokens.
>
> - With rank 2 (Δ_n), we take *two* copies of the n tokens of rank 1, organized as triangles.
>
> - With rank 3 ($\widehat{\Theta}_n$), we take *three* copies of the triangles of rank 2, organized as tetrahedra.

We will end with a sketch of the next rank, the *Pentahedral numbers*. You will be able to guess how to proceed at each step.

A.3 ⊕ Pentahedral Numbers

It is natural to ponder whether the regular patterns observed thus far will allow us to go to the next rank—to the *pentahedral numbers*

$$\Pi_n \overset{\text{def}}{=} \widehat{\Theta}_1 + \widehat{\Theta}_2 + \cdots + \widehat{\Theta}_n$$

Let's see …

Extrapolating from ranks 1, 2, and 3, we would expect to encounter the following evaluation for the pentahedral number Π_n.

$$\Pi_n = \widehat{\Theta}_1 + \widehat{\Theta}_2 + \cdots + \widehat{\Theta}_n = \frac{1}{4}\widehat{\Theta}_n\cdot(n+3) = \frac{1}{4!}n(n+1)(n+2)(n+3)$$

This chain of equations is valid! The required direct algebraic manipulations—i.e., replacing the $\widehat{\Theta}_k$ by their expressions and simplifying—are a bit cumbersome but definitely doable.

It is not so easy to extend the pictorial construction. However, we know that there should be four copies of $\widehat{\Theta}_n$ to consider. So let us see where intuition leads us.

We organize three copies of $\widehat{\Theta}_n$ within the same triangular pattern in Fig. A.5.

Because each element is a copy of $\widehat{\Theta}_k$, let us decompose them again as shown in Fig. A.6 (for $n = 4$). Now, let us add the remaining copy of $\widehat{\Theta}_n$ in a manner

Fig. A.5 Computing Π_n: The three basic triangle patterns (solid lines represent integers with the same value)

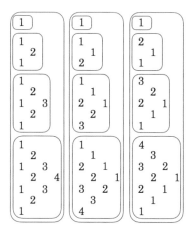

Fig. A.6 Computing Π_n: Organizing the three basic triangle patterns (for $n = 4$)

that exposes an invariant. As we found with lower ranks, we are able to find such an organization: Summing within each row of the structure in Fig. A.7 leads to an invariant: Each row sums to $n + 3$. The final result is now easy to obtain:

- The elements of the first row sum to $n + 3$.
- The elements of the second series of rows sum to $3(n + 3)$.
- The elements of the kth series of rows sum to $\Delta_k \cdot (n + 3)$.

We thereby find that

$$4\Pi_n = (n+3) \cdot \sum_{k=1}^{n} \Delta_k = \widehat{\Theta}_n \cdot (n+3)$$

A.3.1 Can We Go Further?

At this point, even the perspicuous pictorial summations become arduous. We expect—but have not verified—that the hexahedral numbers, obtained by summing the first n pentahedral numbers, will sum to

1		1		1		4		n+3
1		1		2		3		n+3
	2		1		1		3	n+3
1		2		1		3		n+3
1		1		3		2		n+3
	2		1		2		2	n+3
1	3	2	1	2	1	2	2	n+3 n+3
	2		2		1		2	n+3
1		3		1		2		n+3
1		1		4		1		n+3
	2		1		3		1	n+3
1	3	2	1	3	2	1	1	n+3 n+3
2	4	2	1	2	1	1	1	n+3 n+3
1	3	3	2	2	1	1	1	n+3 n+3
	2		3		1		1	n+3
1		4		1		1		n+3

Fig. A.7 Add the last fourth copy of $\widehat{\Theta}_n$ and sum up the rows. The elements of the last column sum to $\widehat{\Theta}_n \cdot (n+3)$

$$\sum_{k=1}^{n} \Pi_k = \frac{1}{5}\Pi_n \cdot (n+4) = \frac{1}{5!}n(n+1)(n+2)(n+3)(n+4)$$

Are you up to the challenge?

Appendix B
$\oplus\oplus$ Pairing Functions: Encoding $\mathbb{N}^+ \times \mathbb{N}^+$ as $\mathbb{N}^+$

Topic-specific references

A.L. Rosenberg (1974): Allocating storage for extendible arrays. *J. ACM 21*, 652–670.

A.L. Rosenberg (1975): Managing storage for extendible arrays. *SIAM J. Comput. 4*, 287–306.

A.L. Rosenberg (2003): Accountable Web-computing. *IEEE Trans. Parallel and Distributed Systs. 14*, 97–106.

L.J. Stockmeyer (1973): Extendible array realizations with additive traversal. IBM Research Report RC-4578.

This appendix extends in three directions the study of pairing functions which we began in Section 8.2.

1. We develop a "generic" version of the construction paradigm of Section 8.2.2, which develops the diagonal pairing function $\mathscr{D}$ of (8.18) in two steps: First, one decomposes $\mathbb{N}^+ \times \mathbb{N}^+$ into "shells". Second, one linearizes the partial order of $\mathbb{N}^+ \times \mathbb{N}^+$ that the "shells" specify. This paradigm was motivated by the "real-life" challenge of devising *efficient* computer-storage mappings for arrays and tables that can be expanded and contracted dynamically.
2. We describe additional specific pairing functions whose structures are specified via the "shell"-based mechanism—and we describe inherent limitations on the "compactness" that is achievable via the "shell"-based paradigm.
3. We exemplify a non-"shell"-based genre of pairing function which is motivated by another "real-life" challenge: how to devise computer-storage mappings for arrays and tables whose rows can be *efficiently* traversed, even after an arbitrary number of dynamic expansions and contractions. One finds one specific scenario that embodies this challenge in [Rosenberg, 2003].

Our motivation here is twofold. On the one hand, the mathematics needed to verify the bijectiveness of the functions we discuss provides valuable practice with some of the tools we have been developing to this point. On the other hand, the many and

© Springer Nature Switzerland AG 2020

A. L. Rosenberg, D. Trystram, *Understand Mathematics, Understand Computing*,

https://doi.org/10.1007/978-3-030-58376-7

varied applications of pairing functions adds value to our excursion into the world of these functions. The detailed survey article [90] provides much more detail than we are able to provide here.

As we embark on our short guided tour, the reader should note that the diagonal pairing function of Section 8.2.2 and the square-shell pairing function of Section B.2 build in essential ways on the L-norms of Section 5.5.1; cf. Fig. 5.4.

B.1 A Shell-Based Methodology for Crafting Pairing Functions

The shell-oriented strategy that underlies the diagonal pairing function $\mathscr{D}$ can be adapted to incorporate shell-"shapes" that are inspired by a variety of computational situations—and can be applied to computational advantage in such situations. We describe how such adaptation can be effected, and we describe a few explicit shapes and situations. We invite the reader to craft others.

Procedure PF-Constructor($\mathscr{A}$)
/*Construct a shape-inspired pairing function (PF) $\mathscr{A}$*/
begin

Step 1.   Partition the set $\mathbb{N}^+ \times \mathbb{N}^+$ into finite sets called *shells*. Order the shells linearly in some way: many natural shell-partitions carry a natural order.

For example, shell c of the diagonal pairing function $\mathscr{D}$ is the following subset of $\mathbb{N}^+ \times \mathbb{N}^+$: $\{\langle x, y \rangle \mid x + y = c\}$. The parameter c orders $\mathscr{D}$'s shells.

Step 2.   Construct a pairing function from the shells as follows.

   Step 2a.   Enumerate $\mathbb{N}^+ \times \mathbb{N}^+$ shell by shell, honoring the ordering of the shells; i.e., list the pairs in shell #1, then shell #2, then shell #3, etc.
   Step 2b.   Enumerate each shell in some systematic way, e.g., "by columns". In detail:

   Enumerate the pairs $\langle x, y \rangle$ in each shell in increasing order of y and, for pairs having equal y values, in decreasing order of x.

end PF-Constructor

Proposition B.1 *Any function $\mathscr{A} : \mathbb{N}^+ \times \mathbb{N}^+ \leftrightarrow \mathbb{N}^+$ that is designed via Procedure* PF-Constructor *is a bijection.*

Proof (Sketch). Step 1 of Procedure PF-Constructor constructs a partial order on $\mathbb{N}^+ \times \mathbb{N}^+$ in which: (*a*) each shell is finite; (*b*) there is a linear order on the shells. Step 2 extends this partial order to a linear order by honoring the inherent ordering of shells and imposing a linear order within each shell. The function constructed via the Procedure is: *injective* because the disjoint shells are enumerated sequentially; *surjective* because the enumeration within each shell begins immediately after the enumeration within the preceding shell, with no gap.   □

Having noted how to use Procedure PF-Constructor to construct pairing function $\mathscr{D}$, we now use the Procedure to design two other pairing functions which produce efficient storage mappings for extendible arrays and tables.

B.2 The Square-Shell Pairing Function $\mathscr{S}$

One computational situation where pairing functions can be useful is as storage-mappings for arrays/tables that can expand and/or contract dynamically.

In conventional programming systems, when one expands an $n \times n$ table into an $(n+1) \times (n+1)$ table, one allocates a new region of $(n+1)^2$ storage locations and copies the current table from its n^2-location region to the new region. Of course, this is very wasteful: one is moving $\Omega(n^2)$ items to make room for the anticipated $2n+1$ new items. On any given day, the practical impact of this waste depends on current technology. However, this is a mathematics text, not an engineering one, so we are exploring whether *in principle* we can avoid the waste. The answer is "YES". If we employ a pairing function $\varepsilon : \mathbb{N}^+ \times \mathbb{N}^+ \leftrightarrow \mathbb{N}^+$ to allocate storage for tables, then to expand a table from dimensions $n \times n$ to $(n+1) \times (n+1)$, we need move only $O(n)$ items to accommodate the *new* table entries; the rest of the current entries need not be moved. For square tables, the following *square-shell* pairing function $\mathscr{S}$ manages the described scenario perfectly. After describing $\mathscr{S}$, we comment on managing tables of other shapes.

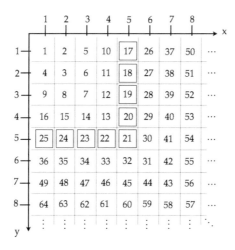

Fig. B.1 *The square-shell pairing function* $\mathscr{S}$. *The shell* $\max(x,y) = 5$ *is highlighted*

$$\mathscr{S}(x,y) = m^2 + m + y - x + 1$$
$$\text{where} \quad m \stackrel{\text{def}}{=} \max(x-1, y-1). \tag{B.1}$$

One sees in Fig. B.1 that $\mathscr{S}$ follows the prescription of Procedure PF-Constructor: (1) it maps integers into the "square shells" defined by: $m = 0$, $m = 1,...$. (2) it enumerates the entries in each shell in a clockwise direction. (Of course, $\mathscr{S}$ has a twin that enumerates the shells in a counterclockwise direction.)

Enrichment note.

Using somewhat more complicated instantiations of Procedure PF-Constructor, the study in [Rosenberg, 1975] adapts the square-shell pairing function $\mathscr{S}$ to: (a) accommodate, with no wastage, arrays/tables of any fixed aspect ratio $an \times bn$ $(a, b \in \mathbb{N})$; (b) accommodate, with only $O(n)$ wastage, arrays/tables whose aspect ratios come from a fixed finite set of candidates—i.e., $(a_1 n \times b_1 n)$ or $(a_2 n \times b_2 n)$ or ... or $(a_k n \times b_k n)$.

B.3 ⊕ The Hyperbolic-Shell Pairing Function $\mathscr{H}$

The diagonal- and square-shell pairing functions indicate that when the growth patterns of one's arrays/tables is very constrained, one can use pairing functions as storage mappings with very little wastage. In contrast, if one employs a pairing function such as $\mathscr{D}$ without considering its wastage, then a storage map would show some $O(n)$-entry tables being "spread" over $\Omega(n^2)$ storage locations. In the worst-case, $\mathscr{D}$ spreads the n-position $1 \times n$ array/table over $> \frac{1}{2}n^2$ addresses, because: $\mathscr{D}(1,1) = 1$ and $\mathscr{D}(1,n) = \frac{1}{2}(n^2 + n)$. This degree of wastefulness can be avoided via careful analysis, coupled with the use of Procedure PF-Constructor. The target commodity to be minimized is the *spread* of a PF-based storage map, which we define as follows.

Note that an ordered pair of integers $\langle x, y \rangle$ appears as a position-index within an n-position table if, and only if, $xy \le n$. Therefore, we define the spread of a PF-based storage map $\mathscr{M}$ via the function

$$\mathbf{S}_{\mathscr{M}}(n) \stackrel{\text{def}}{=} \max\{\mathscr{M}(x,y) \mid xy \le n\} \tag{B.2}$$

$\mathbf{S}_{\mathscr{M}}(n)$ is the largest "address" that PF $\mathscr{M}$ assigns to any position of a table that has n or fewer positions.

Happily, the tools that we have developed enable us to design a pairing function that (to within constant factors) has minimum worst-case spread. This is the *Hyperbolic-shell pairing function $\mathscr{H}$* of (B.3) and Fig. B.2.[1]

---

[1] Details appear in [Rosenberg, 1974] and [Rosenberg, 1975].

Let $\delta(k)$ be the number of divisors of the integer k.

$$\mathscr{H}(x,y) = \sum_{k=1}^{xy-1} \delta(k) \;+\; \text{the position of } \langle x,y \rangle \text{ among two-part}$$

$$\text{factorizations of the number } xy, \text{ in}$$
$$\text{reverse lexicographic order}$$

(B.3)

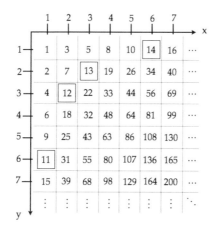

Fig. B.2 *The hyperbolic pairing function $\mathscr{H}$. The shell $xy = 6$ is highlighted*

Proposition B.2 **(a)** *The hyperbolic function $\mathscr{H}$ is a pairing function.*

(b) *The spread of $\mathscr{H}$ is given by $\mathbf{S}_{\mathscr{H}}(n) = O(n \log n)$.*[2]

(c) *No pairing function has better compactness than $\mathscr{H}$ (in the worst case) by more than a constant factor.*

Proof. **(a)** The fact that $\mathscr{H}$ is a pairing function follows from Proposition B.1.

(b) The pairing function $\mathscr{H}$ maps integers along the "hyperbolic shells" defined by $xy = 1,\ xy = 2,\ xy = 3,\ldots$. Hence, when an integer n is "placed" into the table of values of $\mathscr{H}$, the number of occupied slots is within n of

$$\sum_{i=1}^{n-1} |\{\langle x,y \rangle \mid xy = i\}|$$

Elementary calculations show that this sum is $O(n \log n)$.

(c) The optimality of $\mathscr{H}$ in compactness (to within constant factors) is seen via the following argument. The set of tables that have n or fewer positions are those

---

[2] A detailed analysis reveals that the spread of $\mathscr{H}$ is closely related to the *natural* logarithm, whose base is Euler's constant e.

of aspect ratios $a_i \times b_i$, where $a_i b_i \le n$. As one sees from Fig. B.3 (generalized to arbitrary n), the union of the positions of all these arrays is the set of integer lattice points under the hyperbola $xy = n$. It is well known—cf. [85]—that this set of points has cardinality $\Theta(n \log n)$. Since every table contains position $\langle 1, 1 \rangle$, it follows that, for every n, some table containing n or fewer positions is spread over $\Omega(n \log n)$ "addresses". $\square$

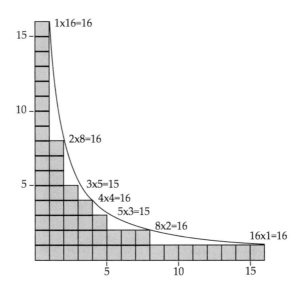

Fig. B.3 The aggregate set of positions of tables having 16 or fewer position. To help the reader understand the figure, we include the curve $f(x,y) = xy$, which provides an upper envelope for the set. A careful look at this curve reveals that it touches the set of positions at the points $\langle x,y \rangle \in \{\langle 1,16 \rangle, \langle 2,8 \rangle, \langle 4,4 \rangle, \langle 8,2 \rangle, \langle 16,1 \rangle\}$, but it does *not* touch the set at the points $\langle x,y \rangle \in \{\langle 3,5 \rangle, \langle 5,3 \rangle\}$

B.4 A Bijective Pairing Using Disjoint Arithmetic Progressions

As we noted earlier, the notion of pairing function has applications to a broad range of computational situations. For some of these (see [90]) it is convenient to have each "row" of the pairing function map onto an arithmetic progression. This can be achieved in many ways. We now exhibit a single such *additive* pairing function, which we call $\mathscr{A}(x,y)$. The reader can find a comprehensive study of such pairing functions in [Stockmeyer, 1973].

For all $\langle x,y \rangle \in \mathbb{N}^+ \times \mathbb{N}^+$, $\mathscr{A}(x,y) \stackrel{\text{def}}{=} 2^{x-1} \cdot (2y - 1)$

We leave to the reader the exercise of verifying function $\mathscr{A}$'s bijectiveness. As an aid to the reader, we provide a prefix of $\mathscr{A}$'s additive bijective mapping of $\mathbb{N}^+ \times \mathbb{N}^+$ in Fig. B.4.

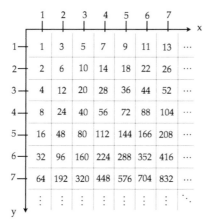

Fig. B.4 A prefix of the additive pairing function $\mathscr{A}$

Appendix C
A Deeper Look at the Fibonacci Numbers

Topic-specific reference

E. Zeckendorf (1972): *Représentation des nombres naturels par une somme de nombres de Fibonacci ou de nombres de Lucas. Bull. Soc. R. Sci. Liège 41*, 179–182.

C.1 Elementary Computations on Fibonacci Numbers

One reason that the Fibonacci sequence has captivated a large community of mathematically oriented fans is the extensive repertoires of identities involving Fibonacci numbers, which are simply presented and gracefully verified.[1] This section is dedicated to three particularly beautiful, unintuitive, identities.

C.1.1 Greatest Common Divisors of Fibonacci Numbers

Our first identity exposes the marvelous fact that the *greatest common divisor* (GCD) of every pair of Fibonacci numbers, $F(m)$ and $F(n)$, is again a Fibonacci number. This surprising property corresponds to an extension of the Euclid's algorithm to Fibonacci numbers.

In this section, we start the progression with $F(0) = 0$ and $F(1) = 1$. This change, which is needed purely for technical reasons, just augments the standard Fibonacci sequence by prepending a 0. (The augmented sequence begins $0, 1, 1, 2, 3, \ldots$ instead of $1, 1, 2, 3, \ldots$.)

Proposition C.1 *For all integers* $m, n, \in \mathbb{N}^+$,

---

[1] An entire research journal, *The Fibonacci Quarterly*, is dedicated to the mathematics of recursively defined sequences of numbers—the genre epitomized by the Fibonacci numbers.

© Springer Nature Switzerland AG 2020   461
A. L. Rosenberg, D. Trystram, *Understand Mathematics, Understand Computing*,
https://doi.org/10.1007/978-3-030-58376-7

$$\text{GCD}(F(m), F(n)) \; = \; F(\text{GCD}(m, n)) \qquad\qquad \text{(C.1)}$$

To make identity (C.1) at least plausible, we cite a single example:

$$\text{GCD}(F(12), F(18)) = \text{GCD}(144, 2584) \; = \; 8 \; = \; F(6)$$
$$\text{while} \quad \text{GCD}(12, 18) = 6$$

Proof. Say, with no loss of generality, that $n \geq m$. Our proof builds on the development of GCDs in Chapter 4, especially Proposition 4.5:

$$\text{GCD}(F(n), F(m)) \; = \; \text{GCD}(F(m), F(\text{REM}(n, m)))$$

We invite the reader to verify the following technical lemmas.

Lemma C.1. *The following relation holds for any integers n and $k \geq 1$.*

$$F(n + k) \; = \; F(k) \cdot F(n + 1) \; + \; F(k - 1) \cdot F(n)$$

Hint: The proof is a straightforward induction on k for fixed n.

Lemma C.2. *For any integer $k \geq 1$, $F(n)$ divides $F(k \cdot n)$.*

Hint: This lemma can be obtained by induction on k using Lemma C.1 by writing $F((k + 1) \cdot n) = F(n + k \cdot n) = F(k \cdot n) \cdot F(n + 1) + F(k \cdot n - 1) \cdot F(n)$.

Lemma C.3. *For any integer $n \geq 1$, $F(n)$ and $F(n - 1)$ are relatively prime:*

$$\text{GCD}(F(n - 1), F(n)) \; = \; 1$$

Hint: The proof is a straightforward induction.

We are now able to verify identity (C.1) via the following assertions, where $r = \text{REM}(m, n)$.

$$
\begin{aligned}
\text{GCD}(F(n), F(m)) &= \text{GCD}(F(q \cdot m + r), F(m)) \\
&= \text{GCD}(F(m), \; F(m \cdot q + 1) \cdot F(r) + F(m \cdot q) \cdot F(r - 1)) \\
&\qquad \text{(by Lemma C.1)} \\
&= \text{GCD}(F(m), \; F(m \cdot q + 1) \cdot F(r)) \\
&\qquad \text{(by Lemma C.2)} \\
&= \text{GCD}(F(m), \; F(r)) \\
&\qquad \text{(by an easy extension of Lemma C.3)}
\end{aligned}
$$

Applying this process until $r = 0$ completes the proof. The last Fibonacci number we have obtained is $F(\text{GCD}(m, n))$. $\square$

C.1.2 Products of Consecutive Fibonacci Numbers

Our next identity asserts the equality of each product $F(n) \cdot F(n-1)$ with the sum of the squares of the first n Fibonacci numbers.

Proposition C.2 *For all integers $n \geq 1$,*

$$F(n) \times F(n-1) = \sum_{k=0}^{n-1} (F(k))^2 \qquad (C.2)$$

Proof. One can observe identity (C.2) "in action" in Fig. 9.6. We augment this visual validation of the identity with the following induction.

Base case. Instance $n = 1$ of identity (C.2) is valid because

$$F(0) \times F(1) = 1 \times 1 = 1^2 = (F(0))^2$$

Inductive hypothesis. Assume that identity (C.2) is valid for all $n \leq m$.

Inductive extension. Focus on the product $F(m+1) \times F(m)$. Invoking the defining Fibonacci recurrence (9.6) and the inductive hypothesis, we generate the following chain of equalities.

$$
\begin{aligned}
F(m+1) \times F(m) &= (F(m) + F(m-1)) \times F(m) \\
&= (F(m))^2 + F(m) \times F(m-1) \\
&= (F(m))^2 + \sum_{k=0}^{m-1} (F(k))^2 \\
&= \sum_{k=0}^{m} (F(k))^2
\end{aligned}
$$

The induction is thus extended, which completes the proof. □

C.1.3 Cassini's Identity

Our final identity on Fibonacci numbers was discovered in 1680 by Giovanni Domenico Cassini, the then-director of the Paris Observatory.

Proposition C.3 *For all integers $n \geq 1$,*

$$F(n-1) \times F(n+1) = (F(n))^2 + (-1)^{n+1} \qquad (C.3)$$

Proof. We proceed by induction.

Base case. Expression C.3 holds for $n = 1$ because $F(0) \times F(2) = (F(1))^2 + 1 = 2$.

Inductive hypothesis. Assume that expression C.3 holds for a given $n \geq 1$.

Inductive extension. Replace $F(n+2)$ by its recurrent definition in the product $F(n) \times F(n+2)$.

$$F(n) \times F(n+2) = F(n) \times (F(n+1)+F(n))$$
$$= (F(n))^2 + F(n) \times F(n+1)$$

Now, from the inductive hypothesis, we have:

$$(F(n))^2 = F(n-1) \times F(n+1) - (-1)^{n+1}$$
$$= F(n-1) \times F(n+1) + (-1)^{n+2}$$

We thus have

$$F(n) \times F(n+2) = F(n) \times F(n+1) + F(n-1) \times F(n+1) + (-1)^{n+2}$$
$$= F(n+1)(F(n)+F(n-1)) + (-1)^{n+2}$$

Now apply the defining recurrence of the Fibonacci sequence to obtain

$$F(n) \times F(n+2) = (F(n+1))^2 + (-1)^{n+2}$$

The induction is extended, which completes the proof.   □

C.2 ⊕ Computing Fibonacci Numbers Fast

How many integer multiplications does it take to compute $F(n)$, the nth Fibonacci number? If we follow the numbers' defining bilinear recurrence, then this computation requires n multiplications. If we are clever, then we can accomplish the computation using many fewer multiplications. The cleverness resides in computing Fibonacci numbers in successive pairs. We sketch how to achieve this, leaving details to the reader.

Proposition C.4 *For all positive integers n, one can compute the Fibonacci number $F(n)$ in $\log_2 n$ multiplications.*

The proof resides in the following Lemma.

Lemma C.4. *For all integers n:*
(a) $F(2n) = (F(n))^2 + (F(n-1))^2;$
(b) $F(2n+1) = F(n) \times (2F(n-1)+F(n)).$

Proof (Sketch for the Lemma). **(a)** The proof is by induction.
The base case $n = 1$ is true because

$$F(2) = (F(1))^2 + (F(0))^2 = 2$$
$$F(3) = F(1) \times (2F(0) + F(1)) = 3$$

If we assume, for induction, that the property holds for index n, for both $F(2n)$ and $F(2n+1)$, then we can compute $F(2(n+1))$ as follows.

Applying the defining recurrence for the Fibonacci numbers, followed by the inductive hypothesis, we have

$$
\begin{aligned}
F(2n+2) &= F(2n+1) + F(2n) \\
&= (F(n))^2 + (F(n-1))^2 + F(n) \times (2F(n-1) + F(n)) \\
&= (F(n))^2 + (F(n-1))^2 + 2(F(n) \times F(n-1)) + (F(n))^2 \\
&= (F(n) + F(n-1))^2 + (F(n))^2 \\
&= (F(n+1))^2 + (F(n))^2
\end{aligned}
$$

the last step following by again applying the defining recurrence for index $n+1$.

(b) We again begin by applying the defining recurrence of the Fibonacci numbers.

$$
\begin{aligned}
F(2(n+1)+1) &= F(2(n+1)) + F(2n+1) \\
&= (F(n+1))^2 + F(n)^2 + F(n) \times (2F(n-1) + F(n)) \\
&= (F(n+1))^2 + 2(F(n-1) + F(n)) \times F(n) \\
&= (F(n+1))^2 + 2F(n+1) \times F(n)
\end{aligned}
$$

This completes the proof sketch for the lemma. □

By exposing the dependencies involved in computing the nth Fibonacci number $F(n)$, Fig. C.1 provides a pictorial argument to show why this decomposition enables us to compute $F(n)$ in $\log_2 n$ steps.

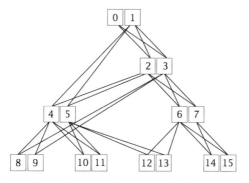

Fig. C.1 The dependency relations for computing the pairs $(F(2n), F(2n+1))$

We exemplify the scheme in Fig. C.2 by using the fast Fibonacci scheme to compute $F(13)$. We observe the computation embedded into a logarithmic-depth complete binary tree of dependencies (with some redundancies).

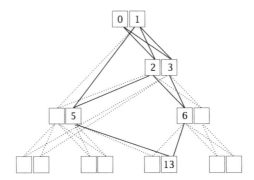

Fig. C.2 Ancestors involved in computing $F(13)$

C.3 ⊕⊕ A Number System Based on the Fibonacci Sequence

This section continues our mathematical-cultural tour of the Fibonacci numbers. Our study of the sequence until now, in Sections 9.2.2 and C.1, has focused "inward", deriving results that are important for understanding the sequence. Now, we focus on an "outward"-looking application of the sequence.

History supplies many examples of nonstandard representations of integers proving significant for reasons that go far beyond the representation of numbers. Even within our exploration of number representations in this text, we have repeatedly benefited from unexpected uses of number representations—particularly from the ability of mathematical objects to *encode* a broad range of situations. (Two simple examples: (*a*) the use of base-2 numerals to count in various combinatorial applications; (*b*) the use of prime-power representations within the realm of coding.) This section is devoted to developing a system for representing integers which is based on the family of Fibonacci numbers.

C.3.1 The Foundations of the Number System

C.3.1.1 Zeckendorf's Theorem

The endeavor of using Fibonacci numbers to represent arbitrary integers is based on the following theorem by the Belgian mathematician Edouard Zeckendorf [Zeckendorf, 1972].

To simplify exposition, we employ two nonstandard conventions in what follows.

1. We index the Fibonacci sequence starting at index 1, not the traditional 0.
2. For integers m and n, we write "$m >^+ n$" to mean that $m \geq n+2$.

Proposition C.5 (Zeckendorf's Theorem) *Every positive integer n has a unique representation as the sum of nonconsecutive Fibonacci numbers with indices $k \geq 2$. In detail, such a representation has the form*

$$n = F(k_1) + F(k_2) + \cdots + F(k_\ell)$$

where $k_1 >^+ k_2 >^+ \cdots >^+ k_\ell \geq 2$.

Explanatory note.

The injunction against employing $F(1)$ in a representation is no real restriction because under this section's convention for the indices of Fibonacci numbers, $F(1) = F(2)$.

Call a representation of an integer n a *Zeckendorf representation* or a *Zeckendorf numeral* if it satisfies the conditions of Proposition C.5.[2]

Illustrating Zeckendorf numerals.

• The Zeckendorf representation of the integer $12,345$ is:

$$\begin{aligned} 12,345 &= 10,946 + 987 + 377 + 34 + 1 \\ &= F(21) + F(16) + F(14) + F(9) + F(2) \end{aligned}$$

• Fig. C.3 illustrates the Zeckendorf representations of the first 26 integers.

C.3.1.2 Proving Zeckendorf's Theorem

A complete, constructive proof of Zeckendorf's Theorem would accomplish three goals.

1. It would specify an algorithm for constructing a Zeckendorf numeral for an arbitrary integer n.

---

[2] The use of the term *Zeckendorf numeral* conforms with the terminology of Chapter 10.

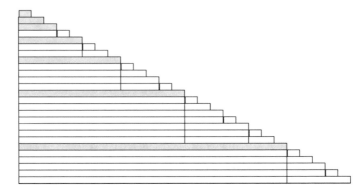

Fig. C.3 The first 26 integers (on the vertical [Y] axis), represented in the Zeckendorf system. The breaks in the horizontal bars represent the summations of Fibonacci numbers. The shaded horizontal bars correspond to pure Fibonacci numbers

2. It would verify that the specified numeral is *valid*—i.e., that the Fibonacci numbers in the representation do, indeed, sum to n.
3. It would verify that the specified numeral is *unique*—i.e., that no other selection of Fibonacci numbers that satisfy Proposition C.5 sums to n.

We present an induction on n which accomplishes the first two goals—producing a numeral and proving its validity. Expanding the induction to accomplish the third goal—proving the uniqueness of the derived numeral—is too intricate for this text, but it should be accessible to the determined reader from the original description in [Zeckendorf, 1972].

Keep in mind our nonstandard indexing of the Fibonacci numbers.

Base cases. The Proposition holds for the cases $n = 2$, $n = 3$, and $n = 4$. That is:

$$F(3) = 2$$
$$F(4) = 3$$
$$F(4) + F(2) = 3 + 1 = 4$$

Uniqueness of these representations is a simple exercise.

Inductive hypothesis. Assume for induction that any integer $m < F(r)$ admits a unique Zeckendorf numeral.

Inductive extension. Focus on an arbitrary integer n in the interval

$$[F(r) \leq n < F(r+1)]$$

We claim that every n in this interval admits a unique Zeckendorf representation.

Of course, this claim is true for the smallest n in the interval, namely, $n = F(r)$, because the equation "$n = F(r)$" obviously specifies a valid Zeckendorf representation of n.

Focus, therefore, on an arbitrary $n \neq F(r)$ within the interval. Write this n in the form $n = F(r) + m$. Note that we must have $m < F(r)$, by specification of the interval and by the defining recurrence of the Fibonacci sequence. Consequently, our inductive hypothesis tells us that the integer m admits a unique Zeckendorf numeral:

$$m = F(k_1) + F(k_2) + \cdots + F(k_\ell)$$

where $k_1 >^+ k_2 >^+ \cdots >^+ k_\ell \geq 2$.

This means, in particular, that the integer n admits a representation of the form

$$n = F(r) + F(k_1) + F(k_2) + \cdots + F(k_\ell)$$

where $k_1 >^+ k_2 >^+ \cdots >^+ k_\ell \geq 2$.

Our final task is to verify that $F(r)$ and $F(k_1)$ are not consecutive Fibonacci numbers, or equivalently, to verify that $F(k) >^+ F(k_1)$. To this end, assume, for contradiction, that $k_1 = r - 1$. Were this the case, we would have (from the Fibonacci recurrence)

$$n = F(r+1) + F(k_2) + \cdots + F(k_\ell)$$

This equality would contradict the fact that $n < F(r+1)$.

Our argument does not verify the uniqueness of n's Zeckendorf numeral, but Zeckendorf's original paper does verify this property as well.

Our inductive proof yields an effective way to compute the Zeckendorf representation of any number n—but there are likely more efficient conversion procedures.

C.3.2 Using Zeckendorf Numerals as a Number System

One can employ Zeckendorf representations as the numerals of a number system for positive integers: As we just showed, each $n \in \mathbb{N}^+$ admits such a representation; moreover, it is shown in [Zeckendorf, 1972] that the representation assigns unique numerals to each integer. Now we can expand the system to the entire set $\mathbb{N}$ by explicitly appending the number 0 to the system. This section discusses some of the pros and cons of the resulting system.

C.3.2.1 The system's numerals

The natural way to represent numerals within the Fibonacci number system is as bit-strings. We employ a somewhat strange system of indexing in order to be consistent with the notation of Proposition C.5.

- 0 is the unique numeral for the number 0.
- For all *positive* integers n: Each length-ℓ bit-string

$$b_{\ell+1} b_\ell \cdots b_2$$

where each $b_i \in \{0, 1\}$, is the unique numeral for the positive integer

$$n \; = \; (b_{\ell+1}b_{\ell}\cdots b_2)_F \; \overset{\text{def}}{=} \; \sum_{k=2}^{\ell+1} b_k \cdot F(k)$$

C.3.2.2 Observations about the system's numerals

We offer a few comparisons between the numerals of the Fibonacci number system and those of the binary (base-2) number system.

The first thing we notice is that base-2 numerals are often shorter than Fibonacci numerals. The integer $12,345$ provides an illustrative example. The *shortest* Fibonacci numeral for $12,345$—i.e., the numeral with no leading 0s—is the 21-bit string

$$12,345 \; = \; 100001010000100000010_F$$

The corresponding base-2 numeral has only 13 bits:

$$12,345 \; = \; 1100000111001_2$$

A second observation is that Fibonacci numerals never have consecutive 1s. This feature endows the Fibonacci system with a modicum of *error detectability*—cf., [87]—which can be a significant, *positive* practical attribute.

One can get a better perspective on numeral lengths and on the impact of not allowing consecutive 1s by perusing Table C.1.

Table C.1 Comparing the base-2 and Fibonacci numerals for integers 1–15

n	Base-2 (4-bit)	Fibonacci (5-bit)	Relevant Fibonacci numbers
1	0001_2	00001_F	$F(2) = 1$
2	0010_2	00010_F	$F(3) = 2$
3	0011_2	00100_F	$F(4) = 3$
4	0100_2	00101_F	$F(4) + F(2) = 3 + 1$
5	0101_2	01000_F	$F(5) = 5$
6	0110_2	01001_F	$F(5) + F(2) = 4 + 1$
7	0111_2	01010_F	$F(5) + F(3) = 5 + 2$
8	1000_2	10000_F	$F(6) = 8$
9	1001_2	10001_F	$F(6) + F(2) = 8 + 1$
10	1010_2	10010_F	$F(6) + F(3) = 8 + 2$
11	1011_2	10100_F	$F(6) + F(4) = 8 + 3$
12	1100_2	10101_F	$F(6) + F(4) + F(2) = 8 + 3 + 1$
13	1101_2	10000_F	$F(7) = 13$
14	1110_2	10001_F	$F(7) + F(2) = 13 + 1$
15	1111_2	10010_F	$F(7) + F(3) = 13 + 2$

C.3.2.3 On incrementing Fibonacci numerals

Our final look at the Fibonacci system illustrates how differently the system's numerals behave when being manipulated, as compared with the numerals from positional number systems. We describe a procedure for *incrementing* numbers—i.e., adding 1—within the Fibonacci system. Of course, one must employ ever-longer Fibonacci numerals as one deals with ever-larger numbers—a simple consequence of the Pigeonhole Principle.[3] It is interesting to observe how the *carries* that embody the lengthening of numerals reflect the underlying Fibonacci recurrences—and to compare this process with the way carries operate within standard positional number systems (which the reader learned in elementary school).

For the sake of precision, what we shall present is a procedure that transforms a Fibonacci-system numeral

$$a_n a_{n-1} \cdots a_1 a_0$$

that denotes the integer

$$(a_n a_{n-1} \cdots a_1 a_0)_F$$

into a Fibonacci-system numeral

$$b_m b_{m-1} \cdots b_1 b_0$$

that denotes the integer

$$(b_m b_{m-1} \cdots b_1 b_0)_F = 1 + (a_n a_{n-1} \cdots a_1 a_0)_F$$

Perhaps the simplest procedure for achieving this transformation—and achieving it *efficiently*—relies on the system of equations (9.9) from Proposition 9.8. We repeat the system for the reader's convenience:

For all integers $n \geq 2$,

$$F(n) = F(n-1) + F(n-3) + F(n-5) + \cdots + 1$$

By combining this system with the fact that the Fibonacci system prominently avoids consecutive 1s in numerals, one verifies that the procedure we describe now correctly computes the *increment* operation. Let us begin with the input Fibonacci-system numeral

$$\alpha = a_n a_{n-1} \cdots a_1$$

Procedure Increment(α)
begin

Step 1   Prepend two consecutive 0s to α, to produce the bit-string

---

[3] The Principle asserts that only finitely many numbers can be represented by numerals of any fixed length.

$$\alpha^{(00)} \stackrel{\text{def}}{=} 00a_n a_{n-1} \cdots a_1$$

Step 2   Find the *rightmost* consecutive positions in $\alpha^{(00)}$ that contain two consecutive 0s. Say that the portion of $\alpha^{(00)}$ beginning with the discovered two consecutive 0s and proceeding rightward from there has the form

$$\alpha^{(right)} \stackrel{\text{def}}{=} 00a_\ell a_{\ell-1} \cdots a_1$$

Because of our prepending step, index ℓ will always exist—i.e., the search will always succeed. Index ℓ will be smaller than index n precisely if the input numeral α contains two consecutive 0s (in one or more places).

Step 3   Replace the bit-string $\alpha^{(right)}$ within the numeral α by the bit-string

$$0100 \cdots 0 \quad (\ell \text{ terminal 0s})$$

end Increment

We extract two illustrations from Table C.1:

$$n_1 = 7 = \quad 1010_F \qquad \text{incr}(n_1) = \quad 8 = 10000_F$$
$$n_2 = 9 = 10001_F \qquad \text{incr}(n_2) = 10 = 10010_F$$

We leave to the reader the exercise of extending our incrementation procedure to devise an algorithm for performing general additions within the Fibonacci system.

Appendix D
$\oplus\oplus$ Lucas Numbers: Another Recurrence-Defined Family

This appendix introduces another family of integers which shares the recursively defined elegance of the Fibonacci numbers and the binomial coefficients. The current family does not share the level of elegance or the range of application that the latter families do, but it does provide valuable experience dealing with recurrences. We remark that yet more recurrence-defined families of integers can be found in the literature: *Tree-profile numbers*[1] represent an example with a somewhat more complex recursive definition.

A continuing preoccupation of mathematicians is to understand why important mathematical structures exhibit their observed properties. A common way to seek such understanding is to perturb the definition of a structure and study the effects of the perturbation. While this stratagem leads to interesting, valuable results only sometimes, it is an invaluable tool in the hands of a gifted mathematician. This section presents a brief survey of such a study by Édouard Lucas.

D.1 Definition

Lucas, who is credited with giving the name "Fibonacci numbers" to the sequence discovered by Leonardo Pisano, investigated the consequences of perturbing the initial conditions, $F(0) = F(1) = 1$, in the classical definition (9.6) of the Fibonacci sequence.

Lucas's approach was simply to replace the Fibonacci sequence's initial values $\langle 1, 1 \rangle$, with the values $\langle 2, 1 \rangle$. It turns out to be much more fruitful—in terms of more striking results and simpler proofs—to make a somewhat more drastic perturbation:

The *Lucas sequence* (or *the sequence of Lucas numbers*) is the infinite sequence of positive integers

$$L(-1), \ L(0), \ L(1), \ L(2), \dots$$

---

[1] See A.L. Rosenberg (1979): Profile numbers. *Fibonacci Quart. 17*(3), 259–264.

© Springer Nature Switzerland AG 2020
A. L. Rosenberg, D. Trystram, *Understand Mathematics, Understand Computing*,
https://doi.org/10.1007/978-3-030-58376-7

which is generated by the recurrence

$$L(-1) = 2$$
$$L(0) = 1 \qquad\qquad\qquad\qquad\qquad\qquad \text{(D.1)}$$
$$L(n) = L(n-1) + L(n-2) \quad \text{for all } n \geq 1$$

Because we conventionally index sequences by *nonnegative* numbers, we henceforth ignore $L(-1)$ and use the following *standard definition* of the Lucas sequence.

$$L(0) = 1$$
$$L(1) = 3 \qquad\qquad\qquad\qquad\qquad\qquad \text{(D.2)}$$
$$L(n) = L(n-1) + L(n-2) \quad \text{for all } n > 1$$

The following finite sequences present the first few elements of both the Lucas sequence (for illustration) and the Fibonacci sequence (for comparison).

n	0, 1, 2, 3, 4, 5, 6, 7, 8, 9, ...
$L(n)$	1, 3, 4, 7, 11, 18, 29, 47, 76, 123, ...
$F(n)$	1, 1, 2, 3, 5, 8, 13, 21, 34, 55, ...

We begin our brief study of the Lucas sequence by noting that just a minor tweak converts the Fibonacci-related identity revealed in Proposition 9.7 into an identity for Lucas numbers, as originally defined—beginning with $L(-1)$.

Proposition D.1 *For all integers $n \geq 0$,*

$$L(n+2) = 1 + L(-1) + L(0) + L(1) + L(2) + \cdots + L(n) \qquad \text{(D.3)}$$

Proof (Sketch). We can literally repeat the proof of Proposition 9.7, with only a change in the induction's base case, which now becomes

$$L(2) = L(-1) + L(0) + 1 = 2 + 1 + 1 = 4$$

The body of the inductive argument holds for the Lucas sequence as well as for the Fibonacci sequence. □

D.2 Relating the Lucas and Fibonacci Numbers

There are several simple equations that relate the Lucas and Fibonacci numbers. We present a few of the most aesthetically pleasing ones,[2] in terms of their exposing an intimate relationship between the two sequences.

---

[2] Aesthetically pleasing, that is, to the authors. As noted by the author Margaret Wolfe Hungerford in *Molly Bawn* (1878), "Beauty is in the eye of the beholder."

Proposition D.2 *For all* $m, n \geq 1$

$$\text{(a)} \qquad L(n) = F(n+1) + F(n-1)$$

$$\text{(b)} \qquad F(n+1) = \tfrac{1}{2}\big(F(n) + L(n)\big)$$

$$\text{(c)} \; F(m+n-1) = \tfrac{1}{2}\big(F(m) \cdot L(n) + F(n) \cdot L(m)\big) \qquad \text{(D.4)}$$

$$\text{(d)} \qquad F(2n-1) = F(n) \cdot L(n)$$

Proof. We consider the identities in turn.
(a) We proceed by induction.

Base case. Equation (D.4)(a) holds when $n = 1$ because

$$L(1) = 3 = F(2) + F(0) = 2 + 1$$

Inductive hypothesis. Assume that Eq. (D.4)(a) holds for $L(2), L(3), \ldots, L(n)$.

Inductive extension. Let us compute $L(n+1)$:

- By definition (D.2),
$$L(n+1) \;=\; L(n) + L(n-1) \qquad \text{(D.5)}$$
- When we apply the inductive hypothesis to both addends in Eq. (D.5), we obtain (after rearranging terms):
$$L(n+1) \;=\; F(n+1) + F(n) + F(n-1) + F(n-2) \qquad \text{(D.6)}$$
- Finally, we invoke the defining recurrence (9.6) of the Fibonacci numbers on the first two addends in Eq. (D.6) and on the last two addends. We thereby transform Eq. (D.6) into Eq. (D.4)(a), which validates the latter identity.

Explanatory note.
Notice that a proof similar to the preceding one yields the identity $L(n) = F(n+2) + F(n-2)$. Similar, but more complicated, identities hold for larger arguments. For the cases $n+3$ and $n+4$, for instance, one can establish the following pair of identities.

$$L(n) = \frac{1}{2}(F(n+3) + F(n-3)) \qquad \text{(D.7)}$$

$$L(n) = \frac{1}{3}(F(n+4) + F(n-4)) \qquad \text{(D.8)}$$

(b) By direct calculation, we derive the desired result:

$$2F(n+1) \;=\; F(n+1) + F(n) + F(n-1) \;=\; L(n) + F(n)$$

(c) This identity is verified via a somewhat complicated induction. We fix parameter n in the argument $F(m+n)$ and induce on parameter m.

Base case. Because $L(0) = F(0) = 1$, the instance $m = 0$ of identity (D.4)(c) reduces to identity (D.4)(b), which we have just proved.

$$F(n+1) = \frac{1}{2}\big(L(n) + F(n)\big) = \frac{1}{2}\big(F(0) \cdot L(n) + F(n) \cdot L(0)\big)$$

Inductive hypothesis. Let us assume that identity (D.4)(c) holds for all $m \leq k$.

Inductive extension. Let us focus on instance $m = k+1$ of identity (D.4)(c). Note first that the classical Fibonacci recurrence (9.6) implies that

$$F(n+k+1) = F(n+k) + F(n+k-1).$$

When we apply the inductive hypothesis to both $F(n+k)$ and $F(n+k-1)$, we obtain the following two identities.

$$F(n+k) = \frac{1}{2}\big(F(k-1) \cdot L(n) + F(n) \cdot L(k-1)\big)$$

$$F(n+k-1) = \frac{1}{2}\big(F(k-2) \cdot L(n) + F(n) \cdot L(k-2)\big)$$

Because both the Fibonacci and Lucas sequences obey the body of recurrence (9.6), the preceding equations combine to extend the induction.

$$\begin{aligned}
2F(n+k+1) &= 2F(n+k) + 2F(n+k-1) \\
&= \big(F(k-1) \cdot L(n) + F(n) \cdot L(k-1)\big) + \big(F(k-2) \cdot L(n) + F(n) \cdot \mathit{l} \\
&= L(n) \cdot \big(F(k-1) + F(k-2)\big) + F(n) \cdot \big(L(k-1) + L(k-2)\big) \\
&= L(n) \cdot F(k) + F(n) \cdot L(k)
\end{aligned}$$

The thus-extended induction verifies identity (D.4)(c).

(d) Identity (D.4)(d) is actually the case $m = n$ of identity (D.4)(c).

This validates our final identity, which completes the proof. □

Proposition D.3 *For all positive integers n, and all $k \leq n$:*

$$F(k-1) \cdot L(n) = F(n+k) + (-1)^{k-1} F(n-k)$$

Proof (Sketch). Let us use induction on k to study the asserted products for fixed n.

- **Base.** The case, $k = 1$, follows by Proposition D.2(a).
- **Extension.** Using an approach similar to the base case, we can observe that

$$L(n) = F(n+2) - F(n-2)$$

Let us see why.

Easy calculations yield the following equations:

$$2L(n) = F(n+3) + F(n-3)$$
$$3L(n) = F(n+4) - F(n-4)$$
$$5L(n) = F(n+5) + F(n-5)$$

Observing that 2, 3, and 5 are consecutive Fibonacci numbers, we obtain the intuition for the general case,

$$F(k-1) \cdot L(n) = F(n+k) + (-1)^{k-1}F(n-k) \quad \text{for } k \leq n$$

which is proved (again) by induction on k.

The base case is straightforward (see case $k = 1$).

Compute $F((k+1)-1) \cdot L(n)$ and apply the definition of Fibonacci number:

$$F((k+1)-1) = F(k-1) + F(k-2)$$

If we replace the two last terms by using the inductive hypothesis, we observe the following.

$$
\begin{aligned}
F(k) \cdot L(n) &= F(k-1) \cdot L(n) + F(k-2) \cdot L(n) \\
&= F(n+k) + (-1)^{k-1}F(n-k) + F(n+k-1) \\
&\quad + (-1)^{k-2}F(n-(k-1)) \\
&= F(n+k) + F(n+k-1) + (-1)^{k-1}(F(n-k) - F(n-k+1))
\end{aligned}
$$

The sought result is finally obtained by invoking the definition of the Fibonacci numbers twice. □

Appendix E
⊕⊕ Signed-Digit Numerals: Carry-Free Addition

Topic-specific reference

K. Hwang (1979): *Computer Arithmetic: Principles, Architecture, and Design.* John Wiley & Sons, New York.

D.E. Knuth (1969): *The Art of Computer Programming, Vol. 2: Seminumerical Algorithms.* Addison-Wesley, Reading, Mass.

Say that you have a box that contains a *counter*. There are two buttons on top of the box: one *red* and one *green*. Each time the green button is pushed, the counter increments—i.e., adds +1 to—its tally; each time the red button is pushed, the counter decrements—i.e., adds −1 to—its tally. Now, in order for the tally on the counter to *always be correct*, you must insert delays between button-pushes: you must always wait until the electronic circuitry inside the box settles into a stable configuration with all necessary carries and borrows up to date. Now, if the electronic circuitry that implements the counter was designed to mimic the binary representation of the number—in technical jargon, the counter implements a *carry-ripple adder*; see [Hwang, 1979] for details—then you observe the following. While you keep pushing only the green (increment) button, you do not feel that the delays between button-pushes are onerous, for the reason exposed in the following analysis.

- Roughly half the time, you have no enforced delays between successive button-pushes: the update of the tally engenders no carry.
- Roughly one-quarter of the time, you have a minuscule delay: the update engenders a carry of only one place.
- Roughly one-eighth of the time, you have a slightly longer delay: the update engenders a carry of two places.
- …Continuing this progression: For each integer k roughly the fraction 2^{-k} of the button-pushes engender a delay commensurate with a carry of $k-1$ places.

Stated differently, the *average* delay you must suffer is bounded by the time required for a two-place carry. In fact, you can even improve on the described pattern of delays, by using more sophisticated circuitry for your counter's adder. If you replace the carry-ripple adder by a *carry-lookahead adder*—see [Hwang, 1979] for

© Springer Nature Switzerland AG 2020   479
A. L. Rosenberg, D. Trystram, *Understand Mathematics, Understand Computing*,
https://doi.org/10.1007/978-3-030-58376-7

details—you can thereby reduce the aggregate number of carries engendered by the first n button-pushes from the $\Theta(n^2)$ carries of the carry-ripple adder to $O(n \log n)$ carries for a carry-lookahead adder.

However, the picture changes markedly as soon as you (or an opponent) start to use the red button as well as the green one. If you wait until the counter has tallied some number of green-button pushes of the form 2^k, hence contains the binary numeral $100 \cdots 00$ with k 0s, then a push of the red button engenders a delay of k carry-units—and a subsequent push of the green button incurs the same delay! In fact, if you (and your opponent) begin to toggle the two buttons—one red push, then one green push, then one red push, then one green push, ...—then you incur a delay of k carry-units with each successive button-push. If you do this for long enough, then your *average* delay starts growing toward k carry-units.

If (the risk of) long delays such as those just described is more than you want to deal with, then you can employ mathematical and electronic technology to replace your counter's adder with one that (almost) *eliminates these delays completely*. The "silver bullet" resides in using a *signed-digit* (positional) number representation to design a counter that implements *carry-free addition*. By changing the form of the numerals your counter uses, you can design adders whose constituent digit-adders can operate *in parallel*. While these digit-adders are roughly twice as complex as those of the more familiar carry-ripple adders, they do guarantee bounded—i.e., $O(1)$—delay between button-pushes, no matter how long you keep pushing a button or which one you choose to push.

The signed-digit number systems which we describe now[1] are positional systems of the same genre as -ary systems: Their numerals are also evaluated by the VAL function of (10.3). These systems achieve their buffered carries via augmented digit-sets that include *negative* digits as well as positive ones. The base-b (or, *radix-b*) redundant number system has the following set of digits.[2]

$$\widehat{B}_b \;=\; \{(-\overline{b-1}), (-\overline{b-2}), \ldots, (-1), 0, 1, \ldots, \overline{b-2}, \overline{b-1}\}$$

For reasons that are purely technical, the simple carry-free adder that we describe requires that the number base b be no smaller than 3. For the smallest relevant base, $b = 3$, one can use the digit-set $\widehat{B}_3 = \{(-2), (-1), 0, 1, 2\}$ (parentheses added to enhance legibility). The "redundancy" in this system is witnessed by facts such as the following: The number 1 is the numerical value of *every* numeral of the form

$$1 \, (-2) \, (-2) \, \cdots \, (-2) \, (-2)$$

because the d-digit instance of these numerals has numerical value

---

[1] There are many such systems and many places to learn about them; see, e.g., the encyclopedic work [Knuth, 1969] on "seminumerical algorithms" and the comprehensive text [Hwang, 1979] on computer arithmetic.

[2] (*a*) Recall that the overline notation, as in "$\overline{b-2}$" is to indicate that we are referring to a single digit. (*b*) We add parentheses when we describe $\widehat{B}_b$ to enhance legibility.

$$\text{VAL}_3\big(1\,(-2)\,(-2)\,\cdots\,(-2)\,(-2)\big)$$
$$= 3^{d-1} - 2\cdot\big(3^{d-2} + 3^{d-3} +\cdots+ 3 + 1\big)$$
$$= 1$$

We now provide a schematic description[3] of (a) a carry-ripple adder and (b) a carry-free adder adding the b-ary integer-specifying numerals

$$x_n x_{n-1} \cdots x_1 x_0 \quad \text{and} \quad y_n y_{n-1} \cdots y_1 y_0$$

In both descriptions, we describe the function of the ith digit-adder, emphasizing the information that it receives from the $(i-1)$th digit-adder and the information that it transmits to the $(i+1)$th digit-adder. Of course, this information is at the heart of the issue of carries or no-carries.

$$
\begin{array}{c}
\phantom{c_{i+1} = }\quad x_i \;\; y_i \\
\phantom{c_{i+1} = }\quad \downarrow \;\; \downarrow \\
c_{i+1} = \big((x_i+y_i+c_i)\ominus \bar{b}\big) \;\longleftarrow\; \boxed{x_i+y_i+c_i} \;\longleftarrow\; c_i \\
\phantom{c_{i+1} = }\quad\quad\quad\quad \downarrow \\
z_i = \min\big(\bar{b},\, x_i+y_i+c_i\big)
\end{array}
$$

Fig. E.1 Digit i of a b-ary carry-ripple adder. The input digits x_i and y_i are summed with the in-carry c_i. If the sum can be represented by a single b-ary digit, then it is the output from this digit-adder, and the out-carry c_{i+1} is set to 0. If the sum is too large, then the maximum b-ary digit $\bar{b}$ is the output from this digit-adder, and the out-carry c_{i+1} is set to the portion of the sum that exceeds $\bar{b}$

A Carry-ripple adder. See Fig. E.1.

- All digits—input digits x_i and y_i, and carry digit c_i (for $i>0$) come from $B_b = \{0,1,\ldots,\bar{b}\}$; $c_0 = 0$.
- Addition is a *single-pass* process. For each digit-index i:

  – Digit-adder i admits inputs x_i and y_i (from the "outside") and carry-in c_i: When $i = 0$, $c_0 = 0$ by convention; for all other indices i, c_i is the carry-in from digit-adder $i-1$.

  – The ith sum-digit, z_i, and the carry-out, c_{i+1}, are evaluated as follows.

$$z_i = \min\big(\bar{b},\, x_i+y_i+c_i\big)$$
$$c_{i+1} = (x_i+y_i+c_i)\ominus\bar{b} \quad \text{where } m\ominus n \overset{\text{def}}{=} \begin{cases} m-n & \text{if } m \ge n \\ 0 & \text{if } m \le n \end{cases}$$

---

[3] Detailed descriptions of both genres of adders—including both operational and implementational matters—can be found in sources such as [Hwang, 1979].

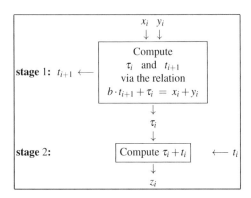

Fig. E.2 Digit-adder i of a $(b > 2)$-ary carry-free adder. The outgoing *transfer digit* t_{i+1} and the *tentative sum* τ_i are computed from the input digits x_i and y_i, using the relation specified in the upper box. These results are transmitted, respectively, to stage 1 of digit-adder $i + 1$ and stage 2 of the current digit-adder i

A Carry-free adder. See Fig. E.2.

- This simple carry-free adder operates in *two stages*. It operates on base-b numerals but requires $b > 2$ (for technical reasons).
- There is no "carry digit". In its place, there is a *transfer digit* $t_i \in \{-1, 0, 1\}$ which is generated in stage 1 of the adder and consumed in stage 2.
- In addition to transfer digits, there are *tentative sum* digits τ_i, which are also generated in stage 1 and consumed in stage 2.

1. Stage 1 of the adder

   a. computes *transfer digit* $t_{i+1} \in \{-1, 0, 1\}$ and *tentative-sum digit* τ_i via the equation

   $$b \cdot t_{i+1} + \tau_i \;=\; x_i + y_i$$

   b. transmits t_{i+1} to digit-adder $i + 1$
   c. transmits τ_i to stage 2.

2. Stage 2 of the adder computes the ith output digit z_i by adding τ_i to t_i (which came from digit-adder $i - 1$).

The important fact to note is that this adder does not have a chain of propagated carries! Digit-adder i computes transfer digit t_{i+1}—which is the only information that it transmits to digit-adder $i + 1$—based only on the ith input digits x_i and y_i. Consequently: *All of the adder's digit-adders can operate in parallel!*

Appendix F
⊕⊕ The Diverse Delights of de Bruijn Networks

F.1 Cycles in de Bruijn Networks

This section is devoted to the following marvelous extension of Proposition 13.12.

Proposition F.1 *Every de Bruijn network $\mathscr{D}_n$ is directed-pancyclic.*
In detail: $\mathscr{D}_n$ has directed cycles of every length from[1] 1 to 2^n.

Proof. We argue by induction on the index n of de Bruijn network $\mathscr{D}_n$.

Base case. A brief perusal of Figs. 12.11 and 12.12 will verify that both $\mathscr{D}_2$ and $\mathscr{D}_3$ are directed-pancyclic.

Inductive assumption. Say, for induction, that every de Bruijn network $\mathscr{D}_m$ with $m \leq n$ is directed-pancyclic.

Inductive extension. Our progress from the inductive assumption to a proof that $\mathscr{D}_{n+1}$ is directed-pancyclic relies heavily on two results from Section 13.2, namely:

- Corollary 13.2: *Each $\mathscr{D}_n$ admits a directed Eulerian cycle.*

- Lemma 13.3: *The line graph of $\mathscr{D}_n$ is (isomorphic to) $\mathscr{D}_{n+1}$.*

Our extension of the inductive hypothesis focuses separately on "small" and "large" cycles.

Case 1. *"Small" cycles in $\mathscr{D}_{n+1}$: lengths $\ell \in \{1, \ldots, 2^n\}$.*
We know by our inductive hypothesis that $\mathscr{D}_n$ contains directed cycles of every length $\ell \in \{1, \ldots, 2^n\}$. Focus on any such cycle, say the one of length k, call it $\mathscr{C}$. Clearly, the line digraph of $\mathscr{C}$ is another length-k cycle, which is isomorphic to $\mathscr{C}$. The proof of Lemma 13.3 therefore tells us that the line digraph of $\mathscr{C}$ is isomorphic to a length-k cycle in $\mathscr{D}_{n+1}$.

---

[1] The cycles of length 1 arise from the *self-loops* on vertices $0 \cdots 0$ and $1 \cdots 1$ in every de Bruijn network.

© Springer Nature Switzerland AG 2020
A. L. Rosenberg, D. Trystram, *Understand Mathematics, Understand Computing*,
https://doi.org/10.1007/978-3-030-58376-7

This additional cycle in $\mathscr{D}_{n+1}$ proves that $\mathscr{D}_{n+1}$ contains directed cycles of all lengths $\ell \in \{1, \ldots, 2^n\}$.

Case 2. *"Large" cycles in $\mathscr{D}_{n+1}$: lengths $\ell \in \{2^n + 1, \ldots, 2^{n+1}\}$.*

Focus on an arbitrary integer $m = 2^n + k$, where $k \in \{1, \ldots, 2^n\}$. We verify that $\mathscr{D}_{n+1}$ contains a directed cycle of length m.

By Case 1 and our inductive hypothesis, we know that $\mathscr{D}_{n+1}$ contains a directed cycle, call it $\mathscr{C}$, of length $M = 2^{n+1} - m = 2^n - k$. By Lemma 13.3, the existence of cycle $\mathscr{C}$ implies that $\mathscr{D}_n$ has a connected Eulerian sub-digraph which contains M arcs. We claim that $\mathscr{D}_n$ *also* has a connected Eulerian sub-digraph which contains m arcs. Once we verify this claim, an invocation of Lemma 13.3 will establish the existence in $\mathscr{D}_{n+1}$ of a cycle that contains $m = 2^n + k$ vertices. Therefore, Case 2 will follow from a proof of the following lemma.

Lemma F.1. *For any $p \in \{0, \ldots, 2^{n+1}\}$, if $\mathscr{D}_n$ has a connected Eulerian sub-digraph $\mathscr{G}$ which contains p arcs, then it also has such a sub-digraph which contains $2^{n+1} - p$ arcs.*

Proof (Lemma F.1). Fix an arbitrary p for which $\mathscr{D}_n$ has a connected p-arc Eulerian sub-digraph $\mathscr{G}$. Because both $\mathscr{G}$ and $\mathscr{D}_n$ are Eulerian, each vertex v of each of these digraphs has equal in-degree and out-degree:

$$\text{IN-DEGREE}(v) \;=\; \text{OUT-DEGREE}(v)$$

Therefore, if we remove from $\mathscr{D}_n$ the p arcs of $\mathscr{G}$, then we are left with a (not-necessarily connected) Eulerian sub-digraph $\mathscr{H}$ of $\mathscr{D}_n$ which contains $2^{n+1} - p$ arcs.

Let $\mathscr{F}_1, \ldots, \mathscr{F}_r$ be all of the maximal connected components of $\mathscr{H}$ which are *nontrivial* in the sense of containing at least one arc apiece. Of course, each $\mathscr{F}_i$ is connected and Eulerian.

If $r = 1$, then $\mathscr{H}$ is the sub-digraph $\mathscr{G}$ of $\mathscr{D}_n$ guaranteed by the Lemma.

If $r > 1$, then we need to do some work to create $\mathscr{G}$ from the components $\mathscr{F}_i$.

Because $\mathscr{D}_n$ is connected and all of its arcs reside in either $\mathscr{G}$ or $\mathscr{H}$, we know that $\mathscr{G}$ must contain some arc $(u \to v)$ where u is a vertex of some $\mathscr{F}_i$ and v is a vertex of some different $\mathscr{F}_j$ (i.e., $i \neq j$). The existence of this arc implies that vertex u has out-degree ≤ 1 in $\mathscr{H}$, even though u has out-degree 2 in $\mathscr{D}_n$. Because $\mathscr{H}$ is Eulerian, it follows that u has equal in-degree and out-degree in $\mathscr{H}$. Moreover, because $\mathscr{H}$ contains at least one arc, we know that u cannot be an isolated vertex in $\mathscr{H}$. Therefore:

vertex u has out-degree exactly 1 *in $\mathscr{H}$.*

It follows that there must be an arc $(u \to w)$ in $\mathscr{F}_i$ for some vertex w of $\mathscr{F}_i$. By symmetric reasoning—using in-degrees instead of out-degrees)—there must be an arc $(t \to v)$ in $\mathscr{F}_j$ for some vertex t of $\mathscr{F}_j$. Because $(u \to v)$, $(u \to w)$, and $(t \to v)$ are all arcs of $\mathscr{D}_n$, there must exist a length-$(n-1)$ bit-string x and bits $\beta, \gamma, \delta, \varepsilon \in \{0, 1\}$ such that

$$t = \beta x$$
$$u = \gamma x$$
$$v = x\delta$$
$$w = x\varepsilon$$

It follows that $\mathscr{G}$ contains an arc $(t \to w)$. This arc resides in $\mathscr{D}_n$ by definition; it cannot reside in either $\mathscr{F}_i$ or $\mathscr{F}_j$ because it would connect these two components which are disconnected in $\mathscr{H}$.

We now transform sub-digraph $\mathscr{H}$ in the following way. We remove from $\mathscr{H}$ two arcs: $(u \to w)$, which belongs to $\mathscr{F}_i$, and $(t \to v)$, which belongs to $\mathscr{F}_j$. We add in place of these arcs the arcs $(t \to w)$ and $(u \to v)$. The resulting new version of $\mathscr{H}$, call it $\mathscr{H}'$:

- *contains the same number of arcs as $\mathscr{H}$ does*;

- *is Eulerian*, because we just exchanged one arc that enters each of vertices u and w for another, and we made a similar exchange for arcs that leave vertices t and u;

- *is connected*, because each of $\mathscr{F}_i$ and $\mathscr{F}_j$, being directed-Eulerian, admits a directed walk that crosses each arc precisely once—and our exchanged arcs connect these directed walks into a composite directed walk through the new component;

- *has one fewer nontrivial maximal connected component than $\mathscr{H}$ does*.

When we iterate the just-described transformation, each iteration yields an Eulerian sub-digraph of $\mathscr{D}_n$ which has p arcs (as desired) and has one fewer nontrivial maximal connected component than its predecessor. After $r - 1$ iterations, we therefore achieve the desired connected Eulerian sub-digraph of $\mathscr{D}_n$. $\square$-Lemma F.1

The m-arc connected Eulerian sub-digraph of $\mathscr{D}_n$ guaranteed by Lemma F.1 implies the existence of an m-vertex cycle in $\mathscr{D}_{n+1}$. Since the number k, hence the number m, was arbitrary, this completes the proof. $\square$

F.2 de Bruijn Networks as "Escherian" Trees

This section is devoted to exposing a mathematically charming connection between a genre of directed rooted tree and the family of de Bruijn networks. The section title acknowledges the "spiritual" relationship between our mathematical connection and the well-known piece "Drawing Hands" (1948) of the Dutch artist Maurits Cornelis (M. C.) Escher.[2]

---

[2] The shared nationality of the artist Escher and the mathematician de Bruijn is an amusing coincidence.

The root of the connection we wish to expose lies in the following algebraic way of representing certain arc-labeled directed graphs. Let us be given a set V (which may be finite or infinite), together with functions F_1, F_2, …, F_k, each F_i being a function from V to V. In our examples, the F_i will be total injections from V to V, but neither of these qualifiers ("total" or "injection") is necessary for the concept we describe. One can generate an arc-labeled digraph $\mathcal{G} = \mathcal{G}(V; F_1, \ldots, F_k)$ as follows.

- The set V comprises the vertices of $\mathcal{G}$.
- For each vertex $v \in V$ and each function F_i, $\mathcal{G}$ will have an arc with label F_i:

$$(v \to F_i(v))$$

We provide three examples, the second two providing the correspondence that motivates this section.

1. The directed infinite binary tree. Let the set V be the set $\mathbb{N}^+$ of positive integers. Define the arc-generating functions as follows:

$$F_0(v) = 2v \quad \text{and} \quad F_1(v) = 2v + 1$$

The rationale for the subscripts of F_0 and F_1 becomes clearer when we point out the effect of each of these functions on the binary representations of the integer vertices of $\mathcal{G}(\mathbb{N}^+; F_0, F_1)$: if x is the binary-string label of vertex v, then $x0$ is the binary-string label of $F_0(v)$, and $x1$ is the binary-string label of $F_1(v)$. We indicate in Fig. F.1 how the described system can be viewed as specifying a graph-theoretic structure. The

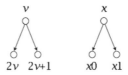

Fig. F.1 The graph-theoretic action of one application of the functions F_0 and F_1: (left) when the vertices of $\mathcal{G}(\mathbb{N}^+; F_0, F_1)$ are viewed as integers; (right) when the vertices are viewed as bit-strings

figure depicts just one tree-vertex and its children. In the system as described, every vertex is a positive integer, and the arcs are generated by the functions F_0 and F_1; when recast into "string-label mode", every vertex is a binary string, and the arcs are generated by the functions "append 0" and "append 1".

2. The root-looped directed infinite binary tree. For reasons that will become clear in the upcoming paragraph 3, we amend the just-described system so that it generates a "cousin" of the directed infinite binary tree. A "prefix" of this tree appears in Fig. F.2. This cousin is the tree with a special vertex, call it r; vertex r has two emerging arcs (as do all vertices): One of the new arcs is a *self-loop* on r; the other

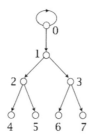

Fig. F.2 The root-looped seven-vertex complete binary tree with special vertex 0

points to the root of the directed infinite binary tree of paragraph 1. One generates the cousin by:

- using the set $\mathbb{N}$ of *nonnegative* integers as vertices, instead of the earlier-used set $\mathbb{N}^+$ of *positive* integers;

- using the same functions F_0 and F_1 to generate the arcs of the new graph.

  Of course, F_0 and F_1 now have an extended domain, but their "action" is un-changed: $F_0(x) = 2x$ and $F_1(x) = 2x + 1$.

In the new digraph, vertex 0 is the special vertex r. The *self-loop* on vertex $r = 0$ occurs because $F_0(0) = 2 \times 0 = 0$. The remainder of this new digraph is the directed infinite binary tree of paragraph 1.

3. The de Bruijn network as the Escherian root-looped directed infinite binary tree. We derive the desired connection between the tree-like digraph of paragraph 2 and the order-n de Bruijn network by presenting the latter digraph algebraically, specif-ically, as the system

$$\mathscr{G}(\{0, 1, \ldots, 2^n - 1\}; F_0^{(n)}, F_1^{(n)})$$

where

$$F_0^{(n)}(v) = 2v \bmod 2^n \quad \text{and} \quad F_1^{(n)}(v) = 2v + 1 \bmod 2^n$$

The system $\mathscr{G}(\{0, 1, \ldots, 2^n - 1\}; F_0^{(n)}, F_1^{(n)})$ thus differs from the infinite system $\mathscr{G}(\mathbb{N}; F_0, F_1)$—which generates the root-looped directed infinite binary tree—by truncating both the set of vertices (by keeping only the first 2^n nonnegative inte-gers) and the arc-generators. The truncation is achieved by reducing all integers modulo 2^n. See Fig. F.3.

To see that the system $\mathscr{G}(\{0, 1, \ldots, 2^n - 1\}; F_0^{(n)}, F_1^{(n)})$ is, in fact, the order-n de Bruijn network, let us observe what the functions $F_0^{(n)}$ and $F_1^{(n)}$ do to an argument integer/vertex v.

- The *number-related* specification

$$v \longrightarrow 2v \bmod 2^n$$

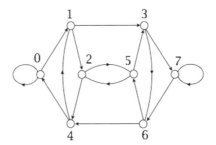

Fig. F.3 A standard depiction of $\mathscr{D}_3$, with vertex-names being the integers corresponding to the usual binary string-labels

corresponds to the *numeral-related* specification

$$\beta x \longrightarrow x0$$

where $\beta \in \{0,1\}$ and x is a length-$(n-1)$ bit-string.

- The *number-related* specification

$$v \longrightarrow 2v+1 \bmod 2^n$$

corresponds to the *numeral-related* specification

$$\beta x \longrightarrow x1$$

where $\beta \in \{0,1\}$ and x is a length-$(n-1)$ bit-string.

The system $\mathscr{G}(\{0,1,\ldots,2^n-1\}; F_0^{(n)}, F_1^{(n)})$ thus specifies the local graph-theoretic structure depicted in Fig. F.4. A look back at Section 12.3.5 will verify that the

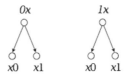

Fig. F.4 The graph-theoretic action of one application of the arc-generating functions $F_0^{(n)}$ and $F_1^{(n)}$ to vertices of the forms $0x$ (left) and $1x$ (right)

system $\mathscr{G}(\{0, 1, \ldots, 2^n - 1\}; F_0^{(n)}, F_1^{(n)})$ is, in fact, (isomorphic to) $\mathscr{D}_n$.

One often reads about connections between mathematics and art. It is probably fair to predict that the connection we have just revealed is not what most references to a mathematics-art nexus are referring to. But there it is!

Appendix G
⊕ **Pointers to Applied Studies of Graphs**

This appendix introduces four graph-related topics that are typically not covered—at least in depth—early in the curriculum, but that are important enough that the reader should at least be aware of them. The topics we chose are motivated by studies in virtually every computational area that benefits from graph-theoretic models. We have tried to present each topic at a level of discourse that will prepare the interested reader to delve more deeply into the material, yet at a level of informality that will make the material accessible to the more casual reader. We thus strive for an intuitive presentation that will not lead any reader astray.

- Section G.1 focuses on the many computations on graphs that can be accomplished efficiently via recursive algorithms that (1) decompose an input graph $\mathcal{G}$; (2) process the resulting subgraphs of $\mathcal{G}$; (3) reassemble the processed graphs.

- Section G.2 introduces the increasingly important topic of graphs whose structure changes dynamically over time. One timely instance of such dynamic evolution is the connectivity graph of the Internet.

- Section G.3 introduces *hypergraphs*—generalized graphs that allow relationships among *multiple* entities/vertices, in contrast to the purely *binary* relationships demanded by graphs' *two-element* edges. Hypergraph-based models find application in areas as diverse as:

  - *social networks:* Hyperedges describe, e.g., collaboration and collusion.
  - *electronic networks:* Hyperedges enable the design of equipotential vertices in voltage-driven technologies such as *VLSI*.
  - *communication networks:* Hyperedges model bus-oriented communication.

- Section G.4 introduces *multigraphs*—generalized graphs that allow multiple edges between each pair of vertices. In contrast to other advanced topics, multigraphs are used more *internally*, as an algorithmic device, than *externally*, as an abstract model of "real" structures. For example, using multigraphs to replace integer-weighted edges with multiple unweighted edges with the same endpoints can trigger algorithmic ideas.

© Springer Nature Switzerland AG 2020
A. L. Rosenberg, D. Trystram, *Understand Mathematics, Understand Computing*,
https://doi.org/10.1007/978-3-030-58376-7

G.1 Graph Decomposition

Topic-specific references

C.E. Leiserson (1985): Fat-trees: Universal networks for hardware-efficient supercomputing. *IEEE Trans. on Computers, C-34*(10), 892–201.

R.J. Lipton, R.E. Tarjan (1979): A separator theorem for planar graphs. *SIAM J. Applied Mathematics 36*(2), 177–189.

A.L. Rosenberg and L.S. Heath (2001): *Graph Separators, with Applications.* Kluwer Academic/Plenum Publishers, New York.

J.D. Ullman (1984): *Computational Aspects of VLSI..* Computer Science Press, Rockville, Md.

The reader will certainly have noted that some "named" graphs are, intuitively, more tightly interconnected than others—with the cycle and clique being the antipodal examples. Even from a purely intellectual vantage point—and all the more so from an algorithmic vantage point—it would be of interest to be able to quantify the tightness of a graph's interconnectedness. Among the various measures that have been proposed for this task, one stands out for its broad range of algorithmic implications: the notion of *graph separator*. In fact, this notion appears in the literature in several flavors. An n-vertex, e-edge graph $\mathscr{G}$ has:

- an α-*edge separator* of size k—where α is a real number and k is an integer with $k < e$—precisely if:

  one can partition $\mathscr{G}$ into two disjoint (not-necessarily connected) subgraphs, each having $\leq \alpha n$ vertices, by removing $\leq k$ edges from $\mathscr{G}$.

- an α-*vertex separator* of size ℓ, where α is a real number and ℓ is an integer with $k < n$, precisely if:

  one can partition $\mathscr{G}$ into two disjoint (not-necessarily connected) subgraphs, each having $\leq \alpha n$ vertices, by removing $\leq \ell$ vertices from $\mathscr{G}$.

We replace the term "separator" with the term *"bisector"* if both subgraphs after a separation operation have $\leq \lceil \frac{1}{2}n \rceil$ vertices.

Commonalities and differences in graphs' inherent separator sizes are often not visually obvious. For illustration, referring to the "named" graphs of Section 12.3:

- It is certainly obvious that cycles are easier to bisect than cliques, as measured by either edge or vertex bisectors.

- It is far less clear that de Bruijn networks and hypercubes are roughly equal in ease of bisection, as measured by vertex bisectors.

Similarity in separation behavior often has very important algorithmic consequences. For instance, the closeness in separation characteristics between de Bruijn networks and hypercubes manifests itself in a large range of algorithmic applications. The range of such applications is hinted at by sources that study the algorithmics of laying out VLSI circuits (cf. [Leiserson, 1985]) and sources that study the ability of a network's interconnections to host a range of communication patterns that enable efficient parallel computation and communication (cf. [3], [Leiserson, 1985], and [Ullman, 1984]).

There is a large literature that develops the algorithmics of finding small separators for computationally significant families of graphs. An early star in the firmament of such studies is the discovery in [Lipton, Tarjan, 1979] of a 2/3-vertex separator of size $\sqrt{8n}$ for n-vertex planar graphs. The dual problem of finding lower bounds on the sizes of graph separators is more sparsely studied but, of course, no less significant. The reader can find a comprehensive exposition of the theory of graph separators in [Rosenberg, Heath, 2001], including both the mathematics that yields lower bounds on separator sizes and the algorithmics that yields upper bounds.

G.2 Graphs Having Evolving Structure

Topic-specific references

W. Aiello, F.R.K. Chung, L. Lu (2000): A random graph model for massive graphs. *32nd Ann. Symp. on the Theory of Computing.*

A.-L. Barabási, R. Albert (1999): Emergence of scaling in random networks. *Science (286)*, 509–512.

B. Bollobas (1985): *Random Graphs.* Academic Press, N.Y.

T. Bu, D. Towsley (2002): On distinguishing between internet power-law generators. *IEEE INFOCOM'02.*

Q. Chen, H. Chang, R. Govindan, S. Jamin, S. Shenker, W. Willinger (2002): The origin of power laws in internet topologies revisited. *IEEE INFOCOM'02.*

M. Faloutsos, P. Faloutsos, C. Faloutsos (1999): On power-law relationships of the internet topology. *ACM SIGCOMM'99.*

S. Jaiswal, A.L. Rosenberg, D. Towsley (2004): Comparing the structure of power-law graphs and the Internet AS graph. *12th IEEE Int'l Conf. on Network Protocols (ICNP'04).*

H. Tangmunarunkit, R. Govindan, S. Jamin, S. Shenker, W. Willinger (2002): Network topology generators: Degree-based vs. structural. *ACM SIGCOMM'02.*

E. Zegura, K.L. Calvert, M.J. Donahoo (1997): A quantitative comparison of graph-based models for internetworks. *IEEE/ACM Trans. on Networking, 5(6)*, 770–783.

A large variety of problems in the area of graph algorithms involve graphs—especially trees—whose structures evolve over time. Such evolution is observed, e.g., in the study of "classical" algorithmic problems such as *Minimum Spanning Tree* and *Branch and Bound*; see, e.g., [36]. What is certain to be more exciting to the reader, though, are the "modern" topics where one encounters graphs with evolving structure, such as *social networks* and *inter-networks* (e.g., *IoT*, the *Internet of Things*).

For "classical" topics, such as the two we have mentioned, the material in Chapters 12 and 13 will provide the reader with the background necessary to deal with graph evolution. Indeed, evolution emerges as an inevitable concomitant of the algorithmics that is superimposed upon the traditional structures of graph theory. The challenge of evolution is to assimilate new algorithmics, not new mathematics.

In contrast, the "modern" topics we have mentioned do require the assimilation of new mathematics. Dealing successfully with the algorithmic issues that arise with social networks and inter-networks requires the reader to understand the structures of the evolving graph-oriented systems and how evolution changes these structures. Among the interesting (and valuable) mathematical questions that one can pose is: If you are a new vertex "applying" to join an evolving network, which vertex in the network is the best one to connect to, in order to best facilitate your interactions or influence within the community. The latter topic leads, e.g., to the study of *power-law* networks.

Explanatory note.

An evolving network *obeys a power law* if there exists a real number $\gamma > 0$ with the following property. For all network vertices having sufficiently large vertex-degrees, the fraction of vertices that have degree k is proportional to $k^{-\gamma}$.

Little of the abstract work on power-law networks would likely be studied in depth in an early course; indeed, the structure of these networks is not yet well understood even in advanced settings. Attempts to understand power laws with rigor have led to a number of competing, rather sophisticated, abstract models—see, e.g., [Aiello, Chung, Lu, 2000], [Barabási, Albert, 1999], [Bollobas, 1985], and [Chen, Chang, Govindan, Jamin, Shenker, Willinger, 2002]—and numerous studies have attempted to understand when the abstract models reflect reality more or less faithfully—see, e.g., [Bu, Towsley, 2002], [Faloutsos, Faloutsos, Faloutsos, 1999], [Jaiswal, Rosenberg, Towsley, 2004], [Tangmunarunkit, Govindan, Jamin, Shenker, Willinger, 2002], and [Zegura, Calvert, Donahoo, 1997].

Explanatory note (cont'd).

We have just cited several sources that deal with the evolving formal notion of power-law graph. This apparent overkill—wouldn't one citation suffice?— is intentional. As new "real-life" concepts emerge which need mathematical modeling—in this case, as social media become an object of increasing study— there is inevitably a period in which competing mathematical models are proposed. All of the models that have emerged thus far capture the central feature of power laws. But every mathematical model embodies "side" features that go beyond the targeted central feature. We must allow multiple models to coexist until we have the opportunity to compare all of the models' side features. Only after this vetting period will the scientific community converge on a single mathematical model of a "real-life" concept such as "power-law graph".

G.3 Hypergraphs

Topic-specific references

F. Amato, F. di Lillo, V. Moscato, A. Picariello, G. Sperli (2017): Influence analysis in online social networks using hypergraphs. *IEEE Int'l Conf. on Information Reuse and Integration.*

D. Liu, N. Blenn, P. Van Mieghem (2010): Modeling social networks with overlapping communities using hypergraphs and their line graphs. Report arXiv:1012.2774, Dec. 2010, `http://cds.cern.ch/record/1314107`.

L. Lovász (1973): Coverings and coloring of hypergraphs. *4th Southeast Conf. on Combinatorics, Graph Theory, and Computing*, 3–12.

A.L. Rosenberg (1985): A hypergraph model for fault-tolerant VLSI processor arrays. *IEEE Trans. Comput. C-34*, 578–584.

A.L. Rosenberg (1989): Interval hypergraphs. In *Graphs and Algorithms* (R.B. Richter, ed.) *Contemporary Mathematics 89*, Amer. Math. Soc., 27–44.

A large variety of modern computing-related topics benefit from the structure inherent in graph-theoretic models but do not comfortably conform to the *binary* relationships imposed by graphs' having *two* vertices per edge. We now discuss a model that retains the general structure of graph-theoretic models while it abandons the *binary* constraint on edge-membership—this is the *hypergraph* model. A hypergraph has vertices that play exactly the same role as with graphs, but in place of a graph's two-element edges, a hypergraph has *hyperedges*, each being a set of vertices whose size is not restricted to 2. A rather general treatment of hypergraphs can be found in the comprehensive graph-theory text [14]; a specialized article that focuses on hypergraph-related versions of several topics from Chapters 12 and 13, including vertex coloring, is [Lovász, 1973].

Because of their inherent conceptual complexity, hypergraphs as graph-theoretic objects are usually relegated to advanced courses. However, the literature contains many studies of hypergraphs that are "fine-tuned" for specific computing-related applications. Several of these studies should be accessible without extensive mathematical background. Sample computing-related applications that benefit from hypergraph-oriented models include the following.

- Bus-connected parallel communication has been part of digital-computer design since its earliest days. The informal picture of such a system is that there are communication channels which multiple agents can retrieve messages from and post messages to. In hypergraph-oriented terms: the vertices/communicating agents aggregate into groups/hyperedges. Each group's agents share "read/write" access to a specific channel. A specialized genre of hypergraph which was invented to study the described scenario is the *interval hypergraph* model developed in [Rosenberg, 1989].

- Modern electronic circuits are implemented using integrated circuit technology, specifically, *VLSI: Very Large Scale Integrated circuitry*; the often-informal book [82] provides an introduction to VLSI for the non-specialist. VLSI technologies are often *voltage*-driven, rather than *current*-driven. Accordingly, much of the attention when designing VLSI circuits centers on the coordination of equipotential points in an electrical network, rather than on point-to-point transmission of signals. Hypergraphs are tailor-made for such technologies.
  A crucially important issue that arises because of the design strengths and weaknesses of VLSI technology is *fault tolerance*—how to cope with the inevitable faulty transistors in massive VLSI systems. A variety of quite accessible mathematical ideas can provide provocative ideas about this important topic; see, e.g., [Rosenberg, 1985].

- Social networks have become so prevalent in society that no one will be surprised to learn that many approaches to modeling the networks' interconnectivity have been studied. In Section G.2, we discussed an interconnectivity model based on evolving graphs and clustering within such graphs. More recently, hypergraph-based models have also been proposed; see, e.g., [Amato, di Lillo, Moscato, Picariello, Sperli, 2017] and [Liu, Blenn, Van Mieghem, 2010].

G.4 Multigraphs and the Chinese Postman Problem

A *multigraph* is a graph-like structure that admits a *multiset* of edges, rather than a set. In detail, a multigraph $\mathscr{H}$ has a set $\mathscr{N}_{\mathscr{H}}$ of vertices and a *multiset* of edges, each edge being a two-element set of vertices. Viewed differently, but equivalently, a pair of vertices in a multigraph can be connected by more than one edge. Figure G.1 depicts a ten-vertex multigraph $\mathscr{H}$. Ten pairs of $\mathscr{G}$'s vertices are connected by single edges; three pairs of $\mathscr{G}$'s vertices are connected by multiedges, each of which (coincidentally) consists of two edges.

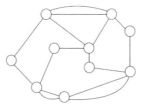

Fig. G.1 A 10-vertex multigraph $\mathscr{G}$ with three multiedges

We noted earlier that multigraphs were usually an algorithmic tool rather than a modeling tool. We now present a sample algorithmic use of multigraphs which allows us to extend the concept of Eulerian tour to a family of graphs that lie outside the domain delimited by Proposition 13.9. Recall that the proposition asserts that a *graph* admits a tour that crosses each edge precisely once if, and only if, each of its vertices has even degree.

From an applied vantage point, *pragmatism is often preferable to purity!* Imagine that one does not have access to an efficient solution for the "pure" version of a practically important problem, because some essential precondition is violated. One is often willing to "bend the rules" if such "distortion" will give one access to an efficient solution to the resulting "not-quite-pure" version of the problem.

The preceding highlighted dictum can be illustrated via the Euler-tour problem.

Proposition G.1 *If a* connected *graph $\mathscr{G}$ has an* even number of odd-degree vertices, *then we can transform $\mathscr{G}$ into a* multigraph *$\mathscr{G}^+$ that admits a tour which crosses every edge precisely once.*

In fact, we can solve this "distorted" version of the Euler-tour problem in a manner that is algorithmically efficient in the following senses.

- *The newly added edges preserve the neighbor relation of $\mathscr{G}$.*

  As we add the edges that convert $\mathscr{G}$'s edge-set to an edge-*multiset*, we never create new adjacencies: Every new edge that we add connects vertices that were already adjacent in $\mathscr{G}$.

- *The algorithm adds the fewest number of new edges possible in order to achieve the described tour.*

  Say that the multigraph $\mathscr{G}'$ is produced by adding edges to $\mathscr{G}$. If $\mathscr{G}'$ has fewer edges than $\mathscr{G}^+$, then $\mathscr{G}'$ does *not* admit a tour that crosses each edge precisely once.

- *The algorithm is computationally efficient: It operates in time polynomial in the size of the graph $\mathscr{G}$.*

Our "distorted" version of the Euler-tour problem is actually a well-known combinatorial optimization problem which is known either as the *Route Inspection Problem* or as the *Chinese Postman Problem*—the latter name in honor of the Chinese mathematician Kwan Mei-Ko who invented the problem.[1] These colorful names arise from stories about a person—either a route inspector or a postal employee—who must traverse all of the roads in a village and who wants to avoid extraneous road traversals.

We illustrate the problem and an algorithm that solves it.

Figure G.2 presents a 10-vertex input $\mathscr{G}$ to our "distorted" Euler-tour problem. Note that $\mathscr{G}$ is not Eulerian because it has four odd-degree vertices—happily, an

Fig. G.2 A 10-vertex graph which is not Eulerian

even number! $\mathscr{G}$'s odd-degree vertices are highlighted in Fig. G.3.

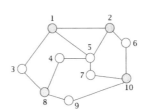

Fig. G.3 $\mathscr{G}$ with its four odd-degree vertices highlighted by shading

We sketch a process that adds a minimally many edges to $\mathscr{G}$ while converting $\mathscr{G}$ into a multigraph that admits a "distorted" Euler tour.

Procedure Edge Augmentation($\mathscr{G}$)
/*Construct a shape-inspired pairing function (PF) $\mathscr{A}$*/
begin

Step 1.   Record, for each pair of vertices $u, v \in \mathcal{N}_{\mathscr{G}}$, the length $\ell(u,v)$ of a *shortest path* between u and v. (Recall that $\mathscr{G}$ is connected.)

[1] M.-K. Kwan (1960): Graphic programming using odd or even points. *Acta Mathematica Sinica, 10* (in Chinese), 263–266. Translated into English in *Chinese Mathematics 1* (1962) 273–277.

Computing each length $\ell(u,v)$, together with a length-$\ell(u,v)$ path between u and v, can be accomplished in low-degree polynomial time; see [36].

Step 2.   Say that $\mathcal{G}$ has m odd-degree vertices $w_1,\dots,w_m$ ($m = 4$ in our illustration). Create a copy of the m-clique $\mathcal{K}_m$ on these m vertices.

Step 3.   Label each edge $\{u,v\}$ of the clique $\mathcal{K}_m$ with the path-length $\ell(u,v)$.

Step 4.   Compute a perfect matching of minimal weight in the edge-weighted clique $\mathcal{K}_m$.

Computing minimal-weight perfect matchings can be accomplished in low-degree polynomial time; see [36].

Figure G.4 illustrates the results of this process on our sample graph $\mathcal{G}$. *Note*: The double-edges that we add make the degrees of all odd-degree vertices even. The double-edge at the top of $\mathcal{G}$ in the drawing makes the degrees of both top vertices even; the length-2 double-edge at the bottom of $\mathcal{G}$ in the drawing makes the degrees of both bottom odd-degree vertices even.

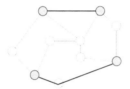

Fig. G.4 A minimum-weight perfect matching labeled by the shortest-distances $\ell(u,v)$

Step 5.   Replace each weighted edge produced by the process by an appropriate number of multiedges.

Figure G.1 performs this last step for the weighted graph of Fig. G.4.

end Edge Augmentation

Our augmented multigraph $\mathcal{G}^+$ now has all even vertex-degrees. We can, therefore, adapt the Euler-tour algorithm from the proof of Proposition 13.9(a) to produce a tour of $\mathcal{G}^+$ that crosses each multiedge precisely once. The required adaptation simply views distinct multiedges as distinct edges for the purposes of the mandate to "leave a vertex by a different edge than you entered on".

The importance of the notion of multigraph in this story is that it gave us a perspicuous bridge between the strongly algorithmic world of edge-weighted graphs and the purely graph-theoretic world that underlies Proposition 13.9. This can be a valuable bridge when computing with graph-theoretic models.

Appendix H
Solution Sketches for ⊕⊕ Exercises

- ## 2.2. Meeting people at a party

  To prove. *Some two attendees shake the same number of hands.*

  The pigeonhole principle is our main tool here. The number of people that each attendee *does not know* belongs to the set $\{0, 1, \ldots, 2n - 2\}$, because each person knows him/herself and his/her partner.

  Because there are $2n$ handshakers (the pigeons) and $2n - 1$ hands to shake (the boxes), some two shakers must shake the same numbers of hands. □

- ## 2.4. Bi-colored necklaces in tubes

  A *bicolored necklace* is composed of $2n = 2(a + b)$ jewels: $2a$ black jewels and $2b$ white jewels. Let us be given such a necklace within a solid tube of the sort depicted in Fig. 2.10, which contains precisely half of the jewels, i.e., n jewels.

  To prove. *There is a way to position the tube so that both inside the tube and outside the tube, there are: equally many jewels, equally many black jewels, and equally many white jewels.*

  We are going to slide the tube clockwise around the necklace, one jewel per step, and count the black jewels and the white jewels within the tube at each step. Of course, if there are currently equally many black jewels inside the tube and outside the tube, then, by simple arithmetic, we are done—we have the desired ideal position of the tube. Otherwise, the tube currently has either too many black jewels or too few. Say, with no loss of generality, that there are too many black jewels inside the tube, namely, $a + c$ black jewels for some $c > 0$; there is, therefore, a complementary number of white jewels, namely, $b - c$. At the end of the process, the tube must contain a black jewels and b white jewels—so the discrepancy c must have been reduced to 0. To see that this will eventually happen, we must see how the discrepancy c changes in a single step of the process.

  The effect of a single shift is to insert a jewel into the tube, at its advancing extremity, and to eliminate a jewel from the tube, at its trailing extremity. The discrepancy can be changed by this shift in precisely three ways.

© Springer Nature Switzerland AG 2020

A. L. Rosenberg, D. Trystram, *Understand Mathematics, Understand Computing*,

https://doi.org/10.1007/978-3-030-58376-7

– If the inserted jewel and the eliminated jewel are of the same color, then the discrepancy c is unchanged by this shift.

– If the inserted jewel is black and the eliminated jewel is white, then the discrepancy c is *increased to* $c + 1$ by this shift.

– If the inserted jewel is white and the eliminated jewel is black, then the discrepancy c is *decreased to* $c - 1$ by this shift.

By the time n shifts have been performed, the discrepancy c will have changed to $-c$, because the tube will be in an antipodal position upon the necklace.

Summing up: Over the course of n shifts, the discrepancy c will have changed to $-c$. The discrepancy changes by ± 1 at each shift. Therefore, there must be some shift among the n when the discrepancy is 0. □

- **2.6. Using *geometric* intuition to sum inverse powers of** 4

We seek a proof of Proposition 2.5 in which the infinite summation

$$S \;=\; \frac{1}{4} \,+\, \frac{1}{4^2} \,+\cdots+\, \frac{1}{4^k} \,+\cdots$$

is analyzed—and ultimately solved—geometrically. The key is to discover how to represent the process of generating successive inverse powers of 4. The nested similar triangles in Fig. H.1 unlock the secret to this representation.

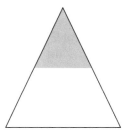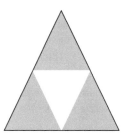

Fig. H.1 Two copies of an isosceles triangle T. The left-hand copy has been partitioned into a top sub-triangle , call it T', whose height is half of T's. The right-hand copy is obtained by adding two copies of T' in the lower half of the left-hand copy. We now have four similar copies of T' within T, which perfectly partition the area of triangle T

– **To prove**. *The four sub-triangles are similar to one another.*

Duplicate triangle T' twice, and put both copies at the bottom of triangle T, in the manner depicted in Fig. H.1(right). The reversed, white, sub-triangle in the middle of the figure has the same sides as the two duplicates of T', and its base is also half of T's. Therefore, it also is similar to triangle T'.

–   **To prove**. *The four sub-triangles are similar to triangle T.*

Sub-triangle T' is similar to triangle T because it is obtained by bisecting both the base and the sides of T.

–   **To prove**. *The common area of each of the four sub-triangles is $1/4$ the area of T.*

This follows from the two previous observations: The four sub-triangles have the same area, and they exactly cover triangle T.

• **To do**. *Assemble the preceding facts into an evaluation of the sum S.*

The sum S is obtained by recursively partitioning the small upper sub-triangles. The process splits the original triangle into layers that each contain three sub-triangles; see Fig. H.2. As the three sub-triangles are identical, the area of the

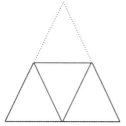

Fig. H.2 First bottom layer of the partitioning

central reversed triangle is $1/3$ of the area of the layer.

The final solution is depicted in Fig. H.3, where the recursive process is applied "all the way". Summarizing the graphical construction. If the area of the original

 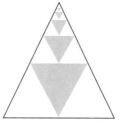

Fig. H.3 The recursive decomposition of triangle T

triangle T is 1, then the area of the grey sub-triangle of Fig. H.3(left) is $1/4$, since it is one of four identical triangles. Then, because each grey triangle in Fig. H.3(right) occupies one-third of the area of its layer, the aggregate area of the grey sub-triangles is $1/3$ of the area of T. One can now apply Fubini's principle to evaluate the sum $\sum_{k \geq 1} 4^{-k}$.

- **3.8. More connections between strings and functions**

  To do. *Craft an argument that predicts the number of permutations, based on the size of set S.*

  – As you create a new string of numbers, in how many ways can you choose *the first number? the second number?* ...

    There are n ways to choose the first element of the permutation. Independently of which first element has been chosen, there remain $n-1$ ways to choose the second element. Independently of which elements are chosen as the first two, there remain $n-2$ ways to choose the third element. And so on ...

  – Based on your answers for the first and second and third numbers of the new string, in how many ways can you choose *the first two numbers—i.e., the first pair of numbers? the next pair of numbers?* ...

    The main message here is that *independent* choices *multiply*. This means that the n choices for element #1 and the $n-1$ choices for element #2 lead to

    $$n(n-1)$$

    ways to choose the first pair of numbers. The persistence of independence of successive selections means that the multiplicative rule continues to hold: There are $n(n-1)(n-2)$ ways to choose the first triple of elements,

    $$n(n-1)(n-2)(n-3)$$

    ways to choose the first quadruple of elements, ...

  To do. *Strengthen your argument by listing all permutations of $S' = \{1,2,3,4,5\}$.*
  The challenge is to convert your mathematical reasoning into a systematic method of enumeration. Here is a natural recursive way of doing this. (i) Fix the first number (among the five possibilities). (ii) Recursively, enumerate, for each copy of the first number, all permutations of the four remaining numbers.

  To do. *Extrapolate from your argument to determine the number of permutations of the set $S'' = \{1,2,3,\ldots,n\}$, as a function of n.*
  $Fact(n) = n \times Fact(n-1) = n \times (n-1) \times \cdots \times 2 \times 1.$

- **4.5. The rationals ($\mathbb{Q}$) and the integers ($\mathbb{N}$) are equinumerous**

  To prove. *Provide a* detailed *proof of Proposition 4.6:* There are "equally many" integers as there are rationals.

  The relevant fact here is that $\mathbb{Q}$ is (isomorphic to) a proper subset of $\mathbb{N} \times \mathbb{N}$. One bijection that witnesses this fact is defined as follows. Map each rational $r \in \mathbb{Q}$ to the unique pair $\langle p, q \rangle \in \mathbb{N} \times \mathbb{N}$ such that:

$$[r = p/q] \quad \text{and} \quad [p \text{ and } q \text{ share no common factor}]$$

Beware: The first of these conditions is satisfied by infinitely many pairs of integers. The second condition is needed to end up with a unique pair, hence with the desired bijection.

Once you have a bijection, you can argue based on definitions.

- **6.4. How to evaluate** $S_1(n) = \sum_{i=1}^{n} i$, **using a "machine" that evaluates** $S_2(n) = \sum_{i=1}^{n} i^2$

To do. *Show how to use the $S_2(n)$-machine in order to compute $S_1(n)$.*

You must exploit how the expressions for $S_1(n)$ and $S_2(n)$ relate to one another. (You need just a little algebra here.) Then you must exhibit how to extract the relation from a "machine" which gives out values in a "black-box" manner.

You can develop the solution strategy as follows: Write the summation $S_2(n+1)$ in two ways. First, isolate the *last* term of the summation; then, isolate the *first* term of the summation:

$$S_2(n+1) = \sum_{i=1}^{n} i^2 + (n+1)^2$$
$$= S_2(n) + (n+1)^2$$
$$S_2(n+1) = 1 + \sum_{i=2}^{n+1} i^2$$
$$= 1 + \sum_{i=1}^{n} (i+1)^2$$
$$= 1 + \sum_{i=1}^{n} (i^2 + 2i + 1)$$
$$= 1 + S_2(n) + 2S_1(n) + n$$

Equating the two final derived expressions for $S_2(n+1)$ and simplifying, we find that

$$S_2(n) + (n+1)^2 = 1 + S_2(n) + 2S_1(n) + n$$
$$(n+1)^2 = 1 + 2S_1(n) + n$$
$$2S_1(n) = (n+1)^2 - (n+1)$$
$$S_1(n) = \frac{1}{2} n(n+1)$$

- **6.6. Evaluating a geometric summation pictorially**

To do. *Use the following figure to evaluate the following summation.*

$$S^{(b)}(n) = \sum_{i=0}^{n} b^i$$

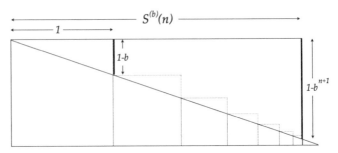

We use Thales's Proposition 6.6 with the two upper similar triangles in the figure.

$$\frac{S^{(b)}(n)}{1-b^{n+1}} = \frac{1}{1-b}$$

$$S^{(b)}(n) = \frac{1-b^{n+1}}{1-b}$$

- **6.7. A direct proof that the harmonic series diverges**
  To do. *Develop a proof of the divergence of the harmonic series*

  $$S^{(H)} = \sum_{k=1}^{\infty} \frac{1}{k}$$

  1. Partition $S^{(H)}$'s terms into groups whose sizes are successive powers of 2
  2. Develop an argument based on the sums within the groups.

  The partitioning step operates as follows:

  $$S^{(H)} = 1 + \frac{1}{2} + \left(\frac{1}{3} + \frac{1}{4}\right) + \left(\frac{1}{5} + \frac{1}{6} + \frac{1}{7} + \frac{1}{8}\right) + \left(\frac{1}{9} + \frac{1}{10} + \frac{1}{11} + \frac{1}{12} + \frac{1}{13} + \frac{1}{14} + \frac{1}{15} + \frac{1}{16}\right) + \cdots$$

  In detail, each group, say the ith group G_i (for $i \geq 1$), is computed from the i numbers

  $$\frac{1}{2^{(i-1)}+1}, \frac{1}{2^{(i-1)}+2}, \ldots, \frac{1}{2^i}$$

  Grouping the terms in the earlier manner, the sum within each group G_i exceeds $1/2$:

  $$\frac{1}{3} + \frac{1}{4} > 2 \cdot \frac{1}{4} = \frac{1}{2}$$

  $$\frac{1}{5} + \frac{1}{6} + \frac{1}{7} + \frac{1}{8} > 4 \cdot \frac{1}{8} = \frac{1}{2}$$

  $$\frac{1}{9} + \frac{1}{10} + \frac{1}{11} + \frac{1}{12} + \frac{1}{13} + \frac{1}{14} + \frac{1}{15} + \frac{1}{16} > 8 \cdot \frac{1}{16} = \frac{1}{2}$$

  $$\vdots$$

Therefore,

$$S^{(H)} > 1 + \left(\frac{1}{2} + \frac{1}{2}\right) + \left(\frac{1}{2} + \frac{1}{2}\right) + \cdots$$
$$= 1 + 1 + 1 + 1 + \cdots$$

This means that the Harmonic sum $S^{(H)}$ is greater than every sum of 1s and hence is infinite.

To do. *Propose an alternative analysis of $S^{(H)}$ which uses groupings whose sizes are multiples of 3*

The aim of this problem is to ensure that you have understood which features of our power-of-2 groupings are inherent to the argument and which are just for convenience.

We argue now in a somewhat different way than with our power-of-2 groupings. We now propose the terms S_i defined by

$$S_i = \left(\frac{1}{3i-1} + \frac{1}{3i} + \frac{1}{3i+1}\right) > 3 \cdot \frac{1}{i}$$

We then have

$$S^{(H)} = 1 + S_1 + S_2 + \cdots + S_i + \cdots$$
$$> 1 + 3 \cdot \frac{1}{3} + 3 \cdot \frac{1}{6} + \cdots + 3 \cdot \frac{1}{3i} + \cdots \tag{H.1}$$

These summations lead to the following absurdity.

If $S^{(H)}$ were finite, then, by Eq. H.1, we would have $S^{(H)} > 1 + S^{(H)}$, which is obviously impossible.

- **6.8. Summations, and summations of summations**

  - a. **To prove**:

$$\widehat{\Theta}_n \overset{\text{def}}{=} \sum_{k=1}^{n} \Delta_k = \Delta_1 + \Delta_2 + \cdots + \Delta_n = \frac{1}{3}\Delta_n \cdot (n+2)$$

The result is obtained by replacing each triangular number Δ_k in the summation by its explicit expression, and then performing the indicated algebraic manipulation. We thereby find:

$$\widehat{\Theta}_n = \sum_{k=1}^{n} \frac{k(k+1)}{2}$$
$$= \frac{1}{2}\sum_{k=1}^{n} k^2 + \frac{1}{2}\sum_{k=1}^{n} k$$
$$= \frac{1}{2}\left(\frac{n(2n+1)(n+1)}{6} + \frac{n(n+1)}{2}\right)$$

$$= \frac{1}{2} \frac{n(n+1)}{2} \cdot \left(\frac{2n+1}{3} + 1 \right)$$

$$= \frac{1}{2} \Delta_n \cdot \frac{(2n+4)}{3}$$

$$= \Delta_n \cdot \frac{(n+2)}{3}$$

- b. **To prove**. *Verify the following identities involving Δ_n.*
  - · i. For all $n \in \mathbb{N}^+$: $\Delta_n + \Delta_{n-1} = n^2$

    You can do the necessary algebraic manipulation, but you can alternatively use a pictorial proof, which is a slight modification of our pictorial evaluation of Δ_n.

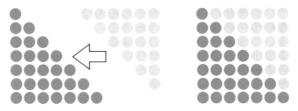

Fig. H.4 Summing Δ_n and Δ_{n-1} to obtain the $n \times n$ square

  - · ii. For all $n \in \mathbb{N}^+$: $\Delta_n^2 - \Delta_{n-1}^2 = n^3$

    Write Δ_n as $n + \Delta_{n-1}$:

$$\Delta_n^2 - \Delta_{n-1}^2 = (n + \Delta_{n-1})^2 - \Delta_{n-1}^2$$

$$= n^2 + 2n \cdot \Delta_{n-1} + \Delta_{n-1}^2 - \Delta_{n-1}^2$$

$$= n^2 + 2n \cdot \frac{n(n-1)}{2}$$

$$= n^2 + n^3 - n^2$$

- c. **To do**. *Derive a closed-form expression for the sum*

$$\widehat{\Theta}_n + \widehat{\Theta}_{n-1}$$

This question is more open-ended than the previous ones since we don't know *what kind of expression* to look for.

When encountering such open-ended problems, you can never go wrong by starting out with the original definitions.

In this case, we express the tetrahedral number $\widehat{\Theta}_n$ as the sum of triangular numbers.

$$\widehat{\Theta}_n + \widehat{\Theta}_{n-1} = \sum_{k=1}^{n} \Delta_k + \sum_{k=1}^{n-1} \Delta_k$$

$$= \Delta_1 + \sum_{k=2}^{n} (\Delta_k + \Delta_{k-1})$$

We now use the expression we derived for the sum of two consecutive Δ_k's:

$$\widehat{\Theta}_n + \widehat{\Theta}_{n-1} = 1 + \sum_{k=2}^{n} k^2 = \sum_{k=1}^{n} k^2$$

We can finally replace this summation by its closed-form sum.

$$\sum_{k=1}^{n} k^2 = \frac{n(n+1)(2n+1)}{6}$$

- **8.2. Discovering fractal-like structure in Pascal's Triangle**
  To prove. *The fractal-like structure that we describe now really occurs.*

We show in Chapter 8 an interesting divisibility-related property of the binomial coefficients within Pascal's Triangle. All of the internal elements of the rows corresponding to a prime p are proportional to p (see Fig. 8.3). This means that all these elements are equal to zero modulo this prime (where each of the triangle's entries is taken modulo this prime). For instance, Fig. H.5 illustrates what happens for $m = 5$. We observe at the top of the figure a small Pascal's Triangle of height 5 and just below, the fifth row with all internal elements at zero bordered by a "1" at both left and right. The next five rows are composed of one reversed triangle whose coefficients are all zeros at the center (in grey) borded by two Pascal's Triangles at the left and the right that are the same as the top small triangle. Thus, the line below is similar to the fifth row. It is composed of a 1 followed by four zeros then, two 1s, then another 4 zeros and a last 1.

Fig. H.5 The first levels of Pascal's Triangle modulo 5

An extended instance of Pascal's Triangle modulo the prime $m = 5$ is illustrated in Fig. H.6. The first m levels of the triangle replicate endlessly, with periodic inverted $(m-1)$-level triangles whose entries are all 0. (In the figure, the inverted triangle of 0s is depicted in grey.) One observes that the original triangle becomes a fractal-like repetitive structure whose pattern of repetitions is dictated by the parameter m.

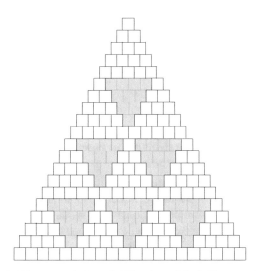

Fig. H.6 The reproducible patterns in Pascal's Triangle modulo 5. The grey-shaded inverted triangles have entry 0 in every position

- **8.4. Divisibility among integers: via the Fundamental Theorem of Arithmetic**

  – **The sieve of Eratosthenes and its implications**
    We formulate the sieve as a regimen for labeling integers with their prime factors. For simplicity, we use the label λ_p to identify integers which are multiples of prime p.

    The multiples of 2, 3, and 5 are labeled in the following table:

1	2	3	4	5	6	7	8	9	10	11	12	13
	λ_2		λ_2		λ_2		λ_2		λ_2		λ_2	
		λ_3			λ_3			λ_3			λ_3	
				λ_5					λ_5			

$\cdots$

To prove.
3. This problem calls for "second-order" insights from the sieve structure.

   Every product of four consecutive integers, $(k+1)\cdot(k+2)\cdot(k+3)\cdot(k+4)$ is divisible by FACT(4).

This result follows by a simple case analysis: Any four consecutive integers contain exactly two even numbers; one of these even numbers is divisible by 4; therefore, the product of the four numbers is divisible by $2 \times 4 = 8$. Finally, every three consecutive numbers contains a multiple of 3. We thus observe divisibility by 3 and by 8, hence (by the Fundamental Theorem of Arithmetic) by $24 = \text{FACT}(4)$.

4. **To prove**. *For every integer n, every product of n consecutive integers,*

$$f(n,k) \stackrel{\text{def}}{=} (k+1) \times (k+2) \times \cdots \times (k+n)$$

is divisible by FACT*(n)*.

We begin by rewriting the product in the following form

$$f(n,k) = \frac{(n+k)!}{k!}$$

We note next that, by definition,

$$\frac{f(n,k)}{n!} = \binom{n+k}{k}$$

Thus, the ratio is a binomial coefficient.

We then make the following observations:

· Being the product of n integers, $f(n,k)$ is an integer.

· By Proposition 9.4, every binomial coefficient is an integer.

The preceding two observations prove that the product $f(n,k)$ is divisible by $n!$. □

- **8.5. The "density" of divisible pairs of numbers**

To prove. *The following assertion is true for every positive integer n.*

If you remove any $n+1$ integers from the set $S = \{1,2,\ldots,2n\}$, then the set of removed integers contains at least one pair p and $q > p$ such that p divides q.

Organize the $2n$ numbers of set S into subsets, based on the largest power of 2 that divides them. Set 0 consists of all odd elements from S; set 1 consists of all numbers that are $2\times$ an odd number; set 2 consists of all numbers that are $4\times$ an odd number; and so on.

When $n = 7$, for instance, set $S = \{1,2,\ldots,14\}$ is partitioned into four subsets:

Subset 0: $\{1,3,5,7,9,11,13\}$
Subset 1: $\{2,6,10,14\}$
Subset 2: $\{4,12\}$
Subset 3: $\{8\}$

Say that each subset-k element $2^k m$ of S is *associated with* its odd divisor m.

Now, since set S consists of the first $2n$ integers, it contains n *odd* integers. Since our challenge begins by removing $n+1$ elements from S, the Pigeonhole Principle assures us that some two of the removed integers are *associated with* the same odd number m. Stated differently: some removed integer has the form $2^{k_1} \times m$ while another has the form $2^{k_2} \times m$.

The smaller of these removed integers divides the larger one.   □

- **8.8. The set $\mathbb{Q}$ of rational numbers is countable**

  To prove Proposition 4.6: $|\mathbb{Q}| = |\mathbb{N}|$

  Hint. The key here is to employ the injections whose existence is guaranteed by Proposition 8.7 to help with this problem. Use the injections in conjunction with the definition of "$|A| \leq |B|$".   □

- **9.4. Karatsuba multiplication**

  1. **To prove**. *The conventional multiplication algorithm uses $\Theta(n^2)$ multiplications to compute the product of two n-bit integers.*

     Hint. Denote by $f(n)$ the number of elementary bit-multiplications used when multiplying two n-bit numerals via the conventional algorithm. We show that $f(n) = \Theta(n^2)$ via the following recurrence.

     $$f(n) = \begin{cases} 4f(n/2) + g(n) & \text{for } n > 1 \\ 1 & \text{for } n = 1 \end{cases}$$

     where g is the cost of the extra arithmetic operations done while merging subproblems. The recurrence arises from the algorithm's two multiplications by powers of 2, plus its three additions. One argues easily that $g(n) = \Theta(n)$. Therefore, an invocation of the Master Theorem (Theorem 9.2) gives the desired solution.

  2. **To prove**. *Karatsuba's algorithm uses* asymptotically fewer than $\Theta(n^2)$ *multiplications to compute the product of two n-bit integers.*

     We need to exhibit a (real) number $\alpha < 2$ such that Karatsuba's algorithm computes the product of two n-bit integers using $\Theta(n^\alpha)$ multiplications.

     Karatsuba's algorithm computes $A \times B$, where A and B are n-bit integers, via the following recipe.

     $$A \times B = (A_1 \times B_1) \cdot 2^n + (A_1 \times B_2 + A_2 \times B_1) \cdot 2^{n/2} + (A_2 \times B_2) \quad \text{(H.2)}$$

     Define the auxiliary number

     $$C \stackrel{\text{def}}{=} (A_1 - A_2) \times (B_2 - B_1)$$

     We discover that

     $$A \times B = (A_1 \times B_1) \cdot 2^n + (A_2 \times B_2) + \big(C + (A_1 \times B_1) + (A_2 \times B_2)\big) \cdot 2^{n/2}$$

This recipe performs three multiplications of $n/2$-bit numerals at each level of the recursion. We incurred the one-time cost of performing two additions and/or subtractions of n-bit numerals as we computed C; and we performed four other additions. We therefore have

$$f(n) = \begin{cases} 3f(n/2) + \Theta(n) & \text{for } n > 1 \\ 1 & \text{for } n = 1 \end{cases}$$

An invocation of the Master Theorem now exposes that

$$f(n) \;=\; 3^{\log_2 n} \;=\; n^{\log_2 3}$$

Thus, $\log_2 3 \;<\; 2$ can play the role of the constant α we are seeking.

- **10.3. The Josephus Problem**

Figs. 10.1 and 10.2 pictorially describe the rounds of the Josephus game for $n = 12$ persons. The circled numbers in each figure are those that are removed at that round. Note that the first round of the process takes $\lceil n/2 \rceil$ steps, while

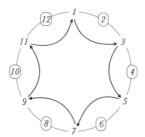

Fig. H.7 The first round of the Josephus erasure process for $n = 12$

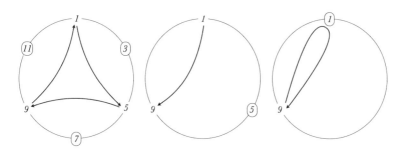

Fig. H.8 The final rounds of the Josephus erasure process when $n = 12$: $J(12) = 9$

each subsequent round takes "half" the number of its preceding round—"half" appears in quotes to emphasize the required rounding-up at each halving.

To do. *Prove the following conditions, which characterize $J(n)$ for particular values of n.*

Condition on n	Value of $J(n)$
(a) For all n	$J(n)$ is odd.
(b) n is even; i.e., $n = 2m$	$J(2m) = 2J(m) - 1$
(c) n is odd; i.e., $n = 2m+1$	$J(2m+1) = 2J(m)+1$
(d) $n = 2^m + k$, with $k < 2^m$	$J(2^m + k) = 2k + 1$

The proof proceeds as follows.

- **(a)** All the even numbers are erased in the first round, so only odd numbers remain thereafter.

- **(b)** This is a simple generalization of **(a)**. If n is even, then the first round leads to a circle containing half as many numbers.

- **(c)** This is the counterpart of **(b)** for the odd numbers.

- **(d)** Observing the successive values for small values of n exposes the relevant pattern. The values $J(n)$ are grouped by successive powers of 2^m. The rule within each group is to start at 1 and then proceed by increments of 2 until you reach the next group ($0 \leq k < 2^m$).

The formal proof that $J(2^m + k) = 2k + 1$ is by induction on n.

Base. If $n = 1$, then $m = 0$, $k = 0$ and $J(1) = 2^0 + 0 = 1$.

Inductive hypothesis. Suppose the formula holds for integers $n < 2^m + k$.

Inductive extension. We branch on k's parity:

· If k is even, then so also is $2^m + k$; therefore, we can write:

$$J(2^m + k) = 2J(2^{m-1} + (k/2)) - 1$$

By the inductive hypothesis, $J(2^{m-1} + (k/2)) = 2(k/2) + 1 = k + 1$. Thus, $J(2^m + k) = 2(k+1) - 1 = 2k + 1$.

· If k is odd, the proof is similar:

$$J(2^m + k) = 2J(2^{m-1} + \lfloor k/2 \rfloor) + 1 = 2\lfloor k/2 \rfloor + 1 = 2k + 1$$

To do. *Provide a closed-form expression for $J(n)$ in terms of n's base-2 numeral.*

Consider the binary numerals for $n = 2^m + k$ and k. We first note that, *numerically,*

$$n = 2^m b_m + 2^{m-1} b_{m-1} + \cdots + 2b_1 + b_0$$
$$k = 2^{m-1} b_{m-1} + \cdots + 2b_1 + b_0$$

where each bit $b_i \in \{0, 1\}$. Therefore, in terms of *base-2 numerals:*

$$(n)_2 = 1b_{m-1}\cdots b_1 b_0 \quad \text{(by definition, } b_m = 1)$$
$$(k)_2 = 0b_{m-1}\cdots b_1 b_0 \quad \text{(because } [n = 2^m + k] \text{ and } k < 2^m)$$

Therefore:

$$(J(n))_2 = b_{m-1}\cdots b_0 b_m.$$

This means that the value of $J(n)$ is obtained by a simple shift to the left of the binary representation of n, with a 1 at the rightmost position.

The pictorial interpretation of this coding is given in Fig. H.9 for $n = 43$.

Fig. H.9 Computing the survivor number $J(43) = 23$

- **10.5. An alternative to Horner's Rule**

  Estrin's method begins with a polynomial of degree d with real coefficients:

  $$P(x) = a_0 + a_1 x + a_2 x^2 + \cdots + a_{d-1} x^{d-1} + a_d x^d$$

  We introduce the recursively defined auxiliary expressions

  $$C_i^{(0)} = a_i + x a_{i+1}$$

  $$\vdots$$

  $$C_i^{(n)} = C_i^{(n-1)} + x^{2^n} C_{i+2^n}^{(n-1)}$$

  The method completes by expressing $P(x)$ in terms of the auxiliary $C_i^{(k)}$

  To do.

  1. *Write $P(x)$ using the auxiliary expressions C_i.*

     We simplify expressions by assuming that $d + 1$ is a power of 2. Thereby, we can illustrate the key manipulations without worrying about floors and ceilings.

     We begin by developing the method for a degree-($d = 7$) polynomial. This will provide a convenient touchstone as we extrapolate to arbitrary degrees.

     $$P_7(x) = a_0 + a_1 x + a_2 x^2 + a_3 x^3 + a_4 x^4 + a_5 x^5 + a_6 x^6 + a_7 x^7$$
     $$= a_0 + x\, a_1 + x^2 (a_2 + x\, a_3) + x^4 (a_4 + x\, a_5)$$

$$+ x^6 \left(a_6 + x\, a_7\right)$$
$$= \left(a_0 + x\, a_1\right) + x^2 \left(a_2 + x\, a_3\right)$$
$$+ x^4 \left((a_4 + x\, a_5) + x^2 \left(a_6 + x\, a_7\right)\right)$$
$$= C_0^{(0)} + x^2\, C_2^{(0)} + x^4 \left(\, C_4^{(0)} + x^2\, C_6^{(0)} \,\right)$$
$$= C_0^{(1)} + x^4\, C_4^{(1)}$$

The generalization for degree-d polynomials follows the same pattern:

$$P_d(x) = a_0 + a_1 x + a_2 x^2 + \cdots + a_{d-1} x^{d-1} + a_d x^d$$
$$= a_0 + x a_1 + x^2(a_2 + x a_3) + \cdots + x^{d-1}(a_{d-1} + x a_d)$$
$$= (a_0 + x a_1) + x^2(a_2 + x a_3) + \cdots + x^{d-3}(a_{d-3} + x a_{d-2})$$
$$+ x^{d-1}(a_{d-1} + x a_d)$$
$$= C_0^{(0)} + x^2\, C_2^{(0)} + x^4 \left(\, C_4^{(0)} + x^2\, C_6^{(0)} \,\right) + \cdots$$
$$+ x^{(d+1)-8} \left(\, C_{d-7}^{(0)} + x^2\, C_{d-5}^{(0)}\right) + x^{(d+1)-4} \left(\, C_{d-3}^{(0)} + x^2\, C_{d-1}^{(0)}\right)$$
$$= C_0^{(1)} + x^4\, C_4^{(1)} + x^8 \left(\, C_8^{(1)} + x^4\, C_{12}^{(1)} \,\right) + \cdots$$
$$+ x^{(d+1)-8} \cdot \left(\, C_{d-7}^{(1)} + x^4\, C_{d-3}^{(1)} \right)$$
$$= C_0^{(2)} + x^8\, C_8^{(2)} + \cdots + x^{(d+1)-4}\, C_{(d+1)-4}^{(2)}$$
$$\vdots$$
$$= C_0^{(\log(d+1)-2)} + x^{(d+1)/2}\, C_{(d+1)/2}^{(\log(d+1)-2)}$$

2. **To do**. *Determine how many additions and multiplications are required to evaluate $P(x)$.*

   As before, we focus on the case $d = 7$ to build intuition for a detailed analysis.

   $$P_7(x) \;=\; C_0^{(0)} + x^2\, C_2^{(0)} + x^4 \left(\, C_4^{(0)} + x^2\, C_6^{(0)} \,\right) \;=\; C_0^{(1)} + x^4\, C_4^{(1)}$$

   – We use two multiplications to compute x^2 and x^4, namely, $[x^2 = x \cdot x]$ and $[x^4 = x^2 \cdot x^2]$.

   – We use four multiplications by x for the 4 products $a_7 x$, $a_5 x$, $a_3 x$, and $a_1 x$. These are followed by four additions.

     Next, we use two multiplications by x^2 for the products $x^2 C_2^{(0)}$ and $x^2 C_6^{(0)}$. These are followed by two additions.

     Finally, we use one multiplication to compute $x^4 C_4^{(1)}$. This is followed by one addition.

   We have, thus, used a total of nine multiplications and seven additions.

   Generalization to degree-d polynomials: We just extend the previous analysis.

   – The first operations to perform are the multiplications by the successive powers of 2: We use $d - 1$ multiplications to compute $x^2, x^4, \cdots, x^{d+1}$.

Thus, $log(d-1)$ multiplications.

– We use $d/2$ multiplications by x for the $d/2$ products $a_{d-1}x, a_{d-3}x, \cdots, a_3x$, and a_1x. These are followed by $d/2$ additions.

Then, there are $d/4$ multiplications by x^2 for the products by the $C_i^{(0)}$, $d/8$ multiplications by the $C_i^{(1)}$, and so on.

We know well the summation of this simple geometric series $d/4 + d/8 \cdots + 1/2 + 1 = \Theta(d)$.

There are the same number of additions.

- ## 11.2. Counting replicated triangles

To do. *Determine how the progression of numbers of triangles contained within successive rank-k triangles T_k grows*

Let N_k denote the number of triangles that reside within a rank-k triangle T_k. We have already observed that $N_0 = 1$, $N_1 = 5$, and $N_2 = 13$.

1. *Compute N_4.*

   We begin with a single rank-4 triangle T_4.

   We next observe Fig. H.10, wherein there are three (partially overlapping) copies of the rank-3 triangle T_3. Next, we inspect Fig. H.11 in order to enu-

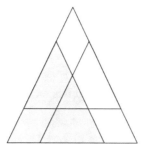

Fig. H.10 The rank-3 triangles T_3 contained within the rank-4 triangle T_4

merate the rank-2 triangles T_2 within T_3. Finally, we focus on the smallest triangle, T_1. Counting them row by row indicates that their number equals the sum of the first k integers (here $k = 4$). This sum is well known, and it is equal to the square of the rank: $4^2 = 16$.

Summing up all levels, we obtain: $N_4 = 1 + 3 + 7 + 16 = 27$.

2. *Develop a recurrence for N_k (i.e., for the general case).*

   The recurrence is obtained by generalizing the analysis of the case $k = 4$.

   For each triangle T_k, we have three intertwined triangles T_{k-1} (which contributes $3N_{k-1}$ to our count). Two of these triangles share a common part, which is a copy of triangle T_{k-2}; this count (which amounts to $-3N_{k-2}$) must

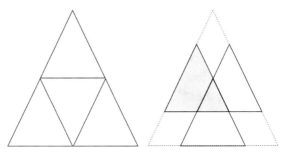

Fig. H.11 four triangles T_2(left) and three triangles T_2(right)

be removed from our count, so that we avoid double-counting. However, there is another part which is common to these three triangles. This part has lower rank and is a copy of T_{k-3} (accordingly we add $+N_{k-3}$ to our count). Finally, we must add the largest triangle, T_k (which augments our sum by $+1$). Finally, when k is even, there is an extra copy of triangle T_{k-2} which appears reversed in the middle of T_k (augments our sum by $+1$ in this case).

In summary, our tally is as follows:[1]

$$
N_k = \begin{cases}
3(N_{k-1} - N_{k-2}) + N_{k-3} + 2 & \text{when } k \text{ is even} \\
3(N_{k-1} - N_{k-2}) + N_{k-3} + 1 & \text{when } k \text{ is odd} \\
5 & \text{when } k = 2 \\
1 & \text{when } k = 1 \\
0 & \text{when } k = 0
\end{cases}
$$

Courageous readers can verify that this recurrence leads to the following explicit expression for N_k:

$$
N_k = \begin{cases}
\frac{1}{8} k(k+2)(2k+1) & \text{when } k \text{ is even} \\
\frac{1}{8}(k(k+2)(2k+1) - 1) & \text{when } k \text{ is odd}
\end{cases}
$$

- **11.6. Monge shuffle: mathematical party tricks**

We introduced two shuffles of a deck of $2n$ distinct cards. Both techniques cut the deck in the middle, creating two n-card decks, and then merge the n-card decks to again obtain a single $2n$-card deck. The versions differ in their merging techniques.

- The *riffle* shuffle alternates the "top" cards from the right-hand and left-hand n-card subdecks.

- The *sophisticated* Monge shuffle operates in stages. At each stage, we rearrange the $2n$-card deck directly, from the middle outward, via the following regimen:

[1] We added the artificial case $k = 0$ in order to simplify the expression.

- Card #1 of the original deck is placed in position $n + 1$ of the shuffled deck
- Card #2 of the original deck is placed in position n of the shuffled deck
- Card #3 of the original deck is placed in position $n + 2$ of the shuffled deck
- Card #4 of the original deck is placed in position $n - 1$ of the shuffled deck
  … and so on, until the original deck is empty.

To prove.

1. *Let p be a prime ($p > 2$), and let $n = \frac{1}{2}(p - 1)$. If we perform the riffle shuffle on a deck of $2n$ distinct cards, then after some number m of cut-then-merge steps, where m divides $p - 1$, the deck returns to its initial form.*

2. *For any positive integer n: If $4n + 1$ is prime, then performing $2n$ steps of the Monge shuffle on a deck of $2n$ distinct cards restores the deck to its original order.*

Toward the end of proving both propositions, let us trace the movements of the cards. For all ranks in the first n, the card at rank k in the deck is shifted to rank $2k$. The n cards of rank $k > n$ retain their relative orders but are moved to respective ranks $1, 3, \ldots, 2n - 1$; in other words, the card at rank k is moved to rank $2k \bmod (2n + 1)$. Continuing in this fashion, after i such rounds, a card at rank k moves to position $2^i \bmod (2n + 1)$.

When $2n + 1$ is prime, we can now invoke Fermat's Little Theorem (Theorem 8.3) to show that the number of rounds needed for each card to return to its initial position must divide $p - 1$. For a standard deck of $2n$ cards, the requisite number of rounds is $2n$. □

- **12.1. Vertex-degrees and the existence of paths**

  To prove. *If every vertex of graph $\mathcal{G}$ has degree $\geq d$, then $\mathcal{G}$ contains a* simple *path of length d.*

  Consider a simple path in $\mathcal{G}$ whose length is maximal. Let vertex v be an end-point of this path. Because the path is *maximal*, all neighbors of v belong to this path, because any unencumbered neighbor could be appended to the path— which would contradict the path's maximality. Because v's degree is no smaller than d, the path must contain at least $d + 1$ vertices. □

- **12.2. Vertex-degrees and their distributions in graphs**

  Prove Proposition 12.2. *For any graph $\mathcal{G}$, the number of $\mathcal{G}$'s vertices having odd vertex-degree is even.*

  Consider the following three sums

$$S^{(all)} \stackrel{\text{def}}{=} \text{sum of degrees of all of } \mathcal{G}\text{'s vertices}$$

$$S^{(even)} \stackrel{\text{def}}{=} \text{sum of degrees of all of } \mathcal{G}\text{'s even-degree vertices}$$

$$S^{(odd)} \stackrel{\text{def}}{=} \text{sum of degrees of all of } \mathcal{G}\text{'s odd-degree vertices}$$

Obviously, every vertex of $\mathcal{G}$ has either even or odd degree, so that

$$S^{(all)} = S^{(even)} + S^{(odd)} \tag{H.3}$$

Proposition 12.2 shows us that $S^{(all)}$ is even. Clearly, $S^{(even)}$ is even, being the sum of even numbers. Eq. (H.3) therefore tells us that $S^{(odd)}$, being the difference between two even numbers, is even. (A sum of k odd numbers is even only if k is even.)

- **12.8. A small graph isomorphism**

  To prove. *The order-4 hypercube $\mathscr{Q}_4$ is isomorphic to the 4×4 torus $\widetilde{\mathscr{M}}_{4,4}$.*

  We represent $\mathscr{Q}_4$ via its associated Gray code: Each vertex is a 4-bit string; each edge connects a vertex v to its four neighbors, which are identified by complementing precisely one of four bits.
  Figure H.12(left) depicts this coding of a vertex and its neighbors.

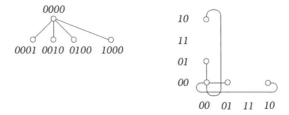

Fig. H.12 Coding schemes for the hypercube and torus-graph

The 4×4 torus $\widetilde{\mathscr{M}}_{4,4}$ is the Cartesian product of two 4-cycles. Each cycle's vertices can be represented as 2-bit strings, and its edges can be specified using the Gray code $00, 01, 11, 10$. Figure H.12(right) shows this coding and its correspondence with the hypercube $\mathscr{Q}_4$.
The complete encoding is obtained by concatenating the torus's two dimensions (horizontal and vertical); see Fig. H.13.

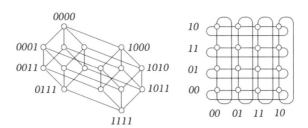

Fig. H.13 The complete coding scheme

The described encoding exposes a one-to-one correspondence between the vertices of $\mathcal{Q}_4$ and $\widetilde{\mathcal{M}}_{4,4}$, and it allows one to observe the one-to-one correspondence between the graphs' corresponding edges.

- **13.2. The aesthetics of de Bruijn networks**

   To do. *Identify, by coloring the edges of $\mathcal{D}_3$, how the network can be viewed as two directed trees which are "embracing" one another.*

   We describe the two "embracing" directed trees separately, but in a way that makes it clear that their arcs partition the arc-set of $\mathcal{D}_3$. Figure H.14 depicts both trees, drawn as though we were drawing $\mathcal{D}_3$. The tree in the left side of the figure is rooted at its leftmost vertex; its isomorphic counterpart in the right side of the figure is rooted at its rightmost vertex. The loops at vertices 0 and 7 have been removed.

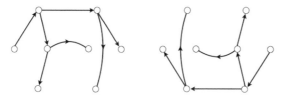

Fig. H.14 Two disjoint trees that cover $\mathcal{D}_3$ in an interleaved manner

As a strong hint at how this covering can be extended beyond order 3, we urge the reader to label all arcs in Fig. H.14 according to the scheme depicted in Fig. 13.14. You will thereby see how the complementary trees in Fig. H.14 are actually encodings of one another, obtained by flipping the bits in their labels from 0 to 1 and vice versa.

Actually drawing the relevant trees beyond order 3 is a daunting task, because drawing any network $\mathcal{D}_k$ for $k > 3$ is a daunting task. This fact reinforces our claim throughout the book of the importance of having many ways to look at the same thing: in this case, graphs and structured arrangements of bit-strings.

Interestingly, the general case of this problem, which is so cumbersome pictorially and so delicate combinatorially, is rather straightforward if one uses the algebraic presentation of de Bruijn networks which we introduced in Section F.2. In that section, we represented the vertices of $\mathcal{D}_n$ as integers:

$$\mathcal{N}_{\mathcal{D}_n} = \{0, 1, \ldots, 2^n - 1\}$$

The arcs of $\mathcal{D}_n$ were then created via the functions

$$F_0^{(n)}(v) = 2v \bmod 2^n \quad \text{and} \quad F_1^{(n)}(v) = 2v + 1 \bmod 2^n$$

Figure H.15 depicts this coding.

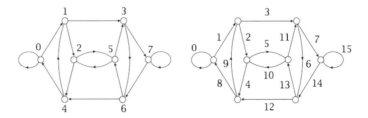

Fig. H.15 The correspondence between vertices and edges in $\mathscr{D}_n$; the case $n = 3$ is illustrated

Leaving details to the interested reader—the details are straightforward applications of the framework in Section F.2—both of the "embracing" trees have vertices $\mathcal{N}_{\mathscr{D}_n}$. One of the trees has arcs that arise from the *partial* functions

$$\underline{F}_0^{(n)}(v) = \quad\ 2v \bmod 2^n \ \text{ for } \ 0 < v \leq 2^{n-1} - 1$$
$$\underline{F}_1^{(n)}(v) = 2v + 1 \bmod 2^n \ \text{ for } \ 0 \leq v \leq 2^{n-1} - 1$$

and the other tree has arcs that arise from the *partial* functions

$$\overline{F}_0^{(n)}(v) = \quad\ 2v \bmod 2^n \ \text{ for } \ 2^{n-1} \leq v \leq 2^n - 1$$
$$\overline{F}_1^{(n)}(v) = 2v + 1 \bmod 2^n \ \text{ for } \ 2^{n-1} \leq v < 2^n - 1$$

The specifications of the four arc-generating functions, $\underline{F}_0^{(n)}$, $\underline{F}_1^{(n)}$, $\overline{F}_0^{(n)}$, and $\overline{F}_1^{(n)}$, encode some interesting lessons within the functions' domains of definition.

- The geometric relations "up" vs. "down" and "left" vs. "right" are elegantly encoded in the patterns of strong ($<$) and weak ($\leq$) relations.
- The strong relations in the specifications—which are the causes of the functions' partialness—remove the self-loops on vertices 1 and 15 in Fig. H.15. This ensures that the subgraphs of interest are, indeed, trees!

Verifying this general algebraic solution also gives some practice with modular arithmetic—a pleasurable exercise you might not have anticipated.

In order to strengthen our comparison of the combinatorial and algebraic approaches to the current problem, we provide the essence of the combinatorial solution for $d = 4$, and we invite the reader to flesh it out with explanation and annotation. Figure H.16 illustrates $\mathscr{D}_4$ as an unlabeled graph. The reader is invited to label the vertices with either their integer names or their names as 4-bit binary strings. (The self-loops are a good place to start.) Then, Fig. H.17 illustrates the "Escherian" trees that are the basis for the current problem. This provides another chance to supply the vertex-labeling.

- **13.4. Fundamental insights into outerplanarity in graphs**

  To prove. *The complete bipartite graph $\mathscr{K}_{3,2}$ is not outerplanar.*

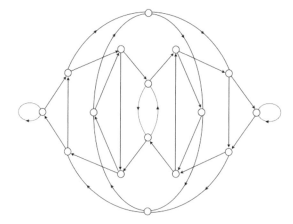

Fig. H.16 The order-4 de Bruijn network $\mathscr{D}_4$

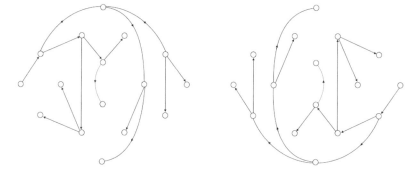

Fig. H.17 The "Escherian" trees that cover $\mathscr{D}_4$.

In contrast to some of our more abstract exercises, this problem requires detailed argument about a specific drawing of a single specific graph.

By definition of "outerplanar", the question at hand is whether one can distribute the two groups of $\mathscr{K}_{3,2}$'s vertices—depicted as three grey vertices and two white ones—along the circumference of a circle, in a manner so that no two of $\mathscr{K}_{3,2}$'s edges cross.

The demonstration that a noncrossing distribution is impossible begins with the observation that there are only two distinct ways to distribute $\mathscr{K}_{3,2}$'s vertices around a circle—in two nonintersecting blocks (left side of Fig. H.18) and interleaved (right side of Fig. H.18). In both cases, the remainder of the verification proceeds by connecting one of the white vertices to all of the grey vertices and noting that these edges block any attempt to connect the other white vertex to at least one of the grey vertices.

Fig. H.18 The two ways to distribute $\mathscr{K}_{3,2}$'s vertices around a circle

Appendix I

© Springer Nature Switzerland AG 2020
A. L. Rosenberg, D. Trystram, *Understand Mathematics, Understand Computing,*
https://doi.org/10.1007/978-3-030-58376-7

Lists of Symbols

Symbols Related to Sets and Their Algebras

Symbol	Meaning / Usage
"iff"	shorthand for "if and only if"
$=$	"equals" — used in all domains
	$1+2+3=6;\ \{a,b\}\ =\ \{b,a\}$
$\overset{\text{def}}{=}$	"equals, *by definition*" — used in all domains
	$S_n \overset{\text{def}}{=} 1+2+\cdots+n$: establishes notation
$\{\circ,\circ,\circ\}$	grouping symbols for listed elements of an *unordered* set
	$\{a,b\}\ =\ \{b,a\}$
$\{\circ\mid\circ\}$	the symbol "|" introduces a specification clause
	$\{x\mid P(x)\}$: the set of x *such that* proposition $P(x)$ holds
$\langle\circ,\circ,\circ\rangle$	grouping symbols for listed elements of an *ordered* set
	$\langle a,b\rangle\ \neq\ \langle b,a\rangle$
$\in$	set membership
	$s \in S$: s belongs to, or is a member of S
$\cap$	set intersection
	$S\cap T$: elements common to S and T
$\cup$	set union
	$S\cup T$: elements belonging to at least one of S and T
$\setminus$	set difference
	$S\setminus T$: elements belonging to S but not to T
$-$	set difference
	$S-T$: elements belonging to S but not to T
$\times$	direct product or Cartesian product
	$S\times T$: ordered pairs $\langle s,t\rangle$ where $s \in S$ and $t \in T$
$|S|$	set cardinality
	$|S|$: the number of elements of set S
$\emptyset$	the empty or null set
	$\emptyset$ is a subset of every set
$\mathscr{P}(S)$	power set
	$\mathscr{P}(S)$: the set of all subsets of set S

Symbols Related to Logic

Symbol	Meaning / Usage
$\wedge$	conjunction or logical product
	$x \wedge y$: this assertion is true iff both x and y are true
AND	x AND y: this assertion is true iff both x and y are true
$\vee$	disjunction or logical sum
	$x \vee y$: this assertion is true iff at least one of x and y is true
OR	x OR y: this assertion is true iff at least one of x and y is true
$\neg$	negation
	$\neg x$: the negation of logical variable x; true iff x is false
NOT	NOT x: the negation of logical variable x; true iff x is false
$\Rightarrow$	Proposition P implies proposition Q (necessity condition)
	$P \Rightarrow Q$: it is true iff $\neg P$ is true or Q is true
$\equiv$	Equivalence relation
	$P \equiv Q$: $P \Rightarrow Q$ and $Q \Rightarrow P$
$\square$	symbolic for "*quod erat demonstrandum*" or *q.e.d.*
	"which was to be proved"
$\forall$	logic quantifier *for all*
	$\forall x \in S$: all elements x of set S
$\exists$	logic quantifier *there exists*
	$\exists x \in S$: there is an element x in set S

Symbols Related to Numbers and Arithmetic

Symbol	Meaning / Usage
$\leq$	less than or equal to $x \leq y$: the number x is no larger (or, greater) than the number y
$<$	strictly less than $x < y$: the number x is smaller than the number y
$<^-$	at least 2 less than
$\ll$	strictly "much" less than "much" is specified by context
$\geq$	greater than or equal to $x \geq y$: the number x is no smaller than the number y
$>$	strictly greater than $x > y$: the number x is greater (or, larger) than the number y
$>^+$	at least 2 greater than
$\gg$	strictly "much" greater than "much" is specified by context
$+$	unary plus sign $+x$: the nonnegative version of number x
$+$	binary addition sign $x + y$: the sum of numbers x and y
$-$	unary minus sign $-x$: the negative version of number x
$-$	binary subtraction sign $x - y$: the difference of numbers x and y
$\cdot$	multiplication sign $x \cdot y$: the product of numbers x and y
$\times$	multiplication sign $x \times y$: the product of numbers x and y
$/$	division sign x/y: the number which multiplied by y gives x in display mode, we write $\dfrac{x}{y}$
$\div$	division sign $x \div y$: the number which multiplied by y gives x
$\lvert x \rvert$	magnitude or absolute value $\lvert x \rvert = $ **if** $x < 0$ **then** $-x$ **else** x
$\displaystyle\sum_{i=a}^{b}$	summation between limits $\sum_{i=a}^{b} c_i$: the summation $c_a + c_{a+1} + \cdots + c_b$ in display mode, we write $\displaystyle\sum_{i=a}^{b} c_i$
$\sqrt{}$	radical (square root) sign $\sqrt{x}$: number whose square is x: $\sqrt{x} \times \sqrt{x} = x$

Symbol	Meaning / Usage
$!$	factorial sign $n! = n \times (n-1) \times \cdots \times 2 \times 1$
$\dbinom{n}{k}$	the binomial coefficient choose-k-out-of-n $\dbinom{n}{k} = \dfrac{n!}{k!(n-k)!}$
$\lfloor \circ \rfloor$	the floor or integer part $\lfloor n/m \rfloor$: the largest integer k such that $m \cdot k \leq n$
$\lceil \circ \rceil$	the ceiling $\lceil n/m \rceil$: the smallest integer k such that $m \cdot k \geq n$
mod	modulo sign n mod m: the unique integer $r < m$ such that $n = \lfloor n/m \rfloor \cdot m + r$.
∞	the *lemniscate curve* used to represent (the point at) infinity $1/n$ tends to 0 as n tends to ∞ Symbolically: $1/n \to 0$ as $n \to \infty$

Symbols Related to Combinatorics and Probability

Symbol	Meaning / Usage
$Pr[E]$	the probability of event E
$Pr[X = x]$	the probability that random variable X assumes value x
$Pr[E \mid F]$	the conditional probability of event E *given that* event F has occurred
$Pr[X = x \mid Y = y]$	the probability that random variable X assumes value x *given that* random variable Y assumes value y

References

1. C. Alsina, R. Nelsen (2006): *Math Made Visual*. Mathematical Assoc. of America.
2. **[H/C]** al-Khwārizmī [Muhammad ibn Mūsā al-Khwārizmī] (ninth century): *The Compendious Book on Calculation by Completion and Balancing*.
3. F. Annexstein, M. Baumslag, A.L. Rosenberg (1990): Group action graphs and parallel architectures. *SIAM J. Comput. 19*, 544–569.
4. K. Appel, W. Haken (1977): Every planar map is four colorable, I: Discharging. *Illinois J. Mathematics, 21*(3), 429–490.
5. K. Appel, W. Haken (1977): Every planar map is four colorable, II: Reducibility. *Illinois J. Mathematics, 21*(3), 491–567.
6. **[H/C]** K. Appel, W. Haken (October 1977): The solution of the four-color-map problem. *Scientific American, 237*(4), 108–121.
7. **[H/C]** K. Appel, W. Haken [with the collaboration of J. Koch] (1989): Every planar map is four-colorable. *Contemporary Mathematics, 98*, American Mathematical Society, Providence, RI.
8. **[H/C]** J. Arbuthnot (1710): An argument for Divine Providence taken from the constant regularity observed in the births of both sexes. *Philosophical Transactions*, The Royal Society,
9. **[H/C]** J.-R Argand (1806): Réflexions sur la nouvelle théorie d'analyse. Manuscript.
10. **[H/C]** J.W. Backus, R.J. Beeber, S. Best, R. Goldberg, L.M. Haibt, H.L. Herrick, R.A. Nelson, D. Sayre, P.B. Sheridan, H. Stern, L. Ziller, R.A. Hughes, R. Nutt (1957): The FORTRAN automatic coding system. *Western Joint Computer Conf.*, 188–198.
11. S.L. Basin (1963): The Fibonacci Sequence as it appears in nature. *Fibonacci Quart. 1*, 53–57.
12. **[H/C]** T. Bayes, R. Price (1763): "An Essay towards solving a Problem in the Doctrine of Chance. By the late Rev. Mr. Bayes, communicated by Mr. Price, in a letter to John Canton, A.M.F.R.S." *Philosophical Transactions of the Royal Society of London 53*(0): 370–418.
13. **[T/R]** E.T. Bell (1986): *Men of Mathematics*. Simon and Schuster, New York.
14. **[T/R]** C. Berge (1973): *Graphs and Hypergraphs*. North-Holland, Amsterdam.
15. J.-C. Bermond, C. Peyrat (1989): The de Bruijn and Kautz networks: a competitor for the hypercube? In *Hypercube and Distributed Computers* (F. Andre and J.P. Verjus, eds.) North-Holland, Amsterdam, 279–293.
16. **[H/C]** J. Bernoulli (1713): *Ars Conjectandi* (opus posthumum). Impensis Thurnisiorum Fratrum, Basel.
17. **[H/C]** F. Bernstein (1905): Untersuchungen aus der Mengenlehre. *Math. Ann. 61*, 117–155.
18. **[T/R]** G. Birkhoff and S. Mac Lane (1953): *A Survey of Modern Algebra*, Macmillan, New York.
19. **[T/R]** E. Bishop (1967): *Foundations of Constructive Analysis*, McGraw-Hill, New York.
20. J. Blazewicz, K. Ecker, B. Plateau, D. Trystram (2000): *Handbook on Parallel and Distributed Processing*. Springer, Heidelberg.

© Springer Nature Switzerland AG 2020
A. L. Rosenberg, D. Trystram, *Understand Mathematics, Understand Computing*,
https://doi.org/10.1007/978-3-030-58376-7

21. M. Blum, W. Sakoda (1977): On the capability of finite automata in 2 and 3 dimensional space. *18th IEEE Symp. on Foundations of Computer Science (FOCS)*, 147–161.

22. [H/C] G. Boole (1854): *An Investigation of the Laws of Thought.* Walton & Moberly, London. Reprinted 2003, Prometheus Books. ISBN 978-1-59102-089-9.

23. [T/R] W.W. Boone, F.B. Cannonito, R.C. Lyndon, eds. (1973): *Word Problems: Decision Problems and the Burnside Problem in Group Theory*, North-Holland, Amsterdam.

24. [T/R] P. Brémaud (2017): *Discrete Probability Models and Methods.* Springer, Cham.

25. [H/C] G. Cantor (1874): Über eine Eigenschaft des Inbegriffes aller reellen algebraischen Zahlen. *J. Reine und Angew. Math. 77*, 258–262.

26. [H/C] G. Cantor (1878): Ein Beitrag zur Begründung der transfiniter Mengenlehre. *J. Reine Angew. Math. 84*, 242–258.

27. [H/C] G. Cantor (1887): Mitteilungen zur Lehre vom Transfiniten. *Zeitschrift für Philosophie und Philosophische Kritik 91*, 81–125.

28. [H/C] A.L. Cauchy (1821): *Cours d'analyse de l'École Royale Polytechnique, 1ère partie: Analyse algébrique.* L'Imprimerie Royale, Paris. Reprinted: Wissenschaftliche Buchgesellschaft, Darmstadt, 1968.

29. G. Chartrand, F. Harary (1967): Planar permutation graphs. *Annales de l'Institut Henri Poincaré B, 3*(4), 433–438.

30. G. Chartrand, S.F. Kapoor (1969): The cube of every connected graph is 1-hamiltonian. *J. Research of the National Bureau of Standards, 73B(1)*. DOI: 10.6028/jres.073B.007

31. N. Christofides (1976): Worst-case analysis of a new heuristic for the traveling salesman problem. Report 388, Graduate School of Industrial Administration, Carnegie-Mellon Univ.

32. [H/C] E.F. Codd (1970): A relational model of data for large shared data banks. *Commun. of the ACM 13*(6), 377–387.

33. J.H. Conway, R.K. Guy (1998): *The Book of Numbers.* Copernicus, Springer, Heidelberg.

34. [H/C] J.H. Conway, J. Shipman (2013): Extreme Proofs I: The Irrationality of $\sqrt{2}$. *The Mathematical Intelligencer 35*(3).

35. S.A. Cook (1971): The complexity of theorem-proving procedures. *ACM Symp. on Theory of Computing*, 151–158.

36. [T/R] T.H. Cormen, C.E. Leiserson, R.L. Rivest, C. Stein (2001): *Introduction to Algorithms (2nd ed.).* MIT Press, Cambridge, MA.

37. M. Cosnard, D. Trystram (1995): *Parallel Algorithms and Architectures.* Int'l Thomson Computer Press.

38. H.B. Curry (1934): Some properties of equality and implication in combinatory logic. *Annals of Mathematics, 35*, 849–850.

39. [T/R] H.B. Curry, R. Feys, W. Craig (1958): *Combinatory Logic. Studies in logic and the foundations of mathematics.* North-Holland, Amsterdam.

40. [T/R] M. Davis (1958): *Computability and Unsolvability.* McGraw-Hill, New York. Reprinted in 1982 by Dover Press.

41. [H/C] M. Davis (1973): Hilbert's tenth problem is unsolvable. *American Math. Monthly 80*, 233–269. Reprinted as an appendix in M. Davis (1982): *Computability and Unsolvability* Dover reprint, 1982.

42. [H/C] M. Davis, R. Hersh (1973): Hilbert's 10th problem. *Scientific American 229*, 84–91. doi:10.1038/scientificamerican1173-84.

43. [H/C] M. Davis, Y. Matiyasevich, J. Robinson (1976): Hilbert's Tenth Problem: Diophantine Equations: Positive Aspects of a Negative Solution. In F.E. Browder. *Mathematical Developments Arising from Hilbert Problems. Proceedings of Symposia in Pure Mathematics, XXVIII.2*, American Math. Soc., pp. 323–378. Reprinted in S. Feferman, ed. (1996): *The Collected Works of Julia Robinson*, American Math. Soc., pp. 269–378.

44. O. Deiser (2010): Einführung in die Mengenlehre. Die Mengenlehre Georg Cantors und ihre Axiomatisierung durch Ernst Zermelo (3rd, corrected ed.), Berlin/Heidelberg: Springer, pp. 71, 501, doi:10.1007/978-3-642-01445-1, ISBN 978-3-642-01444-4.

45. [H/C] C.-J. de la Vallé Poussin (1896): Recherches analytiques sur la théorie des nombres premiers. *Ann. Soc. Scient. Bruxelles 20*, 183–256

46. **[H/C]** R.A. De Millo, A.J. Perlis, R.J. Lipton (1979): Social processes and proofs of theorems and programs. *Comm. ACM 22*(5), 271–280.

47. **[H/C]** A. de Moivre (1718): *The Doctrine of Chances*. W. Pearson, London.

48. G. Estrin (1960): Organization of computer systems—The fixed plus variable structure computer. *Western Joint Computer Conf.*, 33–40.

49. H. Fleischner (1974): The square of every two-connected graph is Hamiltonian. *J. Combinatorics Theory (B) 16*, 29–34.

50. J.E. Foster (1947): A number system without a zero symbol. *Mathematics Magazine 21*(1), 39–41.

51. **[H/C]** G. Fubini (1907): Sugli integrali multipli. *Rom. Acc. L. Rend. (5), 16(1)*, pp. 608–614. In *zbMATH 38.0343.02*. Reprinted in G. Fubini (1958): *Opere scelte, 2*, Cremonese, pp. 243–249.

52. **[T/R]** M.R. Garey and D.S. Johnson (1979): *Computers and Intractability*. W.H. Freeman and Co., San Francisco.

53. **[H/C]** K. Gödel (1931): Über Formal Unentscheidbare Sätze der Principia Mathematica und Verwandter Systeme, I. *Monatshefte für Mathematik u. Physik 38*, 173–198.

54. **[H/C]** J. Hadamard (1896): Sur la distribution des zéros de la fonction $\zeta(s)$ et ses conséquences arithmétiques. *Bull. Soc. Math. France 24*, 199–220.

55. **[T/R]** P.R. Halmos (1960): *Naive Set Theory*. D. Van Nostrand, New York.

56. M. Hazewinkel, ed. (2001): Viète theorem. In *Encyclopedia of Mathematics*. Springer Science+Business Media B.V. / Kluwer Academic Publishers, ISBN 978-1-55608-010-4.

57. **[H/C]** P.J. Heawood (1890): Map-colour theorems. *Quarterly J. Mathematics 24*, 332–338.

58. **[H/C]** D. Hilbert (1902): Mathematical Problems. (Translated by M.F.W. Newson). *Bull. American Math. Soc. 8*(10), 437–479.

59. **[T/R]** P.G. Hoel (1958): *Introduction to Mathematical Statistics* (2nd Ed.) John Wiley & Sons, New York.

60. **[H/C]** W.G. Horner (1819): A new method of solving numerical equations of all orders, by continuous approximation. *Philosophical Trans. Royal Society of London*, 308–335.

61. **[H/C]** W.J. Hutchins [ed.] (2000): Machine translation at Harvard: Interview with A.G. Oettinger. In *Early Years in Machine Translation: Memoirs and Biographies of Pioneers. (Studies in the History of the Language Sciences)*. John Benjamins Publishing Co., p. 86

62. S.L. Johnsson, C.T. Ho (1989): Optimum broadcasting and personalized communication in hypercubes. *IEEE Trans. Computers 38*, 1249–1268.

63. A. Karatsuba, Yu. Ofman (1962): Multiplication of many-digit numbers by automatic computers. *Proc. USSR Academy of Sciences 145*, 293–294. (Translated in *Physics-Doklady, 7* (1963) 595–596.)

64. R.M. Karp (1972): Reducibility among combinatorial problems. In *Complexity of Computer Computations* (R.E. Miller and J.W. Thatcher, eds.) Plenum Press, New York, pp. 85–103.

65. J. Kennedy, R. Eberhart (1995): Particle swarm optimization. *IEEE Int'l Conf. Neural Networks IV*, 1942–1948.

66. S. Kirkpatrick, C.D. Gelatt Jr., M.P. Vecchi (1983): Optimization by simulated annealing. *Science 220*(4598), 671–680.

67. **[H/C]** S.C. Kleene (1936): General recursive functions of natural numbers. *Math. Annalen 112*, 727–742.

68. **[T/R]** S.C. Kleene (1952): *Introduction to Metamathematics*. D. Van Nostrand, Princeton, NJ.

69. **[H/C]** D. König (1936): *Theorie der endlichen und unendlichen Graphen*. Leipzig: Akad. Verlag.

70. H.W. Kuhn (1955): The Hungarian method for the assignment problem. *Naval Research Logistics Quarterly 2*, 83–97.

71. K. Kuratowski (1930): Sur le problème des courbes gauches en topologie. *Fundamenta Mathematica 15*, 271–283.

72. L. Lamport (2012): How to write a 21st century proof. *J. Fixed Point Theory and Applications, 11*(1), 43–63.

73. **[H/C]** P.-S. Laplace (1814): *Essai philosophique sur les probabilités*. Bachelier, Paris.
74. P. Laurent-Gengoux, D. Trystram (1989): *Comprendre l'Informatique Numérique*. Lavoisier Ed.
75. **[T/R]** P.M. Lee (2012): *Bayesian Statistics: An Introduction*. Wiley, Hoboken, NJ.
76. **[H/C]** G.W. Leibniz (Leibnitz) (1674-76): *Sämtliche Schriften und Briefe, Reihe VII: Mathematische Schriften, vol. 5: Infinitesimalmathematik*. Akademie Verlag, Berlin.
77. **[H/C]** J.E. Littlewood (1953): *A Mathematician's Miscellany*. Methuen. Reprinted as *Littlewood's Miscellany* (B. Bollobàs, ed.), 1986, Cambridge University Press.
78. **[H/C]** É. Lucas (1891): *Récréations Mathématiques*. Gauthier-Villars.
79. F. Marchese (1996): Cellular automata in robot path planning. *EUROBOT'96*, 116–125.
80. Y. Matiyasevich (1993): *Hilbert's Tenth Problem*. MIT Press, Cambridge, MA.
81. **[T/R]** C.C. MacDuffee (1954): *Theory of Equations*. John Wiley & Sons, New York
82. C. Mead and L. Conway (1979): *Introduction to VLSI Systems*. Addison-Wesley, Reading, MA.
83. **[H/C]** J. Nash (1950): Equilibrium points in n-person games. *Proc. U.S. Nat'l Acad. Sciences* **36**(1), 48–49.
84. **[H/C]** I. Newton (1687): *Philosophia Naturalis Principia Mathematica* (known popularly as *Principia Mathematica*). Royal Society.
85. **[T/R]** I. Niven and H.S. Zuckerman (1980): *An Introduction to the Theory of Numbers*. (4th ed.) J. Wiley & Sons, New York.
86. **[H/C]** J.A. Paulos (1990): *Innumeracy: Mathematical Illiteracy and Its Consequences*. Vintage Books (Random House), New York.
87. **[T/R]** W.W. Peterson, E.J. Weldon, Jr. (1981): *Error-Correcting Codes*. (2nd Ed.) MIT Press, Cambridge, Mass.
88. M.O. Rabin, D. Scott (1959): Finite automata and their decision problems. *IBM J. Research and Development* **3**(2), 114-125.
89. A.L. Rosenberg (1991): Cycles in networks. Tech. Rpt. COINS-91-20, Univ. Massachusetts.
90. A.L. Rosenberg (2003): Efficient pairing functions—and why you should care. *Intl. J. Foundations of Computer Science* **14**, 3–17.
91. **[T/R]** A.L. Rosenberg (2009): *The Pillars of Computation Theory: State, Encoding, Nondeterminism*. Universitext Series, Springer, New York
92. A.L. Rosenberg (2012): The parking problem for finite-state robots. *J. Graph Algorithms and Applications* **16**(2), 483–506.
93. **[T/R]** S.M. Ross (1976): *A First Course in Probability*. Pearson Education.
94. **[T/R]** J.B. Rosser (1953): *Logic for Mathematicians*. McGraw-Hill, New York.
95. **[H/C]** B. Russell (1902): Letter to Frege. In Jean van Heijenoort, ed. (1967): *From Frege to Gödel*. Harvard University Press, Cambridge, MA, pp. 124–125.
96. Y. Saad, M.H. Schultz (1989): Data communication in hypercubes. *J. Parallel and Distributed Computing* **6**, 115–135.
97. **[H/C]** J.E. Sammet (1978): The early history of COBOL. In *History of Programming Languages*. Academic Press (published 1981).
98. **[H/C]** M. Schönfinkel (1924): Über die Bausteine der mathematischen Logik. *Math. Annalen* **92**, 305–316.
99. **[H/C]** E. Schröder (1898): Über zwei Definitionen der Endlichkeit und G. Cantor'sche Sätze. *Nova Acta Academiae Caesareae Leopoldino-Carolinae (Halle a.d. Saale)* **71**, 303–362.
100. **[H/C]** E. Schröder (1898): Die selbständige Definition der Mächtigkeiten 0, 1, 2, 3 und die explicite Gleichzahligkeitsbedingung. *Nova Acta Academiae Caesareae Leopoldino-Carolinae (Halle a.d. Saale)* **71**, 365–376.
101. J.T. Schwartz (1980): Ultracomputers. *ACM Trans. Programming Languages* **2**, 484–521.
102. **[H/C]** C.E. Shannon (1938): A symbolic analysis of relay and switching circuits. *Trans. American Inst. of Electrical Engineers* **57**(12), 713–723.
103. **[H/C]** C.E. Shannon (1948): A mathematical theory of communication. *Bell System Technical Journal* **27** (July and October), pp. 379–423 and 623–656.
104. I. Shinahr (1974): Two- and three-dimensional firing-squad synchronization problems. *Inform. and Control* **24**, 163-180.

105. **[T/R]** R.M. Smullyan (1961): Lexicographical ordering; *n*-adic representation of integers. *Theory of Formal Systems, Annals of Mathematics Studies 47*, Princeton University Press, pp. 34–36.

106. H.S. Stone (1971): Parallel processing with the perfect shuffle. *IEEE Trans. Computers 2*, 153–161.

107. D. Tonien (2007): A simple visual proof of the Schröder-Bernstein theorem. *Elemente der Mathematik 62*, 118–120.

108. **[H/C]** A.M. Turing (1936): On computable numbers, with an application to the Entscheidungsproblem. *Proc. London Math. Soc.* (ser. 2, vol. 42) 230–265; Correction *ibid.* (vol. 43) 544–546.

109. J.D. Ullman (1984): *Computational Aspects of VLSI.* Computer Science Press, Rockville, MD.

110. **[H/C]** J. von Neumann, O. Morgenstern (1944): *Theory of Games and Economic Behavior.* Princeton University Press.

111. R. Wagner, M.J. Fischer (1974): The string-to-string correction problem. *Journ. Assoc. Computing Machinery (ACM) 21*, 168-178.

112. **[H/C]** H.L. Weber (1891-2): Kronecker. *Jahresbericht der Deutschen Mathematiker-Vereinigung 2*, 5–23.

113. **[H/C]** A.N. Whitehead, B. Russell (1903). *Principles of Mathematics.* Cambridge University Press.

114. A.J. Wiles (May 1995): "Issue 3". *Annals of Mathematics 141*, 1–551.

115. R. Wilson, J. Watkins (2013): *Combinatorics: Ancient & Modern.* Oxford Univ. Press.

116. **[H/C]** V. Yngve (1958): A programming language for mechanical translation. In *Mechanical Translation 5*(1): 25–41. MIT, Cambridge, MA.

117. M. Yoeli (1962): Binary ring sequences. *Amer. Math. Monthly 69*, 852–855.

118. **[H/C]** Zhu Shijie (1303): *Siyuan Yujian.*

Index

© Springer Nature Switzerland AG 2020

A. L. Rosenberg, D. Trystram, *Understand Mathematics, Understand Computing*,
https://doi.org/10.1007/978-3-030-58376-7